Energy for Survival

The Alternative to Extinction

Energy for Survival

The Alternative to Extinction

Wilson Clark
with research by David Howell
with illustrations by James K. Page, Jr.

ANCHOR BOOKS
Anchor Press / Doubleday
Garden City, New York

ENERGY FOR SURVIVAL was also published in a hardcover edition by Anchor Press in 1974.

ANCHOR BOOKS EDITION 1975

ISBN: 0-385-03564-0
Printed in the United States of America

To the Memory of My Grandmother,
Laura Ayers Clark
1882–1973
She provided the spark long ago.

Contents

Preface

Since the international oil crisis of 1973–1974, the economic and social effects of petroleum scarcity and increasing oil prices in the industrialized nations have underscored the essential message of this book—that our high-energy, industrial way of life is extremely vulnerable to energy disruptions, and that our way of life must be changed in order to survive the pressures of the future.

Energy is commonly defined as the capacity to do work. The work of our society is largely done by harnessing sources of energy which are finite. Although we hear much about various future sources of energy (such as nuclear fusion and geothermal power) the work of our civilization is wedded to the fossil fuels—coal, oil, and natural gas—sources of energy that are dwindling rapidly. The institutions of our society that make possible human existence, including medicine, agriculture, transportation, and industry, are all highly dependent on ever-increasing doses of these finite resources.

We are faced with a grave problem. Since the industrial revolution, the industrialized nations have been engaged in a race to increase energy consumption in order to increase population and build great bases of political power. Yet these increases in energy have brought disastrous changes in the social and environmental climate of these nations.

Pollution is an inescapable product of the consumption of energy. Electrical power plants that burn coal release pollutants into the air and require great quantities of water for cooling, leading to thermal pollution of water. Automobiles burn gasoline in ever-increasing quantities, releasing pollutants and heat into the atmosphere. As more energy is consumed by power plants and autos, more pollution is produced. The application of pollution-control technologies—the "Band-Aid" approach—often leads to more energy consumption, though often not at the source of pollution the control technology is designed to eliminate. Automobile pollution-control devices require the expenditure of energy to manufacture, and when they are used, more energy (in gasoline) is consumed to provide motive power for the vehicle as well as to operate the control device.

We must eliminate the sources of pollution, not merely attempt to invent technological "answers" to the pollution problem. A reduction in energy use will solve both the problem of pollution and the problem of conserving our limited stocks of fossil fuels. Although in the early 1970s several U.S. government agencies issued warnings that wide-ranging energy conservation policies should be implemented to help forestall a national energy disaster, the federal government waited until oil shortages actually occurred (in the aftermath of the Middle East war) to initiate fledgling conservation policies.

Energy conservation policies will be necessary on a massive scale to forestall shortages of both energy and material goods in the economy in future years. Better uses can be found for energy supplies than the waste of energy inherent in our present economic system. Using automobiles for transporting people and constructing buildings that "leak" energy in staggering amounts to the surrounding environment are luxuries an energy-short society can ill afford. New, efficient energy-use technologies must be developed and implemented quickly to ease the shock that has already affected the industrial nations.

A currently popular myth is that the energy crisis will disappear if the nation invests heavily in new domestic technologies, such as nuclear fission, and develops "synthetic" liquid and gaseous fuels from deposits of solid hydrocarbons such as coal and oil shale. The purveyors of this myth do not understand the economic relationship between *net* energy and *gross* energy.

Overzealous scientists and planners point to the "unlimited" resources of coal and oil shale, arguing that a massive technological program will enable the United States to rely on its own energy resources for continued economic growth in the future. In fact, the energy required to bring more oil, natural gas, and coal from the earth into use may be more than the value of the energy in the fossil fuel reserves themselves. Though the quantities of fossil energy reserves do indeed appear large, the value of this energy to society cannot be counted until the energy required for processing, concentrating, and transporting energy to consumers is subtracted from the gross energy reserves. The energy that can be used in society is the net energy, which remains after these secondary energy costs are paid. Most government and industry calculations of future energy supplies do not take this factor into account. Consequently, the official calculations of future energy availability are suspect.

The most heavily subsidized energy technology in the United States is nuclear fission, the splitting of atoms of uranium and other heavy metals for electrical power. Not only is the net energy gain from this power process questionable, but also environmental and public protection safeguards appear to be woefully inadequate. Nuclear fission technology is a "fail-safe" technology—as long as the system works perfectly, the public is protected, but if an accident occurs in the nuclear power plant, the waste processing plant, or in the transport of radioactive materials, the public (not to mention other forms of life) may be exposed to the deadliest environmental poisons known, some of which remain lethal for thousands of years.

Much of this book is devoted to discussion of solar energy, our only source of renewable income energy. The use of solar energy offers the promise of a safe, environmentally less destructive technology that can usher in a new era of post industrial development. Little attention has been paid by the current generation of scientists, politicians, and planners to developing a society that operates in a framework of environmental and social stability. In the mad rush to plunder the earth of its dwindling store of non-renewable resources, the planners of today's industrial world have forgotten to make room for future generations. The time for change is not some nebulous future date, but now. Avoidance of environmental and economic collapse is possible if the industrial societies shift the current emphasis of high growth into a transitional phase of planetary maintenance, not exploitation. Gone are the days when the industrial world could engage in resource exploitation (and human exploitation) and think only of economic growth, with no concept of future life possibilities.

The excesses of our current high energy industrial era can be corrected if intelligent and gentle policies of change are immediately implemented to bring about a post-industrial Renaissance.

Acknowledgments

The author is indebted to many people, and it is impossible to recognize them all by name. An examination of the book's Bibliography will identify many of the individuals who provided valuable assistance in the preparation of material. This book would not have been possible without the assistance and perseverance of the author's friend and colleague, David L. Howell, who spent many long hours helping in the planning and writing of these pages. Special thanks is due Congressman Bill Gunter of Florida and Ruth Wallick of the Conservation and Natural Resources Subcommittee of the Committee on Government Operations of the U. S. House of Representatives for assistance in securing Library of Congress books. Monica Andres, the former librarian of the National Parks and Conservation Association (Washington, D.C.) and Sunday Orme of the Washington *Post* library assisted in securing books from other libraries.

While the author lived in Florida, Tom Seay forwarded mail and phone messages from his Washington office. Engineer John Everetts, Jr., reviewed much of the material on architecture and energy use. The staff of the Energy Center at the University of Florida (Gainesville) provided comments as well as office space during a critical period of final writing and review. Special help was provided at Gainesville by Thomas A. Robertson and Chester Kylstra. The Simpson family of St. James City, Florida, made the author's self-imposed exile pleasant and memorable. The illustrator of this book, James K. Page, Jr., provided much inspiration and guidance from the book's beginning to finish. Catherine Lerza of *Environmental Action* magazine helped edit the final manuscript.

Much help and assistance has been provided by the staff of the Environmental Policy Center, Washington, D.C. Special thanks to staff members Joe Browder and Jerry Long for keeping the author aware of ever-changing developments in Washington, D.C., during the sojourn in Florida. Marion Edey of the League of Conservation Voters, Washington, D.C., has provided much help and inspiration. Consultants Richard Sandler and Egan O'Connor have contributed time and valuable ideas during the book's progress. Financial assistance in researching this book has been provided by the Environmental Policy Center, the Scherman Foundation, the John Muir Institute for Environmental Studies, and the Sierra Club Foundation.

Wilson Clark
Washington, D.C.

Energy for Survival

1

The Energy Basis of Our Civilization

From their earliest Earth origins, human beings have continually sought to minimize their labor through the use of tools. During the late Paleolithic period—some twelve thousand to twenty thousand years ago—cave dwellers in southern France and Spain painted images of what may have been the first machine, or sophisticated tool, used by man. On the cave walls are pictured traps used for capturing mammoth and bison. The traps were made in such a way that the great mammoth would enter, release a lever mechanism, and trigger the collapse of a number of tree trunks—felling the beast.[1]

By using this trap, the early hunters were able to brighten the prospects of their existence. By careful planning, they utilized tools—to kill the animals. They stretched their available biological energy, derived from sunlight in the form of food, to get more energy from the bison and mammoth. This process was the daily "energy crisis" faced by our remote ancestors, and the solution they hit upon was the use of technology.

The most vital Stone Age discovery was how to make fire by rubbing together two pieces of wood. This revolutionary discovery—which was almost certainly accidental—has many repercussions in the history of energy. As described by the late economic historian Harry Elmer Barnes:

However it may have happened that man learned to produce fire, the advance from the stage of simply keeping it going to that of being able to relight it if it went out meant an enormous stride forward. For early man fire meant light, heat protection, and a multitude of other things. . . . That which once warmed the bodies of primitive man in the Paleolithic rock shelters reduces to molten form the iron ore in the blast furnace of today, and in the electric torch it cuts steel plates as though they were made of paper.[2]

The rapid advance of human progress that occurred during the following historic periods—from the Neolithic Age (the era of ground and polished stone tools) to what we call the dawn of civilization some three or four thousand years ago—is characterized by the use of solar energy indirectly through the biological system that produced grain and animals for man's survival.

During the later Neolithic period humans used stone tools to harness agriculture. With this step, man managed to couple his own energy—magnified through the tool—with the energy resource of the sun, renewable crops. The first signs of stable human communities developed, with the requirement that housing would be necessary to protect the fragile humans from the vicissitudes of climate. Thus another use of primitive technology was required—the beginnings of "climate control." With their tools, the people constructed houses out of native materials. In some areas, the early dwellings were merely holes dug in the ground and covered with twigs, branches, and vegetation. In other areas during the Neolithic period the beginnings of architecture arose as small communities were developed—housed in dwellings of wood constructed on stilts in northern Italian and Swiss lakes.[3] On other continents, and in other areas, the primitive structures looked different—some were made of earth, some of flimsy bamboo, some of soft wood, some of hard wood. They were all designed by necessity—to meet the climatic needs of the specific areas in which they were located.

This architecture was a clever adaptation to climatic requirements. Success meant survival; failure meant death. Even the form of primitive dwellings was not happenstance.

The great architect Le Corbusier wrote:

Look at a drawing of such a hut in a book on archaeology: here is the plan of a house, the plan of a temple. It is exactly the same attitude as you find in a Pompeian house or in a temple at Luxor. . . . There is no such thing as a primitive man; there are only primitive means. The idea is constant, potent from the very outset.[4]

The basic idea of utilizing building materials and the dwelling design to control the effects of the climate on the structure was an unwritten law in the early cultures. The first recorded analysis of the natural-climate approach to housing was written by the Greek author Xenophon in about 400 B.C.:

> In houses with a south aspect, the sun's rays penetrate into the porticoes in winter, but in summer the path of the sun is right over our heads and above the roof, so that there is shade. If, then, this is the best arrangement we should build the south side loftier, to get the winter sun, and the north side lower to keep out the cold winds.[5]

Early humans had only the natural architectural siting method of dealing with climate, combined with the use of fire to keep warm in winter. As tools for construction and building were developed, early humans succeeded in controlling their microclimate, so that their communities could flourish by remaining stable and in one location. No longer did the population have to migrate with the seasons to escape the effects of climate as unfavorable weather or cold winters intruded.

The technological and societal progress made by Neolithic peoples was translated into the beginnings of modern civilization some four thousand to five thousand years ago, when human society developed into a coherent, organized form in the valleys of Egypt and Mesopotamia. One of the great technical advances of Neolithic cultures was the discovery of spinning and weaving linen, which was developed into a sophisticated art rivaling the much later European workmanship until the days of the Industrial Revolution, when machines were developed for mechanical weaving.[6]

Technological progress consisted of the ability to mine metals and with the help of heat energy—fire—forge them into tools and weapons. First came copper, then bronze (copper and tin). Following the Copper and Bronze Ages came the Iron Age, which may have begun in Africa or the Orient some twenty-four hundred years ago. In Europe, the Iron Age was launched in what is now Austria. It continued from the twelfth to the sixth century B.C.; a later northern European Iron Age lasted from 500 B.C. to the first century A.D.

Before the period of Greece and Rome, the only significant energy sources were natural ones—the power of falling water and the use of wind for ships. In industry, the central energy source was fire—from the burning of wood.

One estimate of the power of the water mills, which were used for basic industries, agriculture, irrigation, and grinding grain, is that in

1066—the time of the Battle of Hastings—there were in England about eight thousand water mills, which served one million people. At 2.5 horsepower per mill, this was about twice the energy of the hundred thousand men who had constructed the Great Pyramid.[7] The other major machine of the period was the windmill, which had been developed in Persia in the seventh century. By the thirteenth century, windmills were common in Europe, with significant advances being made by the Dutch and the English.*

In the Western world, until the advent of the Industrial Revolution, developments in energy use were not as significant as developments in materials use (the production of better metal tools and weapons), the arts, architecture, commerce, and agriculture. The use of energy resources was low, and the extraction of metals did not greatly affect the lot of the average person. Much of the mining and mechanization in early cultures—and certainly in Greece and Rome—was for the purpose of perfecting tools for war. In a very real sense, the liberation of energy through fire made possible the means of large-scale warfare.

The character of life of the earliest people was not much affected by the sophistication of technology. Lewis Mumford observes:

> All the praise of tool-making and tool-using that has been mistakenly applied to man's early development becomes justified from Neolithic times onward, and should even be magnified in evaluating the later achievements of handicraft. The maker and the object reacted one upon the other. Until modern times, apart from the esoteric knowledge of the priests, philosophers and astronomers, the greater part of human thought and imagination flowed through the hands.[8]

Going a step farther, the major source of energy in both Greece and Rome was the energy of human hands, with a considerable part of this energy derived from slaves. Slavery reached a peak in the Roman empire. In one slave mart in Delos during Roman times, as many as ten thousand slaves were sold in a day, and in the period of the Emperor Augustus, Rome alone boasted as many as four hundred thousand slaves.[9]

In view of the fact that the population of Rome was probably not greater than one million inhabitants, the significance of this slave labor, or energy supply, is great. Roman slaves were used not only as servants of the rich, but for maintaining public work, such as the famed Roman aqueducts. For many centuries thereafter, warring Kings would find conquered populations a ready energy source as slaves.[10]

* A more complete discussion of the history of wind power is found in Chapter 8.

The significance of slavery and its historic role in the harnessing of human energy for technology was recognized by Oscar Wilde, when, many centuries later, he wrote:

> The fact is that civilization requires slaves. The Greeks were quite right there. Unless there are slaves to do the ugly, horrible, uninteresting work, culture and contemplation become almost impossible. Human slavery is wrong, insecure, and demoralizing. On mechanical slavery, on the slavery of the machine, the future of the world depends.[11]

"Mechanical slaves" were virtually nonexistent from the Greek and Roman era to the beginnings of the Industrial Revolution. The interim civilizations developed sophisticated tools, but they were powered by men and draft animals, not by the liberation of energy resources by heat to operate engines. In centuries to come, the few natural energy engines in widespread use would be windmills and water wheels.

Yet the idea of the heat engine—a device that would use heat to produce power to turn something—was envisioned as early as 75 A.D. by the Greek inventor Hero. Hero made a toy team "engine," which he called an aeolipile—the "sphere of Aeolus"—which performed no work but functioned well as a toy. It was nothing more than a hollow metal sphere filled with steam, which would spin on bearings as it expelled the steam through vents. The principle of its operation was not known until centuries later, when Sir Isaac Newton formulated the Third Law of Motion: "To every action there is an equal and opposite reaction."[12]

The spread of technology in Europe between the fall of the Roman Empire and the last few years of the seventeenth century took place slowly and was primarily characterized by the development of toolmaking and architecture. Most significant inventions were developed in the fourteenth and fifteenth centuries. During this crucial period—particularly marked by the height of the Italian Renaissance—came the mechanical discoveries of Leonardo da Vinci, discoveries whose application was limited primarily by the absence of readily available energy resources and the technical understanding of harnessing them. The other technical advances that added to the legacy of the windmill and waterwheel were the printing press and the mechanical clock.†

† Lewis Mumford has contributed the observation that the mechanical clock is at the heart of modern mechanized life. He writes that the development of the clock in the fourteenth century ". . . synchronized human reactions, not with the rising and setting sun, but with the indicated movements of the clock's hands: so it brought exact measurement and temporal control into every activity, by setting an independent standard whereby the whole day could be laid out and sub-

But the printing press had to be operated by hand; the clock had to be wound. Civilization still had not found a convenient or concentrated source of energy—even in warfare, where the technology of the cannon took over two centuries to reach widespread use. Cannons were employed in the Battle of Crécy in 1345, but did not displace the bow as the essential English weapon until 1595.[14]

The great Baroque period of European history brought a flood of intellectual and scientific contributions: Galileo's physics, Kepler's astronomy, the philosophy and mathematics of Spinoza, Leibniz, Descartes, Huygens, and Newton. During this period were conceived some of the great laws of nature—as well as the development of natural law—which provided the seeds of decline of the ruling kings of Europe.

During this period, more sophisticated scientific instruments and tools were invented, including the telescope and microscope, the air pump, the barometer, and more. The spirit of the era was exploration and discovery of the world and of applied science. In his utopian *New Atlantis,* Francis Bacon prophesied the future:

> We also have engine-houses, where are prepared engines for all sorts of motions. There we imitate and practice to make swifter motions than any of you have, either out of your muskets or any engine that you have; and to make them and multiply them more easily and with small force, by wheels and other means; and to make them stronger and more violent than yours are, exceeding your cannons and basilisks. . . .[15]

Within the span of two centuries, the Industrial Revolution would be launched—which would ultimately lead to a mechanized future that even Bacon would not have recognized.

To make the Industrial Revolution work, to bring about the era of modern technology, required a new discovery—an engine that could convert an energy source into useful work at higher efficiency than the water mills and windmills. Not that the water wheels and windmills were unsatisfactory. These power sources, in fact, were the central power sources that ushered in the early Industrial Revolution. What was needed was more flexibility. The water mills and windmills were limited to specific sites—the fast-running rivers and the windy hills. The transportation of energy was unknown in the Western

---

divided. . . . Punctuality, ceasing to be 'the courtesy of kings' became a necessity in daily affairs in those countries where mechanization was taking command. The measurement of time and space became an integral part of the system of control that Western man spread over the planet."[13]

world. There were no coal trains, pipelines, electric power lines, or oil tankers.‡

The Steam Engine

By the end of the seventeenth century, the long-awaited engine appeared. Englishman Thomas Savery obtained a patent in 1698 for a machine that he called a "fire engine" because it employed fire to boil water and generate steam in a boiler for use in draining water from mines. His engine was financially successful, even though it was crude and not particularly efficient in pumping water.[16]

Shortly after that, his countryman Thomas Newcomen perfected a more efficient steam engine, which could raise 50 gallons of water per minute from a 156-foot depth. The engine used steam at atmospheric pressure, rather than high-pressure steam, which could have produced more power. Nonetheless, the machine was even more successful than Savery's "fire-engine." After Savery's death, Newcomen purchased the Savery firm, and controlled steam engine sales in England until 1733.[17]

Further contributions to the technology of the steam engine were made by the famed English engineer of the eighteenth century, John Smeaton,* who designed a steam engine with better-fitting parts (including accurately bored cylinders), which made it more efficient.

These early experiments paved the way for a generation of steam engines that would bring the technology closer to the open doors of the Industrial Revolution in the early nineteenth century. The development of the steam engine couldn't have come at a better time in Europe. By the eighteenth century, early attempts at mining coal—particularly in England—were stymied by the fact that the coal mines soon filled with underground water, and pumps of sufficient strength to drain them were not available. By midcentury, timber was essentially unavailable, since the forests had been stripped for fuel wood and to free land for agriculture.

The steam engines enabled the mines to be drained, making possible deep mining of coal, which immediately spurred the development of the English iron industry. Before the development of coal, the iron industry had been limited in production to a mere twenty thousand tons of pig iron a year.[18]

Later improvements on the technology of the engine were made by

‡ In China, however, natural gas had been transported to communities in bamboo pipes for centuries.

* See Chapter 8 for Smeaton's improvements in the design of windmills.

Englishmen James Watt and Richard Trevithick and the American Oliver Evans. The improvements made by these men all increased the efficiency by which the steam engine converted the energy released as heat by fuel (coal or wood) for useful work. Watt patented a vastly improved engine in 1782 that allowed steam to enter both ends of an engine cylinder, allowing the engine to develop two power strokes per cycle of operation.[19] With the development of Watt's engine, the steam engine had arrived as the forerunner of mechanized civilization. Between 1775 and 1800, Watt's company (Boulton & Watt) had sold almost five hundred engines, which were employed for pumping water and supplying mechanical power to textile mills, rolling mills, and flour mills.[20]

Still, all the engines of the eighteenth century used low-pressure steam. They were incredibly inefficient by modern standards. While modern steam engines in electric power plants convert up to 40 percent of the energy content of coal to useful power (still wasting 60 percent of the energy content of the coal, however), the eighteenth-century steam engines did well to convert 1 percent of the energy content of wood or coal to power—wasting almost all the original energy of the fuel.

A significant breakthrough came with the development of engines that could use high-pressure steam. Oliver Evans was convinced that the "elastic power" of expanding high-pressure steam could be used to get more power from a steam engine. Using his life savings, he designed and built a high-pressure steam engine in Philadelphia in 1802. When it proved successful, he built a number of them in the United States for use in various industries.[21]

Independently of Evans, Richard Trevithick, an English coal mine owner's son, built a similar high-pressure engine in England. These discoveries paved the way for the era of steam power. With Robert Fulton's successful operation of the steamboat *Clermont* on the Hudson River in 1807, and Trevithick's use of a steam locomotive to transport coal in Wales, the new industrial era was at hand.[22] Two decades later, George Stephenson of Newcastle, England, manufactured a steam locomotive called the *Rocket*. It won a contest over other locomotives by covering a twelve-mile distance in the astonishing time of fifty-three minutes. This demonstration that the steam locomotive would beat a horse and carriage ushered in the modern era of speed and mechanized transportation.

The inventors and manufacturers of the early steam engines knew little about their precise operations. Today, we can blithely describe the efficiency of this engine or that engine; but until the rise of the science of thermodynamics, little was known about the operation of the engine except that some designs could grind more grain or propel a boat or locomotive faster than others.

The Laws of Thermodynamics

Thermodynamics—from the Greek roots *therme* (heat) and *dynamis* (power)—is an applied science that defines and interprets the relationships among energy, heat, and work. It is the science that makes possible our understanding not only of steam engines, but of the relationships of energy to society and the machines that serve as our technological foundation.

The first scientist to ascertain the relationships among heat, energy, and work was Benjamin Thompson (Count Rumford), founder in 1799 of the Royal Institution in London. Thompson studied the heat produced when a brass cannon was bored. Early theories of energy contended that heat was an invisible fluid called "caloric," which could penetrate and expand all substances as well as dissolve certain materials into a vapor. According to this theory, the chips bored from the brass cannon barrel would be hotter than the barrel because the bore would squeeze the caloric out of the barrel along with the brass chips. Thompson discovered that the heat was at the same temperature in both the bored barrel and the chips, however, and furthermore concluded that the supply of heat must be endless because a vat of water in which he had placed the barrel as it was being bored continued to get hotter and hotter as the horse-powered drilling continued. This led him to speculate that it was *motion,* not the mysterious caloric fluid, that produced heat when the cannon barrel was bored.[23]

Heat, then, was not a material substance, as the caloric theory had held it to be. Later, British scientist James P. Joule established a mathematical measure of the exact relationship between heat and work, which led to the establishment of a principle that is termed the First Law of Thermodynamics: *Energy and matter can neither be created nor destroyed.*[24]

Even more important was the development of the Second Law of Thermodynamics, which was first formulated by a brilliant young French field artillery officer and physicist, Sadi Carnot. Carnot's famous memoir (or technical paper, as it would be termed today), "Reflections on the Motive Power of Heat," was written in 1819, when he was twenty-three years old.† Carnot wrote with rare perception of the operation and significance of the processes governing the engines appearing in Europe. Here are some excerpts from his prescient memoir, published in 1824:

> The study of these machines is of the greatest interest, for their importance is enormous and their use increases day by day. They seem destined to effect a great revolution in the civilized world. Already the heat engine works our mines, propels our ships, deepens our harbors and rivers, forges iron, fashions wood, grinds our [grain], spins and weaves our fabrics, pulls the heaviest loads, etc. It seems as though one day it will be the general motive power which will over[turn] animal power, waterfalls, and wind currents. . . . Iron and fire are recognized as the sustenance and support of mechanical industry. Perhaps there is no single factory in England whose existence is not founded on the use of these elements and in which they are not abundantly applied. If England today were to be deprived of her steam engines, she would also be robbed of coal and iron; all her sources of wealth would be cut off and all her means of development would be destroyed; this would mean the ruin of this vast Power. . . .[25]

After his prefatory comments, Carnot addressed the question of the operation of the engine, commenting:

> The phenomenon of the generation of motion by heat has not been regarded from a sufficiently general point of view. It has only been investigated in engines whose principles have not allowed the full development of which it is capable. In such engines it appears, so to speak, limited and incomplete, so that it is difficult to recognize its principles and to study its laws.[26]

Carnot's analysis of these general principles led him to the conclusion that a heat engine operates on a certain *cycle;* that is, that a fluid in the engine receives and rejects heat during a series of steps in the engine. The fluid, he conjectured, would always return to its original state. He outlined the cycle of operation of a steam engine: Water absorbs heat in the engine's boiler as water evaporates to steam. The steam expands and performs mechanical work; then it is condensed back to water as heat is removed from it in the condenser. Then the cycle is repeated as the water is pumped back into the boiler and converted to steam.

Thus he said there would always be three conditions in a heat engine: heat added, work performed, and heat rejected. Carnot additionally defined the *efficiency* of a heat engine, which he said de-

† Carnot's tragic death came only thirteen years later, during a cholera epidemic.

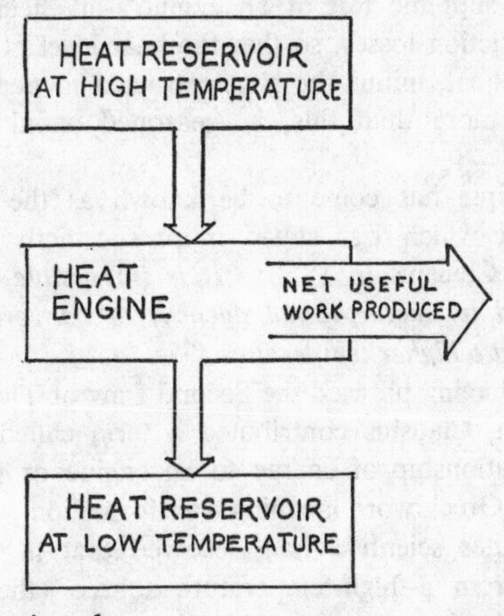

Heat engine (steam) cycle

pended on the temperature relationship of the heat received by the engine's boiler and the heat rejected by the engine's condenser. For the most efficient engine, he stated certain propositions, including:

1. Heat should be supplied at the highest possible temperature from the heat *source*.
2. Heat should be rejected at the lowest possible temperature into the heat *sink*.

A third proposition was that to get full power from a steam engine, not only should the heat temperatures be widely separated, but the steam additionally should be expanded to the lowest possible pressure at the end of an engine's cycle of operation.

Thus Carnot established several concepts, two being the concept of the cyclic operation of the heat engine and the concept of efficiency—based on the temperature relationships of the fluid used in the engine's cycle.

A third, and most important, principle was his demonstration of a "reversible" heat engine. He theorized a perfect engine, which, if it could be built, would work in the following way:

> Whatever amount of mechanical effect it can derive from a certain thermal agency [heat source], if an equal amount be spent in working it backwards, an equal reverse thermal effect will be produced.[27]

Here was the supreme test of an engine—in effect an engine that would have no friction losses, so that the heat level at each end of the cycle would be equal, minus the heat extracted to perform mechanical work. Anything more than this, he reasoned, would constitute perpetual motion.

Carnot's principle has come to be known as the Second Law of Thermodynamics, which was stated more succinctly by the German physicist Rudolf Clausius in 1850: *"It is impossible for a self-acting machine, unaided by any external agency, to convert heat from one body to another at a higher temperature."*[28]

In addition to having phrased the Second Law of Thermodynamics in a concise fashion, Clausius contributed a term called entropy, which describes the relationship of energy to an engine or process. Entropy is derived from a Greek word meaning transformation.

Carnot and other scientists had observed that in the heat engine, heat is drawn from a high-temperature source (the combustion of fuel); some of the heat energy is used to produce work (turning a shaft); and then the remaining heat is rejected (in a condenser) to water or air.

Clausius set out to define precisely how much heat was used in this process to produce mechanical work and how much was wasted in the theoretically ideal heat engine, where no friction would exist in the engine's components. By sophisticated mathematical calculations, he determined that even in a perfect engine (which could not possibly be constructed), there is a constant degradation of the available energy. In other words, even though the First Law of Thermodynamics states that energy can neither be destroyed nor created, another process operates that causes available energy to continuously change to a lower, less-usable form.

A heat engine will convert some of the original energy content of fuel to work in the engine and will dissipate waste heat. So while no energy has been destroyed, some—the waste heat—has been transmuted to a useless form. Energy, then, is constantly moving from a higher to a lower state, Clausius reasoned—not only in available heat engines, or even in the perfect heat engine, but in the universe itself. He called his discovery the Principle of the Increase of Entropy, a corollary of the Second Law of Thermodynamics: *"The energy of the universe is constant, but the entropy of the universe increases towards a maximum."*[29] Today, the Second Law of Thermodynamics is called the Entropy Law, and it is used not only to describe the processes of energy conversion in engines, but to describe the circumstances of many other systems, including those of economics, physics, urban development, biology, ecology, and the nature of the universe itself.

In a general sense, the laws of thermodynamics instruct us that on our planet—and in our universe—all activities of nature and its creatures are subject to two conditions:

1. Energy and matter can neither be created nor destroyed. In other words, "there's no such thing as a free lunch"; and "you can't get something for nothing."

2. Since energy constantly changes from a higher to a lower form, it is destined that "we can't break even." It's impossible to turn around and redo something that has been undone with the same energy that was used to do it in the first place. Once an act is accomplished, the energy required to accomplish it has degraded to a lower level. The entropy law describes the existence of a great natural law—a sort of "energy gravity" of the universe: Energy is constantly moving from a higher level to a lower level. In the same way that gravity influences water flow, energy can't be made to run uphill.

The First Law precludes the notion of perpetual motion. It doesn't exist in our universe. The noted scientist/author Isaac Asimov says that the First Law is the most powerful and fundamental generalization about the universe ever made by science. On perpetual motion, he says:

> You can't get around it. . . . Through centuries inventors have been trying to beat this game and invent perpetual-motion machines. With rolling balls and swinging weights and spinning wheels they have tried repeatedly to make a machine that would run on, once started, getting just a little energy from gravity or some other source to overcome the little bit of friction in its bearings. Such a machine if you could make it, would be a perpetual-motion machine of the kind that would violate the laws of thermodynamics. Nobody has succeeded; nobody ever will; very few, if any, are still trying.[30]

Adding insult to injury, we find that we have to act quickly to even attempt a slight jump on the thermodynamic laws. Economist Nicholas Georgescu-Roegen of Vanderbilt University underscores the real significance of the Second Law of Thermodynamics, or the Entropy Law:

> The Entropy Law teaches us that the rule of biological life and, in man's case, of its economic continuation is far harsher. In entropy terms, the cost of any biological or economic enterprise is always greater than the product.

What does this mean in practice? He continues:

> Every time we produce a Cadillac, we irrevocably destroy an amount of low entropy‡ that could otherwise be used for producing a plough or a spade. In other words, every time we produce a Cadillac, we do it at the cost of decreasing the number of human lives in the future. Economic development through industrial abundance may be a blessing for us now and for those who will be able to enjoy it in the near future, but it is definitely against the interest of the human species as a whole, if its interest is to have a lifespan as long as is compatible with its dowry of low entropy. In this paradox of economic development we can see the price man has to pay for the unique privilege of being able to go beyond the biological limits in his struggle for life.[31]

Another way of understanding the Entropy Law is to visualize the world's entropy process as constantly moving toward disorder. If a room or house isn't periodically cleaned and swept, dirt and dust will accumulate—i.e., disorder. This is nothing more than a demonstration of entropy. A deck of cards held in one's hand represents order; flung to the floor, the deck scatters into fifty-two individual cards in a random fashion. The intact deck is a stock of potential energy that is in a state of low entropy; once the cards are scattered, a person will have to expend more energy in picking them up and arranging them into a deck than the energy originally required to scatter them. Scattered on the floor, the cards are high entropy.

The significance of the discovery of the laws of thermodynamics has not yet been fully appreciated by any human society. The task of interpreting these important laws and coordinating action in accordance with them is vital to our civilization's survival. In our history of energy, the discovery of these laws took place halfway through the nineteenth century, in the midst of the Industrial Revolution in Europe and America. At this point, the Western world had not yet entered the Age of Energy, which spawned the Age of Technology and the Age of Automation. With the application of the thermodynamics laws, the efficient harnessing

‡ The terms "high entropy" and "low entropy" as used in this book are largely to designate available energy for use in a given system (be it a power plant, an automobile engine, or an ecological system). Available energy is always low entropy, and can be used for something; a pound of coal, for example, can be burned to produce electricity in a steam engine-operated electric power plant. The coal is low entropy. On the other hand, the heat rejected from the power plant can't be collected and burned to produce more electricity. It is high entropy. The survival of our species is based on the ability to collect low entropy and use it. Survival would not be possible without the availability of low entropy. High entropy is energy that is simply no longer available for use.

of energy in engines made it possible for engineers and scientists to develop the mechanized arts of the coming age, with its skyscrapers, automobiles, airplanes, telephones, radios, and televisions. But in the middle of the nineteenth century, this had not yet occurred, and life for the average individuals of Europe and America was still relatively placid and rural.

The coming mechanized era was only possible on paper. The real contribution of Carnot, Thompson, Clausius, and the other thermodynamics scientists was that energy could be understood for the first time in history. The Greeks had ordered the universe around fire, earth, air, and water—with fire having the most prominent place—but they didn't understand the nature of energy. James Watt built his steam engine without understanding the nature of the energy that allowed its operation.

More powerful and efficient machines could be built only through applying the science of thermodynamics, whose laws have frequently been called the Laws of the Conservation of Energy. In application, this means that engines could be built to maximize energy by converting a significant fraction of it into useful work instead of wasting 99 percent of it, as had the early steam engines.

The Importance of Energy Conservation

The significance of the "energy conservation" of the steam engine was learned by a British company that made the first "great iron ship," called the *Great Eastern*. The ship was designed and built in the 1840s in order to prove that the age of steam made round-the-world voyages possible, and the nineteen-thousand-ton (deadweight) vessel was designed to hold twelve thousand tons of bunker coal to fuel two steam engines. Unfortunately, the builders had little knowledge of thermodynamics, and their calculations of the coal thirst of the inefficient engines were off so badly that it was found that in order for the *Great Eastern* to make a round trip to Australia, the ship would have needed three times more coal than it could carry—more coal than the weight of the ship! The company folded, and the ship was broken up for scrap metal in 1888.* A ship of its size did not reappear until the next century, when engineers had acquired the necessary knowledge to construct more efficient steam engines that could convert more of the fossil fuel energy to work to turn the propellers—drastically lowering the coal requirement.[32]

The *Great Eastern*'s failure demonstrated another important lesson

* The *Great Eastern* achieved a brief period of success before this, when it was employed to lay the first transatlantic telegraph cable.

in the growing knowledge of energy. The ship exceeded what biologists in another sense call "carrying capacity." It couldn't meet the expectations of its owners and builders because it required more energy to operate than that which it could carry. The same constraints are imposed on human civilization. When a society no longer contains within it the energy supplies necessary to perform the work necessary to sustain it, that society cannot survive.

The history of Western peoples up to the nineteenth century largely chronicles a struggle over chemical energy: food. Famines, droughts, and natural disasters lowered the carrying capacity of many areas in terms of agricultural energy. When there was not enough food, there would be a reduction of population.

The first observer of the effects of food supply on the population was a Scotsman, the Reverend Thomas Malthus, who contended in his 1798 "Essay on Population" that while population increased geometrically (1, 2, 4, 8, 16, etc.), the food supply could not be made to increase on more than an arithmetic ratio (1, 2, 3, 4, etc.). Malthus in his "dismal theorem" foresaw a massive starvation for the future of England's rapidly increasing population, because he concluded that food supplies couldn't keep up with the population growth. The checks to population growth he divided into "positive" and "negative" areas. "Positive" checks—i.e., effective checks—would be war, famine, the spread of diseases, and natural disasters; "negative," or ineffective checks were "moral restraint" in bearing children and the postponement of marriage.[33] Malthus' ideas have been widely disputed, but the fact remains that he recognized the major problem of feeding a large population that overruns the natural carrying capacity of its geographic boundaries.

Since Paleolithic times, human populations had conquered the question of growth only by extending human ability to secure energy sources through the application of human energy-maximizing technology. By Rev. Malthus' time, the population of England had exchanged the technology of the mammoth and bison trap for more sophisticated tools, which lowered the amount of human labor and time needed to raise crops and domestic animals, thereby making possible increased agricultural productivity through technology.

The growth of population through advances in technology—technology made possible through the development of energy resources—had wrought quite a change in the world's civilizations by the nineteenth century. The world's population at the time of Christ was between 200 and 300 million people. By 1650, the population was about

500 million. Thus it had taken 1,650 years for the population to double. By 1850, the population of the world was about 1 billion people; the doubling period was only 200 years. And today, the doubling time is only one-sixth that—about 35 years.[34]

The exploitation of more energy from the natural ecological system by one species—*Homo sapiens*—came at a price: It lowered the capability of the system to support him. The most stunning example was the felling of trees to expand available agricultural land and to use wood as fuel. Before the age of the steam engine, trees were felled and used for fueling fireplaces (climate control), for small industrial applications such as firing forges, and for construction of homes and buildings.

Countless forests have been denuded as humans cut the trees across the globe. Three thousand years ago, Chinese forests were virtually eliminated in many areas to provide agricultural land. Cutting down the trees devastated the landscape to such an extent that annual floods came, as well as droughts in summer. In Europe, as well, country after country was stripped of forests. Author Gordon Rattray Taylor has described the effects of this process in Italy:

> The great floods of the Arno, such as that which bathed Florence in mud in 1967, have been experienced ever since the fourteenth century. When the woodlands round Florence were cut down and used as pasture for goats and sheep, these nibbled so close that the grass died and the ground turned to sterile baked clay. Brooks and wells dried up. By the eighteenth century, the woolen mills of Florence had to import wool and hair from elsewhere. The earliest flood on record was 1333. . . . Since then there has been a flood in Florence every twenty-four years, and a major flood every hundred years.[35]

The path was the same across Europe; England had so few trees left by the time of the Spanish Armada that lumber had to be quickly imported from Norway to build fighting ships. In America, the swath of the pioneers was indelibly marked by the systematic destruction of great forests.

What occurred when the forests were cut was a lowering of the over-all energy-carrying capacity of the regions directly affected. Without trees, there was no fuel. Wood, then, had to be imported. This required energy for transport, which was provided before the Industrial Revolution by the work of animals for land transport and sailing ships for ocean travel. With the reduction of energy resources in one area, the only means of survival for the population was either to leave or to import resources from a distant locale.

Energy Sources

In mid-nineteenth century, the development of technology in America had not yet reached the point where the energy sources required to fuel it involved much more than prairie windmills, water mills, and wood. Until 1865, wood served as *the* energy source, providing between 80 and 90 percent of the fuel for all energy requirements between 1850 and 1865.[36]

As with the railroads and steamboats (both of which primarily burned wood until after the Civil War), the American attitude toward comfort was far different from Europe's. R. G. Lillard described the pioneers' use of firewood:

> All cabin dwellers gloried in the warmth of their fireplaces, exploiting their world of surplus trees where a poor man, even a plantation slave, could burn bigger fires than most noblemen in Europe. . . . The kind of hospitable settler who burned a whole log in order to boil a kettle of tea didn't consider his fire psychologically good until he had crammed a quarter of a cord into a space eight feet wide and four feet deep and had a small-scale forest fire roaring in front of him.[37]

The origins of the U.S. "energy crisis" are firmly rooted in the American ethic, a curious mix of the love of the engine, which brings speed and power; a fixation with the practical, commercial aspect of life; and a childlike and hedonistic desire for comfort at any price. The consequences of ravaging the forests for fuel wood and charcoal are obvious today. The great forests of America are little more than memory. And in this century, few have yet begun to realize that the same fate is in store for the other fuels so prized for the advance of technology.

Behind this American ethic of little respect for resources—particularly energy resources—lurks more familiar attitudes, ones that have characterized humanity since the Stone Age: the preoccupation with domination over nature; the joy of the tool; and the insistence that the use of the tool bring power over the natural system.

This attitude of dominion is firmly embedded in the fabric of our institutions, and most conspicuously in the Western Judeo-Christian tradition. As Lynn White, Jr., observes:

> Our science and technology have grown out of Christian attitudes toward man's relation to nature which are universally held, not only by Christians, but also by those who fondly regard themselves as post-Christians.
>
> Despite Copernicus, all the cosmos rotates around our little globe. Despite Darwin, we are *not,* in our hearts, part of the natural process.

We are superior to nature, contemptuous of it, willing to use it for our slightest whim. The newly elected governor of California [Ronald Reagan], like myself a churchman, but less troubled than I, spoke for the Christian tradition when he said (as is alleged), "when you've seen one Redwood tree, you've seen them all." To a Christian a tree can be no more than a physical fact. The whole concept of the sacred grove is alien to Christianity and the ethos of the West. For nearly two millennia, missionaries have been chopping down sacred groves, which are idolatrous because they assume spirit in nature.[38]

Diversity

The art of survival—without the transfusion of imported energy—is based on adaptation and diversity. The idea of diversity is as important to the understanding of energy as are the laws of thermodynamics.

Diversity is not only a fundamental principle, but the supreme catalyst of the natural environment. Only through the interplay of countless millions of species is stability possible in our world. The unwritten law of diversity was learned by human populations through agriculture. Until the eighteenth century, agriculture was practiced by the "medieval system," which consisted of an early variant of crop rotation. A given plot of land was farmed two years out of three. This was done to secure a steady supply of nutrients in the soil, which were replaced by organic means—the manure of domestic animals and composting of left-over organic material from the crops.[39] Different crops were planted in the fields in successive years to insure that the soil would not be depleted by the steady growth of a single crop.

Thus was the principle of diversity learned. A single crop (or monoculture) grown year after year would soon deplete the soil by robbing it of its capacity to support growth. A succession of different crops would constantly replenish the soil with nutrients, ensuring a balance that would yield long-term gains.

This lesson was learned, not from theory, but from bitter experience of thousands of years of trial and error. Earliest European agricultural science was written as a tribute to this natural system of crop diversity, which flourished in the Low Countries of Europe during the medieval and later periods.†

Since the agricultural system supplied the principal source of energy in the world prior to the Industrial Revolution, the source of food for

† The lesson of crop rotation was passed on to the English by Sir Richard Weston in 1650 in his encyclopedic work on agricultural technology, *A Discourse of Husbandry Used in Brabant and Flanders, showing the Wonderful Improvement of land there, and serving as a pattern for our practice in this Commonwealth.*[40]

animals and men was carefully studied by the scientists of the seventeenth and eighteenth centuries. Medieval agriculture could support about a hundred persons to the square mile[41]—the absolute limit before the development of more sophisticated crop practices. These limits were expanded as the result of several influences, the primary ones being the increase of trade with other nations, which allowed a population increase by adding foreign sources of energy (primarily as food); the introduction of new crops and improvement of old ones with higher yields; and the introduction of better agricultural tools.

In the European Low Countries, agricultural science had reached considerable sophistication. Soil was carefully replenished with rich topsoil periodically added to the fields. The dairy industry was quite efficient, allowing for export of butter and eggs to other European countries. The fallow, or nongrowing, year (one out of three) of medieval farming was gradually eliminated. Organic fertilizer was collected not only in the form of manure from the fields, but the human and animal wastes from surrounding towns as well.[42] In short, the system of energy supply was sufficient to feed the population and work animals, and allow some surplus for export. The carrying capacity of the agricultural system found a natural balance.

The population of Europe increased by two hundred million people in the nineteenth century.[43] Energy support of this population through native means was clearly out of the question. There simply was not that much available land. Had not a massive transport industry developed to bring animals, grains, and other foodstuffs to Europe from the Far East and America, Malthus' "dismal theorem" would have proved entirely correct.

By this point in history, the Europeans had already exceeded the agricultural carrying capacity of their land—and consequently the natural energy-carrying limit of the population. Thus, from the nineteenth century on, Europe has been reliant on overseas support to meet even the most basic energy requirement: food.

The importance of the role of diversity emerges again and again in the course of the history of energy. By the addition of energy through diversified agricultural practices, the population of the Low Countries increased and surpassed the limits of other nations. Yet the size of townships was still limited by the agricultural base. Towns and cities require enormous amounts of energy to support them; and since there is no agricultural land within their limits, they rely solely on imported food energy.

The relationship of diversity to the city is also an interesting phenomenon. The city requires imported energy, which, through agriculture, must come from a diversified biological system. Yet the law of diversity also applies to the growth and economic development of the city itself.

The significance of this has been pointed out by Jane Jacobs in several books.‡ In *The Economy of Cities,* she details the history of two great English cities of the nineteenth century, Birmingham and Manchester. Manchester was thought at the time to be the greatest city in England, with its mighty textile industry, which served as the dominant source of industry and income. During the Industrial Revolution's surge of the 1840s, Manchester was believed by many observers to be the hallmark of the future because it had become powerful by shedding its agricultural past and developing a single industrial base. Borrowing another term from the biological sciences, Manchester represented a *monoculture,* the dominance of a single system—in this case, a single industry.

On the other hand, Birmingham represented the old sort of city, a city of countless cottage industries—what most observers of the time considered to be a throwback to the antiquated guild system of former days. Much of the work of the city was carried out by small companies of craftsmen, numbering no more than a dozen to each shop. The diversity of production ranged from buttons, decorative glass, and jewelry to small machines, guns, steel penpoints, and toys. The failure rate was high in Birmingham. Some of the small businesses made it commercially; many others didn't.

Yet by the beginning of the next century, Birmingham had survived and flourished, while Manchester had declined into stagnation. The diversity of Birmingham's industries had kept it from decay. Manchester, on the other hand, was destined to sadly fail as other cities developed more productive textile mills, which eventually outcompeted Manchester.

"Was Manchester, then, really efficient?" Jane Jacobs asks. It was indeed efficient, she concludes, while Birmingham was not. It was, in fact, Manchester's enviable efficiency, "the efficiency of a company town," that destroyed it when the "company" faltered, while Birmingham's "inefficient" diversity saved it for another century.[44]

Another tragic example of the failure of a monoculture was the great Irish potato famine little more than a century ago. When Ireland's single agricultural crop (or monoculture) failed, the famine nearly resulted in the destruction of the nation. Only through imports of food from Amer-

‡ Jacobs' analysis of the origin of cities first, agriculture later, is tenuous; but her extensive chronicling of the primacy of diversity in cities is without equal.

ica and other countries with energy-rich agricultural systems was the country able to survive.*

The danger of monoculture, then, is a lesson learned from biology—or more precisely, ecology, the science of integrated biological systems. Scientist/author Barry Commoner has modified this definition to "the science of planetary housekeeping."

In his book *The Closing Circle,* Commoner outlines the four basic laws of ecology:

1. *Everything Is Connected to Everything Else*
2. *Everything Must Go Somewhere*
3. *Nature Knows Best*
4. *There Is No Such Thing as a Free Lunch*

From the preceding review of thermodynamics, the bases for Commoner's second and fourth laws are readily apparent. His second law, in fact, is the First Law of Thermodynamics: Everything *does* go somewhere. Matter and energy can't be destroyed or created. We can change the form of something—iron ore, for instance, into a tool; but we haven't *created* a tool. We have merely changed the ore into a useful object at the expense of considerable energy in the process.

Barry Commoner's fourth law—"there is no such thing as a free lunch"—is a restatement of the Second Law of Thermodynamics, which makes it clear that in the process of changing iron ore into finished metals, we have lost—forever—a stock of low-entropy energy, and in the process transformed it to high-entropy (unusable) energy. On the Earth's ecological system, Commoner says that:

> Because the global ecosystem is a connected whole, in which nothing can be gained or lost and which is not subject to over-all improvement, anything extracted from it by human effort must be replaced. Payment of this price cannot be avoided; it can only be delayed.[46]

In his first law—"everything is connected to everything else"—we find the principle and necessity of diversity as opposed to monoculture. An ecological system might be a farm pond, a forest, a few inches of soil, or the entire Earth. The governing principles remain the same. In order to evaluate any system, Commoner says that a new mode of thinking is required: cybernetic thinking. Cybernetics† is concerned

* The great potato famine was caused by the attack of a specific fungus, *Phytopthorax infestans.* In the first four years of the famine, 1.5 million people died, fully one sixth of the Irish population. Millions more emigrated from Ireland, and the country has never been the same again.[45]

† The term "cybernetics" was coined by American scientist Norbert Wiener (1894–1964), who developed the science of cybernetics to study the principles of

with all the cycles and movements of a system. In other words, everything is seen as a loop—where each part of the system is related to all other parts. Everything feeds back to everything else.

Here's how Commoner describes this feedback system:

> The helmsman is part of a system that also includes the compass, the rudder, and the ship. If the ship veers off the chosen compass course, the change shows up in the movement of the compass needle. Observed and interpreted by the helmsman, this even determines a subsequent one: the helmsman turns the rudder, which swings the ship back to its original course. When this happens, the compass needle returns to its original, on-course position and the cycle is complete. If the helmsman turns the rudder too far in response to a small deflection of the compass needle, the excess swing of the ship shows up in the compass—which signals the helmsman to correct his overreaction by an opposite movement. Thus the operation of this cycle stabilizes the course of the ship.
>
> In quite a similar way, stabilizing cybernetic relations are built into an ecological cycle. Consider, for example, the fresh-water ecological cycle: fish—organic waste—bacteria of decay—inorganic products—algae—fish. Suppose that due to unusually warm summer weather there is a rapid growth of algae. This depletes the supply of inorganic nutrients, so that two sectors of the cycle, algae and nutrients, are out of balance, but in opposite directions. The operation of the ecological cycle, like that of the ship, soon brings the situation back into balance. For the excess in algae increases the ease with which fish can feed on them; this reduces the algal population, increases fish waste production, and eventually leads to an increased level of nutrients when the waste decays. Thus, the levels of algae and nutrients tend to return to their original balanced position.[48]

Commoner's third law is that "nature knows best," with its corollary that "the major actions of humans in the natural systems are almost always detrimental to the system."

A study of the ecological systems of the natural environment shows that all species on this planet have evolved in such a fashion that, through natural-selection processes, environmental systems thrive on diversity. Each species interacts in such a fashion with all others that the waste products of one become food for another. If one species be-

---

computers and relate the workings of information systems not only to machines but to life processes and human society. The word comes from the Greek *kubernetes,* which means "helmsman" or "steersman"; it is also the root of the English word "governor." Wiener concluded in his *The Human Use of Human Beings* that automation and mechanization in high-energy societies could spell destruction and dehumanization as well as technological benefits.[47]

comes dominant, either through natural processes or the involvement of humans in the natural system, a possible result is destruction of the entire system through the development of the dominant species—the monoculture. The monoculture lesson pervades not only the science of ecology, but every facet of our lives. A monocultural city is doomed to failure and decay. A monocultural crop eventually will spell the end of the centralized and limited agricultural system. A monoculturally governed society will inevitably disintegrate.

The Accumulated Experience

Up to this point in our history of energy and technology use, we still have not yet entered the twentieth century. We are stalled at the end of a great process in history, which brought human civilization out of the cave and into the midst of a great global enterprise characterized by exploration, discovery, technological innovation, and tremendous population growth. At this point, slightly over a century ago, the science of thermodynamics had just been codified; and ecology, the science of the environment, had just seen the publication of Charles Darwin's *The Origin of Species* (1859). Both Darwin and the thermodynamic scientists dealt with energy in their respective efforts. Darwin described the interaction of species in their daily battle for the energy to survive, and left us with a vital heritage—the description of the process of natural evolution and selection—which illustrates the way in which all species adapt to the energy flow of the Earth's environment.

Darwin's ideas, however, were attacked in the nineteenth century with a vicious fervor. He was accused by other scientists of having linked human life with "lower" organisms, and by Christians for having subverted Holy Writ; finally, his ideas were twisted into the political grab bag called "Social Darwinism." This philosophy, as originally expounded by Herbert Spencer and further enlarged upon by many others, held that man's "dominance" in nature justified not only the destruction of nature itself, but also racism, jingoist wars, and the extension of the capitalistic system as a "natural course" of evolution.[49]

On the other hand, the thermodynamic sciences enjoyed a meteoric rise. The ideas of Carnot and Clausius provided the technical means by which bigger and better engines could be built, which meant bigger and better vehicles, machines, buildings, and the other elements of mechanized high-energy society.

Without the technical contributions of these scientists to the understandings of the workings of heat engines, technology as we know it today would not be possible. As was the case with Darwin, however, some of the most critical ideas contributed by the energy scientists have never been fully accepted and applied to the pressing problems of modern society. The concept of entropy has only recently been applied to economic and environmental problems—and much is left to be learned about the flow of energy both in nature and in the human technological interactions with nature.

The lessons of thermodynamics and ecology provide us with a unique set of symbolic tools with which we can measure the evolution of human culture and of the machine, as well as the energy systems—both biological (food) and synthetic (fuel for technology)—that make that evolution possible.

Rise of the American High-Energy Society

At the midpoint of the nineteenth century, America was a sprawling, energetic nation much preoccupied with growth and the pioneering tradition. Alexis de Tocqueville, the young French writer who toured America in the 1830s, described the fresh character of the American lifestyle in his *Democracy in America*. Looking at the United States from a European perspective, De Tocqueville remarked on the practical quality of philosophy in this country. He noted that, as opposed to Europe, where much of the thinking was in a theoretical realm nurtured by centuries of tradition, Americans were more interested in applying theory and science to getting things done. "These very Americans," he said, "who have not discovered one of the general laws of mechanics, have introduced into navigation an engine which changes the aspect of the world."[50]

So it was. Europe was the cradle of the Industrial Revolution, but it was America that served—and continues to serve—as the crucible of the technological age. The history of the United States appears to us today as a migratory history. From the origins of the Republic, the population surged westward into the interior, leaving behind a network of trails, post roads, railroads, towns, and farms.‡ From the beginnings, the pioneers were preoccupied with speed—getting across the country as quickly a possible. As James K. Page, Jr., and Richard Saltonstall, Jr., observed in their recent book, *Brown-Out and Slow Down,* much of our history can be viewed as a succession of transport modes and increasing speed. They detail this history as follows:

‡ In this process, the inevitable first toll of human "progress" was the annihilation of the native Indians. De Tocqueville noted that, in the 1830s, only 6,273 Indians were left in the 13 original states.[52]

> One of the most romanticized eras of our history is the time when thousands and thousands of courageous people stripped themselves to the bare essentials, packed those bare essentials into wagons, and set off from one of several booming mid-Western towns across the deserts and mountains to settle the far West. The wagons they used were uncomfortable, light, and inexpensive. They were designed for speed, for the quickest possible trip through dangerous territory. What was left of the wagons after the journey was often simply thrown away or broken up and used for firewood.
>
> Steamboats plying American rivers literally raced. Urged on by passengers who not only wanted to get somewhere fast but who also simply enjoyed the sensation of speed, riverboat captains would pour on the steam, strain the boilers and *make time.*[51]

In addition, historian Daniel Boorstin has pointed out that the same was true of the early American railroads. The English built their early nineteenth-century trains of sturdy materials, with strong, heavy engines to pull them. Roadbeds as well were laid out carefully in anticipation of many years of use. On the other hand, American trains were built with lighter, more expendable engines, and the roadbeds were often built in poor terrain, with steep grades and sharp curves. The intent and method of execution was obvious: The Americans' clarion cries were speed and expediency. The Europeans could worry about tradition; Americans were busy getting the job done—quickly.[53]

Energy and War

The American identification with speed and power didn't stop there. The American Civil War, one of the bloodiest conflicts in history, marked the rise of new technologies on the battlefield and on the seas. The inventor of the ironclad ship *Monitor*, John Ericsson,* told President Abraham Lincoln of the importance of war technology:

> The time has come, Mr. President, when our cause will have to be sustained not by numbers, but by superior weapons. . . . such is the inferiority of the Southern States in a mechanical point of view, that it is susceptible of demonstration that, if you apply our mechanical resources to the fullest extent you can destroy the enemy without enlisting another man.[54]

Both the technologies of peaceful transport and of war, however, were entirely dependent on the development of concentrated energy sources.

Ericsson's remark to Abraham Lincoln is typical of the technolo-

* A prolific Swedish inventor and American immigrant, also known for his work in the development of solar-powered engines. [Chapter 6].

gist's monocultural thinking. Although the main concern of this book is with the application of energy resources to the peaceful development of civilization, history shows that the search for concentrated energy has largely been for military purposes. Witness the argument of Admiral Hyman Rickover, the "father" of the nuclear-powered submarine, who says that the two high-energy technologies employed in medieval Europe—sailing ships and gunpowder—wrought a great revolution:

> With ships that could navigate the high seas and arms that could outfire any hand weapon, Europe was now powerful enough to pre-empt for herself the vast empty areas of the Western Hemisphere into which she poured her surplus populations to build new nations of European stock. With these ships and arms she also gained political control over populous areas in Africa and Asia from which she drew the raw materials needed to speed her industrialization, thus complementing her naval and military dominance with economic and commercial supremacy.

Applying this analysis to historical trends, Rickover concludes that "when a low-energy society comes in contact with a high-energy society, the advantage always lies with the latter."[55]

There is a great fallacy in the analysis of Rickover and Ericsson, though: the fallacy of the superiority of high-energy monoculture. Applied to wars and military technology, high-energy societies have not always advanced. For example, the history of the eighteenth-century American struggle for independence and the Indochina wars of France and America in recent years show that low-energy societies have emerged victorious from the conflict. In the case of the American Revolutionary War, the well-equipped "Redcoats" were defeated, not by the superior firepower of a high-energy society, but by the application of time, perseverance, and quasiguerrilla warfare techniques. In Indochina, the example is clearer. Years of struggle taught the strategists and the indigenous soldiers of the Viet Minh and the Viet Cong that the path to victory was assured by application of guerrilla techniques, playing by the diversified rules of the jungle, not by the high-energy-technology rules of the French and the Americans. Superior military technology fueled by an enormous outlay of energy did not help the American armies even adequately *identify* the enemy in Vietnam, much less defeat him. The guerrillas played by the rules of diversity and decentralization, not by the high-energy rules of military monoculture.

In today's world, the alternative proffered by high-energy warfare, to paraphrase Herman Kahn, is akin to the introduction of a new rule to the game of chess: victory by kicking over the board and scattering the pieces, which is precisely what high-energy nuclear warfare has done. With the capability of the extermination of life on whole continents, if not the Earth itself, modern spoils of war are nonexistent. High-energy monocultural victory would eliminate the victor as well as the defeated enemy.

Fossil Fuels and American Civilization, 1850–1900

The backbone of high-energy civilization is the availability of concentrated, readily transportable sources of energy. The energy-intensive lifestyle characteristic of the Western industrialized world today was born with the first oil well and the sudden exploitation of coal about the time of the Civil War.

America has set the pace for the lifestyle of the high-energy civilization. It was here that oil was first drilled and exploited on a large scale. It was here that coal and steam condensed a continent into a few days' train journey. It was here that the electric light and the electric distribution system were invented. It was here that abundant energy and mass production made the cheap automobile ubiquitous. It was here that the potent fire of the atomic nucleus was harnessed to produce still more energy—and destruction.

And it is here that the inevitable collision between unlimited energy demands and limited energy resources is first occurring.

The second half of the nineteenth century was marked by the emergence of the fossil fuel energy society in America. Up to this time, the young nation had been running largely on muscle power and renewable resources—wind, falling water, and firewood. In 1850, about two thirds of all the horsepower-hours produced came from the work of animals—about half of that from horses and one eighth from human beings.[56] Two thirds of all mechanical work in terms of horsepower was done by windmills and falling water.[57]

Coal

Ninety percent of the fuel burned in 1850 was wood, with coal accounting only for the remaining 10 percent, despite the fact that coal was plentiful and a vast reservoir of experience had been developed with coal in Europe.[58]

Furthermore, coal had been known and used by the American Indians. The Hopi tribe in Arizona was mining coal by around A.D. 1000 and burning it to heat their homes and fire their pottery. About a hundred thousand tons had been consumed by the time the Spaniards found their way to Arizona.[59]

There was a reluctance in most parts of the United States to utilize this more concentrated fuel. It was with the extensive cutting of the forests, which raised the price of wood in the East and increased the distance to which it had to be carried to the growing cities, that demand for coal began to rise. Its consumption tripled between 1850 and the beginning of the Civil War in 1861.[60] Demand for it continued to grow rapidly with its increasing replacement of wood as the fuel for steam locomotives and for manufacturing—especially the fast-growing iron and steel industry. In 1885 coal surpassed wood as the dominant fuel and launched the United States into the fossil fuel economy, whose cheap and abundant energy was to propel the nation to affluence and the energy crisis. In that first year of the American fossil fuels era, 40 percent of the coal went to fuel the railroads and 13 percent went to coke production for the steel industry. The remaining 47 percent went to other miscellaneous industries and to homes for residential fuel. The use of coal continued to grow, and its share of the nation's fuels had spiraled from 9 percent in 1850 to 65 percent in 1895. Coal was now King, and it was to remain the dominant fuel well into the twentieth century.

Oil

Oil entered the American economy on a large scale in 1860, when five hundred thousand barrels of crude oil were produced. The nation had long been searching for a good cheap fuel for lamps and lanterns. Substitutes for whale oil had so far been generally unsatisfactory. Camphene, an oil distilled from turpentine and alcohol, produced a bright flame, but it was explosive and burned with a disagreeable odor. Lard oil and candles produced insufficient light. And they were all expensive. Pressure for better lighting was increasing as more people learned to read and as the embryonic factory system began demanding better illumination.[61]

England had developed an impressive coal gas pipeline network for lighting homes, factories, and street lamps, but this approach had not caught on in America—except in a few highly concentrated population centers like Baltimore—because the scattered population made such a system for the most part impractical. Another reason was the still underdeveloped coal industry. Baltimore was still importing coal from England for its coal gas system as late as 1850.

However, another coal product was being received more enthusiastically. An English coal mine operator in Derbyshire had devised a method of producing oil from coal, which he called "coal oil" or "kerosene." By the late 1850s, there were fifty to sixty kerosene plants in

Pittsburgh and the East Coast area, and the nation seemed to have found its answer to the illumination problem.[62]

It was at about this same time, however, that more and more people began noticing the resemblance between the new kerosene and a largely useless "rock oil" that oozed and bubbled from springs and salt wells in western Pennsylvania. Also called "Seneca oil" because the Seneca Indians mixed it with their facial paint, the substance had been touted as a medical panacea. Settlers along Oil Creek and in Venango County, Pennsylvania, insisted that it helped cure their arthritis.[63]

In 1854, a small group of Pennsylvania entrepreneurs organized the Pennsylvania Rock Oil Company with headquarters at nearby Titusville, Pennsylvania, and hired a Yale chemistry professor to analyze a barrel of their rock oil. His report, in 1857, concluded that the substance furnished as much light as any fuel that had ever been tested; that it burned more economically; that it did not turn gummy, rancid, or acid upon prolonged exposure; and that it held up well in very cold environments. It was a raw material, the chemist reported, from which some "very valuable products" might be manufactured.[64]

Bolstered by these findings, the Pennsylvania Rock Oil Company hired a former railroad conductor to undertake the job of getting the rock oil out of the ground in sufficient quantities to be commercially profitable. It quickly became obvious that drilling, rather than digging, was the best means of getting to the substance; and in September 1859, at a depth of sixty-nine feet, the Pennsylvania Rock Oil Company struck oil.[65]†

It was not the first time in history that the potential value of petroleum had been recognized. An oil well was dug on the island of Sante in the Ionian Sea in the year 400 B.C., and its oil also distilled for lamp fuel. The Burmese were drilling for oil in A.D. 100, and it had been sought in other sites around the world.[67] But it was only in the Western industrial world of the 1850s that the confluence of science, technology, and society were at the juncture necessary for oil to be seized upon as the concentrated fuel that would replace coal.

The popularity of the new lamp fuel (the "kerosene" label of the old coal oil stuck with the new petroleum product, through its spelling was commonly revised to "kerosine") rose rapidly. From 500,000 gallons in 1860, the year after the first well came in, crude oil production

† The Titusville area was deluged with people seeking ways to cash in on the much-heralded oil strike. Many of the newcomers were whalers who were now pursuing the whale oil substitute as they had once pursued the whale itself. Richard O'Connor has reported that so many whalers showed up, in fact, that one reporter was prompted to observe—rather prophetically, as it turned out—that "now they have come to harpoon Mother Earth."[66]

rose to 1.7 million gallons five years later, and to more than 2 million gallons by 1870—most of it kerosine. By 1880, crude oil accounted for 13.2 percent of all mineral fuels consumed, a level that it maintained until well into the first decade of the twentieth century. As late as 1900, kerosine still accounted for two out of every three gallons that left the refinery.[68]

Some of the by-products that evolved during this period were lubricants, which solved another major problem for embyronic industry and mechanization; fuel oil; and an apparently useless substance called gasoline, which did find some early market as an additive to improve the combusting qualities of coal oil.[69]

The market for fuel oil was slow in developing—primarily because the equipment required to properly burn it had not been developed, but also because its relatively high price left it uncompetitive with coal and wood as a fuel. During the Civil War the Navy undertook some experiments aimed at testing the potential of fuel oil to drive ships. They proved that oil could be burned in steamships, that oil-fueled ships were faster than coal-burners, that the fires could be started and extinguished much faster, and that the ships could remain at sea two to three times as long without refueling.[70]

However, dangers of explosion, high losses through volatilization, oil's foul, permeating odor, and the very high temperatures created in the engine room appeared to rule it out as a maritime fuel.[71]

The use of crude oil in transportation and manufacturing can be attributed to the resourcefulness and tenacity of Standard Oil promoters, who were charged with the responsibility of unloading the growing surplus of oil pouring from the wells in the Lima, Ohio/Indiana fields that were tapped in the mid-1880s. It was not suited for refining into lamp oil, but some disposition had to be made of the rapidly growing petroleum inventory. The solution was to launch an extensive campaign to persuade industries that oil was better than coal for firing boilers for steam generation. The low price of Lima crude (sixty cents a barrel delivered in Chicago) added considerably to its attraction, and by 1889, fuel oils, with sales of thirty-five million barrels per year, accounted for 35 percent of total petroleum sales.[72] Standard Oil Company's success with the Lima crude had paved the way for acceptance of petroleum as a fuel that could compete with coal.

Invention of the Automobile

Related events were taking place that were to profoundly affect not only the petroleum industry and the fossil fuel economy, but the very fiber of the American way of life. As was noted earlier, the energy pioneers of the eighteenth century had laid the foundation for the Industrial Revolution with the invention and development of the steam engine—first for pumping water out of coal mines. The application of steam to transportation in the form of railroads and ships in the nineteenth century was a gigantic step forward, but the problem of more efficient transportation modes for short distances still had not been solved—in part because coal, the only concentrated fuel available up to that time, was too bulky and inefficient to make coal-fueled steam locomotion practical except on a very large scale.

What the world wanted was a smaller, more mobile engine.

Illuminating gas, benzene from coal, and even gunpowder were being tested—all without success—as possible fuels for operating a gas engine developed back in the seventeenth century. Toward the end of the eighteenth century the concept of compressing an air-gas mixture in the engine before igniting it was developed. In 1870, inventors began testing the engine with gasoline, a still essentially useless by-product of the kerosine refining process.[73] In 1877 Nikolaus Otto's gas engine experiments paid off with the first workable four-stroke gasoline-fueled engine in which the gas was compressed before ignition. Ten years later, this engine, the forerunner of all internal combustion engines in operation today, was first adapted to vehicles, and the first Benz automobile was patented.[74]

By 1900, forty-two hundred automobles had been built in the United States—most of them still steam-driven or electrically powered. Only one fourth were internal combustion. However, the emerging petroleum industry had provided both the fuel and the lubricants required to make the gasoline-powered automobile light, maneuverable, fast, and competitive in cost. Olds switched his Oldsmobile engine from steam to internal-combustion gasoline in 1900. Three years later, Henry Ford introduced his gasoline-powered automobile and announced the formation of the Ford Motor Company.[75] His mass-production techniques would not only revolutionize industry, but also irrevocably anchor the gasoline-powered internal combustion engine as the basic mode of transportation for America—and the world. Once again, as the new energy source became available, the nineteenth-century technology for exploiting it was not far behind.

A more efficient petroleum engine was patented in 1893 by German engineer Rudolf Diesel. Diesel's major purpose was not to develop an engine for vehicles, but to maximize heat engine efficiencies described by Sadi Carnot, the pioneering thermodynamic theorist. In his own account of his invention, written in 1913, Diesel referred back to a lecture on thermodynamics by Professor Linde at Munich Polytechnic in 1878:

I wrote in the margin of my college notebook: "Study whether it is not possible to realize in practice the isothermal curve?"‡ Then and there I set myself the task! This was no invention, not even the conception of one. Thereafter the wish to materialize the ideal Carnot cycle dominated my being. I left College, went into practice, was obliged to win myself a position in the world. The thought pursued me incessantly.[76]

He worked first with superheated ammonia vapor to try to wrest a greater efficiency from his engine, but finally hit upon using superheated air under high pressure, while gradually introducing finely atomized fuel, "and to let the air expand simultaneously with the consumption of the fuel particles, so that the utmost possible amount of heat engendered should be converted to work."[77]

Electricity

With the growing awareness of the ready availability of concentrated fossil fuels came a bewildering barrage of machines and technologies designed to capitalize on them. Many of them proved to be milestones of modern civilization: the automobile, for example; and the telephone, patented by Alexander Graham Bell in 1876. Of perhaps even greater significance were the electric light in 1879 and the subsequent creation of a distribution system to permit its common use.

"It would be hard to overestimate the economic, social, and intellectual significance of these improved modern lights," wrote economic historian Harry Elmer Barnes, adding:

> Their development has made possible many phases of modern industrial life and social concentration. Production, both manufacturing and agricultural, has been enormously increased by making it possible for people to work under conditions and during hours that would otherwise be impracticable for productive labor. Contemporary transportation would be virtually impossible without artificial light. Again, the significance of gas and electric lights for modern recreation, concentrated as it is chiefly in the hours after 6 P.M., is at once obvious. Their educational significance, in the broadest sense of the term, is apparent to anyone who will attempt to estimate the importance of the theater, movies, night schools, university extension work, evening lectures, illustrated lectures, and the like.[78]

Edison himself, however, saw his own greatest achievement not as the light itself, but the world's first electrical power-generating and dis-

‡ The "isothermal curve" refers to the theoretical performance of the Carnot cycle steam engine. What Diesel wanted to accomplish was to develop a more efficient cycle, in practice—which he did.

tribution system, which he initiated in the heart of New York's downtown Pearl Street financial district in 1882. The generating station was located in a building of ironwork structure on a site one hundred by fifty feet, and was designed to serve an area about one half mile square. Four boilers were located on the ground floor, and six two-hundred-horsepower dynamos, or generators, capable of lighting twelve hundred lamps each, were installed on the second floor. Edison himself supervised the workmen who dug the trenches and laid the cable after securing permits from the city of New York. The fact that the enterprise was underwritten by a group headed by financial giants Cornelius Vanderbilt and J. Pierpont Morgan no doubt facilitated the necessary bureaucratic approval.[79]

Edison recalled the venture in these words:

> The Pearl Street station was the biggest and most responsible thing I had ever undertaken. . . . There was no parallel in the world . . . all our apparatus, devices and parts were home-devised and home-made. What might happen on turning a big current into the conductors under the streets of New York no one could say. . . . All I can remember of the events of that day is that I had been up most of the night rehearsing my men and going over every part of the system. . . . If I ever did any thinking in my life it was on that day.[80]

The electric light and the electric power distribution system required to sustain it were to change the world as have no other technological developments. Electricity made it possible to deliver the work potential generated by the steam engine to distant sites cheaply and without the noise, smoke, size, or other characteristics of the steam engine at the generating site. In effect, it put the steam engine, with all its enormous capability, at the disposal of virtually every home, business, and industry in America and much of the Western world.

Edison himself designed Pearl Street Station's six electric generators. They were based on the well-established principle of using a reciprocating steam engine to rotate an armature—a coil of wires around an iron core—in a magnetic field to produce electricity.[81] This is the exact reverse of the principle of the electric motor, which had been invented in 1831 by Michael Faraday.[82] The principle of the motor was that electric current could be made to turn a shaft to perform work. The principle of the dynamo was that a machine, the reciprocating steam engine, could be made to perform work to generate electricity.[83]

The reciprocating steam engine used by Edison and others of the day was very inefficient, and much more efficient methods were quickly devised. The best-known and most widely adapted was developed by Sir Charles Parsons, a British engineer who arrived at the concept of placing several turbines on one shaft, with each turbine powered by steam exhausted from the preceding turbine. His concept, called "pressure staging," has replaced the reciprocating steam engine altogether in the generation of electricity, raising thermal efficiencies from less than 7 percent in 1900 to nearly 40 percent in the most efficient steam electric plants today. When Parsons died in 1931, he had seen his turbine increase in capacity from a 7½ kilowatt unit in 1884 to units of 200,000 kilowatts.[84] Like many other thermodynamic innovations of this period, the Parsons turbine is still in extensive use today.

Hydroelectric Power

Close on the heels of the electric utility industry's September 4, 1882 debut with the Pearl Street Station came another source of electric power generation: hydroelectric power. Harnessing waterfalls and rapids to turn wheels for grist mills and other industrial applications had been a major source of power for pre-1850 industry, and this early experience was now being turned to a new application. Just 26 days after Pearl Street Station went on line, water wheels also began generating the nation's first hydroelectric power on the Fox River in Appleton, Wisconsin. The facility was designed to light 250 bulbs of 50 watts each.[85]

Hydroelectric power expanded relatively rapidly, but it was to remain at a disadvantage in comparison with fossil fuels for electric power generation because most of the suitable sites for harnessing river flow for electric power production were relatively isolated from the population centers that needed the power. The first major hydroelectric station was erected by the Westinghouse Company near Niagara Falls to furnish electricity to the nearby city of Buffalo, New York. It began operation in 1896.[86] By the turn of the century, hydroelectric power accounted for about 2.5 percent of all U.S. energy sources,[87] but about 57 percent of all the electricity generated.[88] This share quickly dropped to about one third, where it was to remain through most of the first half of the twentieth century.

Labor-saving Devices

As electricity began to be more widely available, so did the means of tapping into it to perform work that up to that time had been performed by musclepower. It was at about this time—around 1900—that the term "labor-saving device" was coined,[89] and electrical appliances may well have served as the inspiration for it. Reyner Banham in *The Architecture of the Well-Tempered Environment* has chronicled the barrage of electrically powered devices that quickly became available to the consumer:

> By 1900, manufacturers' catalogues listed and illustrated most of the cooking vessels (kettles, skillets, etc.), with built-in heating elements that are with us today, albeit in rather primitive forms, also toasters, roasters, hot-plates and ovens, radiant-panel heaters, convecting heaters, coffee-grinders, immersion heaters, and such period hardware as electric cigar-lighters and curling-tong heaters. By the time of the General Electric catalogue of 1906, the most prized of all domestic electric equipment, the electric flat-iron, was well established, as was the electric coffee-percolator and the fore-runner of the electric blanket heater. . . .[90]

Still more sophisticated appliances were soon to appear as the result of the small electric motor patented in 1888 by Nikola Tesla. Tesla, an engineering genius who had immigrated to the United States from what was then Austria and is now Yugoslavia, developed the polyphase, or alternating-current, motor.[91] Electric motors up to that time had been direct-current devices, because what current had been available came from batteries. When commercial electric power generation came into being, it was discovered that direct current created major transmission problems—i.e., transmission losses were high, reducing both the efficiency and the profitability of electric power generation. This problem did not occur with alternating current—i.e., current that changed its direction of flow many times per second at a rate corresponding to the speed of the rotating armature. Thus it was alternating, rather than direct, current that was coursing through an increasing number of American cities and villages of the late nineteenth and early twentieth centuries.[92] The significance of Tesla's alternating-current electric motor was that it enabled every home, store, and factory to tap this new energy delivery system and to use it in many ways that had not been envisioned up to that point.*

A year after Tesla patented his motor, he joined forces with George Westinghouse to market an electric fan driven by a one-sixth horsepower alternating-current motor.[93] It signaled the beginning of the mechanization of the home, and ultimately made possible the electric

* Alternating current was not without its ardent opposition. It was condemned—largely, it is suspected, by direct-current interests—as "the killing current," a reputation that ironically was quickly sustained by one of alternating current's first applications: electrocuting condemned men at New York's Sing Sing Prison.[94]

refrigerator, vacuum cleaner, washing machine, dishwasher, and many other home appliances and devices.

Mechanization of Agriculture

The latter half of the nineteenth century also marked a series of inventions leading to farm mechanization that were as momentous as they were timely in the development of the agricultural base necessary to support the high-energy society that was rapidly evolving.

As Professor Siegfried Giedion pointed out in *Mechanization Takes Command,* the idea of the reaper—"the most crucial development in the mechanization of agriculture"—had been devised by 1783, but not until half a century later, in 1834, did Cyrus McCormick obtain his historic patent. Twenty years later found it still among the shelved inventions. Giedion quotes an 1854 Philadelphia account that observed that "scythe and cradle continue to be the principal instruments in use for the cutting of hay and grain in Europe and in America. . . . All the attempts to introduce machinery have failed, more, it may be stated, from the disinclination of the public to encourage them, than from want of merit."[95]

Nevertheless, the reaper did begin to find acceptance around this time in the open Midwest prairies, and in 1858, Illinois farmers C. W. and W. W. Marsh invented the harvester, which did away with the need for raking the grain cut by the reaper. In the same year, John F. Appleby invented the twine binder, which automatically bound with twine the sheaves raked by the harvester. Thus were completed the inventions necessary to mechanize the prairie farmland and feed the ever-increasing agricultural demands of the growing nation. Developments after this time were largely improvements in weight, design, efficiency, and stability, in which steel frames played an important role.[96]

Building Design and Technology

One sector of the American economy, building construction and design, has been especially sensitive to energy availability. The foundation of today's glass-encased skyscraper is a series of inventions and developments that were the direct and logical consequence of increased availability of energy, which suddenly rendered them possible and functional.

It was cheap steel that made the skyscraper possible, and it was the Bessemer process and the abundance of coal to fuel it that in turn made possible cheap steel. The economic incentive for taller and taller buildings was the rising value of construction sites in such burgeoning young cities as New York and Chicago. E. G. Otis had in 1854 designed the hydraulic elevator, a singular invention without which the skyscraper would simply never have come into being. In 1889 the fast, smooth, and silent electric elevator made many-storied buildings practical.[97]

Still it was steel that was the *sine qua non* of the skyscraper. Without it, multistoried office buildings were bulky and inefficient, with prefabricated wrought-iron floors and cast-iron fronts supported by masonry walls. The fourteen-story Pulitzer Building constructed for the New York *World* in 1890 required nine-foot-thick walls at the ground floor, where need for space was especially great. Steel beams permitted the walls to function merely for shelter and privacy, rather than as foundation and support for the structure as well.[98]

Chicago took the early lead in skyscrapers. Steel beams were first used in upper stories of an iron-framed Chicago office building in 1884, and in 1890 the first complete steel frame office building was erected there. In 1892, Chicago's 21-story Masonic Temple was billed as the largest building in the world and one of its seven wonders. New York's first skyscraper didn't appear for two more years; but by the turn of the century, New York had surpassed Chicago, with 29 skyscrapers to Chicago's 16, and a maximum height of 386 feet to Chicago's 300 feet.[99]

There were other developments, interrelated and mutually dependent, that were required before the giant skyscrapers could be considered for mass construction. One of the most important was climate control—the regulation of temperature, air movement, and humidity—which was ultimately to lead to space heating and air conditioning. Logically enough, space heating has deeper historic roots than air conditioning. Benjamin Franklin in 1742 first set forth the principle of heating a room with warm air flowing from a stove.[100] Steam engine inventor James Watt, exploring other applications of steam, heated his own office with it in 1784, and by the first year of the 1800s a public building in Leeds, England, became the first entire building to be heated by steam. By the 1860s, most buildings of significance—both public and private—were heated either by steam or by hot water.[101]

In 1876, Birdsill Holly, of Lockport, New York, acclaimed in heating engineering circles as the "Thomas Edison of central station heating," ran an underground pipe from a boiler in his home to a barn at the back of his property, and later to an adjoining house. A year later, he built the first experimental central station or "district heating" plant at Lockport, which successfully heated many homes, offices, and commercial buildings the following winter.[102]

The concept of forcing hot air out into a room or building, rather than merely letting it find its own way, goes back at least two centuries. The word "ventilator" itself was invented by J. T. Desagulier to describe the man who was to turn the crank of the centrifugal fans with which he proposed to furnish air to the lower decks of ships and the chamber of the English House of Commons in 1736. Such a cumbersome arrangement was obviously impractical for most applications, however, and forced-air heating had to wait until the development of effective mechanically driven fans. A primitive fan furnace was advertised in an 1860 catalogue and, in the decade that followed, fan-forced ventilation gained wide use in industry for providing air for mining and shipping and for drying tea. Power for the fans was provided by the steam engine, and much later, by the gas engine.[103]

The ventilating concept remained static from the 1860s to the end of the century, when Nikola Tesla's alternating-current motors and the advent of electric power suddenly made fan ventilation a totally feasible concept. In fact, Edison's Pearl Street Station had barely started generating electricity when the first mentions of electric fans for cooling New York hotel rooms appeared.[104]

Technology, following on the heels of energy, had now completed the basic groundwork for the skyscraper: artificial heat, light, and ventilation. There were, of course, in the words of architectural historian Reyner Banham, "a gaggle of other devices" that were "equally necessary in order for business to proceed at all—and without ability for business to proceed, skyscrapers would never have happened." These were, of course, the telephone, the elevator, and the flush toilet emptying into a city sewage system—not to mention the revolving door, one of the rare simple, low-energy mechanical innovations falling into the category of "great achievements" in a rapidly developing high-energy society.[105]

1900 to World War I

Energy use jumped dramatically in the first decade of the twentieth century. Coal production nearly doubled; petroleum production more than quadrupled; natural gas production doubled, as did the amount of electric power produced by hydrostations.[106] The economy was soaring also. The Gross National Product increased by 50 percent. Manufacturing production went up 58 percent; production of pig iron, 89 percent; and steel, 125 percent. Power equipment employed in manufacturing almost quintupled—from 9.8 million horsepower to 48.5 million horsepower.[107]

The number of automobiles was also increasing at a startling pace. From a base of 8,000 in 1900, the automobile population had grown to 194,000 by 1908.[108] If the internal-combustion automobile needed a boost (which seems questionable from the statistics), it got it that year in the form of a bizarre motor car competition that has gone down in history as "The Great Automobile Race." It was to start in New York, maneuver across the continent to the Pacific Coast, then on to Japan and through Siberia to Moscow, then to Berlin and finally Paris. For months daily headlines carried progress of the unlikely adventure. The stream of news accounts drilled again and again into the minds of awe-struck readers that the automobile could go anywhere in the world and do anything.[109] It stirred an impressive fury of automobile buying. From 194,000 registered automobiles in 1908, the car population tripled to more than 600,000 3 years later.[110] Standard Oil saw to it that the fuel they needed was there, with a string of service stations throughout the United States.[111]

The automobile spiral continued unabated. By 1920, there were more than eight million; by 1930, more than 23 million; and by 1941, the year the United States entered World War II, the number had reached 29.5 million.[112] The car was to quickly establish itself as the measure of the American middle class and, as such, to institutionalize the corner filling station as just as wistful a part of Americana as the skycraper.

Petroleum

Gasoline still did not constitute a major refinery product in the early years of the century, although by the end of the First World War, it represented about 20 percent of total oil consumption.[113] Fuel oil was the petroleum product that was most rapidly increasing in popularity for the first few years of the century, although coal was still the dominant fuel. And even fuel wood, declining as it was, still outranked the petroleum products in 1910, when it accounted for a little more than 10 percent of the total fuel consumption—a complete reversal from its 90 percent position of 1850. That same year, 1910, coal production exceeded 400 million tons for the first time in the industry's history, a level it was to retain throughout the first half of the twentieth century except for one year during the Depression and one year in the early 1950s.[114]

By 1910 fuel oil had displaced kerosine as the first-ranked petroleum refinery product, and kerosine was fading into the background with the rapid advances of both fuel oil and gasoline. From one eighth of the total refinery output in 1904, fuel oil had risen to one third by

1909, and by the outbreak of World War I it accounted for almost half of the petroleum industry's total output.[115]

Much of this shift can be attributed to the aggressive marketing and promotion practices of Standard Oil Company, which almost completely dominated the petroleum market, although some competition was developing with the Shell and Gulf Oil companies. In 1903, the year Henry Ford founded his motor company, the Wright brothers fulfilled man's age-old dream of flying. And the Standard Oil salesmen were there—with barrels of gasoline and cans of lubricants.[116]

With the spectacular gusher that brought in the legendary "Spindletop" well in West Texas in 1901, and major new finds in California and Oklahoma, oil again flooded the market. Recalling their success with Lima crude in the 1880s, the Standard Oil promotion men took to the road again to once more persuade factory owners, shipping companies, and railroad magnates to switch from coal to the more versatile and concentrated petroleum fuel.[117]

A Gulf, Colorado, and Santa Fe Railroad oil-burning locomotive on a test run of 450 miles burned 42 barrels, the equivalent of 12 tons of coal. Four years later, the Santa Fe boasted 227 oil-fueled locomotives. Shipping, too, was rapidly switching to oil, encouraged by the record-setting voyage of the United Fruit Company's oil-burning steamer *Breakwater* in 1902.[118]

The shift from coal to oil fuel in maritime transportation was accelerated when the U. S. Navy—retrenching from its nineteenth-century position—began converting Navy vessels from coal to oil. This prompted the question of security of supply under wartime situations. In response to this potential problem, several reserve oil fields on public lands were set aside to meet any such need. They included reserves at Elk Hills, California; Buena Vista, California; the infamous Teapot Dome, Wyoming; Point Barrow, Alaska; and three oil shale areas in Colorado and Utah.[119]

The concept of federal involvement in the nation's fuel resources stems from early federal oversight of mineral resources of lead for security purposes (ammunition) in 1807.[120] The Mining Act of 1866 in the wake of the California gold rush had declared public lands to be "free and open" for exploration for minerals and for mining. Special policies were developed to deal with problems of fuel minerals, including coal, oil, gas, and, more recently, uranium. Lands containing high coal deposits were sold freely in small tracts for mining purposes. The conservation movement under President Theodore Roosevelt at the turn of the century succeeded in pulling back from the leasing schedules more than sixty million acres of known coal lands in the public domain.[121]

Problems also arose with petroleum exploration activities, which resulted in the Taft administration's also withdrawing oil lands in 1909 after a U. S. Geological Survey warned that the government would find itself buying back the same oil it would have been virtually giving away by leasing public land to oil developers at token rates.[122]

The Mineral Leasing Act of 1920 established the contemporary policies of issuing prospecting permits and leases for development of the coal, oil, and gas resources on public lands. Uranium was added by the Atomic Energy Act of 1946.[123]

Electric Power

Electric power generation was also increasing at an astronomical pace after 1900. From 1902 to 1907 it increased almost 2½ times, and by 1917 electric power consumption was more than seven times what it had been at the turn of the century.[124] This growing demand came primarily from two sources: homes and factories. Residential demand was growing at the rate of 1 to 2 percent per year.[125]

The major impetus for electric power expansion during this period was the factory, which found electric power—particularly with the availability of the alternating-current electric motor—uniquely adaptable to the growing trend toward assembly line manufacturing. At the turn of the century, electric motors driven by either purchased or self-generated electricity represented less than 5 percent of total installed horsepower in U.S. factories. By 1909, the number of motors had increased ten times, and their total share of manufacturing horsepower had risen to 25 percent. The number more than tripled in the next decade, and the share of the total horsepower rose to 55 percent. After 1910, the number of electric motors increased faster than the total horsepower installed, suggesting that this reflected not only installation of new motors but that steam-driven machinery was being replaced.[126]

By powering the nation's growing assembly lines, electricity nourished itself in a unique way. It made possible the mass production of appliances, which in turn required electric power to operate. Increased demand also permitted larger power stations with higher efficiencies, hence lower electric power costs to the consumer, which still further stimulated the use and growth of electric power.

The early manufacturing assembly lines in which electric motors played such a major role were patterned after the conveyor lines first introduced into the packing industry in the 1860s and '70s. Oliver

Evans, the same young American who had vastly improved steam engine technology, also set the pattern for mass production with his innovative mechanization and continuous assembly line in an early-nineteenth-century grain milling operation.[127] Oddly enough, the railroads had a strong influence on assembly lines. Rails were laid on the floor of a Manchester, England, factory by Johann Georg Bodmer in 1833 to move goods from one point to another. In the American slaughterhouses of the 1860s and '70s, the tracks were placed overhead, an arrangement Henry Ford adapted when he introduced mass production to automobile manufacture in 1913.[128]

The Social Impact

Assembly lines introduced division of labor and factory regimentation, with their far-reaching societal and economic implications. Professor Fred Cottrell has succinctly outlined the relationship between division of labor and the high-energy society:

> . . . the use of high-energy converters to replace men involves a division of tasks into minute specialized operations which machines can be set up to perform. The specialists who are required to build and operate these machines depend in turn upon the makers of instrument and control devices, which involve further specialization. In consequence, high-energy technology requires many times the number of socially approved occupational alternatives than were acceptable in most low-energy systems.
>
> Moreover, as a result of the fact that the materials needed are, in their natural state, widely dispersed, specialized communities must be set up to process them. Accordingly, whole communities appear which are devoted to the assembly of the specialized products of other communities; we see communities engaged in the production of a single product, such as tin, copper, lead, or mercury, or the raising of a single crop such as wheat, rice, sisal, tobacco, cotton, or coffee, and others devoted to the manufacture of products made of a single commodity, such as paper, glass, or rubber.[129]

The introduction of the assembly line in industry and the increasing specialization of people and communities are both representative of a broad shift in the lifestyle of the American high-energy society.†

† The terms high-energy society and low-energy society are relative. A society that uses agriculture in combination with animal husbandry is a higher-energy society than one that has only harnessed solar energy for crops. However, in the context of this book, the term high-energy society designates a society that harnesses many sources of energy, including solar energy (principally in agriculture), fossil fuel energy, hydroelectric and nuclear power, etc. The characteristics of a high-energy society include strong centralized energy supply system, as well as centralized government and economic systems.

In the nineteenth century a highly centralized social order developed, designed to accelerate the use of the concentrated energy sources. A characteristic of this centralization was the development of the central business corporations to facilitate the flow of energy and finished goods in the economy. The economic mechanism of high-energy society in the Western world evolved quite naturally into these corporate structures. Even the Communist bureaucrats of postrevolution Russia recognized the amazing rapidity with which the West adopted industrialization, and their "Five-Year Plans" represented an attempt to introduce a highly centralized socialist structure to duplicate the feats of the corporate structure. They have not yet been able to duplicate the feats of the capitalist world, for the profit motive provides a better inducement for energy exploitation than does the regimentation of the coercive Soviet state.‡

In fact, government institutions of the West now in many ways resemble the form—if not the substance—of the Soviet state. John Kenneth Galbraith, in *The New Industrial State,* says that during this period of rapid fossil fuel development (late nineteenth to early twentieth centuries), a shift of power occurred from decentralized America to centralized America, a power shift that brought on "a new factor in production." It represented the "association of men of diverse technical knowledge, experience or other talent which modern industrial technology and planning require." This group, which Galbraith calls the "Technostructure," is composed of persons both in the corporate structure and the state structure, both of which comprise the industrial system or industrial state. "The industrial system," he adds, "is inextricably associated with the state. In notable respects, the mature corporation is an arm of the state. And the state, in important matters, is an instrument of the industrial system."[130]

The place of human values in high-energy society is subordinate to the value of the industrial machine, the value of competition, the value of technology. In low-energy societies, values arise from the interactions of individuals, families, tribes, villages, and communities, because work and life experiences are precious and, as such, are shared. High-energy society disrupts this decentralized social organization and replaces many of the functions of man with the energy-subsidized machine. As Fred Cottrell notes, ". . . the community

‡ In recent years, the Communist and non-Communist worlds have grown closer together in attempts to facilitate higher synthetic energy flows (fossil fuels, nuclear fuels, etc.) through centralized industrialization. The only anomaly is Communist China, which has preserved a cultural legacy of decentralization common to low-energy societies.

ceases to be a self-contained unit in which the codes that are produced operate upon one another in such a manner as to make them mutually at least tolerable. Instead it [the community] becomes a locus of interaction which may continuously generate local conflict."[131]

The sweeping changes in American society that have been variously identified as resulting from "technology" or "social conflict" or "the loss of religious values" or "the dissolution of the nuclear family" are in reality consequences of the change in the amount of energy available to American society. All the other changes are subsidiary to the availability of cheap, concentrated energies—primarily the fossil fuels—and the resulting societal institutions developed to exploit them. Technology and change follow the liberation of energy. The lifestyle of contemporary America was destined by the development of fossil fuels in this seminal era.

Rise of the Oil Economy

The post-World War I energy era was marked by a leveling off of total energy use and by a continuing shift in the roles of the major fuels. The war marked a watershed in the fuels mix, with coal emerging from it in a much weaker position than it had entered. While production of coal was to remain at about 400 million tons per year, coal's share of the total fuels market was embarking on a long slide.[132] In 1918, coal's share of the national energy consumption was almost six times that of petroleum products. But by 1925, it was only three times higher than petroleum; and by 1930 the ratio had dropped to only about two to one. Petroleum finally overtook coal just after World War II, and has continued to claim an increasing share of the national fuels market since. By 1965, oil accounted for 40 percent of energy use in the United States; natural gas, 30 percent; coal, 23 percent; hydroelectric power, 4 percent; and liquid natural gas, 3 percent.[133]

Several developments accounted for this changing pattern in fuel consumption. First of all, the importance of railroad travel was to markedly decrease with the growing use of automobiles in America; second, many of the trains that remained were converted from coal to fuel oil; and third, coal was to lose more and more of its residential heating market to fuel oil and—later—to natural gas.

The demand for fuel oil pushed it to comprise 50 percent of all oil products in the 1920s, after which it declined to about 40 percent in the 1930s and to about one third in 1955 as gasoline became a more and more important petroleum product.

As the number of automobiles rapidly increased, so did the demand for gasoline to fuel them—but even faster. From 20 percent of all oil consumption in 1920, gasoline's share of the petroleum market rose to twice that by 1930, a share that it retained for the next several decades. From 1925 to 1940, the number of motor vehicles in use increased by almost 70 percent, but the amount of gasoline they consumed increased by 160 percent. Annual gasoline consumption per vehicle for all passenger cars, trucks, and buses during this period averaged 473 gallons in 1925, but rose to 733 gallons by 1940. In the post-World War II period, this level reached 800 gallons.[134]

Two nonhighway uses of fuel also increased sharply during the period of the 1930s, '40s, and '50s: aviation and farm equipment. From 1935 to 1955, civil aviation gasoline consumption increased 30 times.[135]

The number of tractors rose from 80,000 in 1918 to twice that number in 1919. By 1939 the number had risen to 1.6 million; to 3.6 million in 1950; and to more than 4.5 millon in 1955.[136] The rapid increase in farm mechanization was reflected in the sharp rise in agricultural productivity. Siegfried Giedion, in *Mechanization Takes Command,* outlined the significance of farm mechanization:

> In 1880, it is estimated, 20 man-hours were needed to harvest an acre of wheatland. Between 1909 and 1916, this number was reduced to 12.7 man-hours, and between 1917 and 1921—that is, with the advent of full mechanization—to 10.7 hours. The following decade cut the figure to 6.1 (1934–6), that is, to almost the same extent as in the four decades preceding.
>
> This leap in productivity was brought about from the outside: one mobile power, the small electric motor, made possible the mechanization of the household; another mobile power, the internal combustion motor, made possible the full mechanization of agriculture . . . [when] around 1905 the first tractors appear.[137]

Miniaturization and compactness, the next phase of mechanization, made an efficient one-man farm possible. The "combine," which harvested, threshed, cleaned, and bagged grain, was first built in 1836, but did not become practical for the family-size farm for another century. The "midget combine," introduced in 1939, cut a 40-inch swath of grain, cost little more than a grainbinder to buy, and made the small farmer competitive with the giants—but only for a few brief years.[138]

Commercial diesel engines were built in the United States as early as 1898. It was only after the 1930s, however, when an improved fuel-injection system became widely available, that diesel engines began to be in demand. Consumption of diesel oil increased 10 times, from

16 million barrels in 1935 to 169 million in 1955. In 1935, demand for diesel was about 3.5 percent of the combined gasoline/diesel market. By 1955 this had increased to more than 10 percent.[139]

The oil industry has been very flexible in accommodating the rapidly changing patterns of petroleum consumption over the years. At the turn of the century, for example, each barrel of refined oil produced about 24 gallons of kerosine, 6 gallons of fuel oils, and 5 gallons of gasoline. By 1930, this had shifted to 18 gallons of gasoline, and kerosine accounted for just 2 gallons, with fuel oils and residuals accounting for the remaining 15 gallons.[140]

While the petroleum product mix was shifting from kerosine to fuel oils to gasoline, the geographic areas of production were also changing. The oil industry remained focused in Pennsylvania, with some spillover into Ohio and West Virginia until the turn of the century, when these three states accounted for 85 percent of total oil production. In the next decade, however, the focus shifted to California and Oklahoma, so that by 1910 the three eastern states accounted for only one sixth of total production, and California and Oklahoma together produced more than half of the total. These two states continued to dominate the oil scene until the late 1920s, when the gigantic eastern Texas oil fields were discovered and Texas began producing more than either of the other states. By 1930, it was producing more than the other two combined. Louisiana also overtook California and Oklahoma after World War II to move into the second-ranking U.S. oil position among the states.[141] Throughout this time, the nation was on a continuing oil bonanza, with one huge oil field after another coming into production.

Natural gas, which frequently accompanied new oil discoveries, remained largely wasted until the late 1920s and early 1930s, when pipeline technology was advanced to the point that leakproof pipes could be laid, and the heavy equipment was developed to lay them. By the mid-1930s, pipelines were carrying gas for distances beyond 1,000 miles, and gas first began to account for more than 10 percent of the total U.S. energy consumption. Convenience, absence of polluting products, and cheap price prompted rapid growth, which increased even more rapidly after World War II. Between 1945 and 1960, pipelines in the United States increased from 218,000 to 608,000 miles. Natural gas took over from coal what little of the residential and commercial heating markets oil had not already assumed, and then began taking over much of oil's market as well. By 1960, natural gas had become the nation's primary household fuel, although industry, including power generation, was accounting for about two thirds of the total annual natural gas use. Gas has made considerable inroads into the electric power generation market at coal's expense.[142]

Since World War I, rapidly growing requirements for gasoline not only were fully met by U.S. refinery output, but in every year from 1918 to 1949, the United States had surplus petroleum to export. Since that time, the nation has been importing crude oil in increasing amounts.[143]

From the earliest days of oil, the ability of the giants of the industry to generate capital out of their own earnings has given them tremendous economic power, because it has provided the enormous sums of money that have enabled them to afford a powerful Washington lobby to help pass favorable legislation and to prevent laws that would work against the industry. The most flagrant and controversial example of legislation blatantly and unfairly favoring the oil industry is the oil depletion allowance. Written into the first federal income tax law in 1913, it permitted certain "extractive industries"—including oil and mining—to be exempt from taxes on 5 percent of their gross income under the specious argument that because the natural deposits they were extracting were being depleted by production, they should be compensated for the continuing decrease in their capital assets.[144]

The man largely responsible for it was Senator Boies Penrose, political boss of Pennsylvania and a close associate of Andrew Mellon of the Mellon Bank, Alcoa, and Gulf Oil. Senator Penrose was heir to a large fortune and did not have to resort to accepting bribes from the oil industry, but acted as a channel for money from others to his fellow legislators. He distributed twenty-five thousand dollars in this way from John Archbold of Standard Oil, and late in 1918 was successful in getting the 5 percent allowance amended upward by successfully arguing that oil well and mine depletion should be based on market value instead of cost. "As he explained it," Richard O'Connor wrote in *The Oil Barons,* "under the new provision, when a ton of coal was sold part of the excess of the price over cost must be treated as 'a repayment of what was invested.' Only seven enlightened senators opposed him."[145] The year after the amended depletion allowance was passed, Mellon's Gulf Oil had a depletion allowance 449 percent greater than its net income. The oil depletion allowance has continued to serve the oil industry with varying degrees of generosity—up to 27.5 percent until the 1972 tax bill reduced it to 22 percent over strenuous protests of the industry, which maintained it was essential if oil exploration were to continue. Critics of the industry pointed out, however, that despite a generous 27.5 percent depletion allowance, the oil industry had been both exploring less and finding less for several years.

In the meantime, O'Connor estimates "conservatively" that depletion allowances probably took twenty billion dollars in pretax oil income off the tax rolls during the 1960s.[146] Every dollar the oil companies did not pay in taxes was paid by some anonymous American wage earner.

Electricity

Throughout the first half of the twentieth century, the nation's electrical capacity grew steadily. In the first years of the twentieth century, average residential consumption increased about 1 to 2 percent per year, but after World War I, as the number and variety of electric appliances began to proliferate, growth began accelerating rapidly. Since the mid-1930s, annual growth rate per electric customer has averaged 7 to 9 percent—except for the years of World War II. From the beginning of World War I, total electric power demand has been doubling every decade.[147]

While coal's share of most fuel markets was declining during the 1920s and 1930s, it continued to provide fuel for about two thirds of the electric power generation into the late 1930s, with hydroelectric power accounting for the remaining one third. By 1940, however, hydroelectric's share had dropped to 27.5 percent; coal had also dropped to about 60 percent, while natural gas accounted for a little more than 8 percent, and residual fuel, a little less than 5 percent. This trend has continued, and in 1970 hydroelectric power accounted for about 17 percent; coal had dropped to about 45 percent; natural gas accounted for almost 25 percent; residual fuel, about 12.5 percent; and nuclear sources, a little more than 1 percent.[148]

Introduction of Nuclear Power

The first tentative steps toward nuclear fission, the splitting of atoms, as a means of providing the heat for a steam electric power station came at the turn of the century when scientists began asking themselves why Madame Curie's radium gave off its powerful rays. In 1905, Albert Einstein, still in his native Germany, demonstrated mathematically that the energy content of a substance is a function of, or dependent on, its mass. The formula that he evolved—$E=mc^2$—has since become commonplace in the lexicon of modern man, and the power it describes may yet alter the course of mankind.

It was decades before this theory could be proved. Scientist/author Ritchie Calder, in *Living with the Atom,* explains why:

The full implications of this eluded experimental scientists although in chemical reactions large amounts of energy are released as heat. For instance, in the internal combustion engine of an automobile, 100 grams of hydrocarbon fuel undergo reaction with the resulting liberation of a million calories. That sounds like a lot, but, in terms of Einstein's equation, the loss of mass equivalent to the energy involved is less than one part, by weight, in a billion. Such a decrease could not be detected by even the most sensitive chemical balances. So Einstein's remained a paper equation for a long time.[149]

The elusive energy referred to in the equation was stored in the nucleus of the atom, which didn't help much because scientists still weren't sure what the nucleus of an atom was. One concept was that the atom was like a miniature solar system, with the nucleus as the sun and the negatively charged electrons orbiting it like planets, kept in orbit by the magnetic attraction of positively charged protons in the nucleus. This still left some mass in the atom unaccounted for.

Sir James Chadwick of Cambridge University identified in the atom another particle, which had the same mass as the proton, but no electrical charge. He named this new particle the neutron. On the basis of the discovery of the neutron, scientists in the 1930s developed not only the theory of nuclear fission power, but successfully demonstrated that neutrons from radioactive materials could be used to bombard other atoms, releasing the tremendous energy of the atom.

Einstein's formula for energy suggested that the energy was present in the atom, and the pioneering nuclear physicists in Britain and Germany demonstrated it. Yet a controlled reaction, in which the splitting or "fissioning" of atoms releasing more neutrons, etc., had not been demonstrated. This was changed when, on December 2, 1942, a group of scientists headed by Dr. Enrico Fermi gathered under a squash court at the University of Chicago, where the first controlled nuclear reaction was generated in the first reactor, called an "atomic pile," because the uranium fuel used was surrounded by an enormous pile of graphite.

It was actually a twofold experiment: Could a chain reaction be initiated; and if it could, could it be controlled? The experiment was conducted with a heap of several tons of graphite carefully intermixed with lumps of uranium 235, which had been determined to have the most unstable atomic makeup so that it would split easily.[150] The graphite was present because it acts to slow neutrons down, to keep too many atoms from splitting, which might cause the reaction to get out of hand.

In addition to the graphite, rods of cadmium were also inserted in the pile to further slow the reaction. The plan was to initiate the chain reaction by withdrawing all the cadmium rods except one, which would be gradually withdrawn and reinserted as the means of control.[151]

The experiment was successful, and the scientists proved that an atomic chain reaction could be initiated, and that it could be controlled. The public learned about the historic incident almost three years later, following the release of the atomic bomb by U.S. forces over Hiroshima, Japan, on August 6, 1945. From that time on, atomic scientists have been continuously engaged in a controversial struggle to safely harness that awesome energy for the generation of electric power.

The promises of unlimited nuclear electric power as a substitute for fossil fuel-generated electricity that the atomic energy establishment made in the 1950s and '60s had the effect of discouraging development of additional coal mines and generally casting a pall over the coal industry. However, the failure of nuclear power to fulfill that promise has been instrumental in sparking a resurgence in coal production and the construction of coal-fired steam electric plants.

Architecture and Building Technology

To encourage greater electrical mechanization of the home and thus create greater appliance markets, General Electric in 1932 and Westinghouse Electric in 1934 sponsored special cooking institutes in which engineers, chemists, architects, nutritionists, and cooks studied every aspect of the kitchen with an eye toward streamlining it to the ultimate degree for the housewife. In the process of studying how the housewife relates to the kitchen, corporations began to realize that the reorganization of the kitchen also had an impact on the design of the entire house. The trend that grew out of this was the open kitchen; combination kitchen/dining room/living room; combination kitchen/family room, etc. An interrelated development—particularly apropos in the years during and following the Depression—was the design of the "servantless household," in which mechanization and good work-area design became especially important to the newly self-sufficient upper middle-class homemaker.[152]

The two trends—flexible arrangement of rooms and maximized mechanization—soon ran into conflict. The costs of mechanization began to account for up to 40 percent of the building costs in some of the new free-and-open experimental designs. The architectural response was the "mechanical core" of the house, which would include the kitchen, bath, laundry, heating, wiring, and plumbing, and which could be fabricated and assembled in the factory and brought intact to the building site.[153] To a great degree this has been achieved in most contemporary homes.

The mechanical core concept became even more important following World War II, when a wide range of new and more power-intensive appliances became commonplace: electric water heaters, home freezers, clothes dryers, dishwashers, and garbage disposals. And while not part of the "mechanical core," air conditioning and electric panel heating soon appeared on the scene as the latest home appliances to constitute a major drain on electric power.

During this period, prefabricated mechanical equipment and housing was designed by Buckminster Fuller, a leading advocate of mechanization in the household.

It was Willis Carrier who achieved the final environmental breakthrough with his patented air conditioning machine back in 1911. For several decades, however, it remained a curiosity confined largely to hotel dining rooms, theaters, Pullman cars, ballrooms, and other commercial operations in which the high cost of cooling and moisture control could be offset by the increased profit gained from drawing customers who might not otherwise have patronized them.

The air conditioner, though a relatively complex piece of equipment, is based on a simple weather phenomenon: Lowering the temperature both cools the air and removes the moisture from it. In his later life, Carrier recalled the day he walked through fog at a Pittsburgh, Pennsylvania, railroad station in 1902, and came up with the answer to the air conditioner:

> Here is air approximately 100 per cent saturated with moisture. The temperature is low so, even though saturated, there is not much moisture. There could not be, at so low a temperature. Now, if I can saturate the air and control the temperature at saturation, I can get air with any amount of moisture I want in it.
>
> I can do it, too, by drawing the air through a fine spray of water to create actual fog. By controlling the water temperature I can control the temperature at saturation. When very moist air is desired, I'll heat the water. When very dry air is desired, that is, air with a small amount of moisture, I'll use cold water to get low-temperature saturation. The cold water spray will actually be the condensing surface. I certainly will get rid of the rusting difficulties that occur when using steel coils for condensing vapour in air. Water won't rust.[154]

It was the simplest of grade school science concepts: Moisture-laden air has a dew point, the temperature at which the water in it will condense into droplets and fall from the air. Carrier set about to harness it for commercial application on a large scale.

His solution is still the one widely used today. Fresh air and recirculated air are mixed in a chamber, the relative amounts controlled by valves. The air is filtered, then preheated by passing over heat pipes and fins heated with steam or hot water. Some of the excess moisture is removed at this juncture by cooling in the cooler of the air-conditioning plant. Both the moisture content and the temperature of the air coming out of the cooler are determined by the temperature of the cooler. This air is then mixed with more air from the preheater to provide an air mix of the correct temperature. The air conditioner is instrumented so that if there is not enough moisture in the air, water is added by a fine spray. This causes some cooling, after which the air is again passed through a reheater to perfect the air temperature before being ventilated into the room or building by means of a fan behind the reheater. The entire cycle—particularly the reheating phase—consumes extravagant amounts of electricity.[155]

Air conditioning did not become widely available for home use until the early 1950s. In 1952, according to one account, dealers sold out $250 million worth of air conditioning equipment and still had to turn away 100,000 customers.[156]

Although the air conditioner was Carrier's invention, he steadfastly resisted calling it by that name, which had been coined in the first decade of the century by Stuart W. Cramer, a competing inventor. Carrier preferred to refer to it in terms of "man made weather," which he felt more accurately described what actually transpired within the confines of the air conditioning unit.[157]

Urban Architecture

The trend setters for today's glass skyscraper were the United Nations headquarters, built in 1950 under the supervision of Wallace Harrison, and Lever House, headquarters of Lever Brothers, designed by the architectural firm Skidmore Owings and Merrill and constructed in New York in 1951.

Both buildings were very similar: a glass skin stretched over a skeleton of curtain walls and suspended ceilings and entirely dependent on high-energy artificial environment control whose mechanisms are well hidden and present no unsightly view. The two buildings are without doubt the two most important buildings of twentieth-century America. Their style and mechanical systems have been copied and constructed from coast to coast in the ensuing years.

The United Nations building was ostensibly designed by a consortium of ten of the world's greatest architects, but the final product bears the unmistakable imprint of the French architect Le Corbusier, who was the most renowned and influential of the ten. "Here in New York," noted Reyner Banham, "he accomplished his dream of creating a great glass tower in an urban setting, and here in New York he also encountered the talents of the one man, in all probability, who could make it work: Willis Carrier." The Conduit Weathermaster system installed in the building was regarded by Carrier as the greatest achievement of his career.[158]

"Le Corbusier's vision of the Cartesian glass prisms of the slab skyscraper," Banham noted, "and Carrier's practical technology for solving any environmental problem that offered an honest dollar had met, literally, in the UN Building, and the face of the urban world has been altered."[159]

At last, man had achieved the ultimate in the unnatural setting. As an anonymous commentator observed in the magazine *Architectural Forum*, ". . . air conditioning and venetian blinds are pitted against the powerful sun. . . ."[160]

Since he was known as a critic of the megalopolis, the famous American architect Frank Lloyd Wright was never a serious contender for the job of designing the United Nations building. One wonders what might have transpired if the UN had taken Wright's suggestion that the grounds for the United Nations building in New York City be grassed over instead.

"Buy a befitting tract of land, say a thousand acres or more, not too easy to reach," he urged. "Sequester the UN. Why does it not itself ask for good ground where nature speaks and the beauty of organic order shows more clearly the true pattern of all peace whatsoever?"[161]

Unfortunately, the UN chose the high-energy society value system and rejected organics in favor of industrial order.

Energy Use Since World War II

Today, energy-consuming technology has found its extension in almost every facet of American life.

The electric power consumed by the American public has more than sextupled, and electricity now accounts for one fourth of the nation's energy. The electric utilities industry is now consuming about 4 times as much coal per year as was consumed in 1944, the next-to-last year of the Second World War; more than 12 times as much fuel oil; and almost 10 times as much natural gas. In addition, hydroelectric power has also increased substantially, and a newcomer, nuclear power, now accounts for more than 15 billion kilowatt-hours per year.[162]

In terms of total fuel consumption, from 1940, the last year before the United States' entry into World War II, to 1971, coal consumption increased by a little more than 20 million tons; crude oil consumption more than tripled; and whereas 36 million barrels of oil were exported in 1940, 385 million barrels were *imported* in 1971; natural gas consumption soared from 2.6 trillion cubic feet to more than 22 trillion cubic feet; and liquid natural gas (LNG) increased by an even greater margin, from 2.2 billion gallons to 24.8 billion gallons.[163]

During this period, railroads switched almost completely from coal to diesel power; an entire jet fleet of passenger airplanes developed; the auto population experienced phenomenal growth. Air conditioning, television, central heating, and clothes washers and dryers made the transition from luxuries to necessities; a mechanized agricultural industry began using 16 billion gallons of crude oil a year; and monstrous glass towers inspired by Lever House and the United Nations building came not only to dominate the skylines of the cities, but to sprout up even in the small cities of America.

By another measure, the nation's prime movers—the various engines that perform the work of American society—grew from a total of 2.8 billion horsepower in 1940 to 21 billion horsepower in 1971. This rate of growth far exceeds the growth in population, which went from 132 million people (44 people to the square mile) in 1940 to 203 million people (57.5 to the square mile) in 1971. Thus the number of people increased by 54 percent during the three decades, while the prime movers of high-energy technology increased by 750 percent![164]

The Bureau of the Census has recorded a decreasing rate of gain in the population in recent years. In 1973, the population topped 210 million people, but in the two previous years the actual rate of gain was less than 1 percent—dropping in 1972 to the lowest rate increase in U.S. history.[165] However, while the country's population moves toward "zero population growth," the increase in the accumulation of high-energy technology continues to increase. Even with the slowing down in population growth, more and more energy is being used, because the per capita use of energy is constantly increasing.

During the 31-year period from 1940 to 1971, the population was also shifting radically from the farms to the cities. The over-all farm population in 1940 was 30.5 million people; in 1971, only 9.4 million. By 1970, almost 70 percent of the total population was urbanized—i.e., clustered in or around American cities of 50,000 or more. So from nearly one fourth of the population—23 percent—in 1940, the percentage of American people living on farms dropped to less than one twentieth of the over-all population—4.6 percent—in 1971.[166] While the number of farms was shrinking from 6 million to 2.8 million, average farm size was increasing from 167 acres to 839 acres.

The American Association for the Advancement of Science (AAAS) in 1970 commissioned a team of scientists led by Dr. Barry Commoner to review the combined effects of change in population and technology on America's environment. The final report was published in 1972 by the U. S. Commission on Population Growth and the American Future.[167] Assisting Commoner in selecting data and writing the report was Michael Corr, the executive secretary of the AAAS Committee on Environmental Alterations; and Paul Stamler, the committee's research assistant.

They discovered a number of changes in the rates of growth for the production of various consumer goods; the rates of development and extraction of various natural resources, including energy resources; and the patterns of American consumption. Following are some of the growth changes they noted:*[168]

Item	**Percentage Increase from 1946 to 1968**
Population	43
Electric power	276
Motor fuel consumption	100
Plastics	1,024
Synthetic organic chemicals	495
Pesticides†	217

Several conclusions are apparent from this analysis. While the U.S. population grew by a total of 43 percent, the consumption of energy resources grew at a rate many times greater. One notable feature stands out: The common denominator in the list of electric power, motor fuel, plastics, synthetic organic chemicals, and pesticides is that *they are all energy resources*. While it is obvious that electric power is generated primarily by fossil fuels and that motor fuel is gasoline refined from crude fossil fuels, it is not apparent to many that plastics, pesticides, and synthetic organic chemicals are also energy resources in another form. The fossil fuel resources—coal, oil, natural gas—that are used for the production of power in our society have increasingly been diverted to the manufacture of such synthetic materials.

The inevitable result of more oil and gas drilling and coal mining to supply America's economic demands for consumer products (made by and from the fossil fuels) has been to build the economy on a fossil fuel subsidy. Whereas the natural products of solar energy—such as agri-

* This list adapted from Commoner, Barry; Corr, Michael; and Stamler, Paul J., "The Causes of Pollution," *Environment* (April 1971), pp. 1–19.

† Pesticide consumption was measured from 1950 to 1968.

cultural commodities ranging from foods to clothing (cotton from plants, wool produced by grazing sheep, etc.)—once served the economy, the onward rush of fossil fuel energies to society's marketplace eliminated the former solar-derived products and replaced them with synthetic substitutes (plastics, etc.), which were cheaper in narrow day-to-day economic terms, although crucially linked to a finite resource: the stored fuels of the Earth.

As becomes apparent in forthcoming pages, the development of fossil fuels has been a boon to humanity—but a curse as well. All the wonders of today's technology are based on the fossil fuel energies, but the sophisticated technology of high-energy American society has failed to develop a source of energy to fuel itself when the fossil fuels are depleted.

2

Energy: The Limit to Growth

It is apparent from the foregoing pages that the United States is now looking toward an uncertain energy future with a spiraling energy demand that is already by far the highest in the world. The nation's consumption of energy resources in 1970 was half again as much as all of Western Europe's, even though 75 percent more people live in that area of the world; and 2½ times the U.S. energy consumption in 1920.[1]

Oil is now supplying the nation with 43 percent of its total energy requirements. It powers the 100 million motor vehicles that make many of the highways nearly impassable; it fuels the thousands of diesel railroad engines, most of the maritime transportation, and all the planes of the civilian and military fleets. It fuels many millions of small engines, from power mowers to water pumps, and provides the fuel for 12 percent of the nation's electricity.[2]

Its sibling hydrocarbon, natural gas, accounts for one third of the nation's energy consumption. Coal supplies one fifth of the nation's energy; and hydroelectric and nuclear fission power each account for about 2 percent.

One fourth of all primary sources of energy in 1970—oil, natural gas, coal, and nuclear sources—went to generate almost 1.7 trillion kilowatt-hours of electricity, more than was produced in the world's next 4 greatest consumer nations—the Soviet Union, Great Britain, Japan, and West Germany—combined.[3]

The Outlook for Oil

As is readily apparent, where once Coal was King, hydrocarbons—coal, petroleum, and natural gas—now reign supreme. More than one fourth of the land area of the United States is under lease for oil and/or gas exploration.[4] At least up until gas shortages took their toll of many independent entrepreneurs in the summer of 1973, there were 42,000 oil businesses in the country in addition to 181,000 service stations. Nearly 2 million people are employed in exploration and production, transportation, refining, sales, and promotion of oil.[5]

Given the major position oil and natural gas hold in the U.S. energy economy, it is not surprising, then, that the energy crisis began first to be felt in this sector. It is also not surprising that the crisis is frequently couched in terms of availability—or more correctly, nonavailability—of oil and natural gas, even though the crisis falls across every fuel resource.

A sharp increase in demand for fuels—particularly for oil and natural gas—seems inevitable. The most comprehensive U.S. energy analysis available—the *U.S. Energy Outlook,* prepared by the National Petroleum Council (NPC) in December 1972—predicts a doubling of demand for energy in 1985, with two thirds of it to be met by oil and natural gas.[6]

Half of the United States' estimated 200 billion barrel oil heritage has already been consumed, and at the 1970 rate of oil consumption, it would be completely depleted in another 20 years.[7] The nation's oil binge has reduced its share of the world's petroleum from nearly 40 percent in 1937 to 6.7 percent by 1968, and it is continuing to drop as consumption increases.

A recent estimate of proved reserves made by the State Department's Office of Fuels and Energy breaks down world reserves as follows:

Country or Area	**Billion Barrels**
Arab Middle East and North Africa	350*
Iran	60
United States (including Alaska)	40
Venezuela	15
Indonesia	15
Canada	10
Other Western Hemisphere	10
Other African	5
Others	5
Total world reserves	510 billion barrels[8]

* This is a conservative estimate, giving Saudi Arabia credit for 150 billion barrels, although some experts believe the figure may be twice that.[10]

American oil wells pump out almost 10 million barrels of oil per day, but it is only two thirds of what the United States needs to maintain its economic position. And that capacity is steadily falling. Oil production peaked in the United States in the early 1970s, and the present 10 million barrel per day level represents a 10 percent drop from the peak capacity.[9]

While more oil remains to be found, both on land and offshore, the increasing per capita demand will drain it off faster than it can be produced. The amount of oil actually remaining to be discovered in America is unknown. Up to now the oil industry has continuously assured U.S. consumers that there is plenty of oil yet to be found if the incentives—i.e., depletion allowances and other tax favors—are there to induce the oil companies to go out looking for it. However, almost two million drill holes have been sunk in the surface of the United States by now, and many question whether oil reserves in America are still an unknown quantity. There is a pervading suspicion that perhaps the oil industry knows something the rest of the population doesn't know about U.S. oil reserves—or the lack of them—especially since oil companies are now expending almost all their efforts in looking for oil offshore, in Alaska, or other locations off of the coterminous United States.[11]

Today's oil men are finding no more of the great domestic oil finds of the 1920s and 1930s. There has been a sixfold drop in their discovery rate since then, according to scientist/author Ralph Lapp, and each well is considerably more expensive than in the heyday of oil discovery. Some experts have estimated that ultimate U.S. recovery will be about 200 billion barrels of crude oil, as noted earlier. Others, however, estimate the figure as low as 165 billion barrels, and others as high as 300 billion.[12]

The National Petroleum Council estimated that the demand for oil would range from 20.5 million barrels per day to 29.7 million barrels per day by 1985, of which, depending on other fuel availabilities, 3.6 to 19.2 million barrels per day would have to be imported.[13]

In the shorter range, 1980, the National Petroleum Council study predicts demands ranging from 19.6 to 25.3 million barrels per day, with imports of from 5.8 to 16.4 million barrels per day required to meet that demand.[14]

Two years prior to the NPC analysis, a presidential task force headed by then Secretary of Labor George Schultz (now Secretary of the Treasury) predicted that by 1980 the United States would be consuming 18.5 million barrels per day, an increase of only 4 million above 1970. However, in the words of James E. Akins, former U. S. State

Department energy specialist and now Ambassador to Saudi Arabia, "a funny thing happened on the way to the '80s." By March of 1973, when he testified before a House of Representatives Science and Astronautics Subcommittee hearing, the United States was already consuming 18 million barrels per day—approximately what the NPC study anticipated would be the 1975 level of consumption.[15] One third of it was imported, and half of that one third, or one sixth, was coming from the Middle East.[16]

In the early 1970s, Akins reported, the State Department, the Interior Department, and the petroleum industry all agreed that "we should never, never import more than 10 percent of consumption from the Eastern Hemisphere" because the supply was too insecure, and greater dependence than that would render the United States an easy prey of Arab blackmail in the Israeli-Arab tug-of-war. However, "the 10 percent figure was reached last year," Akins said, and "this year [1973] it will be 15 percent." Assuming continuation of the present U.S. oil policy, the United States would be consuming—as a "modest figure"—25 million barrels per day by 1980, for a total annual consumption of more than 9 billion. Half of it would have to be imported from the Middle East.[17]

What would happen if the United States were consuming 25 million barrels of oil per day, half of it from the Arab countries, and the Arabs should suddenly present an ultimatum to the United States to stop support of Israel or lose the Arab oil? Although some Arab countries did stop shipment in 1973, what would the United States do if all Arab countries cut off shipments?

"I don't know," Akins mused. "Nobody does; but the choices seem to be very limited." Those choices, he said, include (1) to accept the sudden oil cutoff, "which would cause something close to economic collapse in the United States"; (2) go to war against the Arabs; or (3) accede to the Arabs' demands.

"All three are absolutely intolerable," Akins continued. "Not one could be considered by the United States." The only possible course of action, he warned, is to make sure the predictions of oil consumption don't come true.[18]

In the meantime, however, the Middle East situation is growing more and more tense, and the possibility of the United States having to choose among intolerable alternatives has grown more and more real. The President of Libya, Col. Muammar el-Qaddafi, in June of 1973 nationalized the Nelson Bunker Hunt Oil Company of Dallas, Texas, as a rebuke to the United States for its pro-Israeli policy. It was no accident that he announced his seizure in the presence of President An-war Sadat of Egypt, who reportedly would also like to use his nation's oil to persuade the United States to force Israel back to its pre-1967 borders.[19]

However, "only one country counts" when discussing U.S. oil imports from the Middle East, Akins told the subcommittee; and that country is Saudi Arabia, because of its enormous oil reserves.† "The other countries can do anything they want. Saudi Arabia can break the price of the world oil; it can overproduce and drive the price down, or restrict production causing world shortages."

As for Saudi Arabia using oil to influence U.S. foreign policy, Akins told the subcommittee that "Saudi Arabia has never made a threat. The King has always said exactly the opposite: that he will never use it as a political weapon." King Faisal's is now, however, a "lone voice," Akins said. In 1972, he added, fifteen different threats were made by various Arabian leaders to use oil as a weapon against the United States. Kuwait joined the chorus in March 1973. "In extremus," Akins said, "all Arab countries could use oil as a weapon in a confrontation with Israel."

Saudi Arabia's "lone voice" has since grown weaker. In April 1973, Sheik Zaki Yamani, a high Saudi government official, informed the United States that it will not expand its oil production to meet America's 1980 oil requirements unless Washington changes its pro-Israeli position. The United States has been relying heavily on Saudi Arabia's pushing its present production of 7.2 million barrels a day to 20 to 25 million by 1980 to accommodate U.S. needs. Sheik Yamani stressed that there is still a "good possibility" of complying with the U.S. request provided the United States in turn creates "the right political atmosphere." He made it clear that he was referring specifically to America's relationship with Israel.[20]

By October 1973, when war broke out between the Arab nations and Israel, the months of rhetoric quickly crystallized into the Arab "oil weapon." By the month's end, the Arab states, under the leadership of King Faisal, began restricting oil exports to Japan and Europe, and banned all exports to the United States and the Netherlands, the countries judged to be closest to the Israeli cause.

In the first few months of application of the Arab oil weapon, numerous "leaks" appeared in the embargo, but the primary purpose

† In view of the United States' intensified public show of friendship and cordiality with Iran, and recent agreements to sell Iran planes and ammunition, it would appear that the United States no longer believes Saudi Arabia is the "only . . . country [that] counts." As the chart on page 60 indicates, Iran's 60 billion barrel oil reserve is substantial.

of the embargo was accomplished: Several economic penalties to Western nations (as well as to economically disadvantaged Third World countries) were incurred, resulting in widespread shortages and escalated prices for all petroleum-based products, ranging from petrochemicals (from plastics to agricultural chemicals to end products, such as gasoline).

In the United States, the oil embargo lent an entirely new tone to the energy crisis. Suddenly the warnings of previous years became the stark reality of economic recession and the familiar snaking lines of autos waiting for limited supplies of gasoline. A new figure in government suddenly garnered headlines—the nation's "energy czar," William E. Simon, head of the Nixon administration's overnight creation, the Federal Energy Office (FEO).

Under Simon's direction, the FEO initiated a series of voluntary and mandatory national programs to allocate scarce fuels and enhance the conservation of energy. Simon told a meeting of the National Academy of Sciences that the effectiveness of Arab embargoes had hastened the day of change for American society. "The U.S. economy must adapt from a low-cost, abundant energy base to a high-cost, scarce energy base," Simon told the meeting, adding that the boycott of oil had accelerated economic change. He concluded: "It is important that the American people realize that the current shortage is not a temporary aberration. Scarce, high-cost energy will be the rule for many years if not indefinitely. The occurrence of the boycott merely means that we must reorder our priorities and modify our life styles now, and not a few years from now."[21]

Even though the Arab oil boycott was lifted in March 1974, the economic effects of the boycott and associated escalation of petroleum prices were keenly felt by the United States and other industrialized nations. Consumers were especially hurt by energy-triggered inflation and commodity scarcities.

An even greater side effect of high oil prices was analyzed in a private memorandum circulated among the Directors of the International Bank for Reconstruction and Development (the World Bank). The memorandum suggested an international monetary collapse as the coffers of the oil-producing states swelled while oil-consuming nations' import bills soared. The World Bank memorandum forecasted an amassing of royalties and oil taxes by the Arab states within the decade. Saudi Arabia, Qatar, Abu Dhabi, Kuwait, and Libya collected about $10 billion from oil royalties in 1973. The World Bank sees the figure escalating to about $60 billion by 1980 and $120 billion by 1985. The net result on the world's economy might be economic havoc for Third World countries, severe hardships for developed nations, and an economic bonanza for the Middle East nations, who would hold most of the world's oil, as well as most of the world's money![22]

In a sense, however, the 1973 Arab oil boycott and the ensuing changes in the international economy may prove to be a boon rather than a disaster, as the energy-consuming nations of the world recognize the paralyzing effects of energy shortages on all sectors of society. Consider what would happen if the high energy nations were confronted with a complete cessation of Middle East oil exports during a political confrontation in the future, when even higher levels of Arab oil would be imported by the consuming nations.

If this should happen, and the United States, Japan, and Western Europe were forced to rely on their own combined reserves of 120 billion barrels, America's oil would be depleted in 10 years—and even less time than that if rising oil demands were taken into account.

A phrase penned in 1960 by two eminent energy authorities may well go down in energy history as among the most famous of last words: "Imports involve no foreseeable problem of supply, either in availability at the source or in transportation."[23]

One probable result of an Arab boycott in the future, if the United States became more dependent on Arab oil, would be military intervention in the Middle East. If the United States were importing half its oil from the Middle East, war would be almost inevitable in the event of a boycott. Numerous political figures have mentioned the possibility of U.S. military intervention in the Middle East, including Senator William Fulbright, chairman of the Senate Foreign Relations Committee, and Nixon Administration Secretary of Defense, James Schlesinger.

The Outlook for Coal and Nuclear Fission Power

The National Petroleum Council estimates that shortages between energy demand and supply through 1985 will be met by coal and nuclear sources—nuclear to the extent that it is available for electric power generation, and coal to fill in whatever gap nuclear power cannot supply. The NPC also predicted that by 1985 from 47 to 339 million tons of coal would be supplied per year for synthetic oil and gas.[24]

Coal is the one fossil fuel the United States is not short of. It represents about 88 percent of the proven reserves of all U.S. fuels. Of total coal reserves of 3.2 trillion tons, some 150 to 200 billion tons are estimated to be recoverable with present technology.[25] If the 1970 rate of consumption were to hold steady, the life span of coal in the United States would be an estimated 1,700 years. It would be 300 years if only coal were to be used to provide the world's energy and if the average energy use in other nations were raised to a par with U.S. consumption today and held constant there.[26]

However, for environmental reasons as well as shortages of water in rich coal-bearing regions of the western United States, much of this coal may be inaccessible. Much of the U.S. coal reserves will require elaborate processing with water, which is simply unavailable in many areas; and since much of the coal is of low quality, with high sulfur concentrations, it may prove undesirable to burn. Combined with these limiting features is the fact that much of the coal lies under rich agricultural and grazing lands, which may prove more productive to society left undisturbed than disrupted for mining coal.

Coal and nuclear power combined, the NPC anticipates, will make up about 30 percent of the total energy picture by 1985.[27] To do this will require production of from 1 billion to 1.6 billion tons of coal per year by 1985, compared with 590 million tons produced in 1970. It also assumes development of coal gasification and/or liquefaction on a commercial scale, which is far from certain.

The most critical assumption made by the NPC relates to the availability of nuclear power on a massive scale. Predictions of the NPC range from 450,000 megawatts of installed nuclear (electrical) capacity in 1985 to 240,000 megawatts of available nuclear power. In the case of high nuclear growth, there would be 64 times as much installed nuclear power as in 1970; or in the lowest assumption, there would be 34 times as much installed nuclear power. All in less than 15 years![28] This level of development, which the NPC largely borrowed from the Atomic Energy Commission's assumptions of nuclear advancement, is predicated on the twin assumptions that enough uranium will be available to fuel nuclear plants as well as the development of a nuclear breeder reactor, which will create additional useful nuclear material. The problem is the rapid burnup of available U.S. uranium reserves, which may not last into the 1980s from such a high level of nuclear growth.‡

The NPC believes that coal and nuclear fuels will comprise the bulk of the fuel for generating electric power, which the council predicts could account for almost half the energy in the nation by 1985, if electric power from coal and nuclear sources could meet more space heating needs for homes and process heating for industry. This would require a growth rate for the electric utilities industry of 8.8 percent per year, compared with 6.7 percent per year projected for an "inter-

‡ A further discussion of the issues surrounding nuclear power is presented in Chapter 5.

mediate" demand situation, and 7.9 percent per year, the utilities' growth rate in the late 1960s.[29]

The Outlook for Natural Gas

The National Petroleum Council has predicted that the demand for the other hydrocarbon keystone of U.S. energy supply, natural gas, would range in 1985 from 22.1 trillion cubic feet per year to 40.3 trillion cubic feet per year, with imports ranging from 5.9 to 6.6 trillion cubic feet per year[30]—compared to consumption of 23 trillion cubic feet in 1970.[31]

The outlook for natural gas is as bleak—perhaps even bleaker—than for oil. In 1950, U.S. consumption of natural gas was 7 trillion cubic feet per year, and 12 trillion cubic feet of new reserves were being discovered annually. In 1971, while consumption had increased to 23 trillion cubic feet, new reserves were still at only 12 trillion cubic feet.[32]

The natural gas supply crisis, as Ralph Lapp recently pointed out,[33] is both man-made and natural. It is man-made on the short-range basis because natural gas companies have been holding back exploration and development pending Federal Power Commission approval of rises in natural gas prices, which are federally regulated in much the same way as electric rates. In the longer term, however, the crisis is very real. Supplies of natural gas appear quite limited. The National Petroleum Council estimated that, at the end of 1970, 1,178 trillion cubic feet of gas remained to be found.[34] At the 1971 rate of consumption, that is approximately 50 years of U.S. supply before there simply is no more in this country, and imports and/or synthetic gases will have to be relied on entirely for gaseous fuels. In actuality, there may be less than two decades left before the natural gas era comes to an end, due to the tremendous amount of energy required to pump gas out of the ground and get it to the consumer. A good fraction of the remaining natural gas must be devoted to "energy pumping" use, leaving even less for ultimate consumption than indicated by reserve figures.

Dr. Earl Cook, Dean of the College of Geosciences at Texas A. & M. University, points out that the cost of drilling an oil or natural gas well in Texas doubles every 3,600 feet of depth. Up to 1970, all the natural gas found in Texas was no more than 10,000 feet underground, and the average total cost of getting to the gas was less than $100,000. Today, however, Texas' estimated natural gas reserves are found at depths averaging 20,000 feet and beyond. At 1970 prices, the cost of sinking these same natural gas wells would be more than $1 million. According to Dr. Cook, the cost of tapping the gas from Texas reserves will escalate the natural gas price from future reserves by four times, assuming the other costs of production do not increase at all.[35] In 1971, the average cost of natural gas at the well was $0.16 per thousand cubic feet. Extrapolating this cost by the fourfold increase of future production plus the costs of transporting the gas to consumers brings the delivered costs up to $0.75 or more. Already, consumers in many areas of the nation, particularly the northeastern United States, are paying more than this for natural gas. Some natural gas is transported as a supercold liquid—liquefied natural gas (LNG)—from northern Africa at an ultimate cost to consumers of between $1 and $1.50 per thousand cubic feet.

As the costs of all the fossil fuel energy sources increase due to lessening availability and the necessity to use more energy to extract dilute deposits, it is becoming clear that the fossil fuel era is almost over for the United States. As the fossil fuels dwindle, the cost of developing alternative energy sources constantly increases, since the newer energy sources must be developed with the remaining stocks of expensive fossil fuels.

Energy and Productivity

Many attempts have been made to correlate the increasing use of energy in advanced civilizations with the rise in economic productivity. Few would disagree that increasing use of energy contributes to increased economic growth; but no one has yet identified the precise cause-and-effect relationship, nor determined if there is an optimum energy/growth ratio that should not be exceeded if the best interests of society are to be served.

A common belief is that the more energy resources a society uses—in the form of nuclear power, fossil fuels, hydropower, or other natural forms—the better off that society becomes. An example of this philosophy is the following comment of Congressman Chet Holifield (D., Calif.), for years a powerful force on the Joint Committee on Atomic Energy and a supporter of energy growth: "Our standard of living is the highest of any nation in the world because we use the highest ratio of mechanical power vs. manpower. One man can do the work of 350 men."[36] Holifield's comment does indicate that mechanization through intensive energy use has accelerated the economic output of each person in the society, but it does nothing to indicate the quality of that person's life in the society.

The relationship between the Gross National Product—a monetary measure of all goods and services produced in a nation—and the level of energy consumption in many countries has been studied by scientists

Arjun Makhijani and A. J. Lichtenberg of the University of California at Berkeley. They found that high levels of energy use (from fossil fuels and other nonsolar energy sources) did not necessarily bring higher incomes. They found that Great Britain, Belgium, Australia, Germany, Denmark, Norway, France, and New Zealand had levels of Gross National Product within a 10 percent range of each other; but in the case of New Zealand, less than half the amount of energy was consumed to sustain a higher GNP than Great Britain. They also noted that per capita energy consumption and per capita GNP "represent averages and say nothing of the distribution of income levels. If a wide disparity in income levels—and consequently a large population segment that is poor—exists in any country, then per capita GNP can hardly represent the standard of living in that country."[37]

Though Makhijani and Lichtenberg chose South Africa as a good example of the case where high GNP is present in the face of great imbalance and poverty, the same rule may be applied to the United States. A favorite argument of the energy industries and boosters of high rates of growth is that more energy should be made available to raise the living standards of the poor. In reality, the problem of the poor has much more to do with the distribution of wealth already available in society. Raising the over-all GNP and increasing levels of energy use without redistributing wealth can only lead to the ultimate situation where the rich get richer at the expense of the poor, for it is the rich who have inherited the high-energy societies. The best, least polluted areas are inhabited by the rich, not the poor. The fruits of technology are primarily enjoyed by the rich, not the poor. The poor are left, not with the bounteous *energy* of society, but with its *entropy*—the waste products of energy production, the pollution. In his short and penetrating book *Replenish the Earth,* ecologist G. Tyler Miller, Jr., points out that rich and middle-class Americans are the "megaconsumers and megapolluters, who occupy more space, consume more of each natural resource, disturb the ecology more and pollute directly and indirectly the land, air, and water with ever-increasing amounts of thermal, chemical and radioactive wastes." He calculates that the pollution engendered through the overuse of resources by the average middle-class American family with three children is equivalent to the pollution that would require thirty to fifty children of a poor American family to produce.[38]

What of the rest of the world? Is there any basis for assuming that American-style consumption and spiraling energy growth holds any answer for the future fate of the poor in other countries? The average American consumes 1,100 times the amount of energy that a resident of Burundi consumes. Even one of our small neighbors in this hemisphere, Haiti, has an energy consumption level of about 68 pounds of coal per person per year in contrast to more than 23,000 pounds of coal consumed per person in the United States. The UN has calculated that the whole world would have consumed more than 38.4 billion metric tons of coal in 1969 if consumption levels across the globe had equaled the U.S. level, whereas the actual consumption was 6.4 billion tons.[39]

There is no industrial machine capable of mining and processing so much energy except the United States. Even West Germany, a country whose per capita GNP exceeded that of the United States in 1973, has a per capita level of energy consumption less than half that of the United States. The world's energy reserves would not last long if other countries were as gluttonous in their use of energy and other key resources as the United States. Is this consumption morally justified?

Scientist/author Ralph Lapp in his book *The Logarithmic Century* singles out the United States as an "island of affluence" in a crowded world that someday will recognize the contrast of resource utilization.

Lapp predicts that someday the world's impoverished billions will sense this disparity: "Through satellite communication and a probable global education-television hookup soon to be deployed, almost every village will become aware of the depths of its poverty and will sense the futility of hoping for human parity."[40]

Studies of the energy/growth relationship in the U.S. economy have been made by economists of the Ford Foundation-supported Washington, D.C., thinktank Resources for the Future, Inc. (RFF). Research performed by RFF in the 1950s led to the publication of *Energy in the American Economy: 1850–1975,* which is now considered a standard work in this area.[41] It analyzes the trend of the relationship between the growth of the Gross National Product (GNP) and the growth of energy use in the United States up to 1960, with projections to 1975. One of the principal RFF investigators, Dr. Bruce C. Netschert, has extensively revised this earlier work and extended the analysis of the energy/GNP relationship into the 1970s.

Essentially, the RFF studies of the 1950s found an obvious trend—beginning in 1920 and extending to 1955—of decreasing energy consumption relative to the growth of the Gross National Product. The relationship was calculated by measuring the growth of the GNP against the total energy input into the economy from primary energy sources (fuels, hydroelectric power, etc.) measured in BTUs. The percentage of primary energy used in the economy up to 1955 averaged between 2.2 percent and 3.3 percent per year. This means that the actual dollar value of primary energy sources (not including their processing, transportation, etc.) constituted an average of less than 3 percent of the Gross National Product during that period.

From 1920 to 1955, the rate of energy consumption relative to GNP declined by an average rate of 1.2 percent per year.[42] This decline at first appears astonishing. How could the economy continuously expand, while at the same time requiring relatively less energy each year to sustain the rising economic productivity? One might assume the opposite would be true—that more energy would have to be injected into the economy per unit of GNP growth.

An interesting aspect of the energy/GNP study helps explain this seeming paradox. As might logically be assumed, with the rise of mechanization, much more energy has been required to take over many functions of the worker in industry. But while more energy was used per worker, the work output per man-hour simultaneously increased substantially with the advent of automation and mechanization. Between 1879 and 1919, the number of man-hours worked in the United States increased by 220 percent, and the horsepower of various machines increased by 160 percent. Yet the actual work output per man-hour increased by only 86 percent. During the following period—1919 to 1955—the man-hours worked increased by 29 percent, the horsepower installed per man-hour increased by 196 percent, and the work output per man-hour increased by a whopping 190 percent—more than twice that of the previous period.

While the advances in mechanization replaced much human energy with fossil fuel energy, it seems likely that there was an additional psychological factor at work, which comes into play at a certain stage of development in high-energy technology and mechanization. And that is that it required some decades for workers to grow accustomed to the regimentation of the industrial machine. The startling increase in work output during the 1930s and 1940s is not surprising. An increase in work productivity is commensurate with increasing mechanization of high-energy society: i.e., mechanization of machine; mechanization of man.

Resources for the Future cited as the primary cause of productivity expansion the increasing use of more efficient technologies for extracting, transporting, and using energy so that continuously less energy was required to produce the same output. Two of the most important expressions of this increased efficiency were (1) the substitution of diesel engine-driven locomotives for the old coal-burning steam locomotives, and (2) the improvements in efficiency at central power plants. Estimates of the over-all efficiency of energy source conversion in the economy vary. J. F. Dewhurst, in *America's Needs and Resources: A New Survey,* estimated in 1955 that the over-all efficiency of converting fuels (wood, coal, oil, etc.) and the energy of falling water into mechanical work in America rose from 1.8 percent in 1850 to 3.2 percent by 1900, and by 1959 was about 13.6 percent.[43] While this latter figure seems abysmally low—since it indicates that even by the middle of the twentieth century, more than 86 percent of energy supplies were wasted in the process of converting them to work—it nevertheless helps explain the fact that less energy was needed to sustain a continuing growth of the GNP. According to Dewhurst's figures, the average efficiency of all areas of energy conversion had quadrupled between 1900 and midcentury.

The increase in efficiency of the electrical power process was especially dramatic. Power plants required almost seven pounds of coal to produce one kilowatt of electricity in 1900, but less than one pound in the most modern coal plants by the late 1950s.[44]

Thus the culmination of all man's technological goals was exemplified in these two complementary trends: (1) the steady growth in the use of fossil fuel energy resources (electricity has doubled each decade since Thomas Edison's Pearl Street Plant); and (2) the increasing efficiency of the use of energy by the available conversion technologies. Factory equipment was being consistently modernized; power plants burned less fuel to get the same amount of electricity; and automobile, train, and other vehicle engines were being improved.

The Significance of These Trends

The significance of these coexisting trends can hardly be overemphasized. Americans were using more and more energy, but because they were using it increasingly efficiently to produce more economic goods, America's energy use and the economy remained in balance. The long-term decline in the energy consumption/GNP ratio abruptly reversed in the late 1960s, as shown in the following chart, which demonstrates the relationship between energy use and the GNP in the United States between 1947 and 1970.[45]

The chart shows the BTUs of energy consumed in the United States for each dollar of Gross National Product. In 1947, more than 105,000 BTUs of energy were expended to produce one dollar in GNP. In 1966, less than 90,000 BTUs of energy were expended to produce one dollar of the GNP. According to National Economic

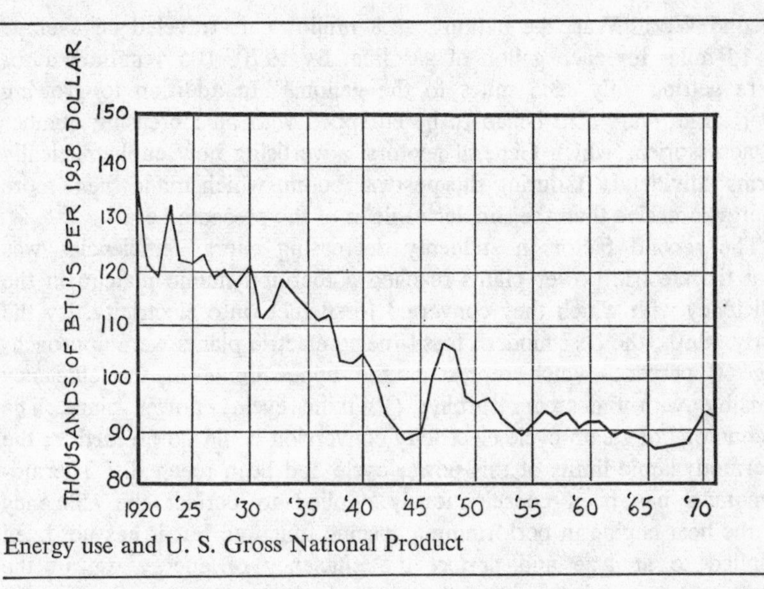

Energy use and U. S. Gross National Product

Research Associates, that was the year that the energy/economy honeymoon ended.*

The reasons are straightforward enough: The laws of thermodynamics at last caught up with the surging economy and its underlying bedrock of cheap energy and easily tapped resources.

As explained by National Economic Research Associates, a number of technological factors played a role in the sudden reversal of the trend toward increasing energy efficiencies. The two primary technologies responsible were the automobile and the electric power plant. Another factor must not be overlooked, however; and that is the increasing nonenergy use of energy resources—such as the dizzying rise of production of synthetic materials and plastics made from energy resources, the fossil fuels. Expressed another way, many important changes in America's economic structure that are just now becoming obvious to trained observers are due to the relationship of energy use to the production and use of manufactured products.

In the case of the automobile and decreasing efficiencies, the underlying factors were the increasing size and horsepower ratings coupled with a simultaneous car population explosion. At the end of the

* Similar conclusions were reached independently by economists at the Bureau of Mines in 1968 in *An Energy Model for the United States, Featuring Energy Balances for the Years 1947 to 1965 and Projections and Forecasts to Years 1980 and 2000* by W. L. Morrison and C. L. Readling, BuMines Information Circular No. 8384, U. S. Department of Interior, Washington, D.C.

Second World War, the nation's 25.8 million cars traveled an average of 15 miles for each gallon of gasoline. By 1970, 105.4 million autos were getting only 13.5 miles to the gallon.[46] In addition to growing ever larger, cars also began to be equipped with an increasing number of accessories (which General Motors' advertising now euphemistically terms "dividends") during the postwar boom, which made them more energy intensive than the simpler vehicles of the preceding era.

The second factor in suddenly decreasing energy efficiencies was that the electric power plants reached a thermodynamic plateau in the efficiency with which they converted fossil fuels into electricity. By the early 1960s, the best modern fossil-fueled electric plants were approaching 40 percent, which represents the upper limits of the efficiency possible with the steam turbine (Rankine cycle) power plant. The technology of steam cycle electricity conversion could go no farther; the thermodynamic limits of this power cycle had been reached.[47] Thermodynamics has been conscientiously applied to perfect the efficiency of the heat engine in performing a specific function; but it has not been applied to analyze and perfect the efficiency of energy use in the entire system that is broadly defined as contemporary society.

To understand this concept, we can examine the way in which energy is used to provide domestic water heating in the home—using, for instance, natural gas as the fuel. There are two common methods by which natural gas, which now supplies more than one third of the available energy in the United States, can be used to heat domestic hot water. The "old-fashioned" means is to use the natural gas from a pipeline in a simple gas burner in the home. The "modern" and fastest-growing way of heating water (as well as performing many other heating tasks) is to burn the natural gas in the boiler of a power plant to produce steam for the generation of electricity. The electricity thus generated is transmitted to the home, where the electric current is converted back to heat in resistance coils, which then heat the water.

Even in the most efficient new electrical power plants, only 40 percent of the original energy content of the natural gas can be converted to electricity, and the national average energy conversion efficiency of all electrical power plants is considerably less—between 25 and 30 percent. For the hypothetical electrical power plant, one can assume an efficiency of energy conversion of 32 percent. This means that one third of the energy of the natural gas burned in the plant's boiler will actually be converted to electricity. The remaining two thirds of the original gas energy winds up as unusable high-entropy or waste heat and is rejected to the water source (a local river or lake) used for cooling purposes at the power plant. When the electricity is transmitted in power lines and distributed through the electrical network or utility power grid, 10 percent or more of the remaining power may be lost before it gets to the home where the electrical water heater is located, so that when the electricity reaches the home, less than 30 percent of the original gas energy is left to use for heating the water. Thus, even though the electric water heater can convert almost 100 percent of the electricity it receives to heat, there's less than 30 percent of the original gas energy left to convert by the time it reaches the electric hot water heater in the home.

A far more logical way to heat domestic hot water is to use natural gas directly by distributing it in a pipeline to the home, where a natural gas water heater—nothing more than a simple burner—can apply the fuel to produce heat directly. The average natural gas water heater converts 62 percent of the energy content of the natural gas into hot water heat at the home, wasting only 38 percent of the energy content of the fuel.[48]

Thus the old and familiar natural gas burner turns out to be a far superior way of getting heat for domestic hot water than the symbol of modern energy technology, the electric power plant. The reason for this can be understood in the light of thermodynamic science. Each time energy is transformed from one form to another, less is left than had originally been available, in accordance with the Second Law of Thermodynamics. By minimizing the steps used to produce and use power, the loss of energy would be minimized.

Moreover, the process of converting fossil fuels to electricity in the power plants being constructed today has already reached the upper efficiency limit (about 40 percent). Further construction of these plants will waste fossil fuels that might be considerably better applied.

In 1970, the average all-electric home in the United States consumed about 20,000 kilowatt-hours of electricity, and the Federal Power Commission (FPC) predicts that annual consumption in the all-electric home will reach 33,000 kilowatt-hours per year by 1990.[49] The magnitude of this level of energy use may be gauged by the amount of coal necessary to produce one kilowatt-hour of electricity in an electric power plant. At current levels of power plant efficiency, about 1 pound of coal must be burned, meaning that the average all-electric home of today consumes 20,000 pounds of coal per year.†

† In the 1970–71 data book of the Edison Electric Institute, *Questions and Answers About the Electric Utility Industry,* the average national figure of .88 pound of coal needed to produce 1 kilowatt-hour of electricity is cited. This figure may or may not be accurate, depending on the operation of each given electric

For the past decades, the electric utilities have been able to justify encouraging ever more rampant and wasteful use of electricity because they have been able to couch it in terms of "cheap electric power." But as noted earlier, the steady decline in energy costs has halted, and costs for fossil fuel—the source of more than four fifths of all electric power in the United States today—have begun to skyrocket. Whereas in 1969 the utilities were paying an average of less than $0.23 to

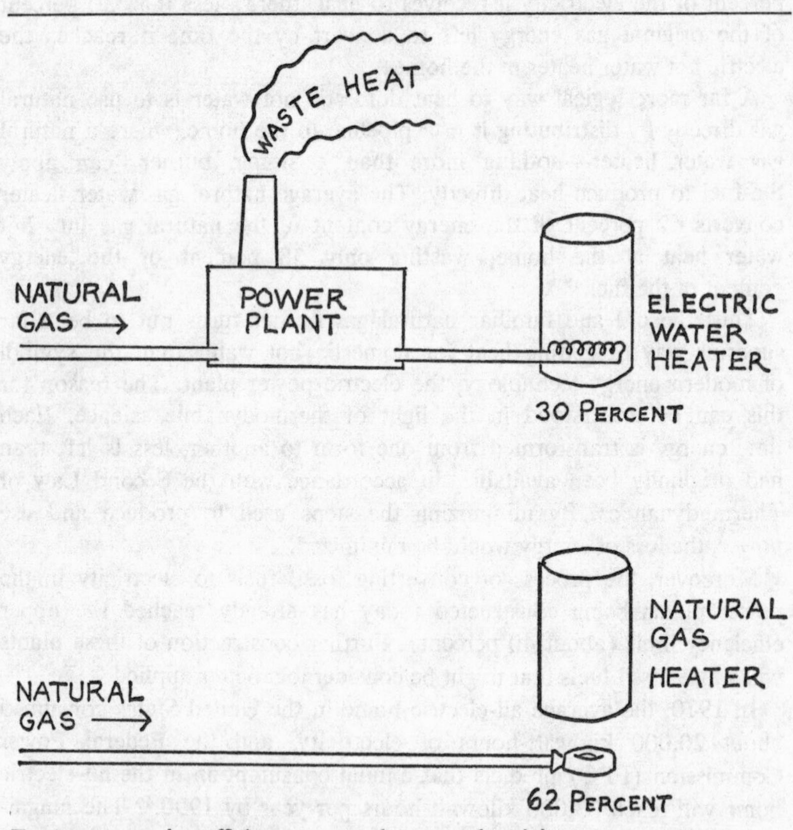

Energy conversion efficiency: natural gas vs. electricity

power plant. If the plant operates at its rated capacity, it consumes less fuel per kilowatt-hour (kwh) of electricity than when it operates at a greater or lesser load. In addition, the full thermodynamic accounting of efficiency of the electric plant's fuel cycle would show that much more input energy was needed to produce the electricity. For example, how much energy was required to mine the coal, ship it, process it? How much energy was required to build the mining, transport, and power plant equipment? Questions such as these alter the glib assumptions often made about the efficiency of any single energy conversion process, which must be analyzed in the context of the whole energy conversion system.

produce 100 kilowatt-hours of electricity, they are now back up to the pre-1960 price levels of $0.33, with costs still rising.[50]

Other factors have contributed to this increase, including the implementation of the Federal Coal Mine Health and Safety Act of 1969 (which has raised operating costs of coal mine operators), and environmental protection legislation (both state and federal) requiring that power plants use antipollution devices; but the essential reason is the combination of increasing costs of extracting fuels from the earth and the simultaneous decline in the average efficiency of the power plants and the electric power distribution system.

As a consequence, millions of buyers of all-electric homes and electric hot water and space heaters and other high-energy appliances in a period of cheap electricity are going to be using them in a period of expensive electricity. The time is foreseeable when the average wage earner can no longer afford to operate them, as relative costs of electric power double and triple over the coming years.

Limits to Growth

A number of economic studies have attempted to project changes in the trend of the energy consumption/GNP ratio into the future. However, fluctuations have appeared since 1970 that make it difficult to forecast a future trend.[51]

Other studies challenge the conventional wisdom of comparing gross energy use in the economy with the value of goods and services. A recent study made by energy specialist Chester Kylstra and economist/accountant Jesse Boyles at the University of Florida indicates that the methodology of the conventional energy/GNP ratio is incomplete and misleading. Their study applies values for *net energy* rather than *gross energy* in the economy, with net energy defined as the amount of energy the consumer receives in the final product (an automobile, for example) rather than gross energy (the energy content of the barrel of oil before processing for use in manufacturing, fuels, etc.). The study concludes that over-all productivity in the U.S. economy (defined as the ratio of net energy/gross energy) is *declining,* even though the U. S. Government calculations show that output per man-hour is increasing. In addition, they find that true *inflation* in the economy (defined as the ratio of dollars of GNP/net energy) is rising at a rate 1 to 2 percent per year higher than the government's inflation calculations, which are based on money supply values rather than net energy/money values.[52]

The statistical message of the analysis of the trend in energy use

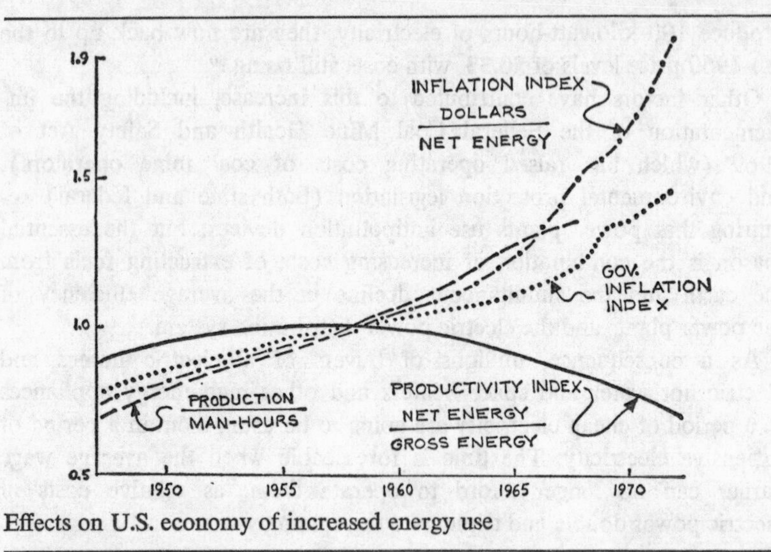

Effects on U.S. economy of increased energy use

and the rise in Gross National Product can be summarized in one sentence: The party is over.

Neither nuclear nor any other source of power can do much to re-create the era of cheap, readily available energy resources to which the nation's consumers have become accustomed over the past century. The golden age of mineral resource discovery and use is largely behind us; the richest, most plentiful ores have already been mined in the United States. We are returning to the politically volatile "Fertile Crescent" of the Middle East for a major fraction of our energy resource supply.

The increases in material prosperity that have accompanied the growth of American industrial society in this century were due largely to a confluence of technological and societal forces, all of which were made possible and shaped by abundant supplies of cheap energy. Accompanying the rise in energy use has been not only the availability of more goods and services, but the presence of greater entropy—expressed as pollution and culminating in economic instability—from the massive development of energy.

The Energy Cancer

Viewed from another perspective, the development of energy resources accompanied by tremendous industrial growth has caused destructive changes to many elements of American society. These changes might be likened to the carcinogenic effects of a powerful chemical. In this analogy, the affected organism is American society;

the carcinogenic chemical is energy—primarily in the form of fossil fuels; and the cancer is the present high energy/high entropy orientation of this society.

The United States social, political, and economic system is now at the ultimate crossroads, the crossroads of energy. For some years, the nation has pressed as a primary goal the exploitation of energy resources for the benefit of the technological machine. But now a diagnosis presents the early warning signs of cancer, and the major overriding question is whether and how the patient is going to be able to face up to the stark reality it presents.

As in the development of certain forms of cancer in living organisms, American society's energy cancer is characterized by the rapid, metastatic growth of wildly uncontrolled cells in widely separated parts of the organism. In contemporary American society these colonies are represented by the erratic growth of certain forms of energy use—electric power, automobiles, the petroleum-based synthetics, and plastics—each of which is characterized by tremendous waste and largely undifferentiated growth. As will be observed in the following pages, no institution of this society has been spared the infestation of the energy cancer. The entire social organism has been exposed to the carcinogenic chemical, and the economic warnings of its destructive presence are just beginning to be felt.

These warnings are best understood within the context of the energy/GNP ratio. The development of energy at low costs tended to encourage the development of assorted technologies to maximize the use of cheap energy—rather than encourage its conservation and wise use. We are now living in the period in which the economics of these inefficient systems are beginning to be felt in the form of severe restraints against continuing the energy policies of the past. The following pages present a more elaborate examination of the energy waste that has precipitated this socioeconomic disease.

Exponential Growth

Several well-publicized studies have been made of the consequences of continuing high levels of economic growth. The most prominent is the computer-modeled study of world growth patterns sponsored by the Club of Rome, an international organization of scientists, economists, government officials, and industrialists. Initial computer models of the escalating levels of growth completed by the Club of Rome were compiled, synthesized, and interpreted in *The Limits to Growth,* the controversial first product of the club's work.[53]

The work has been criticized by some for drawing an apocalyptic conclusion based on too many variables that are too inexact in nature. Nonetheless, it is a valuable study because it has asked the right questions, and in attempting to answer them has raised the very real specter of the disaster that threatens at the end of the road down which the industrialized world is racing at the present time.

The Limits to Growth presents computer models of what the club considers to be the five basic factors that determine and limit growth of all kinds on Earth: population, agriculture, natural resource development, industrial production, and pollution. The study found that in all levels of human endeavor, dangerous curves of exponential growth were apparent. Exponential growth can be defined as the continuous rapid doubling of population, resource use, and pollution within very short periods of time. One current example is the doubling of electrical capacity every seven years.

The exponential growth dilemma has been compared to an old riddle popular among French children, concerning a farmer, a pond, and a water lily:

> The lily is doubling in size every day. In thirty days it will cover the entire pond, killing all the creatures living in it. The farmer does not want that to happen; but being busy with other chores, he decides to postpone cutting back the plant until it covers half the pond. The question is: On what day will the lily cover half the pond? The answer is: on the twenty-ninth day—leaving the farmer just one day to save his pond.[54]

The basic question for our civilization is: Are we now in our twenty-ninth day?

The Club of Rome study noted that the world's population is now doubling each thirty-three years, at an annual growth rate of 2.1 percent. On the other hand, industrial output is increasing at an average rate of 7 percent per year, concentrated in the advanced technological nations: the USSR; Japan (with a per capita GNP growth rate of 9.9 percent per year in 1968, and now one of the highest in the world); Western Europe; and the United States. The study found that to continue this growth rate for much longer, i.e., beyond the year 2000, would severely overload the world's support systems for feeding, clothing, and housing its continuously increasing numbers of inhabitants, while simultaneously stripping most of the Earth's mineral and fuel reserves to supply the industrialized countries. The charts (pp. 84–87) from *The Limits to Growth* outlines the Earth's natural resource limits.

Although the Club of Rome study notes the limits to the current spiraling levels of use of natural resources, it does not take into account the full impact of the complex energy structure that determines not only the exploitation of all resources—from iron to water—but the carrying capacity as well of the Earth or any section of it for the support of human populations. It is the development of energy resources that determines the limits of all growth. As has been noted, it was the development of solar energy in agricultural technology that enabled the first human stable societies; and later periods saw the exploitation of wind power, water power, the fossil fuels, and nuclear energy, which have proved to be the underlying determinants of resource exploitation and population growth.

The authors of *The Limits to Growth* did recognize one essential limiting factor that energy contributes to the exploitation of natural resources: The availability of cheap energy resources and fuels determines the cost and—ultimately—the wisdom of mining low-grade energy or mineral resources.

However, some technological advocates of certain energy sources, most notably of nuclear power, insist that this is not true. They argue that "cheap" energy sources such as breeder reactors will open an era of unlimited, available energy, which will make available low-grade mineral ores to our civilization—the extraction of ores from granite, for instance. Geologist Thomas S. Lovering offers a sobering refutation to this claim:

> Those who are familiar with geological and metallurgical limitations on the mineral industry are not impressed with the suggestion that the advent of cheap nuclear energy will transform common rock to "ore" and supply virtually unlimited quantities of all the metals needed by industry. . . . Cheaper energy, in fact, would little reduce the total costs (chiefly capital and labor) required for mining and processing rock. The enormous quantities of unusable waste produced for each unit of metal in ordinary granite (in a ratio of at least 2000-to-1) are more easily disposed of on a blueprint than in the field. After crushing, the rock volume increases by the volume of the intergranular pore space, which ranges from 20 to 40 per cent. Nor does leaching in place seem to offer an economical substitute for mining except under very special conditions. To recover materials sought, the rock must be shattered by explosives, drilled for input and recovery wells, and flooded with solutions containing special extractive chemicals. Provision must then be made to avoid the loss of solutions and the consequent contamination of ground-water and surface water. These operations will not be obviated by nuclear power.‡[55]

‡ The myth of cheap nuclear power, whether from nuclear fission reactors or nuclear fusion, has been promulgated and disseminated by the Atomic Energy Commission and others—some unwittingly—for years. Note this comment in a 1973 *Business Week* article on mineral resources: "A giant technological step

It is largely just in this century that mineral resources have been tapped to provide the base for phenomenal industrial and economic expansion. Lovering points out that the metal consumed in thirty years at present growth rates is equal to that used in all previous history.[56]

However, this growth is limited by the nature of the resources now available for exploitation. Most of the world's best stocks of mineral resources have been known for some time, and the United States and other countries have mined them for years. This mining has required little energy compared to the enormous drains that would be required to get at low-grade, dispersed minerals in the future. The National Commission on Materials Policy recently concluded that the international competition to exploit high-grade mineral energy deposits "may evolve into a mutually destructive race for resources." According to the director of the commission, James Boyd, "in the U.S. and many parts of the world we have already mined or are mining the richest and most easily accessible ore bodies. Increasingly, we are going to have to look farther, dig deeper, and find ways to mine lower grade ores." This, he said, would mean tremendous capital investments years ahead of anticipated production.[58]

We may, in the words of Mr. Boyd, have to look farther and dig deeper, but this does not mean that we will be able to economically recover the resources that lie in dilute deposits or on the ocean's floor. Many technologists would have us believe that we can turn to the oceans as a source of new mineral resources, as well as food resources. They forget the laws of thermodynamics, which indicate the price that is to be paid for the search. In many cases, it requires more energy to mine these diffuse mineral resources than the energy content of the resources themselves! Society will pay a bitter price for wasting valuable energy reserves in this latter-day attempt to achieve perpetual motion.*

---

such as the harnessing of nuclear fusion for cheap energy would change the outlook radically. 'You could boil granite and distill off the metals and materials you wanted,' says Frank Yans, metals specialist at Boston's Arthur D. Little & Company.'" The possibilities of either "boiling granite" or developing "cheap" nuclear fusion are precluded by several factors—most notably, the laws of thermodynamics.[57]

* Likewise, technologists who argue that intensive cultivation of the seas will provide vast new food supplies forget that most of the ocean is a desert, insofar as biological productivity is concerned. An average area of ocean roughly the size of California provides one half the world's fish supply; the biologically productive areas of ocean are the coastal estuaries, fertile areas that have taken millions of years of natural evolution to adapt into the best system converters of sunlight into chemical energy—the food products of fish and other marine animals and plants.[59]

While the technologists blithely assume that a new march to the sea's frontiers will bring forth resource and food bounties, the productive estuaries are being filled in, polluted, and destroyed, cutting off the only *assured* marine source of food on which society depends.[60]

Financial and Other Limits to Growth

There are a number of ways in which energy growth may find itself being limited. One of them will be increasingly experienced by the many companies involved in energy extraction and utilization who may find it difficult to finance future energy schemes—from constructing power plants to building oil tankers. The National Petroleum Council has predicted that the total capital requirements of the energy industries from 1971 to 1985 will range from $451 billion to $547 billion. Of this, the NPC predicts that the costs of developing, processing, and distributing fuels will represent from $215 billion to $311 billion (of this figure, exploration and production of fuels will consume up to $172 billion); and another $235 billion will be spent to build electrical power plants and transmission facilities.[61] These amounts are staggering, and some industry leaders have repeatedly expressed fears that raising the capital for this level of construction will be extremely difficult, *if not impossible.*

Consider what exponential growth means to electric utilities. Dr. Jerome Kohl, of North Carolina State University, has calculated the past, present, and future demand for electricity in North Carolina and has compared it with the growth of the state's population. Between 1940 and 1970, electricity use in the state went up 8 times (800 percent) and the use per person went up 6 times (600 percent); but the state population went up only 40 percent. On the basis of this electrical doubling time in the past three decades, the largest electric utility in the state, the Carolina Power and Light Company, has announced that the trend will continue on into the future. Dr. Kohl calculates that for this utility to meet the year 2000 electrical capacity (the fifth doubling time of the utility), more money would be needed for capital investment than the total present income of all residents of the state. In 77 years, by the time of the eleventh doubling of electrical power, $768 billion would be needed to construct the plants. By that time, *the total 30,000 square mile area served by the utility* would be required just to site the power plants —not including the transmission lines. The magnitude of the problem can be appreciated when one considers that the 30,000 square mile service area covers half of North Carolina and one fourth of South Carolina.[62] Obviously, within 77 years, land for power plants would

Nonrenewable Natural Resources

1	2	3	4			5	6
Resource	*Known Global Reserves* [a]	*Static Index (years)* [b]	*Projected Rate of Growth (% per Year)* [c]			*Exponential Index (years)* [d]	*Exponential Index Calculated Using 5 Times Known Reserves (years)* [e]
			High	*Av.*	*Low*		
Aluminum	1.17×10^9 tons [j]	100	7.7	6.4	5.1	31	55
Chromium	7.75×10^8 tons	420	3.3	2.6	2.0	95	154
Coal	5×10^{12} tons	2300	5.3	4.1	3.0 [k]	111	150
Cobalt	4.8×10^9 lbs.	110	2.0	1.5	1.0	60	148
Copper	308×10^6 tons	36	5.8	4.6	3.4	21	48
Gold	353×10^6 troy oz.	11	4.8	4.1	3.4 [l]	9	29
Iron	1×10^{11} tons	240	2.3	1.8	1.3	93	173
Lead	91×10^6 tons	26	2.4	2.0	1.7	21	64
Manganese	8×10^8 tons	97	3.5	2.9	2.4	46	94
Mercury	3.34×10^6 flasks	13	3.1	2.6	2.2	13	41

7	8	9	10
Countries or Areas with Highest Reserves (% of world total) [f]	*Prime Producers (% of world total)* [g]	*Prime Consumers (% of world total)* [h]	*U.S. Consumption as % of World Total* [i]
Australia 33 Guinea 20 Jamaica 10	Jamaica 19 Surinam 12	U.S. 42 U.S.S.R. 12	42
Rep. of S. Africa 75	U.S.S.R. 30 Turkey 10		19
U.S. 32 U.S.S.R.-China 53	U.S.S.R. 20 U.S. 13		44
Rep. of Congo 31 Zambia 16	Rep. of Congo 51		32
U.S. 28 Chile 19	U.S. 20 U.S.S.R. 15 Zambia 13	U.S. 33 U.S.S.R. 13 Japan 11	33
Rep. of S. Africa 40	Rep. of S. Africa 77 Canada 6		26
U.S.S.R. 33 S. Am. 18 Canada 14	U.S.S.R. 25 U.S. 14	U.S. 28 U.S.S.R. 24 W. Germany 7	28
U.S. 39	U.S.S.R. 13 Australia 13 Canada 11	U.S. 25 U.S.S.R. 13 W. Germany 11	25
Rep. of S. Africa 38 U.S.S.R. 25	U.S.S.R. 34 Brazil 13 Rep. of S. Africa 13		14
Spain 30 Italy 21	Spain 22 Italy 21 U.S.S.R. 18		24

1	2	3	4			5	6
Resource	*Known Global Reserves* [a]	*Static Index (years)* [b]	*Projected Rate of Growth (% per Year)* [c]			*Exponential Index (years)* [d]	*Exponential Index Calculated Using 5 Times Known Reserves (years)* [e]
			High	*Av.*	*Low*		
Molybdenum	10.8×10^9 lbs.	79	5.0	4.5	4.0	34	65
Natural Gas	1.14×10^{15} cu. ft.	38	5.5	4.7	3.9	22	49
Nickel	147×10^9 lbs.	150	4.0	3.4	2.8	53	96
Petroleum	455×10^9 bbls.	31	4.9	3.9	2.9	20	50
Platinum group [m]	429×10^6 troy oz.	130	4.5	3.8	3.1	47	85
Silver	5.5×10^9 troy oz.	16	4.0	2.7	1.5	13	42
Tin	4.3×10^6 lg. tons	17	2.3	1.1	0	15	61
Tungsten	2.9×10^9 lbs.	40	2.9	2.5	2.1	28	72
Zinc	123×10^6 tons	23	3.3	2.9	2.5	18	50

7	8	9	10
Countries or Areas with Highest Reserves (% of world total) [f]	*Prime Producers (% of world total)* [g]	*Prime Consumers (% of world total)* [h]	*U.S. Consumption as % of World Total* [i]
U.S. 58 U.S.S.R. 20	U.S. 64 Canada 14		40
U.S. 25 U.S.S.R. 13	U.S. 58 U.S.S.R. 18		63
Cuba 25 New Caledonia 22 U.S.S.R. 14 Canada 14	Canada 42 New Caledonia 28 U.S.S.R. 16		38
Saudi Arabia 17 Kuwait 15	U.S. 23 U.S.S.R. 16	U.S. 33 U.S.S.R. 12 Japan 6	33
Rep. of S. Africa 47 U.S.S.R. 47	U.S.S.R. 59		31
Communist countries 36 U.S. 24	Canada 20 Mexico 17 Peru 16	U.S. 26 W. Germany 11	26
Thailand 33 Malaysia 14	Malaysia 41 Bolivia 16 Thailand 13	U.S. 24 Japan 14	24
China 73	China 25 U.S.S.R. 19 U.S. 14		22
U.S. 27 Canada 20	Canada 23 U.S.S.R. 11 U.S. 8	U.S. 26 Japan 13 U.S.S.R. 11	26

Nonrenewable natural resources

[a] U. S. Bureau of Mines, *Mineral Facts and Problems, 1970.* Washington, D.C.: U. S. Government Printing Office (1970).

[b] The number of years known global reserves will last at current global consumption. Calculated by dividing known reserves (column 2) by the current annual consumption (U. S. Bureau of Mines, *Mineral Facts and Problems, 1970*).

[c] U. S. Bureau of Mines, *Mineral Facts and Problems, 1970.*

[d] The number of years known global reserves will last with consumption growing exponentially at the average annual rate of growth. Calculated by the formula

$$\text{exponential index} = \frac{\ln\left((r \cdot s) + 1\right)}{r}$$

where r=average rate of growth from column 4
s=static index from column 3.

[e] The number of years that 5 times known global reserves will last with consumption growing exponentially at the average annual rate of growth. Calculated from the above formula with 5s in place of s.

[f] U. S. Bureau of Mines, *Mineral Facts and Problems, 1970.*

[g] UN Department of Economic and Social Affairs, *Statistical Yearbook 1969.* New York: United Nations (1970).

[h] *Yearbook of the American Bureau of Metal Statistics 1970.* York, Pennsylvania: Maple Press (1970). *World Petroleum Report.* New York: Mona Palmer Publishing (1968); UN Economic Commission for Europe, *The World Market for Iron Ore.* New York: United Nations (1968); U. S. Bureau of Mines, *Mineral Facts and Problems, 1970.*

[i] U. S. Bureau of Mines, *Mineral Facts and Problems, 1970.*

[j] Bauxite expressed in aluminum equivalent.

[k] U. S. Bureau of Mines contingency forecasts, based on assumptions that coal will be used to synthesize gas and liquid fuels.

[l] Includes U. S. Bureau of Mines estimates of gold demand for hoarding.

[m] The platinum group metals are platinum, palladium, iridium, osmium, rhodium, and ruthenium.

ADDITIONAL SOURCES:
Flawn, P. T., *Mineral Resources.* Skokie, Illinois: Rand McNally (1966); *Metal Statistics.* Somerset, New Jersey: American Metal Market Company (1970); U. S. Bureau of Mines, *Commodity Data Summary.* Washington, D.C.: U. S. Government Printing Office (January 1971).

be out of the question; but much sooner than that the utility would have tremendous difficulty finding the capital to construct its facilities.

Suppose capital were not a problem. The utilities are looking to the oceans as a "new" source of land for siting power plants. Dr. David Anthony of the University of Florida calculates that the exponential growth of electrical power in the state of Florida will have the following effects on the coastline of the western side of the state. By 1981, he says, the Gulf Coast will need 16 power plants spaced 20 miles apart; 32 years later, there would be 256 plants, spaced slightly more than a mile from each other; and 40 years after that, there would be 8,192 plants, only 211 feet apart! Anthony drily notes that this period of time is only 80 years from now, "within the lifetime of our children." The plants "would have to be staggered along the coast because each one takes up more than 211 feet."[63]

Dr. Malcolm Peterson of the National Committee for Environmental Information has pointed out that assuming the country's power needs were presently being met by 300 electric plants of 1,000-megawatt capacity taking up 1,000 feet on a side, that at continuing growth rates "all the available land space in the U.S." would be taken up within two centuries (less than 20 doubling periods) by plants alone—leaving no additional room for transmission lines, or people to enjoy the benefits of the electricity. And by the year 2000, the increased number of power plants would be consuming enough water for cooling to heat the total volume of water that crosses the continent by 20 degrees yearly.[64]

Even the ultimate "limit" to growth in electricity use has been set by California Institute of Technology scientist Jerome Weingart. The grand finale, he says, would be reached "if we were able to stack the new power plants on top of the old ones. After six more centuries of steady growth at 7 per cent per year, the United States would be a solid shaft of power plants whose outer boundary would be expanding with the speed of light! This represents what physicists would call the *relativistic* upper limit."[65]

The point here is unmistakable. The growth rates experienced in the development of energy in the world today cannot be sustained beyond a few more years. Exponential growth rates are indicative of the development of this lethal cancer in the industrialized nations of the world.

Energy Use, Economics, and the Environment

The effects of energy growth in the United States in recent years have been visibly demonstrated in terms of the effects of increased energy use on the nation's environment. As one observer put it:

> The recent upsurge of public concern over environmental questions reflects a belated recognition that man has been too cavalier in his relations with nature. Unless we arrest the depredations that have been inflicted so carelessly on our natural systems—which exist in an intricate set of balances—we face the prospect of ecological disaster.[66]

The author of those remarks, Richard Nixon, is surely no wild-eyed environmentalist, yet he recognizes the obvious presence of environmental disorder in our high-energy society even though he probably doesn't realize that the "environmental crisis"—or disorder in the environment—is *the inevitable result of processes of energy conversion.* As G. Tyler Miller, Jr., indicates, "according to the second law [of thermodynamics] any increase in order in the system will automatically and irrevocably require an even greater increase in entropy, or disorder, in the environment." Miller adds to his observation the paradox of our current ways of looking at the destruction of the environment, that "as we increasingly attempt to order, or 'conquer,' the earth we must inevitably put greater and greater stress on the environment."[67]

As we have seen in the history of energy use in the United States, the concept of waste is not only imbedded in the American dream, but in the religious and social institutions of Western society. What happens in high-energy societies, particularly the United States, which currently is the highest (though Japan is quickly catching up), is the ordering of all societal institutions around the availability of concentrated energy sources. In our society economic growth is made possible only by the availability of cheap energy (mostly fossil fuels), and technologists and planners hope that a new generation of cheap nuclear fuels will allow the high-energy spree to continue.

Understanding the nature of the environmental crisis, which affects all Americans, requires a look at the nature of our economic growth machine, which accelerates environmental pollution as it creates more technological artifacts at the expense of increased energy use. Economist Nicholas Georgescu-Roegen, in a searching analysis of the relationship between the thermodynamic science and economics, points out that economic theory today is completely ignorant of the relationship between energy and environment. He shows that, notwithstanding the "noise caused by the Entropy Law" in physics and the philosophy of science, economists have failed to pay attention to this most basic law of nature.

In the nineteenth century, English economist Sir William Petty popularized the notion that labor was the father, and nature the mother, of economic value, but today's economists have discarded that basic concept, and Georgescu-Roegen says that a student of economics learns of it as "a museum piece."[68] He notes that the result of modern economic education is that most economists believe that the economic process can simply continue to grow without "being continuously fed low entropy."[69] Of course, this is impossible, and at some point the energy costs of developing newer dilute resources alone will add appreciably to economic costs, resulting in increased costs and consequently lower consumer demand, resulting in a slowing of the economic process.

Economics today is pictured as a closed system, in which resources are converted to goods that are purchased by consumers, creating jobs and promoting the general welfare. The effects of this cycle on the environment are not considered by most economists as an integral factor of the equation: the environmental effects of growth (or the conversion of low entropy—mineral ores—into high entropy—automobiles—with resulting environmental deterioration brought about by the addition of incremental high-entropy pollution and heat into the environment) are the "external" costs. The "internal" costs are all measured by dollars circulating in the economy. Georgescu-Roegen and other economists who have noted this appalling discrepancy in economic theory argue that the economic process is based on the production of waste. The value of wasteful products is greater in our economy than the value of products that last longer. The reason automobiles are built to deteriorate in a few years, rather than being built to give years of service, is an economic reason. There's not as much profit in the long-lasting product—although it should be noted that there is far more "profit" to the environment, since long-lasting automobiles and consumer goods take a lessened toll in resources and energy.

Georgescu-Roegen sees the economic process, in its full ramifications, as "a continuous transformation of low entropy into high entropy, that is, into *irrevocable waste* or, with a topical term, into pollution. . . . The economic process is entropic: it neither creates nor consumes matter or energy, but only transforms low into high entropy."[70]

Just as economists generally do not recognize the significance of the waste-creating aspect of the economic structures of high-energy society, environmentalists and well-meaning policymakers rarely recognize the over-all implications of the laws of thermodynamics on environmental policies. There is no *pollution-free* energy in this world of ours.† Pollution is an inescapable by-product of the conversion of energy. In the natural world, a forest fire pollutes, and the by-products of the release of its smoke into the air may be just as contaminating as the burning of fossil fuels in limited amounts.

Dr. Robert Mueller, of the National Aeronautics and Space Admin-

† Terms such as "pollution-free energy" are bandied about by proponents of various energy technologies, including solar and wind power, and these terms will be encountered later in this book. The use of the term "pollution-free energy" refers to the fact that environmental disruption is *minimized* to a great degree by the use of natural energy technologies, but the elimination of entropy is impossible. Natural energy technologies may decrease environmental disruption that would otherwise be caused by the combustion of fossil fuels or the use of nuclear power.

istration's Goddard Space Flight Center, explains that the striking difference between pollution created by our energy technology and pollution in nature is a question of degree. The environmental crisis is a crisis because energy has been exploited to operate such a high level of technology in society that the natural energy systems of the Earth cannot withstand the assaults made on them. As Mueller notes, "the organic world contains limited genetic resilience to respond to technologic innovations of this magnitude."[71]

The following analysis shows the primary effects of energy extraction and use on the natural environment, including the drilling and transportation of oil, mining and burning of coal, heat released from engines and power plants, and climatic effects. The specific environmental challenges of nuclear power and other new energy sources are covered in more detail later.

Extraction and Transportation of Oil

Environmental damage from oil derives primarily from its uncontrolled release to surrounding areas during production, refining, or transport.

The primary problems of oil pollution at the production phase are "blowouts," uncontrolled bursts of flowing oil from a well—often as it is being drilled. As the result of technological developments over the past few decades, oil well blowouts on land are now largely controlled, and environmental contamination from them limited.

However, a blowout or other accident at an offshore well can cause major environmental consequences. Two relatively recent incidents focused the public's attention on the problem of offshore accidents and their devastating environmental consequences. The large Santa Barbara Channel oil spill in 1969 caused extensive damage to shore birds, shellfish, and many other marine creatures and plants. The total range of effects still is not fully known.[72] Other spills of varying magnitude continue to plague the Santa Barbara area.

An even greater disaster occurred in early 1970 when a Chevron Oil Company drilling platform in the Gulf of Mexico near Louisiana ignited and spilled 1,000 barrels of oil per day into the sea near $100 million shrimp and oyster beds. "Compared to Santa Barbara," Secretary of the Interior Walter J. Hickel was quoted as saying, "this is a disaster. There is more oil involved, more pollution, a wider area, and it will take much longer and be much harder to clean up."[73]

Major environmental damage also comes from the routine operations of many oil refineries, mainly in the form of water pollution through the release of petrochemicals and lethal trace elements. The former Federal Water Quality Administration reported that one large refinery dumped up to 2½ tons of toxic lead per day into the Mississippi River in Louisiana.[74] Other waste products released by refineries include heat, oil, phenols and other organics, hydrogen sulfide, heavy metals and other inorganics, solids, and product losses.[75]

The third, and most widespread, source of serious environmental damage from oil occurs during transportation—mostly at sea. The breakup of the *Torrey Canyon* off the coast of England in 1967 caused massive ecological damage and pointed up the seriousness of the oil transportation problem. The *Torrey Canyon* carried 118,000 tons of crude oil, most of which (95,000 tons) was dumped into the ocean. A British Government report on the disaster indicated considerable destruction of sea life and shore life along many hundreds of miles of coastline.[76]

Most oil spills of 100 barrels or more occur on ships during transportation. In 1968, 347 out of a total of 714 incidents in U.S. waters were from sailing vessels; and in 1969, 532 out of 1,007, according to Malcolm F. Baldwin, a Washington conservation lawyer. Of the world's annual oil production, 60 percent, or 1 billion metric tons, is transported by sea, Baldwin wrote, a skyrocketing increase since 1938, when only 85 million metric tons were shipped in vessels.[77]

An estimated 2.2 million metric tons of oil were spilled in the oceans in 1969, approximately 0.1 percent of the estimated 1969 world crude production of 1.8 billion metric tons.[78] At the same percentage rate, approximately 4 million metric tons would be deposited in the ocean in 1980 if annual crude oil production increases by then to the estimated level of 4 billion tons.[79]

By far the largest factor in the 1969 spills was normal oil operations: cleaning the tanks and ballasting of oil tankers; bilge pumping and cleaning in nontankers; and normal offshore production. As indicated in the following table, these routine operations accounted for almost half of the total oil spillage.[80] The highly publicized accidental oil tanker spills, offshore well blowouts, submarine oil pipeline breaks, etc., were estimated to represent only one fifth the oil spills of normal operations. Waste oils from refineries, petrochemical plants, or other industrial plants, and automobiles contributed the remaining 1 million tons of oil to the oceans.[81]

Oil Flows to Oceans in 1969
metric tons per year

Normal ship operations	1,000,000
Normal offshore production	100,000
Refinery operations	300,000
Oil in sewage and rivers	600,000
Accidental spills	200,000
Total direct input by man	2,200,000
Natural seepage (approx.)	100,000
Total	2,300,000

More recently, the Maritime Administration's environmental impact statement on the nation's oil tanker construction program indicated that oil spills from tankers had risen to 1,457,000 tons per year—967,000 metric tons from routine tanker operations and ballasting and cleaning of cargo oil tanks.[82]

Survey ships of the National Oceanic and Atmospheric Administration (NOAA) "have given us an indication of how far gone our oceans are," Barbara M. Heller of the Environmental Policy Center in Washington, D.C., says. According to these surveys, oil contamination covered 50 percent (80,000 square miles) of the survey area along the East Coast continental shelf; 80 percent (280,000 square miles) of the survey area in the Caribbean to the Gulf of Mexico; and 90 percent (306,000 square miles) of the survey area north of the Antillean Chain, which extends roughly 500 miles north and south of the coasts of the Bahamas and the West Indies. Total surface area in which oil contamination was detected amounted to 665,000 square miles of open ocean.[83]

Environmental effects of oil spills take many forms. There is of course the obvious damage: dead shore birds, fish reeking of the smell of oil, and petroleum-coated beaches. A British Government report on the *Torrey Canyon* noted the "sobering thought that each thousand tons of crude oil washed onto a beach 30 feet wide could form a layer half an inch thick for 20 miles," which was the *Torrey Canyon*'s calling card. It would have been at least twice as serious, the report observed, if the spill had not occurred in a relatively auspicious site.[84]

Biological damage from oil contamination in water has been documented to some extent. Dr. Max Blumer of the Woods Hole Oceanographic Institution has conducted a series of experiments on the toxicity of oil in aquatic environments. Other than direct kills of aquatic life through contact with high concentrations of oil, Blumer notes that oil in low concentrations produces significant changes in marine food chains, which have led to the destruction of certain forms of life. Blumer's work has pointed up the cancer-producing effects of oil contamination on marine life and a reduced resistance to infection in marine animals exposed to oil from spills.[85]

Various cancer-causing compounds in oil and petroleum may be passed along the food chain to humans in the same way that DDT reaches human food. The long-standing assumption that fish and shellfish tainted by oil would be edible again after the oily odor has disappeared is probably fallacious. Additional research in this area is needed, as well as on the effects of lubricants which, while small in volume, may be far more toxic than crude oil.[86] Another fact is that oil may interact with sewage, chlorinated hydrocarbons, and other chemicals to create combined effects that would be more harmful than the effects of any pollutant alone.[87]

Adequate measures do not exist for cleaning up oil spills. In some recent spills, including the *Torrey Canyon,* detergent-based dispersing agents were to demonstrate the latest in technological wizardry. The experience revealed, however, that the dispersing agents caused as much or more damage as the oil spills themselves. Scientists at the Plymouth Laboratory in England have documented widespread destruction in marine food chains from the use of chemical dispersing agents. Chemical tests on one commonly used U.S. chemical showed it to be "toxic to anchovy eggs and larvae at concentrations as small as two parts of dispersant per million parts of sea water."[88]

For all its immense damage to the ecology of the British coast, the *Torrey Canyon* was a relatively small vessel. As noted previously, it carried 118,000 tons of crude oil. Today, tankers with capacities of more than 500,000 tons ply the seas,‡ and vessels twice that size are being designed. A *Torrey Canyon*-style accident involving one of these gigantic supertankers could spell disaster for an entire coastal area. To meet the requirements of these vessels, proposals have been put forth for deep-water ports, refineries, and other industrial facilities, all of which could magnify many times the impact of spills occurring during normal operations.

Since 1960, the major oil companies have practiced, under a "Clean Seas Code," new load-on-top (LOT) procedures whereby oily water from ballast and tank washing is kept in a separate tank and allowed to separate. The sea water is then pumped out, along with some oil,

‡ As this was being written, the *Globtik Tokyo,* at 477,000 deadweight tons and with a 180 million gallon capacity, was the largest supertanker in service. If the nation had to import all its energy needs by such tankers, nine of them would have to be offloaded every day—nearly 3,400 shiploads a year.[89]

and the oily mixture above is retained for ultimate refining. New crude is loaded on top. Although some commercial difficulties remain, and the load-on-top scheme is not universally popular, the system is widely hailed and utilized by major oil companies.[90]

However, the LOT system can be applied only to tankers carrying crude oil. It cannot be used for tankers carrying finished products such as gasoline, fuel oil, and other refinery products, because these refined products can't be mixed, even in the small amounts that would be required with the LOT system. "Thus we believe," Barbara Heller told a congressional hearing, "that LOT is a stopgap measure, only partially effective, and that segregated ballast systems should be required on all tankers which would use our port facilities." The U. S. Coast Guard has also recommended such measures.[91]

A Massachusetts Institute of Technology Sea Grant study on the potential impacts of an oil find on Georges Bank off the northeastern United States and southeastern Canada concluded that any attempts to provide an oil containment and collection system against winter spills there "are presently futile, and almost certainly will remain so." This conclusion, Heller contends, "raises serious questions about the ability to control oil spills at potential deep-water port facilities, particularly off the northeast Atlantic coast."[92]

Natural Gas

The environmental problems arising from the production and transportation of natural gas are generally minimal, with the exception of some ecological damage from pipelines and growing threats from the increasing practice of transporting natural gas in its liquefied state in gigantic floating insulated jugs cooled to —259° F.

Supertankers filled with cargoes of liquefied natural gas (LNG) are now entering the nation's ports. Current projections are that the United States will need 130 LNG tankers by 1990—each carrying the equivalent of 2.7 billion cubic feet of gas.[93] It is speculated that 30 LNG supertankers would be required just to transport the liquefied natural gas that the United States has contracted to buy from the Soviet Union over the coming decades.[94]

Liquefied natural gas poses the greatest localized potential for mass destruction and annihilation of any energy technology except nuclear reactors. A preview of what could be in store for a nation with 130 LNG supertankers moored along its coastline occurred in Cleveland, Ohio, in 1944, when 1 of 4 LNG storage tanks ruptured and doused the city with 1.2 million gallons of LNG, the equivalent of about 100 million cubic feet of gas. It ignited, and poured into streets and sewers, burning as it went. Once in the constricted sewer pipes, the vaporizing LNG exploded, taking with it 30 acres of city sewer lines, 10 industrial plants, 80 houses, and 200 automobiles. Three hundred were injured, and 133 killed.[95]

In an analysis of the hazards of LNG supertankers, Timothy H. Ingram, contributing editor of the *Washington Monthly,* reported that land-based LNG storage and freezing plants are now relatively safe as the result of technical advances since the Cleveland explosion. The worry now, he pointed out, is "what happens when you start importing the stuff, not in 10-foot-thick, reinforced-concrete storage tanks, but in thin-walled ships." In 1971 alone, he noted, 140 tanker collisions and "hard groundings" were reported to the Coast Guard.

If a tank aboard an LNG tanker should rupture, the subsequent events would probably follow one of two patterns: It might ignite, creating heat intense enough to vaporize and burn the rest of the liquid gas on the ship and to consume any other ships that might have the misfortune of being nearby. The other possibility is that it would not ignite, but would fan out in the form of a low-hanging vapor cloud that could drift into an adjacent city, ignite, and blow entire sections off the map.

"A tanker which ruptured in, say, New York Harbor might well flood Manhattan with flames," Ingram reported. "Such a fire could begin if nearby vessels ignited the cloud, as happened during a recent naphtha spill near Staten Island," when a passing tugboat was invaded by some of the vapor, which then exploded in the tugboat's small cabin area. The resulting fire flashed back to the ruptured naphtha ship.

According to an Exxon study, a spill of LNG in harbor waters would represent an "unacceptable development." Dr. David Burgess of the U. S. Bureau of Mines told the *Washington Monthly* that if LNG vapor should find its way into a building and then were touched off by a furnace or cigarette, the entire building could explode, detonating the meandering plume in the process.[96] This could mean, he said, "a large tongue of flame, moving through the city back to the tanker," where it could immolate the leaking vessel and perhaps others with it.

In *Environment* magazine, James A. Fay, professor of mechanical engineering at the Massachusetts Institute of Technology, and James J. MacKenzie, a member of the Joint Scientific Staff of the Massachusetts and National Audubon societies, concluded that "the burning of the full cargo of an LNG supertanker would be equivalent to the burning of 100 *Hindenburgs* [the hydrogen-filled zeppelin that ignited and killed

62 passengers over Lakehurst, New Jersey, in 1937]. A conflagration of that size in a major city could be a disaster. . . ."[97]

Extraction of Coal—Deep Mining

Environmental hazards exist for all coal mining. The nature and severity depends on whether the coal is taken from deep mines or by stripping. Deep mining does not make the massive environmental assault on the environment that inevitably accompanies strip mining, but it does cause environmental degradation and poses some additional health and safety hazards that do not usually accompany strip mining.

Long exposure to coal dust in the mines causes black lung disease, or pneumoconiosis, among many workers. A 1965 Public Health Service report indicated that throughout the Appalachian region 10 percent of all active coal miners and 20 percent of all former coal miners showed X-ray evidence of black lung disease.[98]

The National Safety Council has listed coal mining as the most hazardous occupation among 41 U.S. industries.[99] However, it has been demonstrated that coal mining does not have to cause the deaths and injuries that have accompanied it up to now. Mine injuries and deaths appear to be preventable to a considerable degree, as some coal operations have shown.

In an attempt to institutionalize greater mine safety throughout the industry, Congress enacted the Federal Coal Mine Health and Safety Act of 1969. The law is designed to reduce—and hopefully eliminate—roof cave-ins, methane explosions, and other disastrous accidents that have killed more than 1,200 miners and injured about 60,000 others since 1966. Electric utility personnel estimate that the law will eventually cause an increase of $1 to $2 per ton of coal from existing underground mines, and $1 in future mines.[100] However, results so far indicate that the law has less effect on safety practices than long-standing company attitudes and policies.

U. S. Steel, for instance, has a long record of mine safety. In the course of mining 18.7 million tons of coal in 1970, 1 miner was killed and 35 were injured. On the other hand, in the same year Pittston Coal Company accidents killed and injured the same number for every million tons of coal produced. Thus, if U. S. Steel had the same record of carelessness and accidents as Pittston Coal, at least 18 miners would have been killed and 630 injured in the U. S. Steel mines in 1970.[101]

Strip mining is generally considered a less hazardous occupation than deep mining, so that it would be reasonable to assume that the companies with the best 1970 safety records were largely stripping operations, while the worst records would be amassed by companies that were primarily in deep mining. However, U. S. Steel, with a remarkably good safety record, obtained less than 1 percent of its coal from strip mines. On the other hand, Peabody Coal (a subsidiary of Kennecott Copper Corporation), which gets 80 percent of its coal from strip mining, had 17 times as many workers injured in the mines in 1970 as U. S. Steel.[102]

Deep mining is not without its environmental problems. Like strip mining, deep mining is accompanied by the formation of large quantities of sulfuric acids which, as "acid mine drainage," frequently seep into local water tables to contaminate aquifers in mining regions. The Federal Water Pollution Control Administration (now the U. S. Environmental Protection Agency's Office of Water Programs) estimated that the Appalachian region annually loses $7.5 million as a result of water deterioration from mining operations.[103] Clean-up operations will be expensive, since more than 10,000 miles of rivers and streams in the United States are contaminated by acid mine wastes. The American Chemical Society believes that "prevention at the source holds the best long-term hope, but it must be supplemented meanwhile by methods of treating acid (or alkaline) mine water after it forms."[104]

Solid wastes from coal mines also present an environmental problem. They are disposed of in great "gob piles," composed of various coal residues, some of which are combustible and hazardous. A Bureau of Mines survey indicated that in 1963 there were 213 fires from gob piles in West Virginia. These wastes are often stacked hundreds of feet high and may lead to slides, such as the one that caused the tragedy at Aberfan, Wales, in the late 1960s.[105] Or they may be illegally bulldozed in the shape of a dam to hold back mountain water. It was such a dam as this—constructed both incompetently and illegally—that gave way in the early 1970s and drowned entire families in the Buffalo Creek area of West Virginia.

A recent Library of Congress report to the congressional Joint Economic Committee concluded that expansion of deep mining would prove increasingly destructive of some environmental qualities. "Even if it be granted that the coal tipple and other surface appurtenances, along with the generally dreary mine villages are no more offensive to many persons than some nonmining aspects of the environment, there remain generally unsolved problems of acid drainage and of long-burning culm [coal refuse] banks," the report noted.

"Few persons appear to regard coal mining in any of its versions as contributing favorably to the quality of the environment," the report concluded. "To press on to higher levels of production will almost inevitably involve increased 'exploitation' of present coal mining areas, large development of some new areas, and greater conflict with an increasingly critical public."[106]

Strip Mining

While strip mining, which now accounts for more than one half of all the coal mining in the United States, is highly lucrative for the coal operator who practices it, it is destructive in the extreme to the land and water environment.

By 1970 more than 3 million acres—1½ times the size of the state of Delaware—had been strip mined for coal in the United States. More than two thirds of that area, or some 2 million acres, was officially rated as "unreclaimed" by the Department of the Interior. Much of what was officially classified as "reclaimed" was reclaimed in name only, and bore no resemblance to the state of the land before the coal was ripped out from beneath it.[107]

In a review and projection of U.S. strip mining published by the COALition Against Strip Mining, Edwin Cubbison and Louise Dunlap noted that:

> If stripping is allowed to continue, in the 15 states surveyed, 36 billion tons of coal will be produced at the expense of 8,455.1 square miles of land, not including spoil bank and off-site damage. The strippable reserves of these 15 states constitute only 4.8 percent of the total recoverable coal reserves in this nation. Yet the pursuit of this 4.8 percent will create a wake of devastated lands equivalent to a path 2.5 miles wide from New York City to San Francisco.[108]

The extent of the damage to the earth's surface, they pointed out, is dependent on the depth of the overburden (the soil between the coal and the surface of the earth), the thickness of the "seam," or layer of coal; and the type of terrain. In addition, they said, acidity of the soil, the humidity of the climate, and proximity to streams, rivers, and subsurface waters are critical factors in determining the extent of the environmental damage that will be done by strip mining.[109]

Characteristic of the strip, or surface mining, industry is the sheer immensity of the equipment it uses to chew up mountainsides. Gigantic 17-story shovels like the "Gem of Egypt" and "Big Muskie" devour 325 tons of coal-laden earth at a time, taking with it topsoil, rocks, small trees, plants, and wild animals and their nesting places without discrimination.[110]

Strip mining's environmental effects include geological cracks caused by blasting; acid and mineral seepage into the water table; redistribution of subsurface waters; soil bank slippage with resultant loss of soil stability; landslides and mountainslides; siltation; lower water retention capacities of watersheds; flash floods; reduced storm-carrying capacities of rivers; loss of spawning beds in streams and rivers; soil sterility; lost topsoil; and replacement of diverse rich hardwood forests—if at all—by fast-growing but vulnerable plant monocultures.[111] As can be readily seen, these devastating effects may range far beyond the immediate vicinity of the strip mine.

The strip mining operations now under way to support the Four Corners electric power plants in New Mexico provide a classic example of the nature and extent of strip mining, and offer a preview of much of the western United States if strip mining is permitted to go unchecked to support the exponential growth of the electric utilities.

The strip mine leased by the Navajo Indians to the Peabody Coal Company is the most extensive supply contract of its kind. It calls for 117 million tons of coal over a 35-year period to furnish the fuel for 2 generating units that are geared up to burn 5 million tons per year, or 16,000 tons per day.[112] This and other leases are vague in relegating the industry's responsibility for restoring the land. One lease requires only that Peabody leave the land "in as good a condition as received, except for ordinary wear, tear, and depletion incident to mining operations." Malcolm F. Baldwin, who conducted a policy evaluation of the Four Corners operation for The Conservation Foundation, concluded that "in effect . . . the coal companies have been granted *carte blanche* in their reclamation plans" for the strip mining operations that go to furnish coal for the gigantic electric power plant.

Strip mining does not have to entail the tragic devastation to the land that it has up to the present time in the United States. The West German Government has adopted a comprehensive program of planning, oversight, and rigid regulation of strip mining that has resulted in re-establishing the landscape virtually to its former state.[113]

The program involves carefully removing first the topsoil, then the rest of the overburden, with the aid of a gigantic wheel excavator. Both the topsoil and the overburden are set aside for reuse. After the mining (some of the mines are 900 feet deep) is finally completed, the excavations are filled with the sand, gravel, and clay overburden. They are then leveled and the topsoil diligently reapplied. Within two years agricultural crops are again being harvested.[114]

The major reasons why the strip mining industry doesn't emulate this practice in the United States are: (1) it is expensive and (2) nobody forces them to. The actual costs of restoring mined-out lands to their original state in West Germany run from $3,000 to $4,500 per acre. This would eliminate the enormous cost advantage strip mining now enjoys over deep mining in America.[115] J. Davitt McAteer, a coal specialist and consultant to Ralph Nader's Center for Study of Responsive Law, recently observed that deep-mined coal is approximately $4 a ton more expensive to produce than strip-mined coal, but that if the cost of reclamation were added, "it would bring strip-mined coal into a more equitable cost position with that from underground."[116]

Vigorous enforcement is one aspect of West Germany's 4-phase reclamation program, which involves incorporating surface mining into an over-all regional development plan scrupulously overseen by a widely representative planning body, whose recommendations are finally subjected to review in public hearings before the strip mining operations are permitted to proceed. Finally, and most important, an enforcement agency vested with the necessary power enforces the approved plan to the letter.[117]

Legislation prohibiting additional strip mining passed the U. S. House of Representatives in 1972, but did not pass the Senate. Several bills were introduced again in the Ninety-third Congress, but prospects for their passage against the clamor for more electric power are dim. In the meantime "the vast acreage of strippable coal leased before the [current] regulations went into effect in 1969—some 2.4 million acres—are beyond the law."[118]

"Strip mining," aptly observed Congressman Ken Hechler of West Virginia, "is like taking 7 or 8 stiff drinks: You are riding high as long as the coal lasts, but the hangover comes when the coal is gone, the land is gone, the jobs are gone, and the bitter truth of the morning after leaves a barren landscape and a mouthful of ashes."[119]

Energy Conversion in Engines

The environmental effects of energy use that are most apparent are the "end uses" of energy: When energy sources are converted to electricity in power plants or in the great furnaces of industry; when solid wastes are combusted; when gasoline distilled from petroleum is converted to motive power to drive automobiles, trucks, trains and aircraft, etc.

The development of the engines for automobiles and mass transit systems makes possible the cities and suburban areas, but this development also brings inevitable waste heat and pollution from the vehicles to the densely populated areas. Likewise, the development of electricity makes possible greater population densities; and even though electrical power plants may be located many miles from cities, the landscape is scarred by the transmission lines that carry power to the cities.

When energy sources are converted into work, pollution invariably results—in the form of heat or water or air pollution (in accordance with the Entropy Law).*

Air pollution from fossil fuel use is seen in the following table of air pollutants in the United States in 1970:[120]

1970 AIR POLLUTANTS BY WEIGHT IN MILLIONS OF TONS PER YEAR

Source	Carbon Monoxide (CO)	Particulates	Sulfur Oxides (SOx)	Unburned Hydro-carbons (HC)	Nitrogen Oxides (NOx)
Transportation	111.0	0.7	1.0	19.5	11.7
Fuel combustion in static sources, power generation	.8	6.8	26.5	.6	10.0
Industrial processes	11.4	13.1	6.0	5.5	.2
Solid waste Disposal	7.2	1.4	.1	2.0	.4
Miscellaneous	16.8	3.4	.3	7.1	.4
1970 Totals	147.2	25.4	33.9	34.7	22.7
Totals in 1940	85	27	22	19	7

From this table of 1970 air pollutants, it is clear that transportation sources, particularly automobiles, represent the greatest source of air pollution in the nation, followed closely by power generators (both electric utility generators and those of industry).

The comparison of 1940 data with those of 1970 indicates that in certain areas increases in the use of energy have not necessarily generated a corresponding increase in pollutants generated. In fact, an actual reduction in the generation of airborne particulate matter has been accomplished. This is due primarily to the substitution of petroleum fuels for coal. The emission of particulates to the atmosphere comes primarily from coal burning. The increasing substitution of natural gas for coal in power plants in recent decades has resulted in the reduction of this pollutant source, since natural gas—the cleanest fossil fuel—burns with ease and contributes few pollutants to the at-

* It should be noted that *all* air and water pollution result from energy conversion processes. Such statements as "energy systems were the *largest source* of the 264 million tons of pollutants emitted into the air in 1970" (from a report of the Council on Environmental Quality, emphasis supplied) are false, and do not take into account thermodynamics. The web of energy—and entropy—stretches farther than the smokestacks of power plants. All pollution is the result of energy use.

mosphere in contrast to coal. Increasing use of abundant coal as a substitute for scarce natural gas at electrical power plants will reverse this trend in the future.

The following sections discuss some of the primary environmental and health effects of fuel combustion in stationary and mobile heat engines.

Stationary Sources (Electricity)

In 1970, the burning of fossil fuels at stationary sources—which are mostly electric power plants, but also include home and building heating units (not including industrial plants)—released to the air 26.5 million tons of sulfur oxides, 10 million tons of nitrogen oxides, 6.8 million tons of particulates, 800,000 tons of carbon monoxide, and 600,000 tons of hydrocarbons (see the chart on page 146).[121] Combustion of fossil fuels also released an undetermined amount of lead, mercury, cadmium, nickel, beryllium, and other trace elements into the atmosphere.

A 1970 report of the Office of Science and Technology estimated that electric power plants emitted one half of all sulfur oxides, one fourth of the nitrogen oxides, one fourth of the particulate matter, 5 percent of the gaseous hydrocarbons, and something less than 2 percent of the carbon monoxide—in addition to the trace elements.[122]

There are serious health hazards associated with all these pollutants. Sulfur dioxide (SO_2) and particulates in combination begin to reflect increased illness and death when the sulfur dioxide reaches a level above one-fourth part per million (ppm). Health hazards are also thought to result from nitrogen dioxide (NO_2) emissions. Research has shown effects on the lungs similar to those caused by aging, as well as an emphysema-type condition, as a result of NO_2 exposure. One study of health effects on humans showed increased respiratory infections at levels above .06 ppm. A 1972 workshop on Energy and the Environment at Cornell University concluded that "since this level is regularly exceeded in 85 percent of all cities over 500,000 populations, NO_2 presents a potentially serious hazard. . . ."[123]

In the atmosphere the nitrogen oxides, under the influence of sunlight, undergo complex chemical reactions to form photochemical smog and ozone. Ozone causes respiratory mucous membrane irritation in many people at levels of .2 to .3 ppm, and impairs lung function, particularly among asthma sufferers. While there is no firm evidence that the respiratory tracts of healthy people suffer serious damage from prolonged exposure to ozone under normal conditions, the Cornell workshop pointed out that it behaves in the body much like radioactive substances in many respects. "Accordingly, the same toxicologic considerations should enter into setting an acceptable air quality standard for oxidants as are used in setting maximal permissible concentrations for radioactive substances," a report of the workshop concluded.[124]

Deterioration in athletic performance has been observed in smog with levels as low as .03 ppm total oxidant. Although there is no evidence of a permanent health effect on a normal population from this hydrocarbon-oxidant mixture, it can seriously aggravate asthma, emphysema, and bronchitis.[125]

In fogs, nitrogen dioxide may combine with water to form nitric acid, which can cause corrosive damage to physical materials and to plants and can irritate the lungs.[126]

Carbon monoxide, fatal in high concentrations, appear to also lower sensory perception and discrimination at levels of about 50 ppm when exposure is prolonged. These conditions could pose serious hazards in many occupations and in driving a motor vehicle.[127]

Fossil fuel combustion also releases into the air particles of the heavy metals—lead, mercury, cadmium, beryllium, nickel, etc.—in amounts which, alone, might not be significant, but when added to levels accumulated in the body from other sources could cause toxic effects. Because of the elimination of lead from gasoline, thought to be a major contributor to levels of lead in the human body up to now, the release of lead may not be as serious as some of the other trace elements. Recent research linking heavy metals to chromosome aberrations and genetic defects suggests that even very small amounts of these compounds might present a significant danger to health.[128] The Environmental Protection Agency (EPA) has noted that trace elements may also "induce, accelerate or aggravate dangerous chronic disorders including heart disease, renal diseases, lung diseases, and cancer."[129]

Some idea of the volume of pollutants that are produced in fossil fuel combustion can be gained from a review of plants now either completed or under construction in the U. S. Southwest. The best-known of these is the Four Corners Plant at Fruitland, New Mexico, where the borders of New Mexico, Utah, Colorado, and Arizona converge. The Four Corners Plant and the Mohave, Nevada, plant 270 miles to the west now generate nearly 4,000 megawatts as the first 2 units of what is termed the Southwest Power Complex.[130] Three more similar plants at Huntington Canyon, Utah; San Juan, New Mexico; and Navajo, Arizona, are under construction and will produce an additional 3,000 megawatts by 1977. The intention is for these plants, as part of the Southwest Power Complex, to provide the power for continued rapid growth of Los Angeles and San Diego, California; Las Vegas, Nevada; Phoenix and Tucson, Arizona; and Albuquerque, New Mexico.[131]

The Environmental Protection Agency estimates that when the 5 power plants are all completed they will release 77.2 tons of particulates, 839 tons of sulfur dioxide, and 644 tons of nitrogen oxides *per day,* in addition to consuming 100,210 acre feet of water from the Colorado Basin each year for cooling purposes.[132]

The pride and joy of the Southwest Power Complex is the proposed Kaiparowits plant planned for southern Utah's Kaiparowits Plateau. The Kaiparowits plant would be the largest coal-fired power station in the world, generating 5,000 to 6,000 megawatts and consuming 50,000 tons of coal per day—some 15 million tons per year.[133] One utility consultant told a Senate Interior committee that the existing plants at Mohave and Four Corners are already releasing 260 tons of particulate matter per day. He also said that *if* the utilities are able to reduce particulate emissions as required, and *if* new sulfur dioxide removal units at the Navajo plant perform at their best and can be duplicated in the other plants, the total emissions from the Four Corners, Mohave, Navajo, San Juan, and Kaiparowits plants operating at full capacity will be 100 tons per day of particulate matter; 600 tons per day of sulfur dioxide; and 1,140 tons per day of nitrogen oxides.[134]

The Environmental Protection Agency has stated that pollutants from power plant emissions in the Southwest Power Complex can ". . . be expected to have a significant impact on terrestrial biota and on water quality." The agency cited possible damage to water in lakes in the area from sulfuric acid rains from two of the Southwest plants, along with fallout of mercury emissions.[135]

Of particular concern to health authorities is emphysema, which is now the leading cause of death by disease in many states, including Arizona and Colorado, two of the "Four Corners" states. According to Malcolm Baldwin, the statistics "might indicate that people affected by emphysema go there for clean air. However, the circumstantial evidence has led one scientist to surmise that small, submicron particles found in these regions may be the cause."[136]

"If this is true," Baldwin continued, "the proposed power plants, with the small particles they create, will add significantly to the problem. The particles that even the best technology cannot now remove from stack emissions are the small ones—those below 2 microns in size. These are perhaps the most dangerous to human health." Dr. Stanley Greenfield, assistant administrator of EPA for Research and Monitoring, points out that the nostrils prevent particles larger than 1 micron from entering the lungs, but smaller particles enter and are retained there.[137]

It was concern over continuously increasing levels of air pollution and its potentially damaging effects that produced a series of national air pollution laws over the past decade, culminating in the Clean Air Act amendments of 1970, which also created the Environmental Protection Agency. As a result of the authority delegated to it in the act, the EPA in 1971 established a 2-tier set of standards for stationary sources, and required states to submit plans for meeting the standards within specified deadlines. Primary standards, those designed to protect the public health, were to be achieved within 3 years after the EPA approved the states' plans. Secondary standards, aimed at protecting less vital environmental aspects—aesthetics, vegetation, and physical materials—were to be achieved within a "reasonable" time period.[138]

On May 30, 1972, the day before the EPA's deadline to approve or disapprove state implementation plans, a federal district court ruled that a plan could not be approved that allowed significant deterioration of existing air quality in areas where the air already was cleaner thn the EPA standards required.[139]

In July 1973, the Sierra Club petitioned the U. S. District Court to compel the EPA to issue final standards that would prevent significant deterioration of air quality, complaining that the EPA had not complied with the 1972 decision. The Sierra Club suit could halt power plant construction in less-populated areas, including those projects now under way as part of the Southwest Power Complex.[140]

The Environmental Protection Agency believes that technology is now available to remove sulfur oxides from the smokestacks of furnaces in which coal and oil are burned, although industry vigorously denies the assertion. A recent survey indicated, however, that sulfur oxide pollution control technology would be available in 1976.[141]

Development of nitrogen oxide removal technology is even farther away than sulfur oxide control technology. One environmental engineer for the Bechtel Corporation, builder of the 2,310-megawatt Navajo power plant in Page, Arizona, has said that significant reductions of nitrogen oxides from coal-fueled power plants should not be expected until the end of the 1970s.[142]

Thermal Pollution

Heat is the ultimate waste product of electric power production, in accordance with thermodynamic laws. Because electric power plants are relatively inefficient, most of the fuel energy is lost as heat before it can be converted to electricity.[143] This waste heat is commonly referred to as thermal pollution. The most modern and efficient fossil fuel plants convert 40 percent of the fossil fuel energy to electricity. Another

45 percent is lost as waste heat to the cooling water, and the remaining 15 per cent as waste heat lost to the air through the smokestack. Most fossil fuel plants are less efficient than that, with the average at 33 percent or lower, meaning that two thirds of the fossil fuel energy is dissipated to the environment in the form of thermal pollution.

Nuclear power plants are even less efficient—about 30 percent—so that 70 percent of the uranium energy is added to the environmental burden as waste heat.[144]

The Federal Power Commission has predicted that U.S. electric power production will double each decade, and that an increasing percentage of that power will be produced by nuclear reactors. This means that if these projections hold true, the amount of thermal pollution released to the nation's air and water can be expected to more than double each decade, on the basis of present industry practices.[145]

The long-term effects of thermal pollution are not well defined, but current research into the effects of heated water on aquatic life has established certain ground rules. Eugene Cronin of the University of Maryland's Natural Resources Institute states that "temperature is a profound master factor in the aquatic environment. It can kill, it can affect the movement of some species, and it can regulate rates of biochemical and physiological processes, especially those associated with reproduction."[146]

Although many water organisms can withstand gradual rises in temperature, they cannot withstand sudden changes in temperature, such as those caused by routine additions of large quantities of heated water to streams, lakes, and rivers by power plants. Thermal pollution decreases the amount of dissolved oxygen in water, thus hampering the ability of many animal and plant species to survive. Another effect of the added heat is the growth encouragement of populations of heat-resistant organisms, such as algae. When algae thrive, the production of thick mats of slime may clog rivers, cause odors, and radically deteriorate the quality of the water.

The addition of large "thermal blocks" of water from power plants may prevent the normal migration of fish; if the water in the rest of the river becomes hot enough, certain fish species will not survive. Lake trout and salmon, both prized as sport fish, are especially sensitive: Lake trout begin dying when water temperature increases above 77° F, and their spawning is hampered at temperature ranges above 48° F. The Oregon fish commission says that the hatchability of salmon eggs decreases "disastrously" when the normal temperature of the Columbia River is increased by 5.4° F.[147]

Raising the temperature of the water may also affect its use for recreational purposes, and often increases local fogging problems. The increased rate of evaporation resulting from the addition of heat may also increase the salinity or mineral content of the water, especially since minerals will also dissolve more easily in heated water. This can cause economic impacts as a result of increased corrosion of highways, ships, intake structures, and other materials with which the water comes in contact.[148]

Technical alternatives to the engineering of devices to eliminate the problem of disposal of the waste heat to waterways include the use of various types of cooling towers, which are massive structures that recirculate water from power plant condensers to the cooling element, then release it to the waterway. One problem experienced in a common form of tower, the "wet," or evaporative, tower, is the release of clouds of steam into the air around the installation. In addition, a 1,000-megawatt plant with an evaporative cooling tower might require up to 10 million gallons of water per day, an amount of water that would serve a community of 100,000 people.[149]

Wet cooling tower systems cost an additional estimated $5 million to $9 million for fossil fuel plants, and $8 million to $13 million for nuclear plants. This would increase the average residential electricity bill only by about 1.28 percent.[150]

A more sophisticated and still less damaging approach is the use of dry cooling towers, which transfer the heat to the atmosphere without the accompanying moisture. A dry cooling tower can be expected to add $20 million to $25 million to the cost of electric power plants—representing a 1.5 percent to 3 percent increase in the average residential electric power bill.[151]

While this is the most desirable means of eliminating release of thermal pollution to water, it still poses many environmental problems. Since the heat is released to the air instead of to the water, the environment still feels the impact, and because air is not as good a cooling agent as water, a much larger volume of air is required for the same amount of cooling. Similarly, because the dry cooling tower system does not cool the water that recirculates through the plant as much as an evaporation system would, the efficiency of the power plant is also lowered in accordance with the laws of thermodynamics. Finally, the immense towers rising several hundred feet into the air raise aesthetic objections and pose some danger to aircraft. Finally, they have not been tested on large electric power plants, so that it is still not a certainty that the dry cooling towers would prove adequate on such a massive scale.[152]

While billions of dollars to reduce thermal pollution in the nation's waterways seems a high price to pay, the costs of permitting thermal pollution to continue may prove even higher, although the charges may show up on a different balance sheet. Raising the temperature of a body of water substantially lowers the capacity of the stream to assimilate waste materials. When the temperature of the Coosa River in Alabama was raised 9° F above the 77° F summer level, the capacity of the stream to assimilate waste was reduced by 11,000 pounds of oxygen-demanding wastes per day if the established dissolved-oxygen levels were to be maintained.[153] The potential seriousness of the problem is reflected in the fact that if power consumption continues to grow at its present rate, with no significant increase in efficiency, power plants of all types will produce enough waste heat by the year 2000 to raise by 20° F the temperature of all the surface water of the United States.[154]

Transportation Sources of Pollution

About one fourth of the nation's energy is devoted just to fueling the various engines in the transportation sector of the economy. Eighty percent of the petroleum is used for powering cars, trucks, and buses. Airplanes take an additional share.

Moving the nation's 210 million people and the goods they demand accounts for the majority of unburned hydrocarbons, carbon monoxide, and nitrogen oxides released to the U.S. atmosphere. In urban areas, transportation energy use is responsible for nearly all the unburned hydrocarbons and carbon monoxide.

Primary pollution damage comes from the emission of the enormous volumes of pollutants into the air, particularly from automobiles. The pollutants in 1970 amounted to more than 111 million tons of sulfur oxides, 19.5 million tons of hydrocarbon residues, and 11.7 million tons of nitrogen oxides.

The health effects of some pollutants are well understood, but little is known of others. Studies have shown that exposure to 10 parts per million (ppm) of carbon monoxide for eight hours dulls mental performance; yet many cities are subject to levels of 70 to 100 ppm in the atmosphere for varying periods.[155] Air quality is notoriously poor in urban pockets, where most travel is conducted by automobile. In Los Angeles County in 1970, fully 25 years after the enactment of California's first air pollution control laws, photochemical oxidant standards were not met on 65 percent of the days; carbon monoxide standards on 55 percent of the days; and nitrogen oxide standards on 31 percent of the days.[156]

The essential federal strategy to control pollutants from transportation (as well as from fossil fuel combustion at stationary sources) is to control emissions from individual vehicles under the terms of the amended Clean Air Act of 1970. Under the terms of the act, emissions of carbon monoxide and hydrocarbons are to be reduced by 1975 to 90 percent of 1970 levels, and emission of nitrogen oxides reduced to 90 percent in 1976.† [157]

Numerous solutions have been proposed to solve the problem of pollution from transportation energy sources, particularly the automobile. In the case of ground transportation, the options for reducing pollution may be divided into three categories:

1. Use clean fuel, such as natural gas, in vehicles. As in power plants, natural gas and its derivatives—such as propane—burn cleaner and can meet stringent federal standards. The Environmental Quality Laboratory at the California Institute of Technology has recommended that one third of the vehicles in the Los Angeles Basin (primarily fleet vehicles operated by telephone companies, industries, and trucking concerns) be converted to gaseous fuels. The gaseous fuels release minimal pollutants into the atmosphere, meeting California state standards and approaching the federal Clean Air Act standards. Since the cleaner fuels do not clog engine components as does gasoline, savings are realized by owners of gas-operated vehicles. In addition, gaseous fuels are usually cheaper than gasoline, and they are exempted from gasoline taxes by the state of California (and from federal gasoline taxes as well).[158]

Alas, the promise of cleaner fuels for vehicles is short-lived, since supplies of natural gas are not capable of fueling the extensive conversion of America's vehicles. The Environmental Quality Laboratory indicated that future supplies might be adequate for Los Angeles, but available sources would not be adequate for other areas as well, given the current demands on natural gas for heating, industrial uses, and electricity.

Although supplies of natural gas are inadequate to meet needs, the philosophy of using clean fuel is a sound one and points the way toward research aimed at cleaning up other fossil fuels at refineries and special processing plants, so that clean fuels might be made available in the future—especially for mass transit in urban areas.

2. Another option is to utilize devices on each vehicle to control emissions. This is the philosophy adopted by the Environmental Pro-

† It seemed clear, as this book was completed, that variances to these standards would be allowed by the Congress in response to pressure from automobile manufacturers.

tection Agency and the nation's auto manufacturers. The basic antipollution device seized on by manufacturers is the "catalytic converter," a cylindrical device containing a catalyst, such as platinum, that reacts chemically with exhaust pollutants, converting the exhaust into carbon dioxide and other gases. Numerous disadvantages have been linked to the use of these converters, not the least of which is the expense of converting all the nation's automobiles; the National Academy of Sciences estimates that this conversion would add up to $23.5 billion per year by the 1980s.[159] Various cost estimates per automobile have ranged from under $100 to several hundred dollars. Another problem is that the United States must import platinum (and other possible catalysts) from foreign countries. The two countries richest in platinum are South Africa and the Soviet Union. Not only would the U.S. supply base not be particularly reliable, but the U.S. balance of payments would be adversely affected, as one more critical resource would be added to the import list.

Tests made at New York State's Department of Environmental Conservation indicate that the catalytic converters emit tiny metallic particles in use (suspected to be platinum, chromium, and nickel), which may pose a serious health problem. The metals involved are of known toxicity, says the New York agency, and are small enough to enter human lungs.[160] The national strategy to equip cars with the converters to meet the EPA standards involves removing lead from gasoline (not for health reasons, but because lead "poisons" the converter); but the addition of the toxic metals may overrule the environmental benefits to be derived from ridding automotive exhausts of lead.

The catalytic converters tested by New York State also added a 23 percent penalty to gasoline mileage,[161] which is to be expected with the addition of any antipollution add-on device to an engine.

Attempts to foist a political/technological solution to the problem of automobile exhausts, or moving to smaller cars, ignores the real problem. As transportation consultant George W. Brown points out:

> The whole idea of emission standards is a cruel hoax which only delays the inevitable time when we will be forced to change our basic concepts regarding necessary transportation modes. The idea that we can continue to condone and encourage a population explosion of motor vehicles if we can only clean up the exhausts, is ridiculous. Space for roads and streets is finite. Our atmosphere is finite. . . . The industry and the EPA prefer the add-on junk which does nothing except distract attention from the serious population explosion of vehicles and the fact that they are becoming less, rather than more, efficient.[162]

3. The last option for reducing pollution is the most logical one. The source of the problem is the automobile culture: The sheer numbers of cars render clean-up schemes absurd, if for no other reason than the tremendous amount of money and energy involved to accomplish the herculean task. The EPA has recently recognized this dilemma and indicates that stringent transportation controls may be imposed on seventeen urban areas by 1977. The EPA suggested that all auto use in Los Angeles may be barred by that year, and 60 percent of vehicle traffic would be eliminated in northern New Jersey. Other measures suggested by the agency included banning motorcycles (whose two-stroke engines are highly polluting and inefficient) and rationing gasoline to meet Clean Air Act standards. As EPA Acting Director Robert Fri noted: "We are basically attacking the problem by asking people to change their habits—their long-standing and intimate relation to the automobile."[163]

The basic limiting factor to this approach, assuming further political developments will actually result in a change of habits, is that alternatives to car transit are not readily available, and, while the 1973 Federal-Aid Highway Act "opened" the venerable Highway Trust Fund to mass transit expenditures, it limited the scope of the expenditures to about one fourth of the monies that will be devoted to further construction of highways for continuation of the automobile culture.[164]

Energy Use and Climate

The effects of energy use on climate have been intensively studied by scientists for some years; yet the area remains little understood. Man's historic impact on climate has been great. For centuries, practices of grazing domesticated animals have reduced vegetated regions to deserts and semideserts, notably in areas of Africa and Southwest Asia. According to the report of a recent international conference of atmospheric scientists, 20 percent of the surface area of all continents has been radically changed by man's activities, causing changes in the heat and water budgets of the continents.[165]

All life is powered by the input of solar energy; the daily workings of the climatic cycles on Earth are the result of the addition of solar energy. On a worldwide average, about 30 percent of incoming solar radiation is scattered back into space by various aspects of the Earth's atmosphere, such as clouds and the protective ozone layers of the upper atmosphere. Another 20 percent of incoming solar rays is captured in the atmosphere: Some energy is trapped in the ozone layer; some is trapped by dust, water droplets, and vapor in clouds. Of the remaining 50 percent of solar radiation that actually reaches the surface, some is absorbed in land and water, some evaporates water (triggering the global water cycle), and some is reflected back into the atmosphere.[166]

The heat balance of the Earth's surface is maintained by the addition of solar energy. As the surface becomes warmer by the addition of solar heat, the Earth gives off heat at a faster rate than when it is cool. An increase in added heat will cause faster heat release, and a decrease in added heat will slow the process. Climatic balance is based on the variations of the sun's input, combined with the secondary natural energy cycles of winds, water, ocean currents, and tides. On the average, the Earth yields back to space as much energy as is received from the sun each year.

Alterations in this natural energy balance by human actions in the environment will be of great importance in the future. Interventions up to the present have affected the climate in local areas and regions (the "microclimate"), and the continuing development of high-energy societies may alter the Earth's over-all climate (the "macroclimate").

Changes in the Earth's microclimates by human intervention have almost always led to increases in heat releases. This has been accomplished in numerous ways, the most common being the cutting of trees, which changes the Earth's albedo, or capacity for reflecting light. Changes in the albedo have led to distinct regional weather trends, and even the presence of a single dry road in a moist area could provide a heat source in the local environment, accompanied by a convective plume, which leads to the formation of cumulus clouds —thus, precipitation.[167] The significance of the construction of highways alone in the United States, which comprise close to 1 percent of the over-all land mass (70,000 square kilometers—a little more than 27,000 square miles—of road surface) is a significant climatic factor.[168]

The release of particulate matter and exhaust gases from vehicles combines with the change in the Earth's albedo in all high-energy societies to produce significant climatic changes. The precise extents of change are unknown, and atmospheric scientists are currently attempting to analyze data to produce models that can predict the long-term consequences.

The effect of energy use on the climate of urban areas is better understood. The combination of heat releases from power plants, automobiles, and other vehicles, and the changes in albedo from construction of concrete highways and buildings in energy-intensive urban areas lead to distinct climatic changes. The report of the Study of Man's Impact on Climate (SMIC) Conference sponsored by the Massachusetts Institute of Technology in 1971 lists the following measured changes that result from urbanization in high-energy societies:[169]

AVERAGE CLIMATE CHANGES FROM URBANIZATION

Element of Climate	Change in Urban vs. Rural Areas
Contaminants:	
particulate matter (condensation nuclei)	10 times more
Cloudiness:	5 to 10 percent more
winter fog	100 percent more
summer fog	30 percent more
Precipitation (rain and snow)	5 to 10 percent more
Sunshine duration	5 to 15 percent less
Ultraviolet radiation (winter)	30 percent less
Annual mean temperature	.5 to 1.0° C more
Winds	20 to 30 percent less

The energy released from some urban areas is much greater than the worldwide averages in the above table. In many cities, much more heat is released into the atmosphere than the solar energy striking the ground! Leading the list is the venerable symbol of high-energy cultural extravagance: New York City. The average annual solar radiation reaching one square meter of Manhattan's surface is about 93 watts. Energy consumed is about 630 watts per square meter, more than 6 times the sunlight striking the surface area of the city. Expressed in per capita terms, energy use (and waste heat released) is twice that of Los Angeles, which itself is no symbol of frugality in the use of energy, with a full 60 percent of its surface dominated by highways, service stations, and other appurtenances of the automobile culture. Manhattan uses 143 times more energy for an equivalent amount of surface area than the total reported in the 21 other metropolitan areas that constitute the Boston-to-Washington megalopolis! The energy used (and heat released) by New York City is 2,625 times the whole United States average.[170]

If the rest of the country were covered by Manhattan-type examples of high-energy culture, such as skyscrapers and vehicle supercongestion, the climatic effects alone would be serious enough to trigger horrendous natural changes, such as the melting of the polar ice caps, raising the Earth's temperatures to such an extent that America's coastal areas would be inundated in water, forcing the evacuation of much of the population. Manhattan itself would be awash! This is a textbook extrapolation of what *could* happen, assuming the human population were still alive from a bath of concentrated pollution and heat from energy conversion processes. But what makes the prospect chilling is that many atmospheric scientists are agreed that only a slight change in the Earth's temperature might trigger a warming of the Earth that would cause melting of the ice caps and associated natural disasters across the globe. A number of internationally recognized experts in climatic science, most notably Professor M. I. Budyko, director of the Leningrad Geophysical Observatory, say that a 2 percent change in the Earth's energy budget (relative to the net solar radiation absorbed) that would result from increased human energy use could cause a severe planetary climate change.[171] At a recent conference on energy use, the director of the Atmospheric Science Laboratory of the National Center for Atmospheric Research at Boulder, Colorado, Dr. William Kellogg, offered his own scenario. The 2 percent change, he said, were it due to man's heat input, would result in an approximate 2° F change in the Earth's temperature. "However," he admonished, "consider that over the land, which is only about one fourth of the [Earth], that is where most of this energy is going to be generated. Some of it may go into the oceans, but most will be over the land. Then, if we consider that the poles respond more sensitively to change than the rest of the globe, we must imagine that there will be a good many degrees' change at the pole, and it seems a good bet that if we do—if we add this much energy, we will in fact, lose the Arctic Ocean Ice Pack."[172]

Dr. Kellogg went on to say that he didn't believe that the loss of the Arctic ice pack would "necessarily be so drastic." One wonders how he defines "drastic," since he noted that this climate change might cause cumulative effects that would melt the Antarctic ice as well, warming the Earth over-all and raising the level of the oceans on the order of 100 meters (328 feet)!

Just how soon a 2° F change in the earth's temperature might occur is the subject of intense speculation on the part of climate scientists, but there is little agreement on the future of climate or of the exact effects of human intervention. The SMIC Conference noted that, since accurate thermometers have been available, temperature measurements in the Earth's Northern Hemisphere show that temperature increased 0.6° C (1.08° F) from 1880 to 1940, and since 1940 there has been a drop of 0.3° C (0.54° F).[173]

At least one scientific conference has agreed that further increases in the use of fossil fuels will lead to an increase of 1.4° F in the surface temperature of the Earth by the year 2000, which is painfully near the 2° F temperature change that atmospheric scientists think may result in far more drastic climate changes than a warming of the surface.[174]

However, several schools of thought in climate science interpret existing data in different ways. One argues that, instead of growing warmer, the Earth may enter an Ice Age as a result of man-made fuels combustion. The combustion of fossil fuels releases large quantities of particulate matter into the atmosphere, which may reflect sunlight away from the Earth, thus cooling the planet. The other school, which maintains that warming may occur from increased energy use, argues that human heat additions to the Earth, combined with the release of carbon dioxide (CO_2) to the atmosphere, will worsen the "greenhouse effect," which is the natural effect—like a greenhouse—in which the Earth absorbs incoming solar short-wave radiation, but prevents the emission of long-wave radiation. This natural effect is worsened by the addition of CO_2 from fuel combustion, because the extra CO_2 absorbs more solar rays than would be absorbed in the absence of combustion.

Another serious climatic effect of increased petroleum use is the contamination of the oceans with thin films of oil. Oil films on the ocean definitely affect the albedo of the ocean's surface, and experiments have shown that the films result in temperature changes in the surface of about 0.5° C (0.9° F). In addition, the oil slicks reduce turbulent motion of the ocean's surface and reduce the intensity of short surface waves.[175] Just what the long-range climatic effect might be is presently unknown, but increased oil pollution of the oceans is undeniably increasing, and the possibility of climatic alteration increases with it.‡

The only significant area of agreement among international climate specialists is that increasing energy use with concomitant urbanization will raise the surface temperatures of a number of areas on the order of 10,000 to 100,000 square kilometers (3,861 square miles to 38,610

‡ Given the severity of man-made changes in the Earth's surface and climate already, it is distressing that the federal government spends about 99 percent of its resources for weather research on programs to artificially modify weather, hardly any of which have ever proven successful, and only a fraction of its funds to ascertain the effects of human activities in disturbing climate including increased energy use.[176] Not only does the overwhelming bulk of man's problems with climate come from the *inadvertent* modification of climate by increased energy use, but the intentional modification of weather has increasingly been suspected as the causative factor in disastrous local climatic effects, including the Rapid City, South Dakota, flood of June 1972, which caused $100 million in property damages and took nearly 250 lives.[177]

square miles) to the same level of incoming solar radiation.[178] What this portends is unknown.

What is known is that escalating human energy use will alter the Earth's climate. The increase in energy use has several multiplying effects. For example, the transportation of oil to cities means oil pollution of the oceans, which disrupts the heat exchange of the sea's surface. When the oil is combusted, more heat and pollutants are released. As more energy is used to construct urban areas, natural vegetation is destroyed to build structures that change the albedo of the surface. As more energy is used to support high-energy agriculture, trees are destroyed, and diversified farming succumbs to monocultural farming, which further changes the albedo of the surface.

Water and Land Use

The rapid spiral of energy use in this century has had an enormous impact on ground water and land. The use of heat engines, as we have seen, requires the rejection of heat following the course of the power cycle to a "heat sink." The water requirements for power plant cooling are beginning to take a significant toll of the nation's water supply. In addition, the growth of all energy conversion technologies requires large amounts of land. As has been noted, the use of land for electric facilities threatens to take up all available land space within one or two centuries. The fight over land and water use is occurring already in the United States. The two problems are at the core of the nation's energy and environmental crises.

Water use and energy development go hand in hand, since water is the preferred cooling source for steam engines (power plants), many industrial processes, and many hoped-for future energy conversion processes involving the extraction of dilute fossil fuel energy resources, such as oil shales and gas from coal deposits. At the same time that water is being sought in such quantity by the energy industries, consumption of water has far outstripped local supply in many areas.

Increased water use has followed the development of high-energy civilization. In ancient and medieval times, the per capita consumption of water was on the order of 3 to 5 gallons daily. In the nineteenth century, during the rise of the Industrial Revolution, per capita use rose to about 10 to 15 gallons daily.[179] Since the turn of this century, U.S. water use per person has increased about tenfold, as high-energy technology has not only spawned higher industrial and agricultural use, but has introduced mechanized conveniences in the home, such as garbage disposals, dish and clothes washers, and the like. Currently, water use in the United States is about 450 billion gallons per day and is projected to reach 580 billion gallons daily in 1980 and 900 billion gallons daily in 2000.[180]

By the end of the century, water needs for power plants will exceed the available fresh water supply, and power companies are already siting plants near oceans, to take advantage of salt water cooling. Even these sites are limited, however, and forecasts indicate that much of the fresh-water needs will be met by desalinating water. However, desalination of water takes considerable energy, which will further deplete existing stocks of fossil fuels, or increase the rate of nuclear power growth.* Furthermore, the problems associated with brine disposal from the desalination process have not yet been satisfactorily resolved.

While the energy industries and various federal agencies are clamoring for the development of more water resources for power plant and industrial use, the existing U.S. waterways and ground water supplies are rapidly becoming overpolluted. The President's Council on Environmental Quality has estimated that more than 90 percent of the nation's watersheds are "moderately" polluted, and the Environmental Protection Agency has found that 29 percent of U.S. stream and shoreline miles suffer from water pollution.[181] One of the difficulties is ascertaining the exact constituents of water pollution. A primary cause is agriculture, which consumes about half of the nation's water supply. Because of the fossil fuel subsidization of agriculture in America, with its intense energy use, pesticides, and synthetic fertilizers, the waste runoffs constitute a major environmental problem. The increased use of fertilizer and synthetic chemicals in energy-intensive farming adds wastes at an unending pace, since the yearly waste burden is increased by the necessity to use more synthetic chemicals to sustain high crop yields.

The addition of phosphate detergents and the many industrial and agricultural wastes lead to serious oxygen shortages in water, which suffocate fish and other organisms. At one time, it was commonly believed that the flow of water diluted pollution and restored natural balance, but recent research has shown that increased rainfall in many cases intensifies the levels of pollution by washing more pollutants into streams.[182]

In highly urbanized and industrialized areas, this effect is height-

* Water can be desalted in solar-powered plants, as is discussed in Chapter 6, but the construction of these plants on a massive scale would require enormous amounts of land and would require considerable energy to build, which would probably come from existing fossil fuels.

ened. The further development of urbanization and heavy energy use present a great dilemma, as the availability of usable water is constantly decreasing; and only through the application of more energy-intensive technologies, such as sewage treatment devices and water desalting facilities, can more usable water become available. The continuing growth of high-energy technology impinges on the static water resource, with the inevitable result that water supplies will contribute a necessary limiting factor to energy growth.

Just as water is essentially a static resource, the availability of land —which is nonrenewable—poses another dilemma for the continuation of high-energy society. The problem of land availability comes up with each increase in the size of America's energy industries—whether the land needs are for strip mines, new ports for oil tankers, or sites for electric power plants. One of the basic policies of the Nixon administration has been to submit to the Congress power plant siting legislation that would facilitate long-range planning of electric power facilities. The administration has attempted to pass national siting legislation—using the guise of "power crises"—that would exempt nuclear power plants from full environmental review, as required by the National Environmental Policy Act.[183]

The U. S. Senate voted in July 1973 to approve the trans-Alaskan oil pipeline and prevent further court review of the pipeline's impact on the environment, thereby delivering a crippling blow to the National Environmental Policy Act (NEPA), which was signed into being in 1970. Similar bills were approved by the Interior Committee of the U. S. House of Representatives, further eroding NEPA's ability to protect the environment against wholesale assault in the name of energy.[184]

Such moves signal the beginning of a legislative era accelerating the widespread use of all available energy resources, particularly on publicly owned lands. Lands owned by the federal government include not only lands rich in oil, gas, coal, and uranium resources, but also petroleum resources on the continental ocean shelves. Federal leasing programs are already under way to open up the oil shale land to operation by private corporations. More than 60 percent of uranium reserves are on public lands, and about 280 million acres of federal land are underlaid by coal deposits, mostly in the western states.[185]

Of particular significance in the struggle to allocate lands for various uses by society is the philosophy of power plant siting legislation. Power plant siting is based on the notion that estimates of future electricity "need" can be made today, and valuable land committed for power plants in coming years. In essence, the legislation codifies the wishes of the energy industries to the pursuit of untrammeled power growth.

The primary difficulty in deciding how to allocate water and land resources—whether for a new power plant or oil field vs. recreation, parkland, or agriculture, for example—is that the economic system has no means by which values can be placed on "free" environmental resources. Canadian economist J. H. Dales points out that at the center of the problem is the question of public or common-property "ownership" of these valuable resources. *"Common-property* ownership," he says, "is, from an economic point of view, virtually nonownership. A common-property asset is one that can be used by everyone, for almost any purpose, at zero cost." He cites as examples the medieval commons, wild game, freeways, etc., and points out that "empirically, it is clear that if the asset is depletable it will be continuously depleted on the grounds that 'everybody's property is nobody's property': medieval commons were overstocked; modern freeways (but not tollways) quickly become congested; wild animals (but never domestic animals) become scarce or extinct; and the deteriorating quality of our air and water resources has become a matter of widespread concern."[186]

The most encouraging work to date that has attempted to find grounds for placing value on the natural environment as well as the synthetic, technological environment is that of Howard T. Odum and associates at the University of Florida. Odum has devised a symbolic "energy language," which, by using energy values as a "common denominator," relates all sources of energy to the human environment, whether the flow of energy comes from fossil fuels, nuclear power, or natural sources such as solar radiation, water flows, wind, etc. One of the flaws in historic and economic analysis of man's natural and technological supporting structures, Odum points out, is that natural sources of energy have been treated as "gifts" rather than actual energy products of the environment that we incorporate into our structures for survival. "We are not accustomed to thinking of water, fertilizer and air as fuels," he says, "but they are as much chemical reactants as oil." In the case of water use and problems of supply, he shows that the flow of water in a stream—as it is used by society for such purposes as cooling, or diluting industrial, commercial, or households wastes—loses its potential energy value. He points out that water can be more valuable as a "chemical fuel" than as a source for hydroelectric power when dams are built and flow is blocked.

His computations of value in the American economy show that

water can be given a dollar equivalent of $1 per 10,000 kilocalories of potential energy (in water, or other energy sources). Using this money equivalent, 400 gallons of water are worth $1 as "chemical fuel." "Contrast this with the usual cost figures for water," Odum says, which equal 1,000 to 5,000 gallons of water processed for $1.[187]

Other studies using this energy language to designate the flow of energy in the natural environment have shown that an acre of trees is worth more than $10,000 per year, or more than $1 million over a 100-year period, not counting inflation.[188]

In a number of far-reaching analyses and studies, Odum and his associates have pointed out the value of the natural environment in performing work for human society that otherwise would require costly and limited fossil fuels. For example, he argues that we should use natural ecological systems to treat sewage, rather than using high-energy intensive sewage treatment plants, which only increase our dependence on the limited fuels. Why not have the natural system perform the work, rather than constantly destroying it for new urban enclaves, parking lots, and other like examples of "advanced" civilization?

Were Odum's energy language—or some similar method of environmental accounting—used by businesses, planning commissions, and government agencies in arriving at decisions over the growth of synthetic energy-consuming technology, far different results might emerge. The significance of Odum's conceptual tool is dramatic, for no longer can we afford the luxury of paving over productive farmland and overloading valuable aquifers under the banner of technological progress. Now it is possible to begin to compute the true value of vegetation and clean water. The sacred grove, while not fully protected from the ravages of technology, has at least been given a price tag—something that land developers, overzealous Interior Department bureaucrats, and industry planners understand.

Population

The focal point of all environmental concerns is the relationship of the human population to the Earth's environment. Throughout history, civilizations have survived or failed only through their ability to stay within the "carrying capacity" of the land and its resources. As Odum puts it:

> Support of man must include all the necessary works that provide stability, reserves, protection, and all the crisscross controls required to regulate his many complex needs. Either we must retain the old network of hundreds of species of plants and animals which carried on

these functions for man, or we must set up new machinery to perform them on energy subsidy. The carrying capacity of one man is the minimum area into which the work to support him can be concentrated.[189]

The carrying capacity of the current American civilization is precariously balanced on the enormous subsidy of fossil fuels, with small incremental energy subsidies coming from nuclear power, hydroelectric power (indirect solar energy), and forms of solar energy such as photosynthesis in agriculture. The development of the enormous synthetic energy subsidy has brought with it blessings of high-energy technology, ranging from electric appliances to the extension of life span through modern medicine. The subsidy has also resulted in the development of monocultural institutions, ranging from single-crop mechanized agriculture to vast urban complexes.

Many environmental scientists have indicated that the greatest stress in the environmental system comes from the continuing increase in the human population itself, even though, in the past few years, the rate of U.S. population growth has slowed. However, even the slowing in the population's growth rate will not appreciably change the magnitude of the environmental problems our society will experience in the future, since a substantial time-lag is involved. The National Commission on Population Growth and the American Future has pointed out that "even if immigration from abroad ceased and couples had only two children on the average—just enough to replace themselves—our population would continue to grow for about 70 years."[190]

Of even greater consequence is the higher growth rate in the world's population, which is expected to reach 4.7 billion people by the late 1980s. The present world population is about 3.7 billion people. Both the U.S. population and the world's people will be seeking more and more resources and conveniences of high-energy civilization, resources that are in short supply already. The people of the United States cannot expect to continue to indulge in their current disproportionate share of the planet's resources: i.e., the use of 35 to 40 percent of the world's mineral and energy resources to support 6 percent of its people.

Notwithstanding the inevitable future competition over Earth's resources, the U.S. population, which is becoming ever more dependent on the synthetic energy subsidy, may suffer even more from pollution than current experience indicates. The cost of attempting to clean up the environment to maintain a healthy place for the coming generations is great. Not only is the financial cost of supporting the growing population burden enormous—estimated at 4 percent of national income to support each 1 percent increase in population[191]—but the energy cost is staggering as well. Maintaining a high-energy society devoid of the elaborate natural biological networks, agricultural systems, and other facets of the natural system would prove not only suicidal, but impossible.

The population problem in America lies at the heart of the energy crisis. With dwindling supplies of fossil fuel energy and an increasing population accustomed to the luxuries of that energy supply, the civilization confronts its oldest and most persistent problem: carrying capacity. Consider, for example, the lesson of the American economic depression of the 1930s: The toppling economy forced the return of many people to the farms from which they had emigrated. In 1930, more than half the American people lived in rural areas. Today, less than 20 percent live in rural areas, and fewer than 5 percent on farms.[192] If an economic depression were to occur now, there would be no natural agricultural system "buffer" to fall back on, and food sources would dry up.† An abrupt change in the economic structure—perhaps brought on by energy shortages—would reduce the agricultural system in America to a shambles and would create widespread misery and possibly starvation.

The environmental crisis is nothing more than a mirror in which the accelerated use of energy is reflected by accelerated entropy. The continuing injection of increasing amounts of energy to support the American lifestyle will inexorably exact the price demanded by the Entropy Law: increased pollution, waste heat, solid wastes, and social instability. As the society exerts more artificial "control" over nature in the name of convenience and prosperity, the society suffers the risk of greater disorder.

Looking Ahead

There are two possible courses for the development of energy policy in the United States: Either (1) the continuation of the present high-energy economy, with its inherent waste; or (2) the substitution of other lifestyles and methods of getting and using energy. A look into the future will indicate the folly of continuing the full development of the cancer without treatment and changing life patterns required to restore health and direction to our societal organism.

† A fuller discussion of the dependence of the high-energy agricultural system on cheap sources of fossil fuels is presented in Chapter 5.

3

Energy and Society

Tracing the movement of energy in American society is difficult, since the exploitation and use of energy pervade every aspect of life, both in the natural and the economic environment. Without access to cheap sources of concentrated energy, no goods could be produced from mineral ores; no trucks or trains could be either manufactured or supplied with sufficient fuel to move the goods to consumers; and the consumers themselves would have nothing to consume.

Reducing the dependence of America on the fossil fuel energy resources, which are not only running out, but which are depleting the life-support capacity of the environment, requires a close examination of the flow of the concentrated energy sources in America. Tracing this flow involves examining the economic and political control of energy sources, the pricing structure, the advertising of products to encourage energy consumption, and the end uses of energy by American industries and consumers.

Corporate Energy Monopolies

All basic energy resources—fossil and nuclear fuels and hydroelectrically generated electricity (as well as electricity from the fossil and nuclear fuels)—are controlled by very powerful corporations, corporations that were initially developed to facilitate the movement of concentrated energy flows in society. Understanding the flow of energy in America necessitates a basic understanding of the nature of the corporate monopoly.*

Dr. Bruce Netschert of National Economic Research Associates, Inc., recently told a U. S. Senate committee investigating antitrust issues that all twenty-five of the largest oil companies are also involved in the natural gas industry; eighteen of them have investments in oil shale; eleven own interests in coal; eighteen own uranium lands; and seven are in tar sands. Six of the ten largest are in all four of the domestic fuels industries—oil, gas, coal, and uranium—Netschert said.

One fifth of all coal production is in the hands of oil company subsidiaries, he noted; and one, Occidental Oil Company's Island Creek subsidiary, accounts alone for one fourth of all new mine openings now in process in the United States. Netschert said that the two U.S. companies that have coal reserves of six billion tons or more are Humble Oil Company (now Exxon) and Consolidation Coal, a subsidiary of another oil company (Continental Oil Company).[1]

These energy conglomerates wield enormous economic and political power. Forcing thirteen hundred independent gas stations out of business by declining to sell them gasoline, as occurred during the 1973 summer shortage, is a bare muscle-flexing exercise for an industry in which the annual sales and assets of the five largest companies exceed the Gross National Product of all but four countries in the world. The political clout is repeatedly demonstrated; as, for instance, when Richard Nixon's 1973 congressional energy message echoed the requests of the oil industry in calling for deregulation of natural gas prices, increases in federal leases for offshore drilling, higher oil depletion allowances, quick approval of the trans-Alaska pipeline, relaxation of pollution control laws, and higher taxes on imported oil.

Representative Charles Vanik of Ohio reported that eleven major U.S. corporations paid no federal income taxes for 1972. Two of them —Occidental Petroleum Corporation and Pennzoil Company—were oil companies. Two others—Consolidated Edison Company of New York and American Electric Power Company—were electric utilities; and the remaining seven were corporations that are high users of energy: McDonnell Douglas Corporation; Republic Steel Corporation; Burlington Northern, Inc.; Eastern Airlines; Trans World Airlines; United Air Lines; and Northwest Airlines.[2]

* The corporate control of energy sources extends far beyond the U.S. borders; today's multinational corporations control energy and material flows around the globe.

The electric utility industry is 60 percent greater than its nearest rival in the economy in terms of capital assets. It is interesting to note that petroleum refining is second and natural gas, sixth.[3]

Control of politics by heavy financial backing in political campaigns has long been standard procedure for the oil industry; and the tax concessions granted them in turn assure the perpetuation of government subsidies.

Philip M. Stern, a former deputy assistant secretary of state for public affairs and author of *Rape of the Taxpayer,* has reported data showing that five other major oil companies—four of them with pretax profits of more than $1 trillion—have paid considerably less than 10 percent in federal income taxes, with estimated losses to the U. S. Treasury of about $2.5 billion, as reflected by the following table:[4]

Company	Pretax Profit	Percent Paid in U. S. Taxes	Approximate Loss to U. S. Treasury†
Gulf	$1.324 billion	2.3%	$500 million
Texaco	$1.319 billion	2.3%	$495 million
Mobil	$1.153 billion	7.4%	$375 million
Standard/California	$ .856 billion	1.6%	$330 million
Standard/New Jersey	$2.737 billion	7.7%	$885 million

As Stern noted, the oil industry is the largest beneficiary of government welfare in the world. In addition to the billions of dollars' windfall each year from the import quota program, such tax favors as depletion allowances, write-offs of royalties to foreign governments, and drilling and equipment depreciation allowances add still more billions of dollars to the industry's tax-free income.[5]

As James Ridgeway pointed out in his recent book, *The Last Play,* this handful of energy companies—most of them dominated by oil interests—are in a position to almost fully control the development of the remaining fossil fuels: how fast they will be produced, who will get them, and what they will be used for. In what Ridgeway calls a "diversionary tactic against Arab nationalism," the major oil companies are developing large-scale oil and gas deposits in Southeast Asia, where—with the assistance of the Indonesian Government—they are extending supply lines to Japan. They are also moving heavily into the Arctic regions and onto the floor of the world's oceans. In addition, they are working out new relationships with Third World countries, "acting in

† The difference between what the U. S. Treasury did collect from these companies and what it would have collected had they paid 40 percent of their profits to the U. S. Treasury, as is true for U.S. corporations as a whole.

effect as their diplomatic agents with major powers in return for exploiting their oil and gas deposits," Ridgeway said.[6]

An essential function of U.S. energy policy should be to break down the political power base of the major energy companies, either through strict application of antitrust laws, or the reduction of special privileges to the industry, such as oil depletion allowances and specially tailored tax benefits. While overt nationalization is being called for by many politicians and congressmen, it should be remembered that the primary pitfall in nationalization is the problem of raising capital for the increases in productivity of the industry. A more logical end to the era of political and social control by the big energy companies would be antitrust action to reduce the overpowering size of the companies. A diversified new age in which industry and government should encourage decentralized efforts at efficient energy use and social changes in many areas of American society is not consonant with overt centralization of power and nationalization; the latter would only replace the current oil bureaucracy with a government bureaucracy—which, in all probability, would be even less efficient than the present structure.

The Role of Advertising in Energy Consumption

Advertising is one of the biggest businesses in the United States. It accounts for more than $23 billion in expenditures yearly.

The advertising industry has grown exponentially with the development of high-energy American civilization. In 1949, advertising expenditures totaled $5.2 billion; in 1960, $11.9 billion; by 1970, $19.6 billion.[7] In the first three years of the 1970s, total advertising expenditures had increased by more than 17 percent.[8]

In 1969, the last year for which accurate government calculations are available, the following energy-intensive industries spent these amounts on national advertising:[9]

Manufacturing Industry	Amount in Dollars
Chemical and allied products	$ 2.23 billion
Petroleum refining and related products	349 million
Electrical machinery, equipment & supplies	769 million
Transportation equipment (not including motor vehicle advertising)	103 million
Motor vehicles and equipment	575 million
Transportation Industry total	640 million
Wholesale and retail trade	
Auto dealers and gasoline service stations	625 million
Total	$ 5.29 billion

In addition to this list of industries that advertise the sale of energy products and energy-consuming equipment, machines, and motor vehicles, there could be included a host of other industries that advertise more directly on the consumer level—in local stores, for example. The total 1969 receipts for general-merchandise stores was $1.4 billion. On the local level, merchandising of appliances, air conditioners, various engines, and thousands of other energy-intensive products is a major contributor to consumer demand.

As the catalyst of the American economic growth machine, the advertising industry's essential function is to convince the individual that reliance on the technologies of high-energy culture brings a better quality of life than "primitive" human labor and the identification of the human being with nature. In its own curious way, advertising pits its colorful products before the marketplace in a natural setting. For example, the automobile ads on television show the sleek machines winding through beautiful forests or by the seashore—not on the crowded, smog-choked freeways of Los Angeles.

Psychologist Erich Fromm said of this:

> "[In such ads] the general theme is, love is dependent on a gadget. And this is a very important theme of all modern life: the expectation that—not human power, [nor] human effort, not being, but *gadgets* create the good life; and that there is really no limit to what the gadget can do. . . ."[10]

An interesting comparison is the relationship between the expenditures of American industry for advertising vs. the expenditures for research aimed at improving the future efficiency of the industrial systems and reducing their use of energy, developing advanced technologies, and generally bettering the quality and value of products. The research expenditures for the industries listed in the previous advertising table are all below their expenditures for advertising. For example, the expenditures for research in the chemical and allied industries in 1970 amounted to $919 million; yet the expenditures for advertising the year before equaled more than $2.2 billion.[11] The National Science Foundation combines listings of the 1970 expenditures for research in the electrical equipment and communications industries together; the total was $793 million. Yet advertising electrical equipment alone amounted to $769 million the previous year.

In these fields, industry research expenditures represent almost all of the nation's research; federal research in energy is devoted almost totally to the development of nuclear power and related areas. In the comparison of research and advertising spending, it is obvious that the nature of the energy-pricing system and the availability of bounteous supplies have largely precluded any incentive for more research at the expense of advertising.

Not only do the basic energy extraction and chemical manufacturing industries spend more on advertising than on research expenses, but the advertising budgets of the supply industries, particularly the electric utilities, are also greater than those for research.

Taking their cue from the wasteful official policy of the federal government, the nation's energy companies and electric utilities—particularly the privately owned utilities—have actively promoted the unrestrained consumption of more and more power. Senator Lee Metcalf, of Montana, in a 1973 Senate speech, said that the nation's electric utilities spent more than three times as much of their revenues in 1971 (the most recent year for which Federal Power Commission statistics were available) to promote the consumption of electric power through advertising as they did trying to produce a better and more reliable power system through research and development. Senator Metcalf correctly pinned much of the nation's crisis in energy supply to this extensive advertising practice, adding that "another reason for the energy shortage is the lack of research and development, by both energy companies and the government, on non-nuclear methods of energy production."[12]

Out of 193 large electric utilities—both publicly and privately owned—only 7 companies reported spending more on research than on advertising and sales promotions. The full calculation of expenses showed that advertising accounted for $314 million, vs. $94 million for research. Senator Metcalf noted with some pleasure that congressional and public criticism of the electric companies in recent years had had a salutory effect, since in 1969 the electric companies had spent 7 times as much on advertising as on research.[13]

What the advertising figures show is that utilities actively promote the sale of power. Their argument is that they need to balance the seasonal requirements for electricity. In some areas of the country, electric companies promote electric air conditioners because they have a winter "peak" in electrical demand; and by advertising electricity for use in summer they balance their winter peak with a corresponding rise in electric consumption in summer. Other utilities promote winter electricity use such as electric space heating because they have a summer peak. The results of the advertising, however, are not to just smooth out the mountains and valleys of power consumption. Advertising electric power promotes more and more consumption.

The only possible positive use of the advertising media by electric companies is advertising to decrease the use of electricity. A number of electric utilities, straining to meet consumer demand with antiquated generation and transmission systems, have already had to encourage customers not to use as much electricity during periods of excessive demand.

The first U.S. company to engage in a serious campaign to decrease electricity use was the Consolidated Edison Company (Con Ed) of New York, which after years of aggressively promoting the ever-expanding use of electric power suddenly found itself unable to meet the demand it had worked so assiduously to create. Con Ed initiated the "Save-a-Watt" campaign during the summer of 1971. The utility's problem was simple: Its service area—New York City—had been hit in previous years with more blackouts and brownouts (voltage reductions) than any other locality in the nation. Con Ed calculated that it not only might not be able to supply enough electricity to meet the expected demand that summer, but certain equipment in its system might give out, leading to a power disaster many times the scale of the fall 1965 blackout. The magnitude of the crisis was noted by the fact that the New York City Public Service Commission anticipated that New Yorkers would use—during summer peak demand periods—8,105 megawatts of electricity, and the company had only 8,802 megawatts of installed capacity.[14] Part of this capacity included power plants in the city that were more than 50 years old. Other plants were chronically out of service, including a nuclear plant (Indian Point) near the city, and a massive turbine generator called "Big Allis" that had already malfunctioned in the midst of a 1970 power crisis.[15] The Public Service Commission concluded that even with the addition of outside power—purchased from other utilities and possibly Canada—a tremendous shortage could develop, requiring the utility to brownout the entire service area, and cut off power altogether in outlying boroughs of the city.[16]

Con Ed's eleventh-hour response to the situation was the "Save-a-Watt" advertising campaign, which encouraged residents to minimize use of air conditioners; keep lights low or turn them off; operate appliances during the nighttime, off-peak hours; turn off appliances when not needed; and buy more efficient air conditioners, labeled by energy consumption in stores.‡ Undoubtedly, "Save-a-Watt" worked.

‡ Although New York City requires that air conditioners be labeled according to the energy required to operate them (given in BTUs of cooling capacity), the labeling system is confusing to the average person, and has probably not been very significant in reducing energy consumption. Compounding the problem is the fact that people generally buy the cheapest air conditioners, which usually—though not always—require more energy to operate.

The utility calculated that more than 350 megawatts of power—the equivalent in power to that of an average-size electrical power plant—had been saved during peak periods of summer use.[17]

How much did the advertising campaign cost in relation to the power saved? A study made at the Oak Ridge National Laboratory estimated that Con Ed's savings was much more than 100 million kilowatt-hours of electricity between June and September of 1971. Since the advertising campaign cost $400,000, the study concluded that each dollar spent resulted in eliminating the need for 250 kilowatt-hours.

Yet the savings might have been much more, if the American public (particularly those citizens living in super energy-consuming cities) were less dependent on electric power. The Oak Ridge study noted that an energy conservation advertising campaign conducted in Sweden in 1970 resulted in the savings of 1 billion killowatt-hours during a 4-month period. The campaign, conducted by the country's national power corporation, cost $500,000. Compared with Con Ed's 250 kilowatt-hours' saving per dollar spent, the Swedes eliminated the use of 1,600 kilowatt-hours for each dollar spent.[18]

Even if the residents of New York *wanted* to ease the crisis faced by its electric utility by cutting down on use, they would find it very difficult, because the city is totally dependent on the transfusion of massive amounts of energy.

Use of Energy by the Electric Utilities

Grouped together as a single enterprise, the many electric power utilities constitute the largest industry in the nation's economy. Revenues from the sale of electricity are expected to reach more than $30 billion by 1980, and capital investments over $200 billion by the mid-1980s. Everything about the electric utilities is a manifestation of massive centralization.

In the 1948–66 period, private (nonfarm) business grew at an average rate, measured in terms of productivity, of 2.4 percent. The electric and gas utilities of the United States grew at an average rate of 3.7 percent; this was 50 percent faster than other businesses.[19]

What is the reason for this phenomenal growth? Why are the utilities and their operations gargantuan?

Part of the growth may be attributed to the over-all increase in demand for energy and power in America; but in recent years many analysts and economists have cited the utilities' essentially politico-economic nature as the central reason for their spiraling growth. Electric utilities are not ordinary businesses. They are monopolies created by the government, and are regulated by the Federal Power Commission and by state and local public utility and public service commissions.

Utilities are allowed by law to make a guaranteed rate of return on investment capital, which means they make a legally assured profit on every dollar of capital invested in their operations and growth. Electric and gas utilities had an annual average capital growth rate between 1948 and 1966 of 5.5 percent per annum, compared to average business capital growth of 3.7 percent. Thus, their capital growth, like productivity, has been 50 percent above that of average businesses in the country.[20]

It is no wonder then that utility stocks are considered the bluest of the blue chips and carry the highest ratings.

The federal government owns almost 12 percent of the electric power-generating and transmission facilities in the United States. Forty separate systems are involved, which either generate power (largely hydroelectric) or distribute it in bulk to other utilities for resale.

Through the Tennessee Valley Authority (TVA) in the Southeast and the Bonneville Power Authority (BPA) in the Pacific Northwest, the federal government actively promotes the extensive development and cheap sale of electricity. Both BPA and TVA are products of the public works era of the 1930s. Both institutions have been largely responsible for the development of cheap electricity and have encouraged maximum consumer usage. Typical of the federal philosophy is this comment from a TVA pamphlet in which TVA says that of all its "contributions,"

> most important of all to the nation is the influence of TVA's low rate/high use policy on electric rates elsewhere. Private utilities have followed TVA's example in offering low rates to the Federal Government, saving the taxpayer additional millions of dollars annually. Also, TVA's low residential rates have influenced rate-making and, consequenty, power use and costs in other parts of the country.[21]

TVA says that within two hundred miles of its southeastern United States service area exist the lowest electrical rates in the country. The government agency claims that its promotion of the "all electric" home, in which all heating, air conditioning, and appliance operation are provided by electricity instead of direct fuel use, was duplicated by private power utilities near the federal complex. This active promotion policy extends to the bulk sale of wholesale power for use by industries near TVA and the Bonneville Power Authority. This policy has led to the congregation of industries that are high energy users in these service areas. In Tennessee, the growth policy has resulted in the development of industries that produce electrical equipment and, in the Northwest, industries that refine and smelt aluminum. As TVA says, "as a pebble dropped in a pond causes ripples to flow outward to the surrounding shores, the influence of TVA's low rates flows outward to surrounding areas."[22]

The Federal Power Commission

The regulatory rule of thumb has historically been to grant the power companies whatever they wanted. The Federal Power Commission (FPC), established in 1920, is designed to assure to consumers the free flow of natural gas and electricity in the United States. While it has generally met this mission, it has done so at the expense of creating the most closely knit, monopolistic energy enterprise in history.

Typical of outside comments about the FPC is the observation in James M. Landis' report to President-elect John Kennedy on federal regulatory agencies in 1960: "The Federal Power Commission," he wrote, "without question, represents the outstanding example in the Federal Government of the breakdown of the administrative process."[23]

What Landis and many others refer to is the energy industries' domination and virtual control of the FPC, which continues to this day. The 1973 membership of the Commission indicates the extent of industry domination. The chairman, John Nassikas, is a former legal counsel to a gas utility. Of the other members, one is a lawyer from a Texas firm that represents several oil and gas companies,* and the remaining two are Republican politicians who have consistently supported the interests of the energy industries. To fill the remaining seat on the FPC, President Nixon nominated Robert H. Morris, a California lawyer who represented the Standard Oil Company. The Washington *Post* editorialized that the petroleum industry might ask itself if its real interests "are really served by this crude tactic of excluding all dissent from the Commission."[24]

The extent of corruption in the FPC has been suspected for some years. Certain of the FPC practices not only cater to industry, but seem blatantly unreasonable. For example, the FPC has always asked the

* Congressman George Brown of California reported that this commissioner, Rush Moody, Jr., nominated by President Richard Nixon, allegedly obtained his seat on the commission only after a lengthy interview with a Nixon campaign contributor. The contributor—of seven hundred thousand dollars (in 1968)—was William Liedtke, president of Pennzoil United Company, one of the nation's energy giants.[25]

natural gas companies to account for the known reserves of natural gas in the nation, and has never undertaken its own surveys, or particularly pressed the gas companies for accurate details of their holdings. Thus the basis of natural gas pricing to consumers is the industry estimates of supply, not independent estimates. Even though estimates of domestic gas reserves are given in this book, for example, they are based either directly or indirectly on data compiled by the energy industries themselves, which may be far lower—in order to gain favorable federal treatment—than the real supplies of gas. The FPC has been of no help in clearing the confusion.

In 1973, an FPC official in the Natural Gas Bureau attempted the destruction of internal staff documents relating to the reserves of natural gas rather than turn them over to a Senate investigating committee. The Antitrust Subcommittee of the Senate had experienced constant opposition from the FPC in releasing any information on industry gas data for months, and the attempted destruction only added to the unreality of the whole affair. Nonetheless, it is a part of the daily world of the Federal Power Commission.[26]

The Nixon administration moved in 1973 to take price controls off natural gas. Lawrence Stern noted in the Washington *Post* that a 30 percent increase in the price of natural gas would give the energy conglomerates an additional $6.6 billion per year in increased rates from gas sales. The 30 percent increase, which is the average increase the energy companies have sought for the past few years from the Federal Power Commission, would increase the book value of America's gas reserves by $300 billion.[27]

This point made by the petroleum industries in their recent advertising campaign has been that the price of natural gas would increase the conservation of energy, because consumers would be more thrifty in their use of a more expensive energy, which is true. In addition, however, the companies suggested that the added earnings would give them a positive incentive for new drilling of oil and gas wells to go after deeper U.S. reserves, which might triple or quadruple the nation's available domestic gas reserves. This is pure conjecture—unless, of course, the natural gas companies have deliberately been hiding some of their gas reserves with the intention of suddenly "finding" them when the price goes up. In purely economic terms, the industry will search for the cheapest sources of oil and gas. The cheapest future sources will probably not be found in the United States, but will continue to be located in the rich foreign oil fields of the Middle East, South America, and Southeast Asia. The most likely use the industry will find for increased earnings will be to build overseas facilities and supertankers to transport petroleum to the United States, not to open up deep domestic fields.

The impact on the nation? We will become more dependent on foreign sources of energy, which may prove impossible to keep secure in the advent of war or natural disaster. For many months now, the American Petroleum Institute has been intoning in national television ads that "a country that runs on oil can't afford to run short." A more logical one might be: "A country that runs on foreign oil *will* run short."

In the realm of overseeing the regulation of electricity usage and sale in the United States, the Federal Power Commission does little more than publish statistics and keep track of company earnings. The FPC has the authority to regulate the interstate sale of wholesale power, if it finds that prevailing rates are unreasonable or unjust. In addition, the Securities and Exchange Commission (SEC) has the power to regulate the sale of utility securities.[28] Neither of these agencies has ever chosen to deal with the substantive issues of monopolistic growth of the energy utilities.

The real business of regulation is left up to the states. The concept of public regulation dates back to the nineteenth century, when key court decisions and laws established the basis for public oversight of the interactions of public utilities, such as power companies, telephone companies, water utilities, transport companies, and others.†

The regulation of electric companies is little more than a charade in most states. Staffs are small and ill-equipped, either in personnel or finances, to analyze, understand, and guide the actions of the electric giants. A Massachusetts witness before a Senate committee in 1969 summarized his state's difficulties. The witness noted that the Massachusetts Department of Public Utilities' chief accountant (and lawyer) was responsible for the auditing of accounting returns filed by 14 electric companies, 26 gas utilities, 6 phone companies, 63 water companies, 88 bus and streetcar lines, 816 securities brokers, 2,599 moving companies, and 15,055 trucking companies. His entire staff consisted of a clerk and another accountant; his salary, at $11,752 a

† A view of the history of utility regulation that gives some historic insight was provided by Mr. Vic Reinemer, staff director of the Senate Subcommittee on Budgeting, Management, and Expenditures, in a 1973 speech to the American Medical Association in Chicago: "We have all heard the litany that energy companies are regulated by literally dozens of government bureaus," he said. "Let us remember that the 'regulation' of utilities was invented here in Chicago, by Samuel Insull. He devised the scheme so as to appear controlled by government, and thus stop the growth of municipal power systems. Then he sent his lawyers to Springfield to set up the Illinois Commerce Commission."[30]

year, was probably less than the junior personnel of any one of the major utilities he was charged with overseeing![29] It is in this light that utility "regulation" should be examined.

Electric utilities and natural-gas utilities are privileged monopolies under the American system of regulation. They can seize property under the laws of eminent domain, which is a legal means of conferring on them a special status in the allocation of the use of natural resources. The law has historically interpreted seizure of land for electric power generation and transmission as a "social good," whether in fact far better uses can be found for land than to feed the growth of power utilities.

Electric utilities have waged campaigns for years to encourage consumers to purchase electric heating systems, stoves, and the like, which their advertising claims are cheaper and environmentally more desirable than purchase of direct fueled devices, such as gas stoves and gas or oil heaters. In reality, the electric devices are neither cheaper nor cleaner, but the consumer must pay for the waging of the futile propaganda war nonetheless.

In recent years, publicly supported environmental and consumer organizations have begun to take heed of the special status of the electric companies, and have argued for changes in the regulation of the utilities before a number of state and local commissions. One of the most thorough jobs was performed in 1972 by a number of Detroit organizations led by John J. Musial, head social planner for the Detroit City Plan Commission. The groups argued before the Michigan Public Service Commission that the area's electric utility, the privately owned Detroit Edison Company, was promoting excessive electricity growth to the detriment of the state's and the city's environment, and was as well discriminating against the poor residents of the city in its rate structure.[31]

John Musial pointed out that Detroit Edison's rate structure, like almost every other electric rate structure in the country, started with very high rates for low users of electricity, but, as usage of electricity increased, the rates dropped. For industry and commercial businesses, this means the cheapest power, since they are the highest single-party users. The last block of the rate structure, called the "tailblock," is the block reserved for high users who get the lowest rates. As for the consumers, Musial presented statistics that showed that the poorest residents of Detroit paid the highest electric rates, and the wealthiest residents of the city, living in outlying areas, paid the lowest rates. A poor, inner-city family having an average yearly income of $3,700 was charged an average of 3.64 cents per kilowatt-hour of electricity. In the exclusive Detroit suburbs of Grosse Pointe, Livonia, and Bloomfield Hills, where family incomes averaged from $15,000 to $46,000, electricity averaged 2.78 to 2.44 cents per kilowatt-hours.[32]

Musial noted that "in large measure, the inner city Detroit families are actually subsidizing the bills of wealthy suburban families, relieving them of their fair share of plant capacity costs."[33] A resident of the inner city and chairman of the Detroit Model Neighborhood Social Services Committee, Ms. Pepper Jacques, explained the reasons for this inequity.

First, she noted the correlation between pollution and utility economics: "I would like to point out that there are no energy plants in Bloomfield Hills. I have one in my backyard, however. . . . Our pollution levels [largely from electric power production] are well above federal clean air standards. The inner city does its duty to the suburbs, by taking in the plants which supply suburbs with electricity. We help keep suburban air clean and fresh, so that suburban kids can grow up strong and healthy."

As for the subsidization of the suburban residents by the poor consumers of power, she added: "Now we find that our rates are going to have to be raised again to pay for billions of dollars of new energy plants. But the commission should know that the citizens of Detroit's inner city are aware that the plants down in our neighborhood are large enough to supply all of our energy needs. We have paid for these plants years ago in our electricity bills.

"The new plants are needed to satisfy suburban needs, not ours. The new plants are needed to satisfy the needs of the large consumer, with air conditioners, deep freezes, stereos, three TV sets, and all-electric homes, not the energy needs of the small consumer."[34]

Her analysis is essentially correct. The inner-city residents who suffer from utility-generated pollution pay a 50 percent higher rate per unit than the privileged classes who are living in houses served by tailblock rates.

Utility economics specialist Dr. Charles Olson of the University of Maryland, appearing at the Michigan hearings, suggested to the state commission that rates should be restructured to emphasize equity and conservation. He pointed out that the special tailblock rates, which encourage consumer purchase of such heavy power use items as air conditioners, may actually worsen the yearly position of the utility in supplying power, because such items are very seasonal in nature. For example, utilities attempt to keep electrical generating plants in use year round for more economic operation, and the conventional wisdom of the industry has taught that promoting the use of air conditioners in summer and electric heating systems in winter would better the "load factor" of the utility, by keeping a higher demand level year round so that power plants can be operated at full capacity, improving not only the efficiency of the plants (through even operation), but also the profits of the utility. Olson pointed out that the seasonal fluctuations of electric use have brought about the reverse situation, that promoting electricity growth has been associated with *lower* load factors for the utilities.[35]

Historically, by promoting greater use of electricity, the utilities have been able to finance larger plants and more extensive transmission and distribution systems. However, those days are gone. The power plants are no longer increasing in efficiency as they once did, and other economic factors, such as the rising costs of fuels for electricity generation, have brought about higher costs—not lower costs—to consumers of electricity.

The current trend of the electric utility industry toward bigger and more complex power plants and distribution systems will add great economic and environmental burdens to consumers of electrical power, as well as burdens to the environment by concentrating pollution in limited areas.

The promotional policies of the utilities have led to an unstable climate wherein need for power-generating equipment cannot even keep up with spiraling demand. The rapid escalation of electrical power requirements in the past few years (indicated by national growth rates over 7 percent per annum, 7 times the growth in population) has created something of a disaster, because utilities cannot purchase and install enough power stations to meet growing demand. In its 1973 annual survey of the industry, the trade magazine *Electrical World* reported that of the 44,529 megawatts of installed power-generating capacity that the utilities planned for 1972, only 27,827 megawatts of capacity actually was put on line during the year. The magazine noted that the 1972 additions "fell an alarming 19.3 percent under the minimum that *Electrical World*'s 23rd Electrical Industry Forecast [September 15, 1972] had determined to be essential for a suitable margin over 1972 peak loads."[36]

The magazine predicted a dangerous future situation for the electrical industry because it was relying on new nuclear technologies to meet oversold electricity demands.‡

‡ The magazine commented that one of the major problems was the slippage in completion of nuclear power plants, which were off schedule by 54 percent of planned capacity for the year. Reasons for the failure of this technology to meet expectations are discussed in Chapter 5.

Power Waste

As the electric utilities go merrily along, promoting growth under the "supervision" of the nation's state and local public service commissions and other regulatory bodies, demands on the nation's supply of energy and other natural resources to fill the increased growth pattern are escalating. Not only is the very nature of electricity, which requires a primary fuel to generate it, inherently wasteful of energy, but construction of power plants and transmission facilities consumes ever-increasing amounts of energy-intensive metals and materials, which constitute a double drain on the nation's energy resources to produce. Environmental lawyer Edward Berlin of Washington, D.C., says that the regulatory rate-setting process is partly at fault in creating the national maze of transmission lines. "In situations where common transmission facilities can serve two or more utilities," he argues, "considerations of environmental, social, and economic efficiency operate in favor of joint facilities. Rate base/rate-of-return considerations may, however, stimulate duplication."[37]

In other words, utilities are encouraged to waste under the conventional regulatory process, which allows them a "cost-plus" fixed rate of return on investment. They are actively discouraged by economic disincentives from incorporating joint facilities with gas and telephone utilities, which perhaps would save energy, resources, land, and money. In short, the public utility and public service commissions of the land are legally mandated to treat the utilities as ordinary businesses, without looking at them as actual monopoly utilities. The regulatory structure itself is designed for waste rather than conservation. The wasteful process is further worsened by the structure of electric rates, which encourages additional irresponsible use of electricity.

Altering—if not reversing—the structure of electrical rates so that demand is reduced has been suggested to encourage conservation of electricity. The former White House Office of Emergency Preparedness has suggested that demand charges on electricity be placed during peak hours of use, which would provide an incentive for businesses, industries, and individuals to use electricity during off-peak hours. This measure, OEP says, would reduce the need for additional peak power generation equipment, would alleviate much of the brownout (voltage reduction) problem of utilities across the land, and would reduce the over-all capital requirements of electric utilities.[38]

Other measures to reduce demand would include the elimination of regulatory provisions that allow utilities to charge advertising expenses to consumers of electricity. Many utilities have said that they voluntarily curtail advertising expenses, but this is oftentimes misleading. For example, the president of Washington, D.C.'s Potomac Electric Power Company (PEPCO), told the city's public service commission that "we have ceased to emphasize, promote, or advertise appliances [and] air conditioning. . . ." Yet lawyers for the Sierra Club and a local university student organization called SHOCK (Students Hot on Conserving Kilowatts), who were intervening in a rate case involving PEPCO, found that the utility actively promoted electrical air-conditioning, heating, and a host of appliances through a dummy organization called the Electric Institute of Washington, which was engaged in advertising—90 percent of which was paid by the utility.[39]

The interveners argued before the public service commission that PEPCO should not be allowed to pass its advertising and promotional activities' expenses along to consumers. Instead, they contended, any such expenses should be paid by shareholders of the company's stock. In addition, the legal brief argued that $3.5 million of the company's expenditures should be set aside for environmentally oriented research and development, an amount less than 2 percent of the company's revenues.[40]

Attempts by environmental groups and others to change rate structures have not met a receptive audience in the private power companies. Typical of the industry's response to criticism is a statement of Donald Cook, president of the American Electric Power Company, to *The Wall Street Journal.* Cook told that newspaper that our standard of living "has a direct correlation with the amount of energy utilized," and "unless we say that our standard of living is as high as it should be, that the poor people should forever remain poor, we must come to the conclusion we must make available the amount of power our society asks for." Neither of his arguments are supported by facts, as is shown by a survey of the GNP of other nations and a hard look at the way electric utilities treat poor people.[41]

On the whole, the nation's private power companies have become more aware of the relationship between research and advertising, and in 1973 established an industrywide research organization, the Electric Power Research Institute (EPRI), to pursue further research into better and cleaner methods of generating electricity. Funds for the institute, however, are still only one fourth of what the industry spends for promotion. EPRI will spend a total of $160 million in 1973 and 1974 on research and development, most of which is devoted to developing nuclear reactors and bigger transmission lines. Less than one fourth of the budget is devoted to environmental problems of utilities.[42]

In large part, EPRI was established in response to threats that the federal government might take over responsibility for electric power research. Best chances for a federal program in electric research are given to proposed legislation offered by Senator Warren Magnuson of Washington in 1972 and 1973 (S. 357) to establish a Federal Power Research and Development Program, which would be financed by a 1 percent surcharge on electricity consumption paid by consumers.* The Edison Electric Institute (EEI), the private utilities' trade association, has consistently opposed the legislation; but the public and municipal electric utilities have been in favor of it.

The public has no real voice in the financial game played by the electric utilities, excepting the public's voice in the selection of political administrations of states that appoint public service commissioners.

At present, the regulatory system is not geared to properly evaluate the effects of the growth of electrical utilities, and the wishes of the utilities are reflected in the massive rate increases that have been rubber-stamped by public commissions in state after state. In 1972 and 1973, electric utilities obtained more than $1.8 billion in rate increases, and billions more will be sought each year in the future.[43] The actions of the federally owned power projects have been little better than those of state public service commissions.

Like a gathering storm, the signs of change are apparent in the thinking of many state utility commissions, but years may be necessary to enact necessary changes in the electrical rate structure. What is needed is a thorough revamping, to include the full costs—not just the economic costs to consumers—of electrical power to the nation. As Professor Warren Samuels of Michigan State University puts it: "The rate structure increases the total usage by making people insensitive to energy costs. People are charged lower and lower rates when they consume more and more, when in fact the social costs increase."[44]

So far, a few states, including New York, Vermont, Michigan, and Oregon, have made the first tentative changes in the rate structure to accomplish such obvious goals as eliminating the free advertising ride utilities now get, and modifying the tailblock rates. The Federal Power Commission has asked the nation's utilities to provide plans for energy conservation. So far, however, no substantive federal or state actions have been taken to change the whole outmoded system of utility regulation.

* An interesting aspect of Washington lobbying is the change offered in Senator Magnuson's bill between 1972 and 1973. The 1972 bill provided a surcharge based on each kilowatt-hour of electricity use, which would bear down on bulk users. Due to lobbying efforts, the bill was altered so that the surcharge would be based on a percentage of revenues, not consumption of actual amounts of electricity, a provision that is less equitable to small users of electricity.

The stakes in the race to reduce energy consumption in this area are tremendous. In California, the Rand Corporation thinktank reported to the state assembly that the increase in electricity growth predicted by the state government and the state's utilities represented a policy that portended disaster to the state by the year 2000. The Rand Corporation recommended that the state enact immediate policies to slash the growth rate by a tremendous margin in order to reduce from 130 to 23 the number of power plants scheduled by the power companies for installation in the state by that time. In a controversial and far-ranging study, the Rand Corporation suggested that the external social costs of building the plants far outweighed the economic benefits that would accrue from increased electricity growth. Building 130 new plants, Rand said, would bring severe fuel shortages, would overload the environment with pollution, clog the coastline with power plants, and introduce many new nuclear power plants to the state despite the fact that nuclear plants represent a potential danger to the state's residents.

The main weapon in the Rand arsenal to slow this growth is the club of altered electric rates. The study pointed out that the conventional arguments of free-market capitalism suggest that the competitive price mechanisms of the market are sufficient to maintain equilibrium between supply and demand, leading to the assumption that buyers and sellers on the market can predict the future with accuracy. Not so, says Rand, when there is a serious question "whether energy markets are sufficiently competitive or far-seeing to make the necessary adjustments soon enough. . . . It may be advisable to take action to ensure a dampening of the growth rate to a level that avoids the risk of severe shortages."[45]

The measures it suggested included such steps as initiation of energy growth-slowing policies by the state legislature; the launching by the state and power industries (including equipment suppliers) of a citizens' education campaign to inform people how to conserve energy on an individual level; and a state reassessment of the commitment to potentially dangerous sources of energy, such as nuclear power.

The study recommended a number of economic steps, including tax write-offs for heating equipment companies that would manufacture solar heating equipment,† reducing the drain on the state's fuel resources. The Rand Corporation also recommended that the state install solar equipment and other energy-conserving devices on new state government buildings; rewrite the state's codes on insulation of buildings to reduce heat loss from structures; and establish minimum efficiency standards for electric air conditioners and appliances.

† See Chapter 7.

The Rand Corporation suggested too disallowing promotional expenses by utilities, and establishing a small kilowatt-hour tax (0.1 mill per kilowatt-hour) on electricity sold in the state to finance research and development on methods of further energy conservation. Electric utilities would be required to introduce special peak-load rates for all consumers of electricity, from industries to small consumers, to discourage overuse of power during the winter and summer heavy use periods. This would obviate the need for much power plant construction by minimizing the major source of increased growth.

Furthermore, Rand told the state, if these measures were implemented and did not sufficiently slow the electricity growth rate, then the state should impose a series of more stringent measures, including raising the tax on electricity. Rand specifically pointed out that poor people and small consumers should be exempted from the tax, since the corporation's studies of growth showed that small consumers played an insignificant part in the expansion of electricity demand. Further steps would include imposing minimum efficiency standards on industrial use of electricity, and banning use of natural gas by electric utilities. Consumer purchase and installation of electric ranges, water heaters, air conditioners, and applicances would also be banned, until utilities could install sufficient (environmentally approved) generating capacity to meet increased load requirements.[46]

The application of these measures would reduce the current rate of electricity growth in California from more than 8 percent per year to a conservative 3 percent per year.[47] Understandably, the Rand proposals have been greeted with attacks from the state's utilities, but the significance of the Rand report was appreciated by the state government. The state's Assembly established a special Subcommittee on Electrical Energy Policy, which proposed an initial legislative package in 1973 that would establish a state power surcharge (0.2 mill per kilowatt-hour of electricity sold), which could be used to fund research and development into methods of slowing growth and building better, cleaner power plants. The surcharge would provide $32 million a year in revenues for the research fund, while adding only $1.16 to the average consumer's yearly electricity bill. Other provisions of the bill would provide for establishing a new power plant siting law that would provide for a streamlined process for state approval of new power plants. The state would set up energy efficiency standards for electric appliances, insulation, heating and cooling of buildings, and other energy conservation measures. The proposed legislation is only a first step, and does not incorporate the major recommendations of slowing growth that the Rand Corporation suggested. However, the Speaker of the California

State Assembly summed up the significance of the move: "Unlike other problems such as alleged oil shortages, which involves the federal government, imports, international companies, and foreign governments, our electrical problem and its solution can be dealt with within our California borders."[48]

Other states, including Florida and Oregon, have also decided to go it alone, without waiting for the creaking wheels of the federal process, divine intervention, or a radical change in the thinking of private power companies. Other states have considered moves to implement special programs to slow growth, including New York, but one thing appears clear. The goals of slowing electricity growth, and growth of energy use in general, cannot be met by applying the old logic of the Federal Power Commission and the nation's regulatory commissions to the new problem of environmental and social costs incurred by spiraling energy use. Rather than turning to the overcentralized federal government or the old centralized model of utility "regulation," it is encouraging to see the states recognize that the energy crisis can best be curtailed by actions designed to control energy where it is used, not relying on the obsolete policies of constantly increasing supplies.

End Uses of Energy

The following chart sets out the end uses for energy from all sources in the United States in 1968. The chart is based on a recent study made

End Uses of Energy (*consuming sector of the economy in parentheses*)	**Percent of Total**
Transportation (*fuel, excludes lubes, and greases*)	24.9
Space heating (*residential and commercial*)	17.9
Process steam (*industrial*) ‡	16.7
Direct heat (*industrial*) ‡	11.5
Electric drive (*industrial*)	7.9
Raw materials and feedstocks (*commercial, industrial, and transportation*)	5.5
Water heating (*residential, commercial*)	4.0
Air conditioning (*residential, commercial*)	2.5
Refrigeration (*residential, commercial*)	2.2
Lighting (*residential, commercial*)	1.5
Cooking (*residential, commercial*)	1.3
Electrolytic processes (*industrial*)	1.2
Total	97.1 percent

‡ Since these industrial categories account for energy used in space heating, the space heating category may equal or surpass 20 percent.

by the Stanford Research Institute for the National Science Foundation; the end uses in 1968 are not significantly different in terms of the over-all amount of energy in use as the end uses in the United States today.[49]

These twelve applications account for all but 2.9 percent of the nation's total energy use. The remainder of the energy use is spread through a variety of large and small applicances, electricity for elevators, and many other uses.

By taking these uses of energy and separating them into broad use categories or sectors, we find that industry accounts for about 42 percent of the nation's energy consumption; transportation, about 25 percent, and homes and commercial buildings, about 33 percent.

Energy Use in Industry

About 42 percent of the nation's energy use is accounted for by the industrial sector of the economy. The U. S. Bureau of Mines found that fully two thirds of this energy consumption has been concentrated in the six major industries listed below, with their respective levels of energy consumption (in 1968) in quadrillion BTUs:[50]

Industry Group	Total Energy Used in Process—Excluding Material *(in quadrillion BTUs)* *(primary energy and electricity)*
Primary metal industries	5.30
Chemicals and allied products	4.94
Petroleum refining and related products	2.83
Food and related products	1.33
Paper and related products	1.30
Stone, clay, glass and concrete products	1.22
Total	16.92 quadrillion BTUs
All other industries	8.05 quadrillion BTUs
Grand total	24.97 quadrillion BTUs

In terms of the kinds of energy used by industry, the Rand Corporation found that almost half comes from the use of natural gas (46.5 percent). Coal accounts for 26 percent; petroleum fuels, 16.8 percent; and electric power, 10.6 percent.

The Rand Corporation studied industrial energy use patterns from 1946 to 1969 and found that the use of natural gas increased, as did that of petroleum and electricity.* The use of coal by industry, however, experienced a significant decline. In 1946, coal furnished about 56 percent of the energy requirements of industry; but by 1969 its share had dropped to only 26 percent. What occurred during the postwar period was the increasing substitution of other fuels—notably the nation's cleanest and most sought-after fossil fuel, natural gas—for coal. The use of natural gas increased from 20 percent of over-all use in 1946 to 46.5 percent in 1969—the exact reverse of the coal trend.[51]

The preceding table of energy areas in industry shows where much of the nation's energy is poured: into the metals, chemicals, and energy industries themselves. These three groups of industries consume over half of the total amount of energy used by all industries, which means that fully one fifth of the nation's energy is devoted to these three industry groups.

The efficiency with which energy is used for the production of materials by these industries has actually increased since the Second World War, and achievements in advancing the efficiency of energy use have come about largely through the application of better technologies.† The application of better technologies has noticeably improved the efficiency of the steel industry, and more recently, the aluminum industry. In the early 1960s, the steel industry began replacing open-hearth steelmaking furnaces with the newer basic oxygen process furnace. The new method has significantly reduced the level of energy use. At the Bethlehem Steel Corporation's Lackawanna plant, for example, the introduction of basic oxygen furnaces between 1963 and 1967 cut by two thirds the average use of energy to produce a ton of steel.[52] In the United States, more than half the steel furnaces in use are of the basic oxygen process type, and the White House Office of Emergency Preparedness predicts that the ultimate addition of these furnaces will cut the steel

* The use of electricity in industry means that even more energy has been used than shows up in most statistics, due to the fact that electricity is made from *primary* energy sources, with resulting inefficiency. Three to four times as much initial energy was expended to produce the electricity—a *secondary* energy source.

† The only industry groups that now use more energy to produce material than was required in years past are the tobacco manufacturing industry, the apparel/clothing-finished product industries, lumber and wood industries, and the printing and publishing industries. Only 2 percent of the total industrial energy use can be attributed to these industries, however. Their higher energy use is due to increased mechanization in recent years, since they were formerly dependent on more human labor.

industry's energy consumption in half.‡ Currently, the steel industry uses 6 percent of the national energy supply to manufacture steel, and the reduction to a level of 3 percent of the national total will be significant.[53]

In 1973, the Aluminum Company of America (Alcoa) announced that it had perfected a new process for producing primary aluminum that would reduce by up to 30 percent the energy needed for aluminum production. The Alcoa technology would replace the conventional Hall process, which is used to refine alumina (the product of bauxite ore) in electrolytic cells filled with fluoride chemicals. The new process would employ an improved technique that uses chlorine to replace the fluorides. Alcoa's chairman, John D. Harper, said the process would be in limited use by 1975, but that "it should be clearly understood that this does not signal any overnight obsolescence of established facilities for smelting alumina," and that additional aluminum plants would probably be built for some time with the older Hall-type equipment.[55]

The aluminum industry could well use newer technologies; the growth of aluminum use has been dramatic in the United States; energy for the manufacture of this single metal represents 4 percent of all the electricity consumed annually in the United States, and the figure is constantly growing larger.[56]

The chemical and refining industries together account for even more of the nation's energy than the metals industries. All told, these industries consume about one third of the industry total, and represent about 13 percent of the nation's over-all energy use. A number of studies have been made to determine the efficiency by which the industries use energy, but the nature of the industry has largely precluded this. The chemical industries produce thousands of chemical products, adding over 400 new ones each year.* The various chemical companies use a

‡ The application of the basic oxygen process may not constitute the great "technological fix" to the problem of energy use in the steel industry that the federal government thinks it will. The recycling of waste products is as important an aspect of energy conservation as is the initial efficient conversion of raw materials into useful products. The older open-hearth furnaces can be operated with up to 100 percent scrap metal. However, the newer basic oxygen process furnace cannot operate with a high percentage of scrap, and its efficiency may decrease even when loaded with small amounts of scrap.[54]

* The production of chemicals has resulted in the introduction of more than 100,000 synthetic chemical compounds into the environment since the turn of the century. Frank Wallick reports in his book *The American Worker* that little knowledge is available on the toxicity of these substances—either to flora, fauna, or humans. Workers in American industries may be exposed to more than 8,000 chemical compounds, yet safety limits (called threshold limit values, or TLVs) have been set for only 450. Wallick says of the old chemicals as well as the new ones that "some

myriad of technological processes to produce chemicals (some chemicals may be made by numerous processes), and detailing the exact way in which energy is used is a laborious and virtually impossible task. Adding to the complexity is the fact that the major raw material in these industries is also the fundamental energy source of our society: fossil fuels, primarily from petroleum and natural gas, the premium fuels for power generation purposes.

In the petroleum refining industries, it is difficult to determine the exact amount of energy used to refine crude oil and other petroleum feedstocks into higher-grade fuels. Of the more than 260 petroleum refineries in the United States, no two are alike. Each refinery uses different methods of refining petroleum—and each of the many methods requires varying amounts of energy. On the average, upward of 10 percent of the energy content of the fuels being refined is used to operate the refinery processes. While various fuels are used by refineries, half of the energy need is met by the high-grade fuels, natural gas and fuel oil.[58]

Most of the energy required to refine fuels is for heat in various refining stages. Refineries today generally use less energy to refine products than they did 20 years ago, but there are signs that this trend may reverse itself. Energy consumption is related to the complexity of the refinery operation. Simple refineries—i.e., refineries that process crude oils into higher-grade fuels—need far less energy than complex, "integrated" refineries, which produce a number of different products from petroleum, such as lubricating oils and greases, waxes, coke, and sulfur products.[59]

The chemical and refining industries make a double-barreled attack on the country's energy sources: They use fossil fuels both for power generation and for manufacturing the finished products (which may even be other forms of the original fuel). The Stanford Research Institute (SRI) estimates that, in 1967, the following product groups of the chemical industry (not including petroleum refining) consumed approximately two thirds of the industry's energy: chlorine, synthetic soda, acetylene, oxygen, carbon black, ammonia, aluminum oxide, phosphorus, methanol, sulfur, ethylene, propylene, butadiene, aromatics, C_2, C_3, and C_4.

---

are undoubtedly quite harmless—but nobody is really sure. Little testing as to the human health effects is made of chemical compounds; workers who use new chemicals are treated as human guinea pigs. Even worse than that—job-related illness or deaths go unreported, undetected, unnoticed. . . . Pollution on the job is a microcosm of pollution everywhere. [It is] little understood and seldom measured. Yet it unquestionably inflicts far more damage to health than does the admittedly serious pollution which descends upon an entire community."[57]

The actual energy used for the manufacture of these products (832 trillion BTUs) was exceeded by the use of fossil fuel "feedstocks." The petroleum feedstocks represented twice the energy consumption (1,815 trillion BTUs) for these products in the chemical industry as the energy used for the production of chemicals![60]

Outside of the "big three" industry groups, high-energy use in industry is concentrated in the food; paper; and stone, glass, and concrete industries. While advances in technology have reduced the amount of energy per unit of production in these industries compared to 20 years ago, their over-all energy use has nevertheless increased as production has risen to meet demand.

Energy needs of various industries can be reduced by some further improvements in technological processes, and by careful use of energy. To a great extent, this is already practiced by industries, since conserving energy and using efficient processes are to their economic advantage. With the advancing national problems in energy supply, many individual companies have embarked on ambitious programs to cut energy uses.

A good example is the DuPont Company, which has engaged in an energy conservation program for several years. DuPont consultant manager G. Frank Moore announced in 1973 that the company's careful attention to energy conservation had enabled reductions of 7 to 15 percent below previous annual energy use. Moore announced that DuPont would undertake, on a consulting basis, similar analyses of energy losses for other industries. He pointed out that energy losses by various industries "are often tolerated simply because no one knows they exist." DuPont found instances of inefficient combustion processes in certain plants that were costing the company more than $300,000 per year in increased fuel costs.[61]

The former White House Office of Emergency Preparedness (OEP) reported the best way to make industry more efficient in its use of energy is to raise the price of energy through federal or state taxes. Another Office of Emergency Preparedness proposal is to change federal depreciation rates on equipment so that companies will have a further economic incentive to use energy more wisely. The OEP says that the allowable payoff rate for equipment averages only 3 to 5 years, and that a longer depreciation period would encourage the life cycle costing of equipment; i.e., more consideration would be given to lifetime energy use and durability, with less concentration on cheap equipment, which requires more energy to operate.

The current effects of low-depreciation schedules may be gauged by this example offered by the OEP: A refinery has the choice between purchasing an efficient electrically operated pump that requires 10 horsepower (hp) to function and costs $1,600, and another, less efficient pump that requires 13 hp and costs $1,000. If the industry chooses the more efficient pump, it will save money during a 10-year depreciation period because the pump uses less energy, thus reducing its lifetime costs. However, at any depreciation rate of less than 10 years, the industry will lose money by buying the more expensive pump, because it cannot amortize the costs of shorter depreciation and because of higher federal taxes.[62]

The OEP has identified one of a number of federal policies that encourage the waste of energy and resources by ill-planned laws that hinder the development of more efficient processes. A number of industries, such as DuPont, have begun to develop more efficient practices without the benefit of corrective legislation. The reason is the rise of energy prices. With changes in tax structure and depreciation rates, many more improvements can be made that will reduce the energy consumption of industries.

A number of recent studies have indicated that energy use in industry can be reduced from 10 to 40 percent within a few years. A General Electric Company survey of 225 manufacturing plants in the United States and Canada estimated that improvements in existing facilities would save $7 million worth of fuel annually.†[63] The DuPont energy conservation program suggests a savings of *$5 million daily,* if advanced conservation measures were practiced by all industries in the United States. Assuming a cost of $3.50 per barrel of crude oil, this would be the equivalent of 1.5 million barrels of oil per day.[64] The costs of foreign crude oil have already increased beyond this level.

The threat of increasing energy prices, fuel shortages, and electrical blackouts has had a positive effect on the thinking of industry planners. The first effect has been the rethinking of energy use in specific industrial plants, and attempts have been made to redesign and modify outmoded processes so that coming increases in cost will not bear down heavily on production capability. A second, and potentially revolutionary, impact has been a trend toward the decentralization of energy production for all uses in manufacturing plants. As is discussed later in greater detail,‡ a great source of energy waste in the United States is the failure of our technology to adequately use waste heat. Heat from

† Dr. David Large of The Conservation Foundation, Washington, D.C., has written an excellent study of energy use, from which this reference is taken. Dr. Large's *Hidden Waste: Potentials for Energy Conservation* is available from The Conservation Foundation, 1717 Massachusetts Avenue, N.W., Washington, D.C. 20036. It is $3.00 in paperback.[65]

‡ See Chapter 4 on total energy and decentralized energy systems.

energy conversion processes is not widely used for constructive purposes such as heating buildings. Instead, it is rejected in large quantities to air and water, adding a serious environmental problem of thermal pollution. Were industries to build their own small-scale power plants at the sites of manufacturing plants, tremendous engineering opportunities would be made available for saving energy by capturing the waste heat of the power-generating process.

In the Rand Corporation study in California the managers of a number of California industrial plants were interviewed. The Rand researchers found that the general response of plant managers confronted with the threat of the shortage and possible rationing of electricity from the state's utilities would be to install their own electrical generating stations, an eminently logical solution.[66]

The amounts of energy used by industry can be related to the nature of industrial production and the potential energy content of various products. The following table illustrates the amount of energy needed to produce a ton of various industrial materials:

Energy Consumption in Basic Materials Processing Per Ton of Product

Material	Energy used for production, manufacturing equipment use, and transportation of finished goods (*in thermal kilowatt-hours*) *
Steel	12,600
Aluminum	67,200
Copper	21,000
High-grade steel (silicone and metal) alloys	59,200
Zinc	14,700
Lead	12,900
Electrically processed metals	51,200
Titanium	141,200
Cement	2,300
Sand and gravel	21
Inorganic chemicals	2,700
Finished plate glass	7,200
Plastics	2,900
Paper	6,400
Coal	42
Lumber	1.51 per board foot

* A thermal kilowatt-hour is an energy measure in electrical terms that takes into account the efficiency of the electrical production and distribution system. A con-

This table, prepared by Arjun Makhijani and A. J. Lichtenberg at the College of Engineering of the University of California at Berkeley, shows the range of energy intensiveness—the relative amounts of energy necessary to make the products. The production of steel requires about one sixth as much energy as does the production of aluminum; some forms of glass require only half as much energy to make as an equivalent weight of steel. Thus, aluminum is more energy-intensive than steel, and steel more energy-intensive than glass.

The two most important recycling concepts that are important in terms of saving energy in the economy are (1) producing material requiring low levels of manufacturing energy use as opposed to the production of high-energy-intensive materials; and (2) reusing and recycling waste materials rather than discarding them and using up more virgin raw materials.

An example of this is the use of aluminum cans as opposed to returnable bottles. Cans made from aluminum represent a significant percentage of aluminum sales in the United States. Between 1971 and 1972, the use of aluminum cans increased an amazing 31 percent, and 8.7 billion beverage cans were made by the industry. American aluminum companies argue that this production helps the recycling of materials, and is a positive force. Alcoa President W. H. Krome says that "without aluminum scrap, we might never build an economical solid waste recovery system in a city."[68]

Krome and others in the aluminum industry point to the fact that recovering aluminum from the nation's solid waste stream will increase in coming years, because it is ten times more valuable than an equivalent weight of steel. However, the industry neglects to point out the energy waste inherent in their recycling scheme. The remelting of aluminum to produce more cans requires additional energy. Wouldn't it be more logical to use returnable, rewashable, refillable glass containers and eliminate both high energy use and environmental problems?

Professor Bruce Hannon of the University of Illinois' Center for Advanced Computation at Urbana, Illinois, has answered this question with a resounding yes. His elaborate studies of the economics and the energy use in the beverage and bottling industries show that nonreturnable cans and bottles use up three to five times as much energy as returnable glass bottles. Were the state of Illinois alone to switch to returnable bottles and outlaw "throw-aways," the monetary savings to consumers would be $71 million annually.[69]

---

ventional kilowatt-hour is one third of the thermal kilowatt-hour, and represents electricity converted at point of use, not the full energy equivalent of electricity in the system.[67]

Increasingly, the products of industry in the United States are being added to the solid waste stream, a national problem of enormous magnitude, since the costs of disposing of wastes amount to more than $6 billion annually. Spurred on by incessant industry advertising to persuade consumers to buy more and more, the solid waste problem is growing. In 1920, the average American generated 2.75 pounds of solid wastes per day; in 1970, 5 pounds; in 1980, estimates indicate that 8 pounds will be produced daily per American consumer.[70] In energy terms, these wastes represent the familiar thermodynamic problem of entropy. The original raw materials have been transformed into useless entropic artifacts of wasteful, high-energy civilization. The way out of this national quandary rests as much on the actions of consumers of industrial products as on the manufacturers of these products.

Energy Use in Transportation

Moving people and goods around the country has long been one of the great (if not in fact the greatest) driving forces in American society. This kinetic, ever-changing quality of the country prompted George W. Pierson, in his book *The Moving American,* to observe that one of the essential "freedoms" of American democracy is the freedom to move. He persuasively argues that Descartes' famed dictum, "I think, therefore I am," has been upended by the American philosophy and rendered, "I move, so I'm alive."[71]

This movement, appropriately, accounts for the largest single block of energy used in the United States. Without regarding the enormous amount of energy required to make cars, airplanes, buses, trucks, ships, and other transport machines, the energy required just to fuel them represents one fourth of all the energy consumed in the country.

By the same token, the various commercial facets of transportation—from businessmen on Los Angeles-to-New York jet hops to iron ore moving down barges on the Mississippi—comprise the largest single contribution to the Gross National Product. The transporation industries, which with related service industries employ more than 10 million people, are responsible for pouring $200 billion a year into the nation's economy.

As opposed to the *financial* costs of transport, the following table shows how the *energy* costs looked for the various transport methods, or modes, in 1970. This U. S. Department of Transportation table indicates the percentages of energy consumed—as fuel—in the various transport modes that together account for one fourth of the U.S. energy budget (all percentages are expressed as the fraction of the transport total, not of the entire national energy consumption budget):

Transportation Mode	Energy Consumption (%)	
Automobiles	55.0	
Trucks	20.0	
Buses	1.0	
Other (including motorcycles)	.48	
Total highway transportation		76.48%
Airplanes†	9.87	
Railroads (freight)	3.51	
Railroads (passenger)	.06	
Shipping (inland and coastal barges and ships)	4.84	
Fuel pipelines†	5.24	
Total		100.00%

The above calculations are based on a comprehensive study of U.S. transportation made by the Massachusetts Institute of Technology for the U. S. Department of Transportation.[72] From these statistics, it is clear that most of the transportation energy use is due to the movement of materials and people in motor vehicles on the nation's highways, with air transport a runner-up.

What is not reflected in this accounting is the radical change in methods of transportation, or transport modes, over the past few decades. Eric Hirst, an engineer at the Oak Ridge National Laboratory, has analyzed these changes over the period between 1950 and 1970.

Hirst found that the ways in which we now move people and freight are less efficient than the ways in which people and freight were moved in 1950. He calculated the *energy intensiveness*—that is, how much energy is used in each mode of transportation, such as cars, buses, airplanes, etc.—by comparing the amount of energy (expressed in BTUs) required in each transport mode to move either one passenger or one ton of freight a distance of one mile. The term *energy intensiveness* is the reverse of *energy efficiency*. The greater the energy intensiveness of a system, the less efficient its use of energy.

† Various studies give widely conflicting analyses of the total energy consumption devoted to transportation in the United States. For example, Eric Hirst of Oak Ridge National Laboratory has calculated that airplanes used more energy in 1970—about 11 percent of the total transportation energy use. His studies also indicate that fuel pipelines used only 1.2 percent of the transportation energy, rather than the 5.24 percent reported by the U. S. Department of Transportation.

Hirst found that the fastest-growing methods of moving people and goods in the decades of the 1950s and '60s were also the most energy intensive. A glance at the following list shows the energy intensiveness of each transport mode in 1970:[73]

FREIGHT TRANSPORT

Mode	**Energy intensiveness, expressed in BTUs of energy needed to move 1 ton of freight 1 mile**
Airplane	42,000
Truck	2,800
Waterway	680
Railroad	670
Pipeline	450

PASSENGER TRANSPORT

Mode	**Energy intensiveness, expressed in BTUs of energy needed to move 1 passenger 1 mile**
Airplane	8,400
Automobile (urban)	8,100
Mass transit (urban)	3,800
Automobile (intercity)	3,400
Railroad (intercity)	2,900
Bus (intercity)	1,600

Airplanes

From the foregoing transportation statistics, it is evident that the most energy-intensive means of moving goods and resources is by air. Yet this least-efficient method of transportation has grown inordinately since 1950. The airlines' share of total intercity passenger traffic increased by a factor of 5 between 1950 and 1970; and their share of freight traffic, sevenfold. Thus jet engines have given a powerful thrust not only to the movement of people and freight by air, but to the consumption of energy required to do it.

In 1950, the energy-intensiveness of air freight was 23,000 BTUs per ton-mile. In 1970, this had increased to 42,000 BTUs per ton-mile—almost double the 1950 figure. Correspondingly, the energy-intensiveness for passenger planes in 1950 was 4,500 BTUs per passenger mile; but by 1970 it took almost twice as much energy—8,400 BTUs—to move a passenger the same distance. Two trends are responsible for this increased fuel use between 1950 and 1970. One is the obvious growth of the air industry, which was 12 times larger in 1970 and carried 21 times as many people as two decades previous. The second was the increasing energy-intensiveness, or decreasing efficiency of airplanes. The primary reason for this was speed, the time-honored keystone of American philosophy. In 1970, the average speed of commercial aircraft was more than 400 miles per hour, in contrast to 1950, when the average plane speed was only 180 miles per hour.[74]

In terms of efficiency, the growth of the air freight and passenger industry has reflected a "the faster the worse" ratio.‡

Automobiles

The second most energy-intensive form of transport is the private automobile. In fact, when used to move people within the city, it requires almost as much energy—8,100 BTUs—to move 1 person 1 mile as does the airplane, based on 1970 figures. On the other hand, cars moving between cities showed an average energy-intensiveness of 3,400 BTUs per passenger mile. The average for automobiles is 5,400 BTUs per passenger mile. Even at this reduced average figure, the automobile uses more energy per passenger mile than any other form of transport except the airplane.

In the two decades since 1950, the energy-intensiveness of private car transport increased 12 percent. While this does not at first appear as significant as the near-doubling of airplane energy-intensiveness, the sheer number of automobiles more than compensates in terms of actual increased fuel consumption during this period. In 1971, with a population of 207 million people, the United States boasted a registered automobile population of 92 million.[76]

Since 1950, automobile traffic increased 142 percent, but the amount of energy required to make them go increased 171 percent during the same period. In terms of direct fuel use, automobiles consumed 66 billion gallons of gasoline in 1970—the equivalent of 13 percent of all energy used in the country in that year. Two thirds of this fuel was used for the least efficient means of travel—urban driving; the remaining one third was spent to move cars between cities and through the countryside.

The magnitude of the automobile's contribution to use of resources other than energy may be gauged from the following list of metals and rubber, with the percentages of U.S. supply currently earmarked for automobile manufacture:[77]

‡ An interesting facet of the growth of air transport is that the average cost of traveling by air has increased only 8 percent during the 1950–70 period, while bus fares increased 90 percent, and rail fares, 40 percent. This is due in large part to heavy government subsidies to the air industry.[75]

Resource	Percentage of U. S. Supply Devoted to Automobile Manufacturing
Rubber (synthetic)	60
Aluminum	10
Steel	20
Copper	7
Nickel	13
Zinc	35
Lead	50

The reason for the increase in automobile traffic is that the automobile population has more than doubled since 1950, when there were only 40 million registered automobiles in the United States. In fact, according to Eric Hirst, this increase in car population is the single major factor in the drastic increase in transportation energy consumption during that time. The greater traffic that has resulted encourages more stop-and-go driving, which increases energy consumption. The increase in traffic, he says, accounts for 89 percent of the increased auto energy use since 1950.[78]

Nevertheless, there has been considerable decline in energy use efficiency due to the changing characteristics of the automobiles themselves. Like airplanes, cars have been designed for speed rather than energy economy. Automobiles today are larger than in the 1950s and are packed with fuel-guzzling extras such as air conditioning (61 percent of the new 1971 cars were air conditioned), unnecessarily powerful V-8 engines, automatic transmissions, power steering, power windows, power seat adjustors, etc.

The automobile-dominated society was made possible not only by Henry Ford's revolution in production, but also by the efforts of the U. S. Government in ensuring that a No. 1 priority of U.S. land use would be the construction of highways. Since 1944, when the first Federal Highway Act was passed providing federal funds for highway construction, including the planned completion of the 42,500-mile interstate highway system, the process of assuring the development of America around the automobile has become ever more entrenched. When Congress debated the construction of the interstate system in 1944, the prevailing argument was national defense, that the highways were needed to move troops in case of war or emergency. Yet since that time, the interstate system, expanded by subsequent amendments, has grown like Topsy. The Highway Trust Fund,* created under these

* Analyzing the Highway Trust Fund is beyond the scope of this book. The best introductory volume on the subject is Helen Leavitt's *Superhighway-Superhoax,*

laws for road construction, and fueled by federal taxes on gasoline sales, has only recently been broken open for such uses as mass transit, which offer a much more rational solution than the private automobile to many of our transportation problems.

Trucks

The third-ranking energy-intensive consumer in transport, after airplanes and autos, is the truck. In 1970, trucks accounted for about 20 percent of the fuel used for transportation in the United States—second in fuel consumption only to automobiles. Trucks carry almost all urban freight, and about one fifth of the freight carried between cities. While trains carry most long-distance cargo, trucks carry almost all "short hauls"—distances less than 200 miles. The changing pattern that occurred between 1950 and 1970 in the growth of car and airplane transport also held true for truck transportation.

In those two decades, use of truck fuel increased 142 percent; truck mileage, 137 percent. The energy-intensiveness of truck use did not increase drastically; it started high. In 1950, it took an average of 2,400 BTUs of energy for a truck to move a ton of freight one mile. In 1970, the figure was 2,800 BTUs per ton-mile. About 15 percent of the increased energy use in trucking was due to the increased energy intensiveness that came about as the result of the better highways in the country in 1970, which allowed for higher speeds for intercity truck hauls.

Railroads

The key to understanding the phenomenal growth of airplane, automobile, and truck transport in the past few decades is in realizing that these modes of travel have not simply garnered their proportionate shares of an increasing population, but they have increasingly replaced the most efficient form of land transport available today: the railroads.

Today's trains are highly efficient in their consumption of energy for several major reasons: They are built to careful, rugged specifications and are powered by very efficient diesel engines that consume less fuel than either the old coal-stoked steam locomotives or the later electric trains. The electric locomotives suffer from the traditional

Ballantine Books (paper), 1970. Helen Leavitt and her husband, Bill Leavitt, also publish an excellent monthly newsletter on mass transit vs. the automobile: *Rational Transportation,* 4215 37th St., N.W., Washington, D.C. 20008. An active and underfunded citizens group lobbies against the industry interests and publishes a periodic newsletter, *The Concrete Opposition,* available from Highway Action Coalition, Suite 731, 1346 Connecticut Ave., N.W., Washington, D.C. 20036.

limitations of the electric power cycle—i.e., low energy conversion efficiency at the central power station combined with transmission and distribution losses in the process of getting the electric current to the train's electric motor.

Trains carry large loads, which are usually maximized to the requirements and design of the engine. This means that the fuel utilized for pulling the load is used efficiently. In contrast to automobiles, in which fuel is used primarily to move the hulk of the car's body rather than the baggage or passengers within it, most of the train's fuel goes to move the cargo.

Almost 100 percent of intercity trains use diesel engines burning diesel fuel oil, compared to about 90 percent of intercity trucks that are powered with diesel engines.[79] Yet the energy intensiveness of railroads in moving freight was only 670 BTUs per ton-mile in 1970, compared to the truck's energy intensiveness of 2,800 BTUs per ton-mile. This means that trains require less than one fourth the amount of energy that trucks require to move the same amount of freight.

However, trains, unlike trucks, are in a period of sharp decline. In the 1950 to 1970 period, intercity traffic carried by trains dropped from almost half of all freight moved between cities in the United States to just over one third. Passenger traffic carried by trains dropped from 7 percent in 1950 to only 1 percent in 1970!

However, a startling energy statistic stands out. Unlike trucks, automobiles, and airplanes, which require far more fuel to operate today, trains need much less fuel than they did in 1950 to transport the same load of freight or people. Hirst calculates that the energy intensiveness of trains *declined* by 78 percent between 1950 and 1970. Thus trains use an average of 5 times less energy than they did to accomplish the same jobs they performed in 1950.

Buses

Another casualty of the march of "progress" in transportation is the bus. In 1950, buses carried 5 percent of intercity passengers and 8 percent of urban passengers. By 1970, both categories had dropped to less than 2 percent of over-all passenger transportation movement. Yet buses need only 1,600 BTUs to move 1 passenger 1 mile—far less than the energy-guzzling cars and airplanes that have replaced them as the preferred means of intercity travel.

The bus—whether used for urban or intercity transit—belongs in the mass transit picture. Ground mass transit includes subways, trains, and buses used in cities; and buses and trains used for intercity transportation. In the cities, Eric Hirst has calculated that the energy intensiveness of buses and electrified trains averages 3,800 BTUs per passenger-mile—less than half the energy use of the preferred mode of urban transit, the private auto. However, in the 1950–70 period, urban mass transit passenger traffic declined by 57 percent, and its share of energy use dropped correspondingly by 53 percent.

The reason for decline, once again, is the convenience of the private automobile, with the rise in the development of suburban highway belts around cities that are geared directly to the automobile culture. Politically, the development of the Highway Trust Fund has been the major impediment to the development of better transit systems for linking the suburbs to the cities. Not until 1971 did the federal government spend the equivalent of more than a fraction of the funds tied up in the Highway Trust Fund on mass transit systems. That year, $57 million—a mere pittance compared to what is needed—was spent to build mass transit systems under the Urban Mass Transit Act of 1970.[80]

Waterways and Pipelines

The remaining transportation modes—which also reflect the greatest efficiency and conservation of energy use—are waterways and pipelines.

Shipping in the United States on the Great Lakes and on canals and other waterways accounted for 27 percent of all the intercity freight transported—a decline from the 31 percent transported by these means in 1950. During that 20-year period, the energy intensiveness of these systems also declined by 7 percent because of more efficient engines and other technological improvements on the barges and ships involved in transport. The actual traffic of freight shipping grew by 41 percent during the period, which is indicative of some increased use, but not enough to approach the tremendous growth spiral of faster transport modes such as airplanes and trucks.

Pipelines are used almost exclusively for transporting oil, natural gas, and refined petroleum products. Pipelines need very little energy to move liquid resources—only 450 BTUs to move one ton one mile. Due to the use of improved and larger-diameter pipelines, the energy intensiveness declined slightly in recent years. The principal significance of pipelines is that they are a much more efficient and environmentally preferable method of transporting energy than other forms of resource transport. For example, overhead electric power lines slash the countryside with enormous, unsightly rights-of-way, whereas all pipelines are under the ground and out of sight. Pipelines can move energy resources from point to point with almost no loss of energy, in contrast to what happens when secondary energy—electricity—is shipped in power lines in which 10 percent or more of the energy may be lost through transmission and distribution inefficiencies. Yet, the transportation of energy via electrical power lines—not the clean and efficient pipelines—is the fastest-growing energy transport method.

The Transportation Total

The previously cited statistic that transportation energy use accounts for about 25 per cent of the total energy use in the United States represents only part of the transportation total. It reflects only the narrowest definition of energy use, which does not incorporate the energy needed to manufacture all the transport vehicles, pipelines, highways, gasoline stations, and other components of the transportation system.

Dr. William E. Mooz of the Rand Corporation in Santa Monica, California, says that, when all of the energy costs of transportation are added up, the transportation energy consumption figure is boosted to 41 percent of over-all U.S. energy use![81]

Few thorough studies have been made of the extent to which energy is used to both build and operate transport vehicles in the United States. One pioneering work, however, analyzes the energy needed to manufacture automobiles. It is a report prepared by Stephen Berry and Margaret Fels of the University of Chicago Department of Chemistry, for the Illinois Institute for Environmental Quality.[82]

Their report on energy consumption in automobile manufacturing treats the automobile as a part of the over-all energy system and takes into consideration the energy required to mine the various metals that go to make up a car as well as the energy used in industrial processes related to manufacturing it. Berry and Fels take into account the loss of energy in the process of electricity conversion, starting from the built-in heat loss in the process of converting fuels to heat and the heat to electricity. In short, their study is a classic in thermodynamic understanding.

In the Berry/Fels study, a statistically typical automobile was analyzed, one that weighed just under 3,600 pounds and consisted of the following raw materials.†

Iron and steel	3,244 lbs.
Aluminum	74 lbs.
Copper	54 lbs.
Zinc (die castings)	50 lbs.
Laminated glass	24 sq. ft.

† Upholstery and plastic items were not considered in the study; these items constitute about 100 extra pounds of raw material weight.

They found that, taking into consideration all the energy conversion stages in mining and finishing the raw materials used in automobile manufacture, the total energy cost of the car amounted to—in the energy equivalent of electrical power—about 37,000 thermal kilowatt-hours of electricity,‡ or what they term "thermodynamic potential," which is the total energy conferred on a product by manufacturing. This amount of energy is considerably more than the annual gasoline consumption of an automobile.*

The Illinois Institute for Environmental Quality report estimates that 12,000 kilowatt-hours of energy content of the average scrapped automobile could be saved if car hulks were recycled. "If such a savings could be achieved on all the (roughly) 8 million automobiles produced each year," they calculate that "the annual reduction in the energy demand would be almost . . . 100 billion kilowatt-hours [equal to about one thirtieth of the total energy used annually by all road vehicles]. The power requirement would be reduced by about 11,400 megawatts, roughly the rated capacity of ten very large power plants."[84]

A major finding of the Berry and Fels study was that manufacturing automobiles for permanence—i.e., tripling the current lifetime of autos —would result in great energy savings. They calculated that a 5 percent increase in the present total manufacturing energy devoted to autos would be required to build cars that would last 3 times longer. This means that the cars would require 38 percent less over-all energy (energy both for manufacturing and maintenance—i.e., fuels, batteries, replacement parts, etc.) during their extended lifetime.

Many myths abound concerning the wonders of an improved technology that will save the era of the private car. These myths are usually centered around the "revolutionary" new engines that will be highly efficient and will use little fuel.

The following chart illustrates the comparative energy conversion efficiency of most of today's automotive engines:[85]

‡ Thermal kilowatt-hours differ from electrical kilowatt-hours in that the thermal kilowatt-hours reflect the energy required to convert energy resources into electricity. Electrical kilowatt-hours represent the energy value of electricity alone, not the full energy costs of the full energy conversion process of making electricity. Even the figure cited here does not reflect the *full* energy costs (mining, transport, material transport, etc.) in automobile manufacture, which may boost the energy requirements significantly.

* Dr. Peter Chapman, writing in the British magazine *New Scientist,* says that an energy cost of 23,000 kilowatt-hours (electrical) in car manufacturing would be equivalent to the fuel consumed in operating a 3,600-pound auto for 12,000 miles.[83]

Engine Type	Fuel Burned	Cost Per Horsepower	Energy Conversion Efficiency (Fuel to Useful Power)
Conventional Internal combustion (Otto cycle)	Gasoline	$3	32%
Internal combustion rotating engine "Wankel" (Otto cycle)	Gasoline	$4	30%
Internal combustion diesel engine (Diesel cycle)	Diesel fuel oil	$5	35%
Gas turbine engine continuous combustion (Brayton cycle)	Various	$5	30%
Steam engine continuous combustion (Rankine cycle)	Various	$4	28%
Stirling engine continuous combustion (Stirling cycle)	Various	$6	40%

This table indicates the relative possible energy conversion efficiencies of most known vehicle power plants, but does not include the full energy losses (friction, etc.) that are encountered in use; nor does it include electric-powered vehicles. The following chart compares the way in which energy is actually utilized in a conventional car powered by an internal-combustion engine and in an electric car:[86]

Conventional Internal-Combustion Engine-Powered Automobile

Energy input (gasoline)		100%
Engine cooling and exhaust energy losses	−75%	
Idling and transmission energy losses	− 9%	
ACTUAL MECHANICAL POWER OUTPUT		16%†

† See footnote for the following table.

Electric Automobile		
Energy Input (fuel at power plant)		100%
Average loss of energy at electric power plant and some transmission losses	−67.5%	
Battery cycle: Charging and discharging energy losses	−5.0%	
Motor and controls: energy losses	−10.0%	
ACTUAL MECHANICAL POWER OUTPUT		17.5% ‡

From these tables, several facts are apparent. First, no major technological revolution can be wrought by changing the type of engine used in an automobile. Even the best engine type available—the Stirling engine.*—is only 5 percent more efficient in converting fuel to power than the common diesel engine that is already found in cars, trucks, and trains around the world. The Wankel engine, far from being a revolutionary engine in terms of saving energy, actually wastes more fuel than the standard internal-combustion engine produced in Detroit for years. The primary reason the Wankel engine is now being favored by the automobile industry is because it is of simpler mechanical design, making it cheaper to construct on a mass-production basis. The fuel savings attributed to the Wankel and other "revolutionary" designs are not due to the engines used, but to the fact that the cars themselves are smaller, therefore requiring less energy to propel them.

The second conclusion that can be readily drawn from this table is that the major problem in using energy efficiently in vehicles is the problem of energy losses incurred while stopping and starting, and assorted transmission and friction losses in operation.

Third, the electric car offers no salvation because, as noted in the

‡ Even these figures may be too high. Actual power delivered to propel the vehicle, whether gas or electric, has been estimated at lower than 10 percent of the total energy content of fuel supplied.

* The Stirling cycle engine, however, offers the greatest long-range hope for vehicle engine efficiency, particularly for use in mass transit systems. The Stirling engine was originally patented in 1816 by Scottish Rev. Robert Stirling and has been extensively developed in this century by the N. V. Philips Company of the Netherlands and the Ford Motor Company in the United States. Since it is an externally fueled continuous combustion engine, it can operate on a wide variety of fuels, including hydrogen.[87]

table, it carries with it the double penalty of energy losses in actual use as well as the energy losses inherent in the original conversion of fuel to electricity, thereby using energy at about the same rate as the internal-combustion (Otto cycle) engine.

A Strategy for the Future

Changing the orientation of America's transportation systems to reduce energy and resource use and minimize the impact of transportation on the environment will not only require a change of public attitudes toward the "indispensable" character of the private automobile, but will also require the overhauling of major governmental and economic institutions.

A beginning would be shifting the car population of the nation to machines of smaller size, which would significantly alter energy and resource demands.† An 1,800-pound car (the size of small American and foreign import models) requires approximately half the energy necessary to propel it as a typical 3,600-pound American car. Energy used for manufacture and raw material use would also be reduced. According to government estimates, if just one half the cars expected on America's roads in 1980 were to have an average fuel economy of 22 miles per gallon of gasoline consumed (compared to today's national average of 14 miles per gallon), the annual fuel savings would equal 17 billion gallons.[88]

However, at some point, increases in population (of people with cars) would erode the significance of the switch to smaller cars. The nation should immediately face the prospects of developing new transit systems and new cities to maximize the possibilities of life without the car. Today, more than half the trips made by cars are less than 5 miles in length, and fully 82 percent of U.S. commuters travel by car.[89] Numerous alternatives to this mass use of auto travel are possible. People can pedal bicycles on shopping trips or to work, using human energy rather than fossil fuels. New cities can be designed so that car travel is not necessary. Car pools can be better utilized so that workers can share trips to work.

The best hope for reducing the problem of the automobile, however, lies with advance planning. To date, most of the transportation revenues of the federal government, local governments, and industry have been spent on transportation planning that has resulted in the least efficient use of existing transport alternatives. The Highway Trust Fund is the worst offender, and it may be years before its funds are

† Congressman Charles Vanik of Ohio introduced legislation in the Ninety-third Congress that would tax on the basis of the gross horsepower of automobiles.

more equitably distributed in developing mass transit systems. When President Richard Nixon signed into law the Federal Aid Highway Act of 1973, only $6 billion was authorized for mass transit out of a 3-year, $23 billion program. Much of the mass transit money was earmarked for buses, not rail transit. The bulk of the $23 billion was allotted to increasing the size of the nation's concrete and asphalt vehicle arteries. Several new programs have recently been initiated, including an economic growth program for communities under 100,000 population, allowing federal funds for intensive highway development.[90]

The philosophy of the Department of Transportation could hardly be less well-grounded. New communities in the future will require fewer highways and better planning aimed at eliminating use of cars altogether.

A complete reorientation of federal priorities in transportation planning is necessary. The federal transportation budget is devoted almost exclusively to highways and airports, which receive four fifths of the funds. Mass transit systems and railroads get one twentieth of the federal total. Federal transportation research funding seems designed to waste as much energy as possible: Air transport research gets 60 percent of the funds, followed by 17 percent for highway planning. Mass transit and railroad research split 12 percent of Department of Transportation research outlays.[91]

The present network of energy-intensive automobile, truck, and air transport is due in large part to heavy government subsidization of these wasteful modes. The Joint Economic Committee of Congress recently released a compilation of studies on government subsidization in transportation areas. A study of the Interstate Commerce Commission by George W. Hilton of the University of California at Los Angeles pointed out that this agency's policies resulted in massive inefficiency of operation of both the trucking and railroad industries. Hilton estimated that both industries operated at only 50 percent capacity because of the Interstate Commerce Commission's (ICC) often archaic and ill-conceived regulatory rules, which have only succeeded in producing great industry monopolies. He succinctly summarized the effect of the ICC's policies: ". . . The present organization of the transporation industry is in the nature of a major tax on the economy which results mainly in waste, rather than in a subsidy which has major benefits for society or for many individuals."[92]

The Federal Aviation Administration subsidizes commercial airlines in the United States, as well as general aviation (private fliers, and business planes). According to Jeremy Warford of the International Bank from Reconstruction and Development, subsidies to general aviation alone amount to more than $640 million annually. What is amazing about the federal subsidy is that businesses with expensive private jets receive the greatest subsidies. The taxpayer pays more than $38,000 yearly for the operation of each business turbojet in operation in the United States.[93] Thus, the worst "energy offenders"—operators of inefficient aircraft—get a free ride from the unsuspecting public.

Elimination of the elaborate government subsidy structure in transportation will be a significant step in developing a rational transportation system. Future city and community planning should eliminate the built-in dependence on the automobile as the only mode of transport. Only through advance planning can society become less dependent on the tremendous energy devoted to transportation.

Agriculture and Energy Use: From Soil to Oil

All life is made possible by solar energy, and the harnessing of solar energy through agriculture permitted the concept of stable human cultures. As anthropologist Leslie White observed, "after about a million years of cultural development, during which time cultural systems were activated almost exclusively by energy from the human organism, a radical change took place: solar energy was harnessed in the form of plants and animals."[94]

The history of agriculture is a history of various levels of energy use, expressed through technologies that man has developed to maximize his own energy and produce more from the soil.

If we divide the development of agriculture in the Western world into three ages, the first would appear as the use of the direct flow of the sun's radiant energy—both directly, to generate photosynthesis; and indirectly, in the form of food converted to human and animal energy—to subsidize the growth of more food, thus expanding the efficiency of the food-growing process. This age stretches over many thousands of years, from the Neolithic period to the end of the nineteenth century.

The second age is the period of transition to mechanization, a period in which the energy subsidy in agriculture shifted from animate energy to fossil fuels, as advanced farm machinery—powered by fossil fuels—became available. Compared to the thousands of years of evolution of the first great agricultural epoch, this period was a very abbreviated one, stretching only from the late nineteenth century to the middle of the twentieth. It might best be described as the period during which solar energy came to be *supplemented* by stored fossil fuel energy.

The third agricultural age—the one in which we are now living—is only about three decades old, and is in reality the extension or culmination of the second one. This current period, however, represents much more than just the more sophisticated mechanization of its predecessor. It involves supplementing natural energy with synthetic fossil fuel energy through fuel for machines and the accompanying extensive use of synthetic chemicals and fertilizers in the farming cycle.

Until the advent of the third age, technology had not fully invaded the balance of the natural system. Fertilizers were organic; crop rotation was practiced; and pest control was accomplished by natural means—the best method being the planting of diversified crops so that pest attacks would not cripple entire harvests. Until the third age, the only synthetic pesticides used were metal compounds—lead, copper, arsenic, etc.—which not only destroyed pests, but acted as an immediate poison to people and animals as well, so that they were used sparingly and with great caution.

Fossil Fuel Subsidization

The decade of the 1940s proved to be the historic dividing line between the second and third periods. Since then, agriculture has been based on almost total subsidization by fossil fuel energy.

This subsidization takes three major forms: (1) direct use of fuel to power the highly mechanized farm equipment; (2) the massive use of chemical insecticides and herbicides derived directly from fossil fuel resources; and (3) the indirect use of the enormous increments of energy required to produce the synthetic fertilizers upon which modern agricultural production is largely based.

Adding the subsidy of the synthetic energy resources (fossil fuels) to the natural solar energy of the agricultural system has resulted in a phenomenal increase in productivity. Between 1920 and 1940, crop output per acre remained relatively unchanged. Since 1940, crop output has risen, with an annual growth rate of almost 2 percent per year.[95] In 1972 was recorded the highest yield in American history of agriculture. Yield per acre was 96.9 bushels per acre for corn (the 1969 yield was 83.9 bushels per acre); 32.7 bushels per acre for wheat; and 28 bushels per acre for soybeans.[96]

Writing in the journal *Environment*, economist Michael Perelman of California State University at Chico sums up the lessons of the energy changes in agriculture over the past few decades. Perelman notes that agriculture is not the main user of energy in U.S. society, and contends that the equivalent energy content of 150 gallons of gasoline does not seem extravagant to feed and clothe one person. "But," he adds, "agriculture could be an energy-*producing* sector of the economy. The crops we harvest capture the energy of the sun and store it in a useful form so that we can use it to nourish our bodies or to perform some other service for us. And yet our agriculture has become a major consumer of our stores of energy. In fact, agriculture uses more petroleum than any other single industry."[97]

On the other hand, as noted by former Secretary of Agriculture Clifford Hardin, agriculture "does an amazingly efficient job of providing food." He backs this up with the observation that 1 person can raise 60,000 to 75,000 chickens in a modern, mechanized broiler feeding system, that 1 cattle feedlot operator can take care of 5,000 cattle, and that 1 dairyman can manage 50 to 60 milk cows.[98] American crop yields, fiber, food, and domestic animal production are at all-time highs. But it is made possible only by the constant injection of massive doses of highly concentrated synthetic fossil fuel energy—in the form of chemicals, fertilizer, fuels, and electricity.

The real significance of the third age of agriculture is that this great enterprise, which historically has determined the level of population by the organic limits of the land's carrying capacity, no longer plays this role. These population limits are now determined by the ability to inject synthetic energy into the land and the farming process.

In short, agriculture today has shifted in a very real sense from soil to oil.

Agribusiness

A unique but logical product of this third age of agriculture is the introduction of the gigantic American corporations, or "conglomerates," into the role of superfarmer under the label of "agribusiness." It is logical that these giant corporations would turn their interest to agriculture. It is by far America's largest industry. Total value of farm output in 1970, for example, was $53.1 billion.[99]

Agribusiness denotes large-scale farming, which is increasingly controlled by corporations involved in the totality of the process—including the use of large machines for cultivation and harvest; manufacture of nitrogen fertilizers and other inorganic chemicals; breeding of special genetic strains of plants; and production of synthetic pesticides.

This "seedling to supermarket" industry was unknown in America only a few decades ago; but by 1969, the U. S. Department of Agriculture reported that no less than 81 million acres were farmed by corporations. The acreage under corporation management now reaches upward of 10 percent of the total U.S. farmland—and is growing.[100]

Not surprisingly, corporate conglomerates specializing in energy resources are becoming increasingly interested in industrialized farming.

A typical example is the Tenneco Corporation. Tenneco is a leading American energy conglomerate with substantial holdings in oil, gas, and uranium lands. It also owns the Newport News Shipbuilding Company, which builds oil and gas tankers. It has now become a leading agribusiness corporation as well.

Tenneco farms more than 1.6 million acres of land in California, New Mexico, and Arizona; subsidiaries of the corporation manufacture farm machinery, agricultural fertilizers, and chemicals. They package fruits and vegetables and ship them across the country—by jet freight![101] Other "energy conglomerates" have become similarly involved in agribusiness, which is today more an energy industry than an agricultural one.

Agricultural Mechanization

The basic use of energy on the farm is as fuel for the vast array of machinery used in agriculture, ranging from tractors and reapers to massive threshing machines. As economist Perelman points out, while the actual number of farm acres planted in the United States has not changed appreciably since 1950, the total value of farm machinery used in working that land has almost tripled.‡[102]

This use of more and bigger machinery is largely responsible for the trend cited earlier—the withering away of the small farm. In 1940 the average farm size was 167 acres. It is now 400 acres. The small farmer who cannot afford to purchase the specialized equipment is forced to rent it, which puts him at a considerable economic disadvantage. Consequently, he has gradually been replaced by large-scale farmers and agribusinessmen who have access to the large amounts of capital required to compete.

The total fossil fuel use on farms is equal to more than 10 million BTUs per acre under cultivation. Perelman calculates that powering the 5 million tractors in the United States requires 8 billion gallons of fuel. This amount of synthetic fossil fuel energy is equal to the amount of chemical energy contained in the food products of American agriculture. Thus, *the use of tractor fuel alone on U.S. farms has surpassed the energy conversion of sunlight for a given unit of agricultural land.*

Before this third age of agriculture, the sun was King. Farm equipment—whether powered by human muscle, draft animals, or inanimate machines—served as a supplement to the natural energy of the biological system. Now the sun's energy system has become subsidiary to the synthetic energy system.

‡ The value of farm implements and machinery was $12.1 billion in 1950; by 1971 it was estimated at $33.8 billion.[103]

About 2.5 percent of the nation's electricity is devoted to agricultural uses. The Edison Electric Institute (EEI) says that electricity now performs some 400 farm tasks that used to be performed by manual labor. A significant role of electricity on the farms is to control the climate of farm facilities. "In the controlled environment of farm buildings," the EEI contends, "the farmer in a few months is able to do what formerly took a year or so."[104]

The Chemistry of Superfarming—Fertilizer and Poisons

As noted earlier, the kind of farming that agribusiness has developed is characterized by the intense application of chemicals to soil in order to raise productivity.

The most graphic reflection of this age appears in Barry Commoner's table of resource shifts, which shows the use of synthetic organic chemicals and pesticides increasing 495 and 217 percent, respectively, between 1946 and 1968.[105] These new synthetics are all derived from fossil fuel energy resources.

The use of nitrogen fertilizers also increased 534 percent during this period, according to Commoner. This product, too, while not derived directly from fossil fuel, is available only as a result of high-technology, energy-intensive processes fueled primarily by the fossil fuels.

The availability of these energy-based synthetic pesticides, chemicals, and fertilizers has drastically changed farming methods. No longer does farming rely primarily on sunlight and the organic materials provided by photosynthesis.

In *The Closing Circle,* Commoner points out that in 1949 about 11,000 tons of nitrogen fertilizer were used to produce crop yields that required the application of 57,000 tons—more than 5 times the amount initially required—to sustain the same yield in 1968. In Illinois in 1949, 20,000 tons of nitrogen fertilizer produced 50 bushels of corn per acre; in 1968, it took 600,000 tons of nitrogen fertilizer to get 93 bushels of corn per acre.[106]

Two facts are obvious from this comparison of fertilizer use. The first is that much of the nitrogen was not used by the crops, and was leached into the soil and ground water. The excess nitrogen contributes nitrate pollution as a central by-product of the agricultural system—one of the major water pollutants in the country.*

Second, the corn plant (as well as other crops) has a limited capacity for absorption of the chemical fertilizer, and in order to

* Nitrate pollution from another major agribusiness practice—central feedlots—is discussed in the section on solar bioconversion, Chapter 6.

extract more bushels per acre, very high amounts of fertilizer must be added. Commoner points out that the efficiency with which the crops used fertilizer in 1968 compared to 1949 declined fivefold. But the farmer has little choice in the matter; he is competing in an economic system that is based on the massive use of synthetic chemicals. The economic break-even point is about 80 bushels of corn per acre, so the farmer *must* use the triple dose of fertilizer to survive.[107]

This high use of fertilizer to extract marginal yields from the crop is considered unimportant in an era of cheap fuels, an era in which the economy is based on dwindling fossil fuel resources—from which the fertilizers are derived.

The change from small farms to large ones, from natural energy to synthetic energy, is more than a quantitative change; it involves the qualitative change of responses to insect, weed, and fungus control. Big farming means monocultural farming—the planting of a single crop over large land areas.

Denzel E. Ferguson, director of the Malheur Environmental Field Station in Burns, Oregon, says that the monocultural planting "provides virtual cafeterias for our insect competitors. Also a monocultural type of agricultural environment, because of its ecological uniformity, is poorly suited to the proliferation of the natural enemies of insect pests."[108]

The small scattered farms of earlier years that were planted in diverse crops were not susceptible to destruction by a single species of pest. This condition is markedly changed with the emergence of large-scale monocultural farming. Millions of years of natural evolution result in inherent pest protection in diversified systems. Monocultural systems—e.g., a single crop of wheat or corn stretching for many square miles—are deprived of the natural genetic protection. The pesticides and other chemicals are applied in hopes of compensating for natural protection.

What will result from this chemical use? For the present, we have increased crop yields, but what the future will bring as a result is uncertain. Denzel Ferguson speculates that

> . . . the most frightening prospects of long-term changes caused by pesticides involved fundamental ecological processes which affect all life. In terms of geological time, what might be the effects of continued use of pesticides upon the process of soil formation? Every ounce of fertile soil contains millions of bacteria, fungi, algae, protozoa, and small invertebrates such as worms and arthropods. These organisms play essential roles in the maintenance of soil fertility and soil structure. When we realize that all these organisms carry on innumerable

biological processes as they perform their functions, the odds become very substantial that introduction of a biocide will have some disruptive influence.[109]

One well-known effect of the large-scale pesticide, or biocide, application is the fact that many pests become immune to the pesticides, and emerge stronger than ever in the struggle for their own energy in the environment. Ferguson says that more than 224 species have developed reistance to various insecticides in the world, of which 129 are agricultural pests and 97 have medical significance. In Mississippi, the application of pesticides in cotton farming has killed off everything but resistant species of fish living in drainage ditches. Woe to the people who eat these surviving fish, which are literally loaded with potent, concentrated doses of lethal chemicals.[110]

Another problem associated with the use of the chemicals in industrialized farming is the loss of the topsoil itself. According to the National Academy of Sciences, one third of the valuable U.S. farmland topsoil is already gone.[111] Nothing has replaced it. All that is left from the third age of agriculture is a deadly residue of chemicals, resistant pests, and assorted concentrations of pollutants, such as nitrates. Every year a bigger and bigger dose of chemicals is necessary to sustain monoculture crop yield—at the expense of the entire system's stability.

As is the case with other monocultures, success is bought at a high price. It can survive for only a given period before the bitter night sets in. It is not certain how long the fossil fuel energy-subsidized age can prevail in America, but it is known that, when it has ended, there will be little legacy from the land for farmers of future generations to till. The soil will have been wrung into depletion for the windfall gains of a few decades.

Are There Alternatives?

Few studies have been made to determine the economic price of lowering the energy base of modern agriculture. Entomologist Robert van den Bosch of the University of California at Berkeley has analyzed the use of pesticides in California's San Joaquin Valley cotton industry. He thinks that lowering the use of pesticides and substituting a diversified, integrated pest control approach will save the growers and society $10 million to $15 million annually, out of a total crop value of $13 million. Of the failure of monoculture farming, he says that it is "imperative that we abandon the fruitless search for a simple all-encompassing solution to the insect problem. We must instead revert to the more flexible integrated control approach in which a variety of techniques, materials and information are meshed together into systems which maintain environmental integrity, while managing pest populations—in such a way as to minimize their detrimental effects."[112]

On a national scale, agricultural specialist J. D. Headley told a U. S. Department of Agriculture symposium that the planting of 12 percent additional land would reduce the use of insecticides by up to 80 percent, while still maintaining current food output. This 12 percent additional land would amount to about 60 million acres—the farm acreage, he noted, that is now diverted into the federal "land bank" through federal subsidies paid for keeping land out of productivity.

The cost of maintaining the agricultural subsidy program is about $3 billion to $4 billion per year, and the cost of the pesticides applied to 60 million acres would be about $750 million. These costs would be eliminated by planting the soil bank's additional acres using diversified farming techniques, although the total cost of planting the acreage would be more.[113]

Understanding the energy balance of agriculture helps explain many factors, including one that is not obvious to many. Since World War II, the United States has encouraged many other nations to buy agricultural products from us. These exports of grain and agricultural products to other nations are in fact exports of essentially raw energy resources. Since the American monocultural agriculture system is propped up by synthetic rather than solar energy, farm products are simply extensions of fuel resources. In 1971, exports of agricultural products were valued at $7.8 billion, 14 percent of which was under federally subsidized programs—mostly grains.[114] Since 1971, even larger food-export programs have been subsidized by the government, including large Soviet and mainland Chinese grain purchases. In mid-1972, the federal government arranged the sale of $1 billion worth of basic agricultural grains to the Soviet Union. The partially government-subsidized deal was struck at $1 billion, and by mid-1973, as the remainder of the grain was being shipped to the Soviet Union, agricultural experts valued the product at almost $3 billion. Thus, a net loss of $2 billion was incurred.[115] As fossil fuels continue to be diverted to basic agricultural uses, the ultimate effects of subsidized food exports will be felt as a facet of the energy crisis. Until the renewable agricultural system that the United States once practiced is readopted—as opposed to the present system—exporting agricultural products will only be exacerbating the energy supply crisis.

By eliminating subsidized exports of foods and adapting to a less synthetic and less energy-intensive agricultural system, the country will benefit by a more favorable balance of payments, since fewer of our valuable energy resources will be devoted to the agricultural energy base that underlies much of our export program.

The population-carrying capacity of this nation can no longer be met by sun-powered agriculture; only through the input of fossil fuels are these high crop yields possible. Yet we are faced with a curious dilemma:

The monocultural agriculture system that we have so proudly evolved cannot long survive under its present form; soil depletion and chemical pollution pose grave threats. In short, the system is not an ecologically stable one. On the other hand, a return to complete decentralization, eliminating the fossil fuel subsidy (including the chemicals), would cut productivity to such an extent that endemic famine would be a likely consequence.

The only prudent alternative is to apply the evolutionary lessons of diversity and decentralization. The nation must slowly and carefully begin an orderly return to a stable, decentralized system—hopefully before the full impact of the shortages of fossil fuels have plunged our obsolete monocultural agriculture system into precipitous decline. When the fossil fuel energy agricultural subsidy is removed, the system will fail. Today, as fuel prices rise and the effects are felt in the rising costs of food, the beginnings of the end of the third age of agriculture are already being felt.

Flourishing agricultural systems in other countries use the principle of crop diversity and decentralization with great success. Noted scientist and author Dr. René Dubos, of Rockefeller University in New York, points out that his own birthplace, Île de France, has been under cultivation since Neolithic times, but the land is still rich and fertile. The secret is diversified farming, which has not robbed the soil of its nutrients.[116]

The other side of the industrialized farming coin is complete decentralization of agriculture, with maximized organic farming. A nation whose farming practices closely approach this position is mainland China, where pesticides and chemical fertilizers are probably used in far less than 10 percent of the agricultural undertakings. In China, decentralization is the key word; agriculture is based to a large extent on human and animal labor, and in contrast to Western methods, the farms appear more like millions of gardens, rather than great blocks of single crops; but more than 80 percent of the population is involved in agriculture.[117]

It seems inconceivable to many that U.S. agriculture could begin a transition to such a decentralized system, even if it became national policy and intent—which also seems equally unlikely, given the present urban/rural balance and the inherent resistance to change. In 1970, our entire rural population was only 18 percent of the total population; and actual farm population represented less than 5 percent of the total.[118]

Conclusion

What have been the effects on the American consumer patterns of the fossil fuel energy subsidy in this third age of agriculture? Returning to the Commoner/Corr/Stamler study for the American Association for the Advancement of Science (AAAS) (see p. 86), the following shifts in agriculture-related consumption were recorded for the 1946–68 period:[119]

Consumption (1946–68) Per Capita	Percent Change
Calories	–4
Protein	–5
Meat consumption	+19
Eggs	–15
Grain	–22
Milk and cream	–34
Butter	–44
Wool fiber (clothing)†	–61
Cotton fiber (clothing)†	–33
"Natural base" synthetic fiber (made from cellulose)†	–5
Natural fat for soap products‡	–71

From these statistics, it is apparent that, on the average, Americans ate no better in 1968 than they did in the 1940s. In fact, since per capita consumption of both over-all calories and protein declined, Americans seem to be worse off nutritionally now than they were a few decades ago. The only food item on the list to have increased in per capita consumption is meat, indicating that prosperity since the Second World War has brought an increase in meat consumption, but no corresponding increase in the usable energy of food.

The striking items of the list are the decreases in per capita consumption of fats for soap and wool, cotton, and cellulose-based synthetic fibers. All of these items were once a part of our natural, richly endowed agricultural energy heritage. The fertile soil of the United States his-

† 1950–68
‡ 1944–64

torically produced not only the basic foodstuffs for the country, but also the fibers and animal wool for clothing.

With the advent of the current high-energy society, this has been dramatically altered. We no longer wear and use the natural products of sunlight-induced materials, but the synthetic products of fossil fuel energy. For example, compare the following items of change:[120]

Consumption per Capita	**Percent Increase**
Synthetic fibers (1950–68)	1,792
Detergents (1952–68)	300

The natural fibers of wool and cotton have been almost completely replaced by synthetic fibers, which constitute a further use of fossil fuels, since they are made by converting natural gas and petroleum into plastic "threads." Not only have the farms been transformed into appendages of the energy monoculture, but many farm products have been simultaneously replaced by artificial materials that are themselves made from or with the application of fossil fuels!

In the case of detergents, whose use grew 300 percent from 1952–68, these synthetic chemicals replaced the agricultural fats that went to make up soap.

Popularization of the fossil fuel products is taking two tolls:

1. It further exacerbates the energy crisis by draining off stocks of valuable fossil fuels to replace materials that could be harvested from the natural system with the use of sunlight rather than synthetic energy.

2. It creates severe pollution problems, because synthetics cannot be degraded by natural organisms into harmless forms as are the natural products of sun energy in the agricultural systems. Even the so-called "biodegradable" synthetic chemical detergents may be more harmful than the first group of nonbiodegradable detergents that they replaced, since the new detergents contain benzene, which in water is converted to toxic carbolic acid, an agent that is potentially lethal to fish. "In fact," Commoner observed, "the new degradable detergents are more likely to kill fish than the old ones, although they do not produce the nuisance of foam."[121]

Analysis of fossil fuel use just in farming, with the inclusion of fertilizers and pesticides, electricity for henhouses and milking machines, and fuel for tractors, is only a part of the food supply system. The modern way of delivering food products to the consumer involves significant energy use at all stages of growing crops or animals, processing of food, transporting it, and delivering it to consumers.

As has been noted, the food industry is one of the major energy consumers in the United States. Eric Hirst of the Oak Ridge National Laboratory, in analyzing government economic data for 1963, concluded that food-related activities that year consumed 12 percent of all the nation's energy and more than one fifth of the electricity. In 1963, about six calories of fossil fuel were required to get each calorie of food to consumers.[122] The level is much higher today.

Calculating the level of energy use in food production and distribution entails the consideration of all the energy used in transportation, a wide range of industries, heating and air conditioning of buildings (from farm buildings to supermarkets), and many other activities. One unmistakable point emerges: The growing agribusiness industry has transformed diversified agriculture into a great industrial machine that offers cheap food to consumers only at the expense of very high energy use. The system of industrialized farming is only possible through national policies of accelerated economic growth and federal subsidies that enable large agribusiness corporations to grow food and market it more cheaply than small farmers do.

Consider the energy costs involved in a typical transaction in a Florida supermarket in the summertime. The consumer* purchases a plastic-wrapped package of spring onions, identified clearly as having been grown in Mexico, packaged in California, and shipped (most probably by air) to Florida. The agribusiness corporation—in this case a large oil company—is able to engage in this high-energy-use industrial and transportation enterprise and get the product to the Florida consumer at a lower cost than can the local farmer in the Sunshine State! This is borne out by the fact that locally grown green onions at a neighboring local supermarket were more expensive on that particular summer day than those grown in Mexico!

Even the supermarket in which the purchase is made glitters with the artifacts of high energy use: high lighting levels; full air conditioning (even a perpetually open storefront, which attempts to confine cool air in the building with massive air blowers, requiring the use of more electricity both for the fans and for air conditioning); open freezers and refrigerated coolers; and containers of aluminum and plastic representing still more energy conferred on the packages in manufacturing.

Yet this system of corporate farming and corporate marketing is cheaper under the American economic system than the local production and distribution of agricultural products, which could significantly reduce the rate of energy use.

* In this instance, the author.

The potential for conserving precious stocks of limited energy from fossil fuels (and other sources) is great, but it can be realized only by significant changes in the socio-economic system, including changing the present government incentives structure, which rewards waste at almost every level of the food supply system.

American agriculture could become an enormously more productive system—without the present waste of precious resources—were consumers more aware of the price paid for the convenience of plastic-wrapped vegetables and other products shipped from afar. A return to the local greengrocer, who used to be supplied by local farmers before the advent of agribusiness, would be a desirable alternative course for the nation. It would provide more farm employment in many productive areas of the country where employment is now rendered impossible by artifically cheap energy sources manipulated by powerful corporations.

Changing the basis of American agriculture from the highly centralized, monopolistic agribusiness system to a decentralized base will involve sweeping political changes. A recent study of American agriculture by the independent, public interest Agribusiness Accountability Project in Washington, D.C., concluded that at the center of the problem was the government-subsidized land grant college/agricultural complex, comprising both well-endowed university programs and agricultural experimental stations across the nation. The project concluded in its report, *Hard Tomatoes, Hard Times,* that this complex serves "an elite of private, corporate interests in rural America, while ignoring those who have the most urgent needs and the most legitimate claims for assistance."[123]

The philosophy of the federal government is to replace farm labor by increasing mechanization, constantly displacing jobs for people by the energy-intensive machinery owned by the oil companies and agribusinesses. A U. S. Department of Agriculture task force commented that "labor was viewed as a physical rather than a sociological resource."[124] In other words, the physical welfare of farmers was viewed as subsidiary to the values of the nonhuman components of New Agriculture: the machines. There are some signs that awareness of the dilemma is evident in the federal government, even though no change in policy has been suggested. Dr. Don Paarlberg, the USDA's director of agricultural economics, commented a few years ago:

> I think we should reorient our research, working more on agricultural adjustment, rural poverty, and world agricultural development. The invention of new institutional forms that would help more family farms to survive the technological revolution, and the development of

new ways to help farmers preserve their decision-making role seem to me priority items for research and policy.[125]

Localized farming would not only save energy and promote employment but, by encouraging the practice of more diversified farming, might save the nation from a possible collapse of the monocultural farming system. Small-scale, diversified farming is an eminently rational approach, and policies to assure its success should be pursued by individuals, townships, and governments at every level—from federal on down.

A continuation of the present trend toward bigger and bigger industrial farms portends disaster. With the knowledge gained by centuries of agricultural experience, man now knows the reasons for such debacles as the Irish potato famine of the nineteenth century. A major difference today is that a failure of monoculture farming in the United States would find no generous, overproducing countries to come to the rescue—as Ireland found in the United States.

Energy Use in Homes and Buildings

With the advent of cheap, available fuel sources to heat homes and buildings, and the spread of electrically powered devices ranging from elevators to the various appliances in the "mechanical core" of the modern homes, classic standards of construction have given way to methods of building that are characterized by one quality: their total dependence on finite sources of energy.

Both the Houston Astrodome and the mobile home are artifacts of high-energy civilization. In a world without cheap fuels and electricity, neither would be possible. In most countries of the world, both are out of the question today; and, if future sources of power are not found that are as cheap as the sources currently in use in the United States, they will quickly become little more than memories of the high-energy age we are living in today.

The extent to which energy sources are used in the homes and commercial buildings may be gauged by the statistics on energy use. About 30 percent of all the energy used in the country is used to heat, cool, light, and operate machines within buildings and homes of America. This does not include the energy used to manufacture the materials that are used in the construction of the buildings, or the energy required to make the machines that use energy transmitted for their operation—only the energy that daily is used for *maintenance* of the structures.

One of the characteristic historic traits of human civilization was the ability of people across the world to minimize the use of outside sources of energy by constructing homes and buildings that meshed with the prevailing climates. This great architectural art has virtually been lost in the United States, and with the passing of each year homes and buildings show a striking uniformity that bears no relationship to climatic requirements. This subject is taken up in a later chapter in greater detail, but it is instructive to note that the tremendous use of energy in our structures is due to a turnabout in historic adjustment to climate, a turnabout wrought by the rise of high-energy civilization, which substituted the convenience of machines for thinking and design.

Homes

Energy use in the seventy million homes of America is substantial, taking about one fifth of the national total for all sources of energy and about one third of all electricity generated. These figures do not include the energy required to build the houses, or the mechanically operated devices within them.† Regardless of the source of energy delivered to the home, the actual shell of the house—its walls, ceilings, attic, basement, windows, construction material (wood, metal, concrete, etc.)—is important in determining how much energy the home will require to keep heated in winter or cooled in summer. In other words, how well is the home insulated against the effects of climate?

A poorly constructed house with thin walls and little insulating material will require more energy to keep warm in winter due to loss of heat. In summer, the same factors of good construction and insulation save energy in the air-conditioned home. Whether or not a house is air conditioned, good insulation and quality construction will provide a cooler internal environment. The effects are more pronounced with air conditioning, because heat gain (from solar heat striking the house and from loss of cool air through walls and windows) will require more energy to operate the air-conditioning system.

As is true with almost all areas of energy use, the nature of high-energy American society—particularly the economic system—encourages waste through incentives for "first cost" purchase as opposed to "life cycle" costing. An individual interested in low lifetime costs for heating a home will specify quality insulation and construction for many reasons of durability and low over-all costs. The initial purchase price

† The energy required to build homes and machinery is categorized as an industrial use of energy, and energy required to transport materials and machines to homes is attributed to transportation uses of energy. Not discussed here is the energy waste inherent in the disposal of solid wastes, already more than five pounds per person daily.

of a home with good insulation will be higher than the purchase price of a home with less insulation. Over the lifetime of the home, however, more money will be spent to heat and cool the home with poorer quality design and insulation.

Factors affecting the loss of heat through homes in winter have been tabulated by the National Bureau of Standards (NBS).[126] It recommends that homeowners use the maximum possible amount of weather-stripping and caulking around doors and windows. Improper seals on doors and windows can result in the loss of 15 to 30 percent of the heat generated by the home's winter heating system. Storm windows cut the heat loss from windows in half. Not only will storm windows result in lower heating costs, but an investment in storm windows in most areas of the country will be paid off within a decade of use (at 6 percent interest), and thereafter will yield a yearly dividend in reduced fuel costs. NBS suggests that the installation of six inches of insulating material above the ceiling of the home will be a sound investment. In all areas of the country with relatively mild winters (i.e., where the average temperature between October 1 and April 30 is 45° F.), the investment in insulation, NBS says, will pay for itself within one year, and will return a dividend in reduced fuel use each year thereafter.

Another potential insulating feature in a home is the use of draperies. Closing window draperies at night will provide a small savings in energy, and will also provide greater comfort for persons sitting near drapery-covered windows. Air leaks through attic air vents and other attic openings can be a significant source of heat loss. Warm air is lighter than cold air, and flows upward; in a home, this flow can mean a steady loss of heat throughout the winter heating season. All openings from occupied parts of a home to the attic should be closed.‡

Another source of heat loss in homes can be the improper insulation of hot water tanks. Hot water leaks can also mean energy and water losses. A leak that fills an ordinary cup in ten minutes wastes more than three thousand gallons of water per year.

The National Bureau of Standards recommends essentially the same construction and insulation features for increasing the energy efficiency of home air conditioning systems. A home that is mechanically air conditioned faces one great problem: Air conditioning is essentially a mechanical battle against summer heat from solar radiation. In this war against the sun, the adversary puts up a mighty struggle; and solar heat falling on the roof of the home may be ten to twenty times greater—

‡ In addition to loss of energy, the warm air flowing into the attic can condense and deposit as frost, later melting and causing water damage to clothing and objects stored in the attic.

in energy content—than the energy used to power the home's air conditioner.

Important points to keep in mind in house design to maximize the cooling ability of the air conditioner are the colors the roof and sides of the house are painted. Light colors reflect sunlight, and tend to keep the house cooler in summer. Sunlight striking a rooftop can raise the temperature in the attic forty degrees higher than in the occupied rooms of a home. Good insulation between the attic and the occupied space will minimize the heat gain, which interferes with the air conditioning system; installation of an attic ventilating system will also remove much of the heat. All available measures should be taken in summer to reduce the amount of solar heat that can enter the house, including the use of trees to shade the house; the use of mirrored, reflecting glass to repel solar rays on windows; and sun shading and screening devices to prevent solar radiation from penetrating into the rooms. Light, opaque blinds and draperies drawn during the day can reduce solar heat intrusion by 50 percent.[127]

In short, all insulating materials serve to keep heat inside the home in winter and outside the home in summer.

Recognizing the value of insulating materials, such as mineral wool filler and sheets in houses, led the federal government to adopt minimum standards for federally insured housing. For single-family homes, the Federal Housing Administration (FHA) requires an average of 3½ inches of ceiling insulation and 1⅞ inches of wall insulation.[128] These averages are based on houses in most parts of the nation, and the actual formula for insulation is more complex. Standards adopted in 1971 and 1972 by the FHA apply to houses as well as federally mortgaged housing projects.[129]

However, these federal standards do not apply for any privately financed housing projects or houses. In some states, local codes may require more insulation; but usually standards are lower. This translates as economic losses through energy waste to millions of apartment dwellers and homeowners.

Even the federal standards are not the economic optimum for homeowners. John C. Moyers of the Oak Ridge National Laboratory calculated that for a typical home in New York, wall insulation should be increased from 1⅞ inches to 3½ inches, regardless of whether the house is heated with gas or electricity. The economic optimum thickness of ceiling insulation in a New York home heated with natural gas would be 3½ inches, but a home heated with electricity would need to have 6 inches of ceiling insulation. Moyers assumed that the house would also have floor insulation as well as storm windows (storm windows are not required by the federal standards). The energy saved would be 49 percent more than older federal standards (before the 1971 revisions), and the homeowner would save $32 a year in the costs of natural gas. If the home were heated with electricity, the homeowner would save $155 a year, and energy use would be cut by 47 percent.[130]

Financial savings will continue to grow with the impact of higher energy costs. Even Moyers' calculations are based on energy prices of the early 1970s period, and prices have dramatically risen since then.

On a national level, the improvement of insulation in existing houses, as well as adding storm windows and doors, can have a tremendous effect on U.S. energy consumption.* The National Mineral Wool Insulation Institute's technical committee has outlined a national program wherein all new construction would utilize upgraded insulation, and some existing houses would use more insulating material and storm windows. The institute reports that, between 1973 and 1982, the maximum possible energy savings in the nation would equal 2,780 trillion BTUs, equivalent (in current energy prices, to consumer savings of $5.7 billion. While this goal is probably unattainable, due to the difficulties of converting all older and new houses and buildings, the institute believes that half this is a reasonable goal. If 75 percent of all new construction used good insulating techniques, 25 percent of old housing units were upgraded with better insulation, and 12 percent of all older housing units were equipped with storm doors and windows, the nation's energy bill would be reduced by $3.1 billion in 1982. The cumulative savings from 1973 to 1982 would equal $17.1 billion, while the actual cost to the consumers using the insulating techniques would be $6.4 billion.

The cost of the individual homeowner would average $144 for upgrading existing insulation, and $240 for adding storm doors and windows. These costs would be easily amortized by homeowners within a few years, given the rising prices of energy delivered to the home vs. the savings from insulation.

* An essential point concerning energy and insulation is the energy required to manufacture insulation materials. According to the National Mineral Wool Insulation Institute, approximately 15 BTUs are saved through the use of insulation for each BTU devoted to its manufacture. In some cases, more energy may be utilized for the production of insulation (such as Fiberglas) than can be amortized during the lifetime of the insulated structure—particularly in the case of a home or building that is occupied for only a few years or decades. In any case, the value of energy used for making the insulation vs. the energy it saves by minimizing heat loss represents a trade-off that must be judged by the regulatory agency or builder. It would not be wise to require insulation in housing so expensive that it would prevent home purchases by low-income families.

The institute notes that the penalty for delaying the implementation of the insulation plan is great. A three-year delay in the program would eliminate half the proposed savings in the plan. On the other hand, if the plan were successfully implemented in the nation, the energy savings made possible by improved insulation would significantly reduce the gap between supply and demand for natural gas projected by the Federal Power Commission (FPC). The energy saved would equal 70 percent of the gap in gas availability predicted for 1982 by the FPC, and 90 percent of the gap predicted for 1990.[131]

The nature of mobile home construction provides some interesting analogies of energy use in modern construction. Not only do the mobile homes require construction materials that are highly energy-intensive, but they also require a startling amount of "maintenance" energy for heating and cooling. The walls, roof, windows, and door frames of mobile homes are commonly made of aluminum, as is much of its structural support. Aluminum requires high amounts of energy just to manufacture; and once integrated into the mobile home, it loses heat to the outside twice as fast as wood, the conventional frame material. No giant industry lobby exists to promote natural wooden frames or better and more natural methods of construction, but many industries promote the sale of mobile homes. The private electric utilities alone actively promote the sale of total electrically powered mobile homes through a trade association: the Electric Energy Association in New York. A slick brochure published by this group in 1972 called *Home Sweet Electric Mobile Home* jubilantly noted that electric mobile home sales increased in some states by 200 percent between 1970 and 1971![132]

The association did not point out that electrically powered mobile homes are probably the most insatiable energy-guzzling permanent housing structures in the entire world; nor did it warn unwary purchasers that coming electricity price increases might render their homes economically obsolete.

Heating Homes

While the constructed "skin" of the house or housing structure is the most important design feature in determining the rate of energy use, the energy source used in heating or cooling the home is also a vital element.

The use of natural gas directly for heating hot water in a home is a more efficient use of gas energy than is secondary conversion to electricity; so is the use of gas directly for house heating.† Since electricity

† Although gas in a home furnace is a more efficient form of energy use for space heating than electricity, the use of fuel oil may be less efficient than electricity for

is not a primary energy source, but a secondary one, the inefficiency associated with its production at the power plant must be linked to its ultimate efficiency in the home.

The engineering consulting firm Hittman Associates studied the use of energy in a typical house for a family of four in the Baltimore/Washington area. In the study, conducted for the Department of Housing and Urban Development (HUD), they defined the average home as having two stories, about fifteen hundred square feet of space, wooden construction with good insulation (five inches of ceiling insulation), and equipped with storm doors and windows. The study compared an all-electric home with a "balanced" home, which used natural gas instead of electricity for major heating, cooking, and clothes-drying purposes. Otherwise, electricity was specified for all other purposes, such as clothes and dish washing, air conditioning, appliance operation, etc. The home using natural gas used about half the energy that would be required if it were all-electric. The major cut in energy use was in the central heating system, where the direct use of natural gas in a home furnace cut two thirds of the yearly energy use of the electric system. The gas hot water heater, clothes drier, and cooking range used about half the energy of the all-electric system.[133]

Notwithstanding this potential savings in energy, the Federal Power Commission (FPC) continues to advocate the development of all-electric homes, and predicts that construction of electric homes will surge from the 1970 total of just over four million homes to twenty-four million in 1990.[134] At the same time, natural gas has been increasingly allocated, not for direct home energy purposes, but for burning at electric power plants. The FPC prediction that 40 percent of new dwellings in the 1970s and higher percentages thereafter will be all-electric spells energy disaster, as the already overtaxed energy resources of this country are diverted into the production of wasteful electric power.

Energy-Consuming Appliances

The assertion that Americans are reliant on electrical energy is an understatement. Each year, the number of gadgets, home appliances, and heating/cooling devices requiring electricity mounts. The following chart indicates the growth in per capita electricity for selected house-

---

suburban home heating when the costs of transporting oil in tank trucks is taken into consideration. Although the author has not been able to locate precise data on the energy used to transport oil to homes, it is possible that this may require enough energy use to offset the gain of using oil directly. Natural gas is pipelined with little parasitic loss of energy, and constitutes the most efficient form of household heating energy.

hold electricity uses. The first column for each year shows the relative saturation, or percentage of homes with the selected electrical devices; and the second column indicates the electrical consumption per year in kilowatt-hours (electrical) (KWHE):[135]

	1960 Saturation	ANNUAL Per Capita KWHE Consumption	1970 Saturation	ANNUAL Per Capita KWHE Consumption
Electrical resistance heating	1.5%	41	6.1%	190
Electric water heating	18.6%	232	29.6%	416
Electric ranges	35.6%	118	52.7%	197

	1960 Saturation	ANNUAL Per Capita KWHE Consumption	1970 Saturation	ANNUAL Per Capita KWHE Consumption
Air conditioning total	14.8%	59	42.6%	191
Air conditioning (central)	2.0%	11	5.9%	37
Air conditioning (room)	12.8%	48	36.7%	154
Freezers	22.1%	59	29.6%	92
Dishwashers	6.3%	7	23.7%	27
Portable appliances	—	104	—	223
Refrigerators	98.0%	163	100.0% +	310

Clearly, both the number of electric appliances in the household and the electrical consumption resulting from the use of appliances have risen dramatically in recent years.

Scientists Arjun Makhijani and A. J. Lichtenberg of the College of Engineering at the University of California at Berkeley, in a study of residential energy use, found that in modern, middle-class homes, the primary energy use determinant was the presence of electric appliances and heating/cooling systems. For the sake of simplicity, they converted energy use in 4 sample homes to kilowatt-hours (thermal), which indicates not only the rated electrical energy converted into heat or mechanical power at the home, but also the energy content of the fossil fuel at the power plant.

The energy use in the 4 modern homes varied from 58,000 kilowatt-hours (thermal) of energy consumed per year in a home heated with either oil or gas directly and equipped with an electric refrigerator and clothes washer, to a high level of 112,620 kilowatt-hours (thermal) for an all-electric home, with electric appliances, stove, and air conditioner. The electric home used far more energy, even though it was built with better insulation than the home heated with a direct fuel-burning furnace.[136]

Other than the choice of energy source, the real determinant in home energy use is the lifestyle of the occupants. The dependence on appliances and gadgets can total significant energy use over a year's time. A major reason for the increased per capita energy consumption for the same appliances measured today against a decade ago is the development of larger, more sophisticated appliances, which consume more energy. Color TVs consume more energy than black-and-white TVs; "frost-free" refrigerators consume more than conventional refrigerators; "self-cleaning" electric and gas ovens consume more energy than previous models. The reason for the increase in energy consumption is simple: Convenience in the mechanized home is largely defined by work done for human beings by machines. The more work done by machines (defrosting, self-cleaning, etc.), the more energy required in the process.

Consumers have never had significant access to information on the exact amounts of energy consumed by different appliances. Even the magazine *Consumer Reports* has never devoted a great deal of attention to lifetime energy costs or specific efficiencies of appliances in converting electricity into work. The magazine, which is published by Consumers Union, might have represented a strong voice for the consumer by arguing for higher energy efficiency standards and durable appliances. No such voice has been evident.

Research conducted recently has shown that common appliances can be built better—often at little extra cost to the consumer—so that energy loses are low and the efficiency of conversion of electricity (or, in some cases, natural gas) is high. In their study for the Department of Housing and Urban Development (HUD), Hittman Associates found that redesigning the components of ovens and using different insulation materials could cut energy use in half. They found that more efficient energy use could be achieved by modifying the design of refrigerators and freezers, so that waste heat from the machines' condensers would be captured and used for supplementing the hot water system of the house. However, economic calculations showed that the materials for redesigning the system would cost $50, and would yield a $4 savings per year at current electricity prices.[187]

Thus, adding on certain advanced technologies to refrigerator freezers would add substantially to consumer costs, possibly raising the price of the units too high for consumer purchase. In his thesis on electric energy use, Harvard researcher John Neely found that comparing the total costs of operating currently available refrigerator-freezers yielded substantial economic difference to the consumer. For example, the initial costs of 2 comparable refrigerators differed by a $67 margin, yet the total operating costs of the 2 models over an assumed 20-year life showed that the cheaper ($292) refrigerator cost $746 to use over 20 years at prevailing electricity rates, and the more expensive refrigerator ($359) cost only $392 to operate over the lifetime of the unit. Thus, the second and apparently more expensive machine is actually $287 cheaper to own.‡

The real savings to consumers who purchase more efficient appliances are substantial. John Neely extrapolated from the sale of 6.6 million refrigerators and freezers in 1970, that if all purchasers chose more efficient units, the economic savings (based on $1 reduction in monthly use cost) would equal $80 million at the end of the first year, and by the time 100 million new refrigerator-freezers were sold, the economic savings would equal $1.2 billion per year.

Assuming that the fuel that was burned to create the electricity to power the home refrigerators was coal burned at modern, 1-million-kilowatt-capacity electric power plants, the energy comparison between use of the efficient refrigerators and the inefficient refrigerators gives the following results: Consumer purchase of more efficient refrigerators would mean that 6 projected 1-million-kilowatt plants would not have to be built. The savings in electricity would equal 40 billion kilowatt-hours, meaning that 17 million tons of coal would not be combusted, which would prevent the introduction into the atmosphere of 690,000 tons of sulfur dioxide, 25,000 tons of particulates (plus an additional 2.5 million tons of particulates that would be left as solid wastes), and 147,500 tons of nitrogen oxides. Additionally, thermal pollution —high-entropy heat—would not be released from the power plant operations, and over 26,000 acres would be saved from the ravages of strip mining. Another 10,000 acres would be saved for other uses,

‡ Adding to this calculation inevitable increases in the price of electricity would yield a substantially higher advantage to the consumer in the long run.

which would otherwise provide the necessary land for the power plants and transmission facilities. All this in only 15 years from the initiation of an energy-savings program in refrigerator design![138]

Another study compared the actual efficiencies of black-and-white and color television receivers in converting electricity. The comparison of sets sold in the United States in 1970 indicated that, if all color television buyers had chosen the most efficient color set (a 17-inch screen consuming 130 watts), electric power consumption would have been reduced by 1 billion kilowatt-hours, reducing the amount of coal that had to be mined and combusted in electric power plants by about 1 billion pounds.

If all black-and-white television buyers had chosen the most efficient set (a 19-inch screen consuming 110 watts), 300 million kilowatts would have been saved, eliminating the need for about 259 million pounds of coal.[139] Further extrapolations of the data from these surveys of electricity conservation showed significant savings of other "external" social costs, such as reduced deaths from coal miners' "black lung" disease and pollution-related deaths and injuries.[140]

These estimates of potential consumer savings by use of more efficient appliances clearly show the economic and environmental advantages of the use of efficient appliances; yet the appliances just cited represent only a small fraction of the mechanical hoard of the modern American home. Air conditioners consume many times the energy of refrigerators or television sets; and, depending on the type of unit (window or central), they may consume from 10 percent to as much as half of the yearly electricity requirements in the home.

Seemingly minor changes in the design, construction, and use of air conditioners can affect the amount of electricity required to cool the interior spaces of houses. We have already seen that the principal energy-conserving measures that affect the operation of air conditioners are actions to reduce the solar heat striking the home and entering interior space. Another measure that can save energy and resources is purchase of a properly sized air conditioner, rather than an oversized model. An air conditioner that is larger than needed will not properly dehumidify, since it will cool the room quickly and then end its cycle (by thermostatic control) before dehumidification has taken place. This can cause discomfort by leaving the air cool but moist. An undersized air conditioner is a wiser purchase than an oversized one.

Responding to consumer and environmental pressures (and the legal requirements of a few local governments, including that of New York City), the Association of Home Appliance Manufacturers acted in 1973 to require the efficiency labeling of window air conditioning units so that consumers can compute the life-cycle operating costs. Air conditioners are labeled with an EER, or "energy efficiency ratio." The EER expresses the ratio of the cooling capacity of the unit (in BTUs) to the wattage of the air conditioner. For example, 2 8,000 BTU air conditioners are listed, one with an EER of 5.8 and the second with an EER rating of 9.3. The first air conditioner costs $200 and uses 1,375 watts per hour; the second costs $240 and uses 860 watts per hour. The EER is simply the BTUs of cooling capacity divided by the unit's wattage. The higher the ratio, the higher the initial price, but the cheaper the operating costs over the lifetime of the unit. The lower the ratio, the cheaper the unit (usually) but the higher the operating costs. In the example above, the more efficient air conditioner would cost $19 to run for a summer in Washington, D.C. (1,000 hours at 2.3 cents per kilowatt-hour), and the less efficient air conditioner would cost $31 for a summer's operation. The extra investment in the more expensive, more efficient unit could be recouped in about 3½ years.

Several dozen models of more efficient air conditioners were available from most U.S. air-conditioner manufacturers in 1973 that had EERs of 10 to 12 and above, and required half as much electricity to cool rooms as units commonly sold only 2 or 3 years previously. However, almost all the machines were more expensive than the earlier units because they used more metal components, more refrigerant coils, and bigger and more efficient compressor motors to circulate compression fluid.[141]

Assuming that all U.S. consumers might buy the same number of high-efficiency air conditioners in the future as they bought low-efficiency units in the past, the electricity demands expressed by halving the cooling energy requirements would cause a radical shift in the construction of power plants and the need for coal mining, etc.

However, this is unlikely. First, the energy efficiency ratio system is confusing and awkward. Most consumers don't understand it, and store owners are interested in selling profitable air conditioners, not necessarily efficient ones. A better means of labeling units would be to express the energy equivalents in concrete, understandable terms (adopted for each geographic region).*

* A good start would be labeling along these lines, hypothetically adopted for a sample air conditioner, marketed in Washington, D.C.:

"*Warning:* This air conditioner is hazardous to your health and to the environment. This model, used in normal operations in your locale (1,000 hours in the summer months) will require the consumption of 1 pound of coal (or equivalent fuel) per hour of operation. This air conditioner will consume a half ton of coal

High-Energy Society

The basic problem with residential consumption of energy is that high residential energy consumption is an integral evolutionary development of high-energy society. Low-energy societies are characterized by different social structures. In a social situation where cooperation of many individuals in the home and neighborhood provides a human energy supply, tasks can be shared, and outside energy sources to support mechanization are not essential. In a high-energy society, social organization is oriented around synthetic energy sources. Institutions of high-energy society—television advertising, for example—teach the public that synthetic materials and a profusion of energy-consuming gadgets are essential, both physically and psychologically. The public becomes increasingly convinced of this through the mirror of personal experience. With the passing of each year, the erosion of values common to lower-energy society leads to more energy and resource consumption. The smaller, isolated nuclear family cannot rely on the support of others for the sharing of tasks performed by cooperative groups in low-energy society. High-energy society's values and institutions teach that human labor is to be avoided—i.e., that it is bad—and that machine labor is to be desired—i.e., good. Even in the apartment buildings and condominiums of the high-energy society, the transition to high-energy values is apparent. Washing machines, clothes driers, televisions, and countless other appurtenances of household mechanization are not shared by the concentrated human dwellers. High-energy values teach that happiness and success are measured in terms of each dweller's accumulation of machinery rather than in the use of machines that might be shared by all.

Practitioners of low-energy society values are few and far between in contemporary society. Dwellers or urban and rural communes provide a good example of real energy conservation in action.

A study made for the American Association for the Advancement of Science's Committee on Environmental Alterations by Michael Corr, secretary of the committee, and Dan MacLeod of the Minneapolis Public Interest Research Group, found that energy and resource use was lower in urban communes than in urban conventional homes

---

during the summer, most of which comes from strip mining in Appalachia. In addition, the generation of pollutants from the electric power plant where the coal is burned creates dangerous pollutants, which may cause disease and property damage to your family. The Surgeon General (or the Environmental Protection Agency) advises use of trees for cooling and shading, natural ventilation when possible, and alternatively, when necessary for mechanical ventilation, the use of electric fans, which consume one tenth the energy required for this machine."

(small conventional homes at that, with only 900 square feet of living space, compared to conventional middle-class homes of 1500 square feet and above). They studied the lifestyles of 12 communes made up of a total of 116 members in the Minneapolis area. The communes consisted of young working people, religious groups, political activist groups, and others. Compared with average Minneapolis households (900 square feet of space, 2.6 people per house), the communards consumed 40 percent less natural gas; 82 percent less electricity; and 36 percent less gasoline for vehicles.

An interesting finding of the study was that the national average household saturation of (all) appliances is 74 percent, while the communes studied registered 62 percent appliance saturation. However, while the saturation rate *per person* in a household nationally is 24 percent, in the communes it was 6 percent.[142]

The significance of the study is that energy consumption is related to basic lifestyle, yet the lifestyle of the communards was not in any way that of cave dwellers or hoe-wielding primitives. The communes had most of the amenities of modern civilization; yet, unlike the scattered nuclear families of modern America, the practice of cooperative living and sharing produced a more rational and balanced life pattern.

Communal living is perhaps the one significant aspect of low-energy value that is still preserved in high-energy America. Communal living is more resource-conservative than even this study indicates. The actual process of sharing and using materials in a commune leads to reuse and recycling of basic materials, a practice not found in suburban enclaves. Communal, countercultural living saves energy because it is antithetic to the values of the high-energy system.

Many advocates of "energy conservation" in high-energy society argue that the path to less consumption is through technological advance—more efficient air conditioners, ovens, appliances. While it is true that more efficient appliances are better, the high-energy advocates miss the critical point: The savings of energy by better appliances will only buy time until increases in population and the advertising-instilled mechanical desires of new consumers will negate the technological savings. A case in point is the urging by the former federal Office of Emergency Preparedness (OEP) and others that gas pilot lights be replaced by electronic igniters. Gas pilot lights on gas appliances and home furnaces burn continuously throughout the year. Technologists say that eliminating the pilot lights would save the nation an enormous amount of energy. Indeed, the American Gas Association (AGA) estimates that lights on gas water heaters, ranges, and furnaces consume about 560 billion cubic feet of gas each year. However, the AGA's environmental systems research director, Dan Kennedy, points out that even though the gas is burned, it's not wasted. This is because residences are in "negative heat balance 80 percent of the time," meaning that the residences require heating 80 percent of the time. The pilot lights contribute to meeting the heat needs of the house most of the time, and therefore do not constitute waste.[143] The technologist's "solution" to the pilot light "problem" is the electronic igniter, which only adds to the initial home or apartment owner's expense.

The more logical alternative is for consumers to simply turn off the gas-burning pilot lights at the end of the heating season. Honeywell, Inc., engineer William Burt calculates that if the 20 million homeowners with gas heating would turn off their pilot lights at the end of the heating season, the national gas savings would equal 86 billion cubic feet of gas per year—without the intrusion of additional energy-consuming electronic gadgetry. One month's savings at this rate would be enough energy to heat all the homes in St. Louis, Missouri, for an entire year.[144]

Another technological fix offered by high-energy technologists is this suggestion for the operation of air conditioners: "Raise your thermostat setting to the upper limit of the comfort range, say 80 degrees, during the period of highest load. For example, changing your thermostat setting from 75 degrees to 85 degrees will reduce the air conditioning load by 15 percent or more."[145] This suggestion by the White House consumer affairs office has been put to the public in various forms, including a federal television advertising campaign in the summer of 1973 in various parts of the country.

What the message conveys is: "Buy an air conditioner, but don't use it to significantly alter the climate, which is what you bought it for in the first place." Why not advocate instead that consumers build houses using natural methods of ventilation, and use low-energy fans for cooling? In years to come, as energy prices rise, individuals will be faced with a critical dilemma: How to afford winter heating and summer air conditioning? In winter, thermostats can be set lower to conserve energy, and occupants of houses can wear more clothing (as do Europeans and members of lower-energy cultures)—a form of cheap, personal insulation.

But what will happen in the summer months to the residents of today's "modern" homes, whose cheap, anticlimate construction and orientation preclude any possibilities of natural ventilation and are totally dependent on electricity?

The bedrock of home and building construction today is reliance on machinery for human comfort. Little or no attention is given to installing windows, doors, awnings, etc., to take advantage of prevailing winds, solar radiation, or other natural climatic effects. Many of today's construction methods deliberately preclude the use of natural ventilation. This subject is discussed at greater length in the final chapter. The development of better technologies for energy distribution for home use is also considered in the next chapter on decentralized power distribution.

Energy Use in Commercial Buildings

However profligate the use of energy and other natural resources may seem in the homes of America, commercial buildings are so extravagant in their consumption of energy that they constitute a carnival of waste. The development of commercial and business structures is a key part of the evolution of high-energy society, since commercial structures reflect the feelings and culture of the leading corporate and government entities that control high-energy society. The great buildings of all high-energy societies have reflected the extremes of power and grandeur—as interpreted by the values of the builders.

The supreme expression of American builders is the skyscraper; as Frank Lloyd Wright noted in his Princeton lectures of the 1930s:

> Electricity and the machine . . . have not only made super-concentration in a tight, narrow tallness unnecessary, but vicious, as the human motions of the city-habitant became daily more and more compact and violent. All appropriate sense of the space-values the American citizen is entitled to now in environment are gone in the great American city, as freedom is gone in a collision. Why are we as architects, as citizens, and as a nation, so slow to grasp the nature of this thing?

Continuing, he added: "A little study shows that the skyscraper in the rank and file of the 'big show' is becoming something more than the rank abuse of a commercial expedient. *I see it as really a mechanical conflict of machine resources. An internal collision!*" [emphasis in original][146]

The mechanical conflict Wright scorned four decades ago continues undiminished. Witness the struggle of American business to build the greatest and tallest structures on Earth. For most of this century, the record for tallest building was held by New York City's Empire State Building; then came the World Trade Center, which topped the Empire State Building by 100 feet. In 1973 the record was taken by Chicago's Sears, Roebuck and Company Building, a 1,454-foot urban dinosaur.

The logic behind it all? Sears' former chairman and instigator of the Chicago skyscraper, Gordon Metcalf, explains: "Being the largest retailer in the world, we thought we should have the largest headquarters in the world."[147] In energy terms, the Sears Building towers over New York's World Trade Center. The World Trade Center draws some 80,000 kilowatts of electricity (depending on peak load requirements) —more than enough power to feed all the needs of Schenectady, New York, a city of 100,000 people. The Sears Building requires more electricity than the city of Rockford, Illinois, which has 147,000 people; and through the inverted genius of the electrical rate structure, its power costs far less than the electricity purchased by Rockford residents. The Sears structure contains enough concrete for 78 football fields, and has 80 miles of elevator cables.[148] Appropriately, the architects of the Sears Building, the firm of Skidmore, Owings and Merrill of New York City, also designed the first of the generation of glass-walled energy burners, Lever House in New York City, in 1951.

The most important criteria in determining the energy use and requirements of a building are the skin of the building, the amount of insulation in the building, the structural materials, the amount of window space, and the manner in which the building is sited; in other words, the way in which the building is adapted to climate. The primary energy needs in buildings—whether they are skyscrapers or apartment buildings—are for the comfort of the inhabitants.

The rapid rise in building demands since the elevator and cheap steel made possible the skyscraper has been accompanied by constantly increasing energy use in construction technology. Part of this energy thirst has been for highly energy-intensive materials. In low-energy societies, in which buildings hover close to the earth, construction materials are carefully selected of native materials to adapt to climate. As a result, they generally require little external energy in the form of wood or fossil fuels. The skin of the building (or house) is the principal environmental buffer. In America's high-energy construction, the opposite is true. The overriding criterion in building design is not regard for climate, but the maximum use of glittering, aesthetically appealing materials.

The construction and maintenance of America's buildings requires 57 percent of the electricity produced in the country, and lighting them alone takes about one fourth of the nation's electricity. Changes in design and construction can significantly reduce the energy demands. New York architect Richard G. Stein points out that a reduction in energy use can be made at all four stages of a building's life: the production of materials used in construction; the assembly of the materials into finished products; the maintenance and operation of the building; and the building's demolition.[149]

From the outset, he says, the production process is based on waste and overdesign. Steel for building purposes is not made to conform to the specific needs of buildings; rather, it is made for convenience in rolling mills. "Its cross-section is fairly efficient structurally, but it is needed only at the critical point of span. At all other points there is an excessive amount of material in the beam," Stein reports in *Environment*.[150] Additionally, safety standards often are based on overdesign for simplicity's sake, which leads to energy and resource waste in the use of more material than would be necessitated by wiser design. Stein argues that careful design can reduce the material needed in construction and still provide ample safety margins in multistory structures. Better use of cement alone would save 20 million kilowatt-hours of electricity a year in the nation—enough to provide electricity for 3 million homes.[151] The substitution of lower-energy-intensive materials—such as steel for aluminum—would save energy in buildings. An example is the use of steel in the Chicago Sears Building, which would save 1.3 million kilowatt-hours of electrical power (in the manufacturing process) otherwise required for aluminum processing.

The use of the large glass expanses popularized by Skidmore, Owings and Merrill's Lever House in the 1950s is a basic energy-wasting factor in building design. Glass transmits heat readily in winter, causing a drain on the heating system and requiring more energy. In summer, glass walls admit solar radiation, overtaxing the building's air-conditioning system. Glass has increasingly become the most popular external surface for buildings, regardless of the energy-wasting effect, which makes it antithetic to the very concept of insulation.

To attempt to reconcile the conflicting demands for glass-walled buildings with the increasing requirements for using less energy, the nation's glass companies came up with two innovations. One—double-walled glass—dates back to the 1940s and involves the use of an inner layer of air or inert gas as insulating material. The second glazing technology is the development of tinted, heat-absorbing glass and special vacuum-deposited metallic coatings that enable glass to reflect solar radiation. The mirrored surfaces of modern glass-walled buildings may reflect as much as 90 percent of the incoming solar radiation. The use of mirrored glass and double glazing can reduce solar heat gain in summer and provide more insulation to keep heat from escaping in winter. One glass manufacturer, Libbey-Owens-Ford, has advertised that use of their Vari-Tran mirror glass and Thermopane insulating dual glazing on all U.S. buildings between 1973 and 1982 would save 153 billion kilowatts of electricity; enough, they say, "to keep the homes of a great metropolis going for the next decade."[152] The company bases this on an estimate that over 2 billion square feet of buildings' surfaces will be covered with glass of some variety during the period.

What the advertisement did not say is that, even though the specialized glazing technologies help reduce energy consumption, they're still an unsatisfactory panacea for building energy conservation. Even the most sophisticated glass can't beat good insulation and sound design. Solid glass buildings remain symbols of wasteful ostentation.†

Businesses have quickly assured the public of their newfound interest in energy conservation in buildings—chiefly through the use of mirror glass. New York's First National City Corporation announced plans in 1973 for Manhattan's fifth-largest skyscraper, Citicorp Center. A company vice president said that "technology would be put to work to reduce our energy needs" by using double-glazed reflective glass on the glass walls of the building, which comprise 50 percent of the exterior. Simply substituting well-insulated walls for much of the glass area would come much closer to achieving the energy conservation to which the corporation was paying public lip service.[153]

When glass areas are used on buildings, the effects of solar heat gain in summer can be sharply mitigated by the use of shading devices, ranging from awnings over windows to more complicated sun-protection systems. The great architect Le Corbusier pioneered the use of solar control louvers, called *brises-soleil,* which were first used on a building in Rio de Janeiro in the 1930s.[154] Since then, thousands of various shading systems have been designed to integrate with the architectural style of any building and provide solar control and attractive exterior surfaces.

A more important design feature for buildings that use air conditioning systems is the orientation in relation to the sun's daily course across the sky. Fred S. Dubin, president of the New York engineering firm of Dubin-Mindell-Bloome Associates, says that a rectangular building whose long side is 2½ times as long as its shorter side will use 29 percent less energy for cooling if the length of the building runs east to west, rather than north to south.[155] The rectangular structure with a long north-to-south wall does not intercept as much sunlight as conventional buildings. Other variations in the shape of buildings can effect similar large energy savings. A conference on energy conservation sponsored by the government's purchasing agency, the General Services

† In addition, the technology involved in mirrored-glass production is a very energy-intensive process, reflected in the high prices of vacuum-coated metallic glass.

Administration (GSA), and the National Bureau of Standards (NBS) reported that round and cubicle buildings have lower surface-to-interior volume ratios, and therefore use less energy because heating and cooling losses are reduced.[156] Tremendous potential exists for the architects and planners of buildings by adapting to the climatic conditions of specific sites. Buildings may be designed in various unusual shapes, such as inverted pyramids, spheres, and others, to take advantage of prevailing winds and solar radiation for natural ventilation and solar shading.

The use of trees for shading is an important consideration. Architect-conservationist Malcolm Wells stressed in *Progressive Architecture* magazine that trees are the best sunshades of all. "They cool and freshen the air, they drop their leaves in the fall when the sun's warmth starts to feel good again." Trees also improve the immediate external climate of the building, he notes, by adding fresh oxygen to the air and by absorbing solar radiation in the photosynthesis reaction that also cools by evaporation. Trees can screen the wind and control the flow of air around the building by their presence or absence.[157]

Lighting in Buildings

The use of interior and decorative lighting in buildings is a major energy consumer, using 24 percent of all electricity sold in the country. The massive use of artificial lighting in American buildings has been fostered by an organization called the Illuminating Engineering Society (IES), which is a nonprofit group that sets lighting standards. Its membership and financial support come primarily from the manufacturers of lighting equipment. Since the late 1950s, the recommended minimum lighting standards set by IES have tripled. In the process of conforming to IES recommendations, the New York City Board of Education raised the lighting standards for schools from 20 foot-candles (lumens per square foot) in 1952 to 30 in 1957 to 60 in 1971. Libraries have gone from 20 in 1952 to 70 in 1971.[158] The IES standards range from 70 to 100 footcandles for these applications.

No valid scientific evidence has been presented to confirm the necessity for the spiraling lighting intensity "requirements" set by the IES. On the other hand, considerable research has indicated that reducing the current IES standards by half or more would not affect the visual acuity of individuals in offices, schools, or libraries. Leslie Larson, author of *Lighting and Its Design*[159] and an outspoken critic of both the IES and its lighting standards, says that the lighting industry's control of IES has created a conflict of interest that "would take an incredibly naïve person not to perceive." Of the IES standards, he notes that "excessive light is wasteful of energy and money, is destructive of the visual environment, and may well be psychologically and even physiologically damaging: and that which the IES is insisting upon is clearly excessive."[160]

What is astonishing is not that the IES standards have been met, but that in most buildings they have been exceeded—primarily because businesses use lighting for purely cosmetic purposes, for the visual effect. Reductions in lighting can easily be accomplished by methods ranging from using more natural illumination, to designing lighting selectively in offices and interior spaces so that work areas are illuminated without overlighting other areas. In addition, greater substitution of fluorescent lighting for incandescent lamps has been suggested, since fluorescent lamps convert more electricity to light—thus giving off less heat—than incandescent lamps.‡ Even in buildings where high light levels are used, systems called "heat-of-light" recovery equipment can transfer the excess heat from lighting for use in heating and air conditioning equipment. However, energy savings are higher with better design and less lighting.

One building examined by architect Richard Stein used 9 million kilowatt-hours of electricity per year; he estimates that "an ample but selective lighting system would permit all office functions to carry on equally well with 11 percent of the kilowatt-hours now consumed."[162] Energy saved by redesigning existing commercial lighting systems can eliminate much of the load on the electricity demands of major American cities. A priority goal of urban public service commissions and agencies involved in energy decision-making should be to require businesses to reassess excessive lighting. Stein's example of a case wherein 90 percent of the lighting could be eliminated is not unusual.

Climate Control

High-energy civilization has bred mechanized, controlled climate. Since the early days of air conditioning in the 1920s and 1930s, the technologists' dream of controlled, nearly unwavering year-'round mechanized interior climate has been possible. Standards for the temperature and humidity relationships of the mechanized climate have been set by the American Society of Heating, Refrigerating, and Air-Conditioning

‡ Some research studies have shown that cool white fluorescent lighting may cause detrimental health effects to office workers exposed for lengthy periods. A number of physiological and psychological problems have been linked to various kinds of artificial lighting, suggesting that perhaps the best answer is to provide office and factory workers with balanced internal environments with opportunities for time to be spent in the natural environment in balanced solar light.[161]

Engineers (ASHRAE), an association that had its origins in the nineteenth-century Heating and Ventilating Engineers' Association.

Determination of standards for interior climate control by ASHRAE has long been pursued as a quasiphysiological research specialty by their own laboratories and others of government and industry. The basic intent of climate research has been to determine narrow humidity/temperature limits wherein any group of people, from Eskimos to Texans, will be comfortable and happy in a confined space, whether that space is in Alaska or Texas.

After several decades of experimentation and research, ASHRAE has refined its guidelines to the extent that today it offers standards called "thermal comfort conditions," which call for a temperature range of 73° F to 77° F at relative humidities between 40 and 60 percent. This standard is based on work performed by engineers at Kansas State University, who assert that thermal comfort is "that condition of mind which expresses satisfaction with the thermal environment," and that the standards will satisfy the comfort needs of "sedentary and slightly active, normally clothed people in the United States and Canada." The best air conditioning and heating system would be one that would satisfy 95 percent of the occupants of a building, and the Kansas State scientists believe that 80 percent of the occupants in spaces climate-controlled under the new standards will be satisfied.[163]

The development of the standards has a fascinating history. Over a period of several decades, ASHRAE has been reviewing the previous recommendations and guidelines and narrowing the "comfort" area. Whereas air conditioning engineers once thought it reasonable to cool a space in summer to 80° F with 60 percent relative humidity, the standard now calls for a maximum of 77° F with 40 to 60 percent relative humidity. In practice, the real standard is nearer 70° F than 77° F.

A former ASHRAE president puts it this way: "In following out the applied art and science in physiology of the best designs, I feel there has been entirely too much careless engineering foisted on the public at excessive costs in the magic name of *air conditioning*."[164]

The energy waste precipitated by prevailing contemporary attitudes about heating and cooling is monstrous. The General Services Administration/National Bureau of Standards (GSA/NBS) energy conservation roundtable estimated that a 3° change (from 75° F to 78° F) in the summer air-conditioning setting would result in a 16 percent energy savings.[165] With higher settings in summer and lower settings in winter, energy use in buildings could easily be cut by one fourth.

Energy Distribution in Buildings

Excessive amounts of energy are devoted to precisely maintaining lighting levels and interior climate in buildings. Sital Daryanani, chief mechanical engineer with the New York firm of Syska & Hennessy, observes that a person occupying a typical office of 15-foot by 15-foot dimensions with 5 watts per square foot of lighting and 100 square feet of window area, will consume in one year as much energy as a small family car operating for that year. The heat loss through the glass and the energy required for the lighting totals the equivalent of 500 gallons of oil per year. "There is great potential for energy conservation through re-evaluation of the traditional standards," Daryanani says, adding that 30 percent of the energy can be saved just by reducing the office to a 12-foot by 12-foot space. Further savings can be achieved by lowering lighting intensity and reducing the glassed-in area.[166]

A big loss in energy, Daryanani says, is through the mechanical equipment—called "reheat" equipment—commonly used in the air conditioning systems of buildings. The reheat systems chill all the air used for air conditioning in the building down to 55° F. Individual heaters are then supplied in the zones and rooms of the building to heat the air to temperatures required for occupant comfort—usually 20° above the chilled air.

Numerous technical air conditioning/heating systems exist that use energy better than the centralized reheat systems, but have not been commonly used because the central systems are often cheaper initially and less complicated to install. The General Services Administration energy conservation roundtable concluded that

> although window air-conditioning units are generally inefficient, small unitary air-conditioners should be considered for spaces which may have irregular occupancy hours, in order to avoid operating the central system for a small number of people. The use of incremental units has two basic advantages over conventional central heating and air-conditioning systems. Since each unit is thermostatically controlled, there is less probability of energy waste through excess heating or cooling. Additionally, these units—which are usually controlled by a master clock—can all be turned off at a predetermined time. In the event that an occupant works beyond the normal business hours, he can manually restart his individual unit. The energy savings which are realized through the application of this type of system can be significant.[167]

Whether or not buildings use central heating and cooling systems or decentralized office or zone heating and cooling equipment, numerous opportunities exist for making use of more efficient equipment and using waste heat from heating and cooling equipment. Up to one third of the the heating energy and 20 percent of cooling energy could be saved by diverting exhaust air from equipment to heat exchangers and devices called "thermal wheels" and "heat pipes."*

Standards for the ventilation of closed spaces can affect energy usage, since mechanical equipment under heavy use for ventilation consumes more energy than when under less strenuous use. Architect Richard Stein argues that ventilation has its own "mythology," since standards reflect a tremendous variation from building to building and city to city. "Aside from the direct energy required to operate fans, motorized dampers, and other ventilating equipment, there is a major energy loss in dumping heated or cooled air outside of a building and then heating or cooling the replacement air," he says.[168]

A major energy savings in ventilation can be made by introducing outside air into the building, but he reports that "ironically, no one evaluates the quality of the outside air that will be brought in."[169] Most modern buildings, from skyscrapers to small suburban office buildings, have no provisions for opening windows for natural ventilation. The ventilation is totally dependent on mechanical systems.

It is by no means uncommon for a building—particularly a large one —to use the air conditioning or heating system during periods when the outside air is within one degree of the temperature being maintained inside the building—all because there is no provision whatsoever for natural ventilation. Energy waste under such conditions is simply a matter of thoughtless design and overeliance on technology.

In addition, great savings can be made in heating and air conditioning systems by using decentralized direct fuel equipment, rather than purchased electricity.

A report on the relative efficiency of purchased electricity vs. on-site building power made by the New York engineering consulting firm of David Sage, Inc. concluded that electricity for heating would use more than twice the energy required by an oil furnace-heated apartment building within the service area of Consolidated Edison Company. In addition, electricity compared to steam (district)- heated buildings also showed almost a two-to-one disadvantage.[170] Furthermore, the study found that if buildings were now built with electric heat, "and should the subsequent power system or fuel problems make it desirable to heat them by some other means in the future—the conversion could not be made in any economically feasible manner, nor without substantial demolition and remodeling." On the other hand, buildings designed for

* See p. 419

steam heat or oil boilers could be converted to electricity in the future, in the contingency of fuel problems and a desired switch to electricity.

The use of a decentralized power source for buildings and homes is explored in greater detail, and advanced mechanical equipment such as heat pumps and total energy systems are discussed in a later section on decentralized power systems.

Conclusion

The potential for conservation of energy has never been greater, nor—conversely—has the rising use of energy in society ever been greater. Continuation of present policies of increasing the use of energy for transportation, agriculture, and housing may bring about the decline of civilization in the developed areas of the world. As the cheap, readily available sources of energy—the fossil fuels—decline, so will the productivity of the agricultural system and the efficiency of the transportation system.

Houses and buildings that depend on a daily dose of oil, natural gas, or electricity to heat and feed their occupants will constitute poor forms of shelter in an energy-short world. The time for changing these ways of life is now. We must find new sources of energy to carry us into the future; but at the same time, we must realize the limits to the present fuels and change our lives—and institutions—accordingly.

4

The Coming Decentralization of America's Electrical System

It comes as no surprise to most Americans that the electric companies are no longer able to deliver a product to the consumer when it is needed. Americans—particularly in urban areas—are by now as used to blackouts and brownouts as to a smoothly running electrical system. The major reason for the nation's problem in electricity supply is the technology of our electrical system. Decentralizing the system and encouraging the use of new technologies could radically transform the present-day system so that power could be delivered as needed.

Electric Power and Economies of Scale

The proliferation of electricity as *the* primary energy source for stationary applications (relatively little electricity is used for powering vehicles) brings corresponding weakness and system instability.

One way of examining this instability of the electric power networks is to look at the size of electric power plants. At the dawn of the electric power age in America, the maximum rating of a turbine generator was 1,500 kilowatts. In 1930, the maximum size of a generator had risen to 75,000 kilowatts; in 1940, to 100,000 kilowatts. By 1960, the maximum was 450,000 kilowatts, and today *single* power plants have turbine generators producing between 1 million and 2 million kilowatts of electricity.[1]

The escalating size of the electric power network and the escalating capacity of each individual power plant are accompanied by the escalating instability of the entire power system. The most obvious expression of this instability is the chronic disease of the electric utilities: yearly blackouts and brownouts. Since the great blackout that struck the northeastern United States in 1965 first brought this to public attention, failures of the huge electric grids have become frequent.

Here are two recent examples:

In mid-July 1972, a series of failures in the electric distribution cables maintained by New York City's Consolidated Edison Company caused a cascading 2-week series of blackouts that affected 750,000 people. As the New York *Times* observed, "in Manhattan, Brooklyn and Queens, freezers didn't freeze, air-conditioners didn't cool, and elevators sat at the bottoms of their shafts. In some buildings, water pumps went out and there was no water. Meat and vegetables spoiled; ice cream melted; milk went bad. Losses may be in the millions."[2]

On two consecutive days in April 1973, more than 2 million persons were affected by successive electrical blackouts on Florida's "Gold Coast," stretching from Palm Beach to the Florida Keys. The cause of the failure was the automatic shutdown of one power plant rated at 650,000 kilowatts of capacity,* which triggered a series of failures in other power plants in the electric power network in southern Florida. A Florida Power and Light Company official told the Miami *Herald* that the blackout was analogous to the fate of a 20-mule team wagon when 1 mule drops dead. "If you lose 1 mule," he said, "you may still be able to move the wagon with 19, but you also run the risk of working the remaining ones to death."[3]

In both these incidents, which are typical of failures that periodically plague other electric companies as well, equipment problems in isolated locations triggered the total loss of power in heavily populated areas, causing inconvenience and injury to thousands of people. Had the system been decentralized rather than supercentralized, the failure of a single small power plant would not have affected more than the limited number of people it served. Smaller plants are not immune to failure, but the consequences of their failure do not cripple the activities of millions.

At present, decentralization† is not seriously considered as a viable

* The plant was a nuclear power plant. Issues involved in this power source are discussed in Chapter 5.

† A number of prospective ways to decentralize the power system are discussed in subsequent pages.

option for electric power generation and use in the United States. In fact, power companies and the government agencies that regulate energy use look forward to the day when new technologies can increase the size of power stations far beyond the present million-kilowatt units.

Perhaps the lesson of the futility of the overcentralization of the electrical system will be learned when several states are blacked out, not just single metropolitan regions.

As the size of the giant electrical system grows ever larger, and the power plants grow accordingly, waste heat appears destined to be released into the environment in staggering quantities, heat that might be used beneficially. The energy content of the heat represents a sizable energy resource if it could be tapped.

The nation's electric power plants‡ in 1970 wasted energy in an amount equivalent to:

- 200 million barrels of oil
- 200 million tons of coal
- 2,600 billion cubic feet of natural gas.[4]

Many ways exist for utilizing the high-entropy heat emitted by the steam engine's cycle at the power plant. One way is to eliminate the electric power cycle, and use fuel directly at the home or building. Other ways include using the waste heat from the power plant directly for heating. Hot water or excess steam could be pipelined like natural gas to the home. Like the use of natural gas burners, this practice, called "district heating," is an "antiquated" method of delivering heat to the consumer, developed in the United States in the 1870s.*

In this country, many electric companies provided steam heating for homes and buildings in cities as a subsidiary operation of their electric power plants. New York City's Consolidated Edison is the largest distributor of steam for district heating. However, no new plants in the United States have been constructed in recent years to deliver low-grade heat directly; and in Chicago, the steam-heat subsidiary company of the Commonwealth Edison Company, the city's electric utility, which used to operate five large plants in the city, is now operating only one steam plant to serve fourteen buildings. In smaller cities, steam heat companies are a relic of the past. More than twenty companies have closed or gone into bankruptcy in the past decade.[5]

One reason for the decline of the steam heat companies in the United

‡ This includes not only steam-electric plants, but gas turbines as well, which have a lower efficiency than steam engines in the production of electricity—hence greater waste energy released to the environment.

* See Chapter 1 for Birdsill Holly's early application of central heating in America.

States is the fact that the original underground pipes were built sixty or seventy years ago, and the companies made no effort to replace them because of the advance of "modern" energy approaches, such as use of electricity directly for conversion to heat.

Thus the old, efficient systems were not preserved. Instead, the convenient new electric plants began to take over the business of heating buildings. The science of thermodynamics again was used only to sharpen the efficiency of the electric plant's steam boiler, not the efficiency of the total energy network of the city.

In contrast, the Europeans and the Soviets are building and renewing district heating facilities in many of their cities. A team of scientists from the U. S. National Bureau of Standards recently toured apartment buildings in several Soviet cities, and reported that heat for the buildings was supplied by modern, sophisticated district heating plants (steam only) and by combined district heat/electric power plants. The report of the American delegation to the Soviet Union noted a trend directly opposite from the rapid demise of these once-common facilities in America. Many other buildings apparently are not connected to the central heating distribution system, but new facilities are, the delegation found,[6] which suggests that district heating is increasing, not decreasing, in the Soviet Union.

The same is true in other European countries. In Sweden, the city of Vasteras, with a population of 120,000, is served by a combination electrical generating station/district steam plant. Using the normally wasted high-entropy heat produced by the oil-fueled electric power plant, residents of Vasteras enjoy cheaper power rates than are charged in cities where the power plants supply electricity only.[7] Other European countries use district heat extensively, including France (where several large plants serve Paris) and Germany.

There are many legitimate and necessary uses of electric power—uses that were recognized in many cases by Edison and Westinghouse before the turn of the century. The first function of electricity was for lighting—to provide better-lit homes and offices by replacing dirty, smoking gas lamps. Many other essential uses of electricity include the operation of machines and appliances. What is at issue today, however, is not the use of electricity for lighting or mechanical power, but for applications requiring heat.

A number of proposals have been advanced to improve the efficiency and reliability of the electric power service networks in the United States. These proposals may be divided into two general categories: (1) Centralizing the national power system further by building larger and larger power plants and linking the plants together in a national power transmission grid, so that outages (loss of power from specific power facilities) in some areas can be met by the addition of power from other areas through an interconnected grid of electrical transmission lines; and (2) decentralizing the power system by limiting the maximum size of power plants so that communities and even individual homes are served by small plants that convert transported fuels to electricity and heat at the site where they are to be used.

Repeatedly, over the past several years, Senator Lee Metcalf of Montana has introduced in the Congress legislation that would establish a federally financed national grid, which he says—by balancing the national electrical peak load—would eliminate the need for 25 percent of the planned additions of large electric power plants. In his introduction to the proposed legislation in the Senate in 1973,† he said:

> Because of our insatiable appetite for energy we may within a few years see the Great Lakes, the Pacific, the Atlantic, and Gulf coasts ringed, and rivers lined, with electric generating plants. . . . A national power grid system could slow down such plant proliferation by as much as 25 percent. It would mean lower fuel consumption as well. There would be less demand for peak generating facilities. Bulk power could be moved from one section of the country to another to take care of load requirements peaking at different times of day in the various time zones of the country, and at different seasons of the year in the various regions of the country.[8]

Senator Metcalf's proposed national grid might pose a partial solution to the problems of transmitting electricity over long distances, but at present it is stalled by the intense opposition of the private, investor-owned utilities, which constitute a strong congressional lobby.‡

A national electricity transmission grid faces other problems, of a technological character. There are three major technologies either in use or under development for transmitting large amounts (thousands of megawatts) of electricity over long distances, as would be especially necessary in a national grid: (1) overhead UHV (ultra high voltage)

† S. 1025, "A Bill to Improve the Nation's Energy Resources." See *Congressional Record,* February 27, 1973, pp. S. 3397–S. 3402.

‡ The attitude of the private investor-owned electric utility companies seems to be that the grid is not a bad idea, as long as they control it. In the meantime, they are engaged in their own centralization schemes. Practices such as training public relations personnel to propagandize communities, as well as grooming political candidates to influence communities not to operate their own publicly owned municipal power companies are commonplace. Such practices are widely used by such electric companies as Southern California Edison Company, Pacific Gas and Electric Company, Duke Power Company, Florida Power Corporation, and many others.

and EHV (extra high voltage)* transmission cables; (2) overhead direct-current transmission lines; and (3) underground cryogenically cooled transmission lines.

The only technically satisfactory means of long-distance transmission at the present time is the first method, overhead alternating-current (AC) power lines. A number of EHV power lines (carrying a maximum of 765 kilovolts) have been built already in the United States, and discussion of lines carrying more than 1.5 million kilovolts is under way for the future. The disadvantages of these power lines are threefold: (1) they require long and unsightly rights-of-way 200 feet wide; (2) they suffer a significant power loss, which may range upward of 10 percent of the electrical current carried; and (3) they may create a serious local hazard.

The release of electrons from the line creates a "corona discharge," which gives rise to ionized air and the production of atmospheric pollutants. One potentially lethal pollutant that may be spawned by corona discharge is singlet oxygen, the excited state of the oxygen molecule. Singlet oxygen has been linked to human cancer.[9] Corona discharge produces other hazardous chemicals through atmospheric interactions. In her book *Power Over People,* Louise B. Young says that the planned construction of EHV and UHV power lines has been initiated by power companies "without any attempt to evaluate the effects of these chemical reactions on the atmosphere along the rights-of-way where many people will spend a large part of their lives."[10]

Transmitting power by direct-current (DC) overhead lines has been pursued for some time by electric companies, but significant technical problems remain, including the lack of reliable conversion and power flow regulation equipment. For example, the lines are equipped with no effective DC circuit breakers in cases of emergency.[11] Transmission losses would be present in this system also, and great areas of land would be needed for rights-of-way.

The third proposed means of transmitting power—cryogenically cooled underground lines—is far in the future. Just burying the electric power lines underneath the soil is in itself a major problem. Less than 1 percent of today's intercity high-voltage transmission lines are underground. And if underground lines were economically feasible, the technological barriers of cryogenic transmission would still remain.

Several corporations, including Union Carbide, Phelps Dodge, and General Electric, as well as the federal government's Brookhaven National Laboratory, are now working on cryogenic lines, which many

* EHV lines carry between 230 and 800 kilovolts (thousands of volts) of current; UHV lines carry 800 kilovolts or more.

consider the transmission technology of the future. Cryogenics is attractive because when certain metals are cooled to absolute zero (—459.9° F), they lose electrical resistance and become excellent, nearly perfect conductors. U.S. experiments involve the use of liquid nitrogen, hydrogen, or helium to chill the power lines. However, this poses a major stumbling block. To keep the power lines supercold necessitates the use of refrigerating plants spaced along the surface above the lines, which would require significant amounts of power. General Electric predicts that in their scheme refrigerating plants would have to be placed every 10 miles to keep the nitrogen cold, and Union Carbide scientists say that to refrigerate liquid helium in their underground line would require the use of 75 kilowatts of electrical power to get the equivalent of 1 kilowatt of refrigeration for the helium.[12]

Thus the cryogenic lines would require enormous parasitic power just to keep the lines cold. Another potential headache is the shortage of helium, the most promising element for the refrigerant job. Helium, which is extracted from natural gas, is in scarce supply in the United States, and the National Academy of Sciences estimates that we may come close to running out of helium within 30 years.[13]

In addition, the cryogenic lines require the use of sophisticated, energy-intensive metals for their construction, aside from the energy costs of the refrigerant stations and the cooling elements. In thermodynamic terms, these costs add up to energy losses that cannot be avoided.

In short, all of the concepts for long-distance transmission of bulk power suffer from a common problem: the loss of energy in the transmission of electricity. A more efficient solution is to transport fuels to decentralized power plants, where they can be converted into heat and electrical power on the spot.

Author Louise Young asks what is probably the most relevant question about electric power transmission, and raises the most important issue in regard to extracting energy resources from the countryside for shipment to the cities. She writes:

> The attitude of the power industry is that it is permissible to impose these unpleasant effects on rural citizens living along their right-of-way. What if a few farmers must suffer a little inconvenience of this kind in order that the rest of the people may benefit? But this attitude violates the constitutional principle of equal application of the law. The laws enabling the public utilities to appropriate land for rights-of-way make it possible for them to impose environmental damage on rural property owners that would not be tolerated in more densely populated areas.[14]

In fact, the high-voltage lines are almost never routed through populated areas. The energy industries practice a variety of modern mercantilism with natural resources by denuding the scenic, resource-rich areas of the world in order to pump more energy and materials into the energy-intensive cities and heavily populated areas.

New Systems of Generation and Distribution

There appear to be only two ways to solve the problem of energy waste emanating from the inefficiency of the present electricity generation and delivery system: One way is to scrap the system altogether, and use fuels directly; the other is to develop advanced methods of generating electricity from fuels that will waste considerably less energy than the present fossil-fueled power plants, which have a maximum electricity energy conversion efficiency of about 40 percent.

Several leading technologies have been developed as a result of developments in space age technology. Two of the processes for more efficient production of electricity are MHD (magnetohydrodynamic) power plants and fuel cell power plants.

Magnetohydrodynamics (MHD)

MHD generators convert hot gases directly to electricity. They do not suffer from loss of efficiency imposed by the extra energy conversion stage in a conventional power plant. As we have seen, fuel is burned in a conventional plant, giving off heat, which is used to make steam for driving an electrical turbine generator. This is the familiar steam engine energy conversion cycle. The MHD generator is called a direct energy conversion generator because it makes electricity directly from the heat source without requiring the additional stage of steam generation, as in conventional plants.

Thermodynamic science shows that the most efficient heat engines use heat at a high temperature and reject it at a lower temperature. The greater the difference in the two temperature levels, the more efficient the engine in converting energy to useful work (or to electricity). The MHD process operates at far higher temperatures than for conventional power plants. Power plants that burn fossil fuels for electricity generation produce superheated steam at temperatures just over 1,000° F, and nuclear power plants fission uranium to produce steam at temperatures of 650° F.†

† Nuclear power conversion is discussed more fully later in this chapter. In this context, it should be noted that nuclear fuels do not produce electricity very efficiently. Since the power plants fed by nuclear fuel sources operate in lower heat

On the other hand, in MHD plants, fuels are burned to produce gases of temperatures as high as 5,000° F. Because of this very high temperature MHD plants can produce greater fuel efficiencies than conventional plants, reducing the amount of fuel needed to produce the same amount of energy as the conventional process—thus minimizing the waste involved. In contrast to the upper limit of energy conversion efficiency of a fossil-fueled electric power plant of about 40 percent, MHD plants operate in the range of 50 to 60 percent efficiency.

The following diagram shows how the process works:

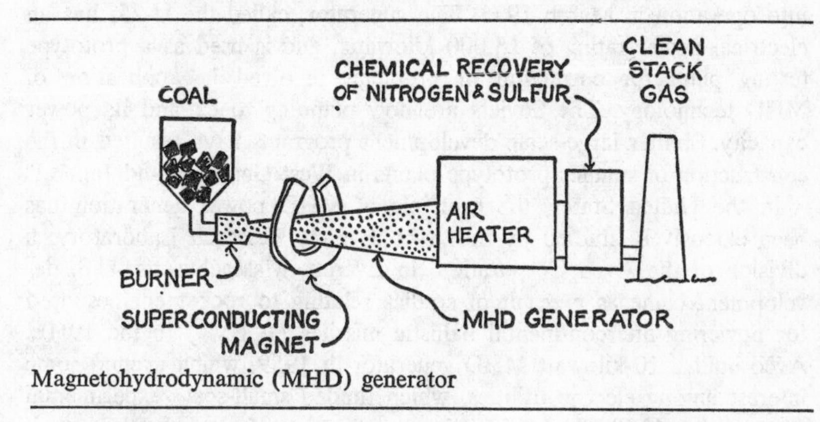

Magnetohydrodynamic (MHD) generator

The MHD generator resembles a rocket engine surrounded by an enormous magnet. At the front or nozzle end of the MHD engine is a burner, in which a fuel—coal, natural gas, etc.—is burned to produce hot gas. The hot gas is then "seeded" with a small amount of an ionized alkali metal—such as potassium or cesium—to make the gas electrically conductive. As the gas expands, it passes through the rocketlike generator, which is surrounded by powerful superconducting magnets, which set up an electromagnetic field. The movement of the electrically conductive ionized gas generates sufficient electrical current within the electromagnetic field that it can be transmitted by electrodes to an external load circuit as direct-current (DC) electricity.[15] It is then converted to alternating current for distribution through conventional electric power lines.

The actual electrical conversion process used in MHD was described by the English scientist Michael Faraday. Faraday discovered the relationship of electricity and magnetism in the early nineteenth century, but until the advent of sophisticated space technology and durable,

temperature ranges than fossil-fueled plants, they are less efficient in converting energy than the fossil fuel plants.

temperature-resistant metals, a process as technologically advanced as MHD was unthinkable.

Only a few research programs have devoted development efforts to the technology in the United States. Consequently, American scientists have very little operating experience with MHD power plants. Japan, the Soviet Union, and some European countries have developed the technology to considerable lengths, however.

In the Soviet Union, a natural gas-fueled power plant equipped with an experimental MHD generator—the largest in the world—was put into operation in March 1971. The generator, called the U-25, has an electrical power rating of 25,000 kilowatts, and is used as a prototype testing plant for correction of problems involved in application of MHD technology. The Soviets are now planning to expand its power capacity. Similar large-scale development programs have resulted in the construction of smaller prototype plants in West Germany and Japan.[16]

In the United States, the principle of MHD power generation has been extensively studied by the Avco Everett Research Laboratory, a division of the Avco Corporation, in Everett, Massachusetts. U.S. development came as a result of studies relating to rocket engines used for powering intercontinental ballistic missiles (ICBMs) in the 1950s. Avco built a 10-kilowatt MHD generator in 1959, which excited some interest among electric utilities, which funded small-scale experimental work in the 1960s that led to the construction of several other small generators. Avco has engaged in other small cooperative efforts with electric utilities, and in 1971 the corporation was awarded $2.4 million by the federal government to build a coal-fired MHD generator.[17]

On the whole, however, the government has never showed much interest in the prospects for MHD generation, even though the President's Office of Science and Technology (OST) in 1969 suggested a large-scale government development program, which OST said would lead to fuel savings of $11 billion by the end of this century through wide-scale application of the efficient MHD plants.[18]

However, according to an analysis of research expenditures for energy development made by the Committee of Science and Astronautics of the U. S. House of Representatives, total U.S. spending for MHD development up to 1971 was about $35 million, about one fifth the amount spent in the Soviet Union just to build the U-25 MHD generator.[19] The scenario has not appreciably changed since then.

Other than the Avco MHD program, an aggressive development program has been pursued by the University of Tennessee Space Institute in Tullahoma, Tennessee. At the space institute, MHD generators have been built that use coal as a power source. According to the director of the MHD program there, Dr. John B. Dicks, MHD power plants producing as much as 2 million kilowatts each can be ready in the 1980s if the government will institute a $410 million development program. He told a subcommittee of the powerful Senate Apropriations Committee in 1973 that, in contrast to nuclear power, the MHD development program would result in efficient power plants that would be safe, would create very little pollution, and would save $40 million to $130 billion in the nation's power bill by the end of the century.[20]

Notwithstanding the tremendous promise that MHD appears to offer, this technology still faces a number of unsolved problems. The tremendous magnets required to produce the electromagnetic field in the generator have, with few exceptions, only been built for laboratory use in high-energy physics. The problems in superconducting magnet construction for small MHD plants may be overcome, but the sizes needed for plants of more than 1 million kilowatt electrical capacity are only on the drawing boards. Another major difficulty is the lifetime of the MHD generator. Experience across the world amounts to only a few thousand hours in laboratory generators. The Soviet MHD plant has not produced more than a few thousand kilowatts of power for short periods.

Potential difficulties may arise from the exposure of metal surfaces to the intense heat of the generator, and from the corrosion of metals and electrodes resulting from increased ash deposits when coal is a fuel source for the plants. Even the use of the seeding materials, potassium and cesium, may interfere with the production of electricity by condensing and penetrating the walls of the generator.

Scientists specializing in the field are confident, however, that research and testing programs will resolve the potential problems with this power-generation technology. They point out that it was once thought that MHD generators would not operate with coal as a fuel, due to problems of coal ash corrosion. Yet, in the past two years, coal has been successfully used as MHD fuel in the University of Tennessee Space Institute tests.[21]

Another problem is that part of the MHD technology involves seeding the hot gas with valuable metals, which must be recovered at the end of the process by means of efficient electrostatic precipitators like those used in conventional fossil fuel power plants as pollution-control devices. MHD specialist and pioneer Dr. Arthur Kantrowitz of Avco Everett Research Laboratory says that the use of the precipitator not only will reclaim the seed metal, but also will effectively trap most pollutants used in the fuel. "The MHD exhaust will be highly cleansed of particulate matter," therefore; and he adds that special pollution-control processes developed at the Avco laboratory drastically minimize the escape of sulfur and nitrogen oxide pollutant emissions.[22] Avco calculations for an MHD plant compared to a conventional fossil-fueled power plant are shown here:[23]

Pollutant Emissions in Tons per Day Based on Use of Coal Having 3 Percent Sulfur Content

	1,000-Megawatt MHD Plant Burning Coal	1,000-Megawatt Conventional Coal Plant
Particulate Matter	3	33
Sulfur Oxides (SO_2)	3	450
Nitrogen Oxides (NOx)	4	80

Use of an MHD plant operating in conjunction with a gas turbine power plant might not need to reject any heat to cooling water; so, in a sense, the thermal pollution of water would be eliminated. Of course, the eventual waste heat from the burning of the coal would go into the air (about 40 percent of the original energy of the coal expressed as heat), but the elimination of cooling water would make possible the siting of MHD plants in many areas where water is not available for cooling purposes.

Assuming the successful development of MHD, there are several ways in which the plants can be used. Since the MHD plant burns fuel at very high temperatures in its operation, and rejects heat at high temperature, it can be used to fuel a secondary power cycle. This is termed a hybrid, open, or "topping" cycle. In the topping cycle, the MHD generator produces electricity, then rejects left-over heat, which is used in turn in another gas or steam turbine to make more electricity. The combination of the two energy converters makes for a highly efficient system that wastes little energy and produces minimal waste products, including heat.

While the exciting possibilities for MHD generation have been recognized in other countries, they have not been recognized in the United States. MHD offers great long-term promise, and can be adapted to the use of many fuels, including all the fossil fuels. The use of MHD for electricity production would require the strip mining of 30 percent less land to produce the same amount of electricity as a conventional plant. Even more exciting is the prospect that MHD plants are less complicated in some ways than conventional power plants—there are no moving parts in the MHD generator, for example—and may be cheaper to construct than other types of power plants.

Dr. John Dicks said that the Soviet approach to MHD power is an all-out, large-scale effort involving the construction of MHD power plant components in a large building at the Moscow High Temperature Institute. As he told the 1973 Senate Appropriations subcommittee, the Soviets "believe that this gamble [in MHD development] does pay off. It is possible that in some programs the United States would be wise to review the Soviet method and perhaps adopt it where it seems advisable, and MHD might be such a case."[24]

It is clear that without sufficient funding the large-scale testing of MHD cannot be accomplished, and we will never know the full potential of this promising technology. To date, the federal government and the power companies have shown no initiative in exploring the potential.

Fuel Cell Power Plants

A promising advanced technology that converts fuel directly into electricity without going through any intermediate stage—such as making steam to turn a turbine, for example—is a space-age device known as the fuel cell.

Like most other energy technologies in use today, the principle of the fuel cell is an old one—discovered in the early years of the nineteenth century. In 1839, English scientist Sir William Grove built a small fuel cell consisting of tubes of hydrogen and oxygen in solution with sulfuric acid. When strips of platinum were connected to the separate tubes, an electrical current was created, which, according to Grove, produced a shock "which could be felt by five persons joining hands, and which when taken by a single person was painful."[25] A few later experiments with fuel cells for electricity generation were attempted with no great success in the nineteenth and early twentieth centuries.

The modern technology was developed in England by Francis Bacon of Cambridge University. Bacon's work began in the 1930s, and in 1959 he reported that a practical cell had been perfected that could be used to power a vehicle or industrial equipment.[26]

This announcement coincided with the origins of the space age, and the National Aeronautics and Space Administration (NASA) as well as several large American corporations became interested in the fuel cell for powering space vehicles. In the early 1960s, NASA spent large sums to develop fuel cell technology, reaching a peak in 1963 when the agency spent $16 million for fuel cell research and development.[27]

Fuel cells made for NASA by the Pratt & Whitney Division of United Aircraft Corporation were used for spacecraft power on the Apollo space flights; and when astronauts Armstrong and Aldrin became the first men to land on the moon in 1969, the electric power required aboard their spacecraft was furnished by three fuel cells supplying 6.6 kilowatts of electricity.[28]

Unlike the conventional heat engine, the fuel cell converts energy into power electrochemically; i.e., it makes electricity from fuel by controlled oxidation of the fuel. The four basic elements of the system are the fuel, the oxidant, the electrodes, and the electrolyte chemicals.‡

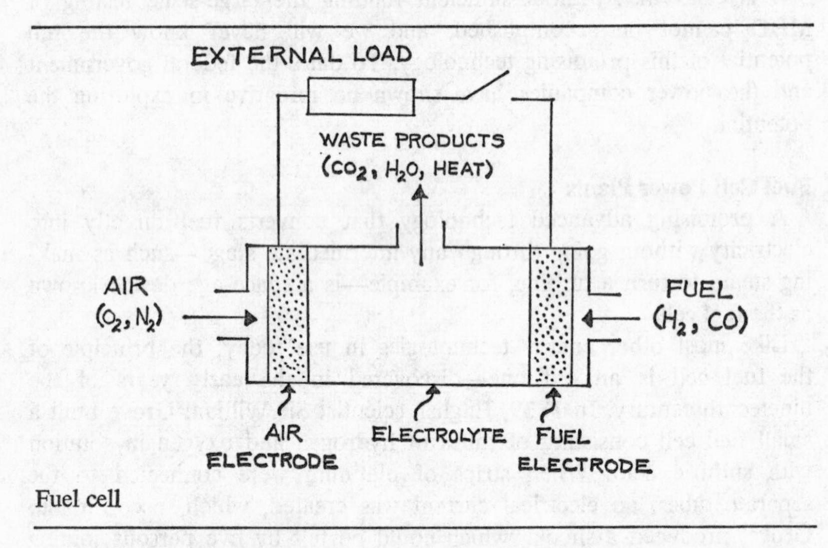

Fuel cell

In the simplified fuel-cell diagram shown above, the fuel is supplied in the form of hydrogen (H_2) and carbon dioxide (CO_2). The two electrodes in the cell are the anode, which is the fuel electrode; and the cathode, which is the oxidant electrode. The electrolyte material that conducts the electric current between the electrodes may be either acid or alkaline, solid or liquid. The cycle of operation begins as the hydrogen and carbon dioxide are fed to the anode, where hydrogen ions are formed, releasing a flow of electrons to the cathode (through the electrolyte medium). The cathode takes oxygen from the air and transforms it into an ionic state in combination with the electrons it is

‡ The human body works like a fuel cell. Fuel in the form of food is oxidized catalytically by enzymes in the body; the body's electrolyte is blood, and some of the food energy we convert to power is electrical, though energy is transformed into useful work primarily through muscle power.[29]

fed from the anode. The oxygen-carrying ions are released into the electrolyte, where they migrate to the anode, completing the process of energy conversion by producing a flow of direct-current electricity and one by-product, water (by combining the hydrogen and oxygen fed to it in the reaction).

The fuel cell's energy conversion process is very basic, but exceedingly difficult to bring about. Its simplicity is obvious: Fuel cells have no moving parts and operate quietly, since the reaction is chemical. The process itself has been described as the reverse reaction of the common high school chemistry experiment of water electrolysis, in which electric current is used to break down water into hydrogen and oxygen. The fuel cell accomplishes the opposite: It combines hydrogen and oxygen to produce electric current and water, and at very high efficiencies in terms of converting fuel to power—as high as 75 percent in laboratory tests.

Though simple and basic, the fuel cell process is subject to certain economic restraints, largely in the use of fuel and the use of rare materials needed as electrode catalysts for the chemical reaction. The best catalysts to date have been noble metals—especially platinum, a metal that is extremely expensive and extremely scarce. Other factors that have contributed to the high costs of fuel cells in the past have been the specialized applications for which they have been designed. NASA, for example, has never cared whether or not fuel cells were economic, only that they would be lightweight, efficient, and suited for space power. Dr. Ernst Cohn, manager of NASA's Chemical and Solar Power Division, says that average fuel cell costs have ranged around $50,000 per kilowatt for experimental uses—and up to $300,000 per kilowatt for space applications. These prices soar far beyond the limits of economically competitive power for conventional use.

Thus, for fuel cells to compete, they must be built cheaply—for a few hundred dollars per kilowatt. Dr. Cohn stresses that this goal is attainable. He told a group of energy engineers at a Japanese meeting:

> We must learn to search for, and to adapt, science and technology that have already been developed over the past 100 years for quite different purposes. And we must learn to think, instead of merely rushing into laboratories, building equipment, measuring, and processing data by computers. The present generation of fuel cells appears to me to be quite unsuitable for normal commerce and industry. But the potential of the fuel cell is great enough to justify a serious drive towards a more competitive and useful kind of system.[30]

Bringing the costs of the fuel cell down to a commercial range has proved to be an elusive goal for many of the companies that have attempted it. Union Carbide Corporation, Allis-Chalmers Manufacturing Company, and the Monsanto Company developed fuel cell-powered motorcycles, farm tractors, and trucks, respectively in the 1960s, but they could not build the fuel cells cheaply enough to compete with conventional heat engines. All three are now out of the fuel cell business. Likewise, Texas Instruments, Inc., and General Electric Company made fuel cells for the military and NASA in the '60s but both these companies have also withdrawn from fuel cell research and development.[31]

However, several companies have persisted, and the end of the road appears to be in sight. The major contenders for the fuel cell market today are the Pratt & Whitney Division of United Aircraft Corporation; Exxon Enterprises, Inc.—a division of Exxon Corporation—in conjunction with a French manufacturer of electrical equipment, Alsthom; Atlantic Richfield Company's Arco Chemical Division, in cooperation with Bolt, Beranek and Newman, a Cambridge, Massachusetts, research firm; and Westinghouse Electric Corporation.*[32]

Pratt & Whitney, as has been noted, built fuel cells for use on the Apollo spacecraft. Largely as a result of its success in the space venture, the company convinced a national coalition of natural gas utility companies to pursue a commercial fuel cell for home use. The TARGET program, as it is called (for *T*eam to *A*dvance *R*esearch for *G*as *E*nergy *T*ransformation) is comprised of more than 30 utilities; and since its creation in 1967 it has been responsible (in conjunction with United Aircraft Corporation) for the expenditure of more than $50 million in fuel cell research at the Pratt & Whitney fuel cell research laboratory, where more than 1,000 specialists are employed in fuel cell research. The TARGET fuel cell produces a maximum of 12.5 kilowatts of electrical power, which the company says is adequate to meet the peak needs for home use of electricity at any time.

The fuel cell is designed to run on natural gas, and the actual home installation, which is about the size of a central air conditioning unit, consists of three components: a reformer unit, the fuel cell itself, and an electrical inverter. Natural gas (methane) is chemically transformed in the presence of steam and a catalyst in the reformer to carbon dioxide and hydrogen, which are pumped into the fuel cell for elec-

*** Other international oil companies that have developed fuel cell technology are British Petroleum Company and Royal Dutch-Shell.**

tricity conversion. The fuel cell produces direct-current (DC) electricity, which is unsuitable for home use, so the electrical inverter makes alternating-current (AC) electricity at the suitable frequency and voltage level for use in the home.

In 1972 and 1973, prototype TARGET fuel cells were shipped to member utilities and were placed in actual use for field tests at 37 locations in homes and office buildings. After a few months of testing, the units were shipped back to the Pratt & Whitney laboratories for performance analysis. Pratt & Whitney has not announced the results of the prototype trials, but it has stated that successful completion of the prototype tests will result in commercial marketing of the fuel cells in the late 1970s.[33]

Another Pratt & Whitney program is the development of larger fuel cells for use by electric utilities. In November 1972, an experimental 37.5-kilowatt fuel cell was installed by Pratt & Whitney for the Public Service Electric and Gas Company in Newark, New Jersey.[34] Utility use of the fuel cell does not offer as great an energy savings as use of the fuel cell in the home. Assuming that the conversion efficiency of the fuel cell is 50 percent, then half the energy of the natural gas pipelined directly to a residence would be converted to electricity by the fuel cell (not considering the loss of energy in the fuel cell's associated steam reformer and DC inversion equipment). A 50 percent efficient fuel cell in use by a central electric utility would convert gas to electricity, and then the electricity would be transmitted to the ultimate users, incurring the additional energy loss in the transmission of the electric current.

However, the fuel cells now in use are not 50 percent efficient. The Pratt & Whitney fuel cell is about 40 percent efficient, the same as a conventional power plant in converting fuel to electricity.[35] Why, then, are utilities interested in the fuel cells for use in their electric system?

There are several reasons: First, the fuel cell maintains a high energy-conversion efficiency even when it is not operating at full load —i.e., when it is producing only half the electricity it is capable of converting from fuel. Conventional steam turbines and gas turbines drastically lose efficiency when they are not operating at their full rated capacity, thus wasting an inordinate quantity of energy when not running at peak output.

The second reason is that a fuel cell can be started up quickly, and reaches full power within seconds, as opposed to conventional power generators, which require warming up for considerable time intervals. This feature of reliability makes the fuel cell a good candidate for a peaking power unit for use in times of great drain on the utility system or as an emergency, or standby, generator to "back up" other equipment.

Finally, fuel cells are potentially cheaper than other types of generating equipment and generate less pollution. Because the fuel cell does not contain moving parts, it can be assembled with greater ease and lends itself to mass production, reducing manufacturing costs. Additionally, there's less to go wrong with the fuel cell than with conventional, complicated power plants, so maintenance costs would be reduced in comparison with other equipment. Due to the fuel cell's high efficiency in converting energy into electricity, it produces less pollution than other power methods—particularly when it is fueled with natural gas, the inherently cleanest fossil fuel in general use.

Dr. Seymour Baron, vice president of Burns and Roe, Inc., a New Jersey engineering firm, has estimated both the costs and the environmental effects of fuel cells compared to conventional power plants. The following table compares the amount of pollutants generated by conventional fossil-fueled plants with pollutants generated by experimental fuel cells for each 1,000 kilowatt-hours of electricity generated by the plants:[36]

Pounds of Pollutants per 1,000 Kilowatt-Hours Of Electricity Generated

Central Power Plant:	Fueled with Natural Gas	Fueled with Oil	Fueled with Coal	Experimental Fuel Cell
Sulfur Dioxide (SO_2)	.3	21.0	28.0	0 to .00026
Nitrogen Oxides (NOx)	4.0	6.8	6.7	0.139 to 0.236
Hydrocarbons	2.8	10.0	20.0	0.225 to 0.031
Particulates	.1	.5	1.0	0 to .00003

As is obvious from the chart, the fuel cell—because of its high energy-conversion efficiency combined with the combustion of clean natural gas—generates only a fraction of the pollutants emitted by conventional power processes. Note that the fuel cell generates only an infinitesimal quantity of pollutants while burning the same fuel that the conventional gas power plant burns. Again, the electrochemical nature of high-efficiency energy conversion provides an ideal solution to environmental effects and plant efficiency. Dr. Baron says of thermal pollution, that "although some heat is generated in the cells . . . the heat from high temperature [fuel] cells can be recovered in the gasification step or the product gases from lower temperature cells can be cooled in air condensers for recovering the water."[37]

In addition, noise pollution from fuel cells would be minimal, since they have no moving parts (except for associated equipment having a few fans and pumps) to generate noise. Combining these environmental characteristics of the fuel cells means that fuel cell power plants can be sited in the heart of urban areas, whereas noisy, inefficient conventional power plants must be located at some distance from the human population.

A unique characteristic of fuel cell manufacture is that they can be produced in small, very efficient units, then stacked together to form a large installation. In other words, 10 of the home-size fuel cells could be linked together to provide a total of 125 kilowatts of electrical power. This combined fuel-cell plant could be located in an apartment building, where the power would be used as needed from 2, 4, or all 10 units, at high operating efficiency.

The chief engineer of the Pratt & Whitney program, William H. Podolny, explains how the modular nature of fuel cell technology could be applied to energy needs of isolated areas:

> For example, the electric power needs of a small rural village could be initiated by the installation of a single 5-kilowatt unit. As the demand for power in the village develops, other units may be added in parallel to supply the growing requirements of the residents. Load paralleling, synchronization, and sharing are accomplished simply and quickly. When operated in this manner, one or more of the parallel power plants may be removed from the system without interrupting service. This feature enhances both the maintainability and reliability of the power supply system.
>
> This modular or building block approach also offers an economic advantage in that equipment and, therefore, the capital investment may be added incrementally only at the time it is required.[38]

Perhaps the most alluring feature of the fuel cell is its estimated price. Pratt & Whitney's estimates for the potential costs of fuel cell power plants for rural areas installed between 1975 and 1980 would be up to 50 percent cheaper than conventional diesel engines used for remote power applications today. Another study, independent of the manufacturers of fuel cells, was made by Hydro-Quebec, Canada's largest public power corporation. Hydro-Quebec estimated that fuel cell plants would save up to 30 percent of the costs of conventional power systems for isolated villages.[39]

Dr. Seymour Baron has studied the relative economics of large fuel cell power plants compared to nuclear plants, and concluded that even if nuclear plants cost $450 per kilowatt in the 1980s,† fuel cells costing $200 to $250 per kilowatt would produce cheaper power. Baron studied two types of fuel cell installations: one installation where large (1 million kilowatt) fuel cell power plants would be located near coal mines and would be fueled by synthetic gas made from coal; and a second type of installation, where dispersed fuel cell power plants of 20,000 to 100,000 kilowatt size would burn natural gas, synthetic gas, or distillate fuel oils. In both cases, he found cause for predicting economic, clean power.[40] He did not compare the relative economics of nuclear energy's social and environmental liabilities, only the cost of installing the power plant, based on current utility and AEC predictions. Had he considered the true economics of the nuclear plants, the fuel cells would have looked even better.

The only other major effort in progress to develop commercial fuel cells is that of Exxon Enterprises and the French Alsthom company. In December 1972 the two companies signed a contract to fund a $10 million development program designed to produce a fuel cell that may be used for homes, remote power applications or even vehicles. In fact, the Peugeot Automobile Company of France has contributed funds to the project with this in mind. In a 1973 letter to *Science* magazine, C. E. Heath of Alsthom laboratories noted that the fuel cell used methanol as the basic fuel, which could be made by the conversion of coal or other fossil fuels. "The fuel cell itself," he said, "is based on the pioneering thin cell concept of Alsthom's Bernard Warszawski, who showed how to pack five to ten times as much active electrode surface into the same volume as found in conventional fuel cells. The Alsthom-Exxon program thus also promises to bring the fuel cell out of space into the hands of industry and the public."[41]

The Westinghouse Electric Corporation is working under a government contract from the Interior Department's Office of Coal Research to develop high-temperature fuel cells that do not need a catalyst (such as platinum) for their operation. However, such fuel cells do require a variety of expensive, rare materials that can stand the rigors of sustained operation at high temperatures (in the range of 1,000° C). This approach may prove to be a long-range answer to problems of catalyst shortages, but only if low-cost materials can be found for fuel cell operation.[42]

A smaller, but significant fuel cell research effort is being conducted

† Reported prices of nuclear power plants in 1973 had reached over $560 per kilowatt, which makes Baron's economic comparison vis-à-vis fuel cells even better.

by the non-Communist world's largest supplier of platinum, the Engelhard Minerals and Chemical Corporation of Murray Hill, New Jersey. The Engelhard fuel cell research is aimed at developing cells that can use ammonia or methanol as fuel. At present, Engelhard is supplying platinum for use in catalytic converters for automobile pollution-control devices. Assuming the widespread use of the limited platinum reserves, which mainly come from Engelhard mine holdings in South Africa, a serious problem could develop in platinum supply. The continued successful development of platinum-catalyst fuel cells and the rapid increase in use of platinum-catalyst auto pollution control devices, required by law beginning in 1975, could precipitate a shortage that could affect the use of platinum for large-scale use in either cars or fuel cells. More likely is the possibility that large-scale use of platinum for auto pollution control might require the exploitation of platinum to a degree that would drive the price of platinum—already high—to an economic level that would prohibit its use in fuel cells.[43]

Although Pratt & Whitney has been relatively silent about the commercial prospects for fuel cells, early orders for fuel cells to be delivered in the late 1970s have been received from a number of electric utilities. Typical of utility enthusiasm for the concept is this statement of long-range energy plans made by New York's Consolidated Edison Company in 1973:

> We are especially hopeful about the large-size fuel cell development program Con Edison and several other utilities plan to fund with Pratt & Whitney, the prime contractor. If this development is successful, we could have in service in the late 1970s virtually pollution-free fuel cells in 26,000-kilowatt units that can be located at many places within our service territory—including existing power-generating stations and substations.[44]

In 1972, 10 of the nation's private electric power companies joined in an effort coordinated by the Edison Electric Institute to design a large fuel cell for integration into existing power systems. The industry trade magazine *Electrical World* observed that, "while it is not expected that multi-megawatt fuel-cell power plants would supplant other power-generation systems, their successful development would give electric companies additional flexibility for providing power where and when needed."[45] Since the coalition was formed, at least one power company has committed a substantial sum to purchase and operate the units. In the spring of 1973, the Southern California Edison Company ordered 15 26,000-kilowatt fuel cell plants from

Pratt & Whitney for operation in 1978. The company agreed to spend $7.5 million to develop fuel cell technology for implementation by that time.[46]

The sudden dramatic interest in fuel cells by the electric utilities is a far cry from only a few years earlier, when the companies considered the fuel cells a way to increase competition between gas and electric utilities. Ray Huse, research director of Newark's Public Service Electric and Gas Company, the first electric utility company to try an experimental fuel cell, told *The Wall Street Journal:* "A lot of our friends in the electric industry think we're crazy for helping the gas industry develop the fuel cell, but we think the gas industry is doing us a favor."[47]

The Future of Fuel Cells

The best possible use for fuel cells has been recognized for some time: putting the fuel cell in the home or building where natural gas can be piped to it. The best use of the fuel cell energy is made when the fuel cells are dispersed and serve the ultimate customers directly, because this eliminates the energy-wasting step of transmitting electricity. In the words of energy economist Dr. Bruce Netschert, the development of the home-size fuel cell would "at one stroke make central electric power generation for home use obsolete."[48]

Netschert's analysis is recognized by the power companies, who see the home fuel cell as a distinct threat to their growing control over home energy use. The reason electric utilities are ordering fuel cells is that they would like to control the future development of this technology and prevent the wide use of fuel cells in the best locations —the homes and buildings, the ultimate sites of power consumption. Fuel cells provide an excellent investment for utilities, because they are potentially cheaper than any conventional power plants, and they can easily meet state, municipal, and federal pollution control laws.

But in terms of efficient energy use, which is what matters most to society, the utilities' use of fuel cells would not be a substantial improvement on the efficiency of present systems. As Thomas Maugh II pointed out in *Science:* "Central station generation of electricity with conventional fuel cells is impractical. Because such cells are no more efficient than the best large turbines, no advantage is gained in their use."[49] While he was referring to the idea of stacking fuel cells together to equal the size of a large fossil-fueled electric plant of 1 million kilowatt capacity, the same observation is valid for the utility's use of the smaller fuel cell plants. In terms of a comparison between use of fuel cells in buildings and homes vs. use of fuel cells for incorporation into the electric utility's system, the advantage clearly lies with the use of fuel cells at the sites of power use. While the successful development of electric utility fuel cells will probably increase the operating efficiency and decrease operating cost of the utility's electricity supply network, society will benefit more by development of the technology for use in its most effective and thoroughly dispersed and decentralized application.

What has astounded many observers is the lack of government funding for fuel cell development. After the successful use of fuel cells for the space program, government funding dwindled to very low levels, effectively starving most of the research effort created in the 1960s. Dr. Seymour Baron termed "shocking" the "complete lack of funding by the U. S. Government agencies in the commercial development of fuel cells . . . considering all [its] positive features."[50]

Possibly in the future the fuel cell will be fueled directly with hydrogen, which can be made by the electrolysis of water. Taking the projected 40 to 50 percent estimated efficiency of today's fuel cells as a base figure, it can be seen that changing the fuel from a fossil fuel (which contains carbon and other impurities) to direct use of hydrogen would increase the efficiency of the fuel cell to levels as high as 60 to 70 percent. In addition, the fuel cell could be made more cheaply, since expensive fuel-reforming equipment would not be needed in conjunction with the fuel cell. As has been pointed out, the Pratt & Whitney fuel cell is equipped with a steam reformer unit, which constitutes a considerable part of the cost of the installation. This reformer unit converts natural gas and other fossil fuels into hydrogen (and carbon dioxide).

The hydrogen-fueled fuel cell would not require the reformer stage. It would also produce even less pollution than today's fuel cells, since the fuel would be oxidized at a higher efficiency, and since the fuel is cleaner than fossil fuels. In fact, the only pollutant products would be small quantities of nitrogen oxides—assuming ordinary air were used in combination with the hydrogen—and the major by-product: water. The fuel cells that were used on the Apollo space flights produced fresh water for the astronauts' personal hygiene. Similarly, fuel cells in homes could produce water to supplement household needs.[51]

Combined Cycle

A readily available electric power plant technology that significantly increases the efficiency of converting the energy in fossil fuels to electricity is the "combined cycle" power plant. The term denotes the combination of a gas turbine power cycle with a conventional steam turbine power cycle in one power plant installation. The combination of the two units allows waste heat from the gas turbine—which only converts about 25 percent of the energy content of natural gas into useful power—to be used to make steam for the operation of a conventional steam turbine.

The principle of operation of the gas turbine was discovered at the turn of the century by a graduate student at Cornell University, S. A. Moss, but it was not until midcentury that the turbines were made efficiently. The following illustration shows the process of energy conversion in the gas turbine:

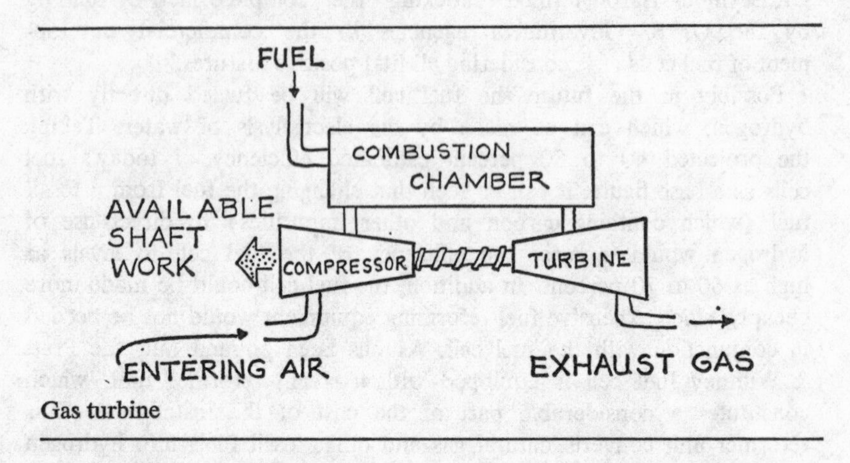

Gas turbine

The three basic components of the engine are the air compressor, the combustion chamber, and the turbine. After air is compressed in the compressor (usually between 50 and 80 pounds per square inch), it is then discharged into the combustion chamber, where it is mixed with fuel and burned. The heated fuel/air mixture flows against the blades to turn the turbine. This motion rotates a mechanical shaft attached to the turbine and provides the work necessary to operate an electrical generator.[52]

The development of the gas turbine for electricity production was originated in the United States during the Second World War, when gas turbines were produced in large quantities for the operation of military warplanes. The technology is not in widespread use today, but immediately after the war a few installations were custom-designed for industrial uses. The first installation was by the Oklahoma Gas and Electric Company to generate electric power from plentiful natural gas. Since then, perhaps two dozen combined-cycle plants have been built, the largest of which is a 300,000-kilowatt plant that supplies electricity and steam for a Dow Chemical Company industrial plant in Freeport, Texas.[53]

At present, the over-all efficiency of a combined-cycle power plant burning natural gas is limited by the ability of the gas turbine engine to withstand high-temperature heat. The units now in operation reach over-all energy-conversion efficiencies that are about the same as modern coal-fueled steam-electric power plants—close to 40 percent. What engineers working in the field are hopeful of is an improvement in the metals and materials used in the gas turbines that would allow far higher temperatures in operation—and thus increased efficiency in conversion of energy to power. Gas turbines producing electricity now operate in the temperature range of about 1000° C (1832° F), but scientists at United Aircraft Laboratories in East Hartford, Connecticut, predict that turbines will be available in the 1980s to operate at 1440° C (2624° F).[54]

At the predicted higher temperature range, energy can be converted at 50 percent efficiency in the units, a 20 percent increase over generator performance today. Notwithstanding the fact that the combined cycle plants convert fuel to electricity at about the same rate as modern fossil-fueled steam plants, they offer several advantages over the big steam plants: First, they are cheaper to build and install—costs for combined-cycle plants are about $100 to $150 per kilowatt of plant capacity, compared to a minimum of $200 to $250 per kilowatt for the cheapest fossil-fueled steam electric plants; second, they require very little cooling water, compared to the voracious needs of conventional electric plants, since their waste heat is largely rejected to the atmosphere rather than to the water supply.

Third, the concept of the combined cycle process may find even greater success by using different types of energy conversion cycles than the gas turbine and the steam engine cycle. For example, Dr. Arthur Fraas at Oak Ridge National Laboratory suggests an improvement called the "multiple cycle," which employs liquid potassium in a closed system to drive a special power turbine. A gas turbine provides heat for the potassium to boil into a vapor—which is expanded in a second turbine. Then the exhaust from the second turbine provides heat to drive a final turbine. In this cyclic process, less heat is wasted, and more is converted to useful power. The result is a process that is 55 percent efficient. While this specific plant concept has not been built, the technology is currently available.[55]

A promising application of the combined-cycle approach has been developed by a large Japanese company, Ishikawajima-Harima Heavy

Industries (IHI). The IHI plant uses a closed-cycle steam turbine similar to that proposed by Oak Ridge National Laboratory. However, their turbine uses a fluorocarbon compound as the heat transfer fluid —a chemical identical to that used in many refrigerators to transfer heat. In fact, the IHI fluorocarbon turbine can be used to make electricity from fuels as well as to drive a special compressor, enabling the fluorocarbon to be used for refrigerating or air conditioning. The company claims efficiencies as high as 80 percent in combined electricity/refrigeration applications. Two systems have been installed in Japan, to power a department store and a chemical plant.‡[56]

The importance of the combined-cycle technology is that it leads the way toward significant achievements in eliminating energy waste by adapting the power plant to specific needs of individual locations. The most efficient combined-cycle installations generate electricity through energy conversion in several stages, finally using the waste heat from the engines for other purposes. L. O. Tomlinson, the manager of an engineering firm (Power Systems Engineering, Inc.) that specializes in power plant engineering, says that the secret of the combined-cycle system is decentralization:

> Combined-cycle systems are well suited for application in decentralized utility systems arranged for serving large industrial customers with electric power and steam or hot gas for process heat. These systems are highly efficient because steam condensed in a process or hot gases used for process heating perform useful functions in place of the traditional problem of dissipating the heat to cooling water or to atmosphere.

Tomlinson backs up his contention with a comparison of the efficiency of a combined-cycle plant in delivering electricity and steam to a factory vs. the energy costs of delivering electricity and steam from a conventional power plant. The conventional process uses 28 percent more fuel and releases 28 percent more heat to the environment than the decentralized combined-cycle plant.[57]

Thus, as is the case with other energy conversion methods, the optimum use of combined-cycle power plants is in specific locations where the maximum efficiency of converting fuel energy can be ex-

‡ The fluorocarbon turbine has been developed by an American inventor, Wallace Minto of Sarasota, Florida, for powering automobiles. Minto's company, Kinetics, Inc., has contracted with Japanese car manufacturer Nissan Motors for the installation of the turbine in Nissan's Datsun automobile. Similar applications of closed-cycle turbines use other fluids, such as isobutane, for transferring heat. Several American and Japanese designs have been applied toward electricity generation using geothermal heat as the heat source.

pressed as a function of electric power generation *and* the use of waste heat for some useful purpose, such as refrigerating or air conditioning a building.

Unfortunately, this is not the use that electric utilities in the United States have in mind for the combined-cycle power plants. By early 1973, utilities had ordered a total of 18 combined-cycle power plants, representing a total of 4,500 megawatts of electric power, about the equivalent of 5 large fossil-fuel power plants.*[58]

These plants will in most cases be used for production of electricity at current efficiencies—just under 40 percent—which makes the combined cycle-plant no more efficient than the conventional fossil-fueled steam power plant. In addition, the fuels that are burned by combined-cycle plants are "premium" fuels—natural gas and distillate fuel oils—both of which are cleaner-burning than heavier oils and coal, and both of which are in short supply. Recognizing this fuel limitation has prompted the federal government to encourage the future use of combined-cycle power plants at sites of coal gasification plants, where electricity can be made from synthetic natural gas made from the coal.

Decentralized Power System

During the famous electric power blackout of November 1965, which left much of the North Atlantic seaboard without power or lights, several newspaper photographs of New York City showed a surprising sight: a colony of lighted buildings in Queens. The lights were on on that night in November because the buildings were in the Rochdale Village cooperative housing project—consisting of apartments and a housing complex—which had been built with its own power plant (of 20,000 kilowatt electrical generating capacity) and was thus independent of the public utility grid.

Back in the nineteenth century, the buildings and homes of many wealthy people contained their own generating plants for supplying

* Combined-cycle power plants using gas turbines as the basic plant component, and gas turbine power plants used alone (with less efficient fuel conversion) are increasingly being viewed as the answer to peak power needs by the nation's electric utilities. Since the plants can start operation within seconds, they provide an excellent source of standby power when utility loads are greatest. New York City's utility, Consolidated Edison Company, announced in 1973 that 880,000 kilowatts of capacity added to its system in the late 1970s would be gas turbine and combined cycle plants. Con Ed already has a number of barge-mounted gas turbines in service in New York City. Adding the newer units will afford greater energy efficiency by using the combined cycle plant in conjunction with the supply of steam for the utility's district steam service to buildings in the city.[59]

heat and electricity. Thomas Edison himself designed a power generator for the Vanderbilt mansion,† and major buildings with independent power systems included the U. S. Capitol, the Chicago Board of Trade, and New York's Park Row Building. The trend was reversed within a few years after Edison opened the Pearl Street central electric generating station in New York; and by the early 1900s, independent plants were eliminated—abolished by the convenience of the central electric system.[60]

"Total Energy" Systems

By 1960, however, the trend toward a return to the concept of independent power generation was beginning under the label of "Total Energy," a term coined by the gas industry.

The Total Energy plant does not on the surface appear to be much different in concept than the plant Thomas Edison built for the Vanderbilts a century earlier. However, the Total Energy plant in operation represents one of the greatest single contributions to the savings of energy in the history of power plant engineering. The reason is a simple one. Throughout the development of electrical power, there had never been a great concern for the waste heat given off by the power plant; only a concern for the minimization of costs incurred in wasting fuel. More efficient electric plant prime movers—steam turbines—used less fuel, which saved money for the plant owners. However, even with the modern fossil-fueled plants, 60 percent of the heat at the facility is still rejected to cooling water. And, as has been pointed out, other engines such as gas turbines burn fuel at only 25 percent efficiency, resulting in the conversion of three fourths of the fuel energy into waste heat.

The Total Energy plant is designed around this one factor: waste heat. In the Total Energy plant, the waste heat from the prime mover is put to work to provide air conditioning and heating for the occupants of the building. In many cases, the useful application of this "waste" results in the total reversal of the central station power plant efficiency ratios. For example, engineers speak of Total Energy plant efficiencies in the range of 85 percent, meaning that 85 percent of the fuel burned in the plant's engine is used for something—either making electricity or heating and cooling rooms.

Thus, the Total Energy plant, which—in a sense—is a *throwback* to the period that preceded Edison's revolutionary central station electrical generator, wastes as little as 15 percent of the available fuel

† Mrs. Vanderbilt had the generating plant removed because she was annoyed by the noise it generated.

energy, compared to the modern coal plant, which wastes 60 percent or more; and compared to the even more sophisticated nuclear power plant, which wastes as much as 70 percent of the energy of the fissioned uranium fuel at the power plant.

At present, there are approximately 600 Total Energy plants in the United States, ranging in size from a few hundred kilowatts of electrical energy to more than 20,000 kilowatts of electrical generating capacity in a few plants like Rochdale Village.[61] A Total Energy system is generally characterized by the following services and components:

1. The plant produces electricity for the building or development at the site of use; fuel for the generating plant is usually natural gas or distillate fuel oil, and is either pipelined directly to the plant or is delivered by other transport means.‡

2. The electrical generating engine in the Total Energy plant is always equipped with heat recovery equipment to distribute heat that may be used for the operation of air conditioning or refrigeration equipment; for a variety of industrial functions (process steam and heat); and for space heating.

3. The plant serves a single site, and the power and heat produced by the plant do not cross public thoroughfares. Electric utility legal franchises in the United States prohibit the passage or sale of power from independent plants across public thoroughfares. Observing this legal constraint, the Total Energy plant may be an installation comprised of a single engine and associated equipment serving a small building, or it may be comprised of a number of large engines linked together that produce power and heat for a housing development or an industrial park.

The lines that divide several energy technologies discussed in this book are often not distinct. For example, an electric utility may own a power plant that not only produces electricity, but also distributes heat or steam to industrial or commercial customers, as in the combined electrical district heat plants. This is nothing more than a scaled-up version of a Total Energy plant. The advantage of the Total Energy plant is a matter of scale: The plant can be carefully built according to both present and future needs of a specific enterprise. The energy savings are largely a function of this scale.

‡ The most desirable form of fuel delivery for a Total Energy plant is fuel that can be pipelined to the plant, such as natural gas. Not only is fuel usually cheaper in this form, but—as has been pointed out—it takes considerably less energy to pipeline fuel than to deliver it via truck or railroad.

Electric utilities may operate advanced power plants, such as combined cycle power plants or fuel cells, in order to produce great fuel economies and effect the delivery of more energy to customers, but unless these devices are equipped so that their waste energy can be used, society will not realize most of the savings derived from greater energy efficiency in the advanced engines. Only through the process of decentralization can maximum energy savings be fully utilized by society at large. Interestingly, this concept has been adopted by at least one electric utility—Southern California Edison Company, which has installed several Total Energy plants for customers, and provides maintenance by its own service crews.[62]

One of the nation's leaders in advancing the use of Total Energy equipment is New York architect and engineer Fred S. Dubin. In a manual on the design of Total Energy facilities written for the Ford Foundation-supported Educational Facilities Laboratories,[63] Dubin analyzed the selection of plant equipment for use in schools that could be adapted for use in other buildings. He pointed out that the selection of engines and related equipment for Total Energy plants must be closely related to the "electrical load profile" of the building or building complex. The load profile is calculated by ascertaining the total electricity demands of the building over a twenty-four-hour period, and over the entire year. Also, all energy needs—cooking, hot water, space heating and cooling, etc.—should be calculated so that the exact size and type of engine can be chosen for the job. This precise energy calculation means the difference between success or failure of the Total Energy plant. For example, the system should be designed so that the engines and electrical generators meet the energy needs of average use most effectively.

Dubin points out that all engines used in Total Energy systems operate most efficiently when they are running at 60 percent or more of their rated electrical capacity. If a 1,000-kilowatt engine generator is only producing 300 kilowatts of electricity, it does not burn the fuel as efficiently as it does at full load. "Since the electric load in schools and colleges varies for different conditions of occupancy, the most efficiently designed Total Energy plant would be one which utilizes generating sets sized so that combinations of the sets are always operating as nearly as possible to full rated capacity," he wrote.

One intriguing aspect of Total Energy plant selection relates to the kind of engines chosen. Engines commonly used in the plants are either gas turbines or reciprocating engines, such as the diesel. As has been noted, gas turbines are not particularly efficient—about 25 percent of the fuel is converted to electricity (at full load), so that the majority is converted to waste heat. Diesel engines are a third more efficient in producing electricity.

However, this very inefficiency makes the gas turbine particularly attractive for Total Energy plants under certain circumstances. Because it is inefficient in converting fuel to electricity, it produces twice as much waste heat as the more efficient reciprocating engine. Dubin points out, for instance, that a gas turbine can produce 7 to 13 pounds more steam per kilowatt generated (at 15 pounds per square inch) than can a reciprocating engine, which will produce 4 to 6 pounds of steam per kilowatt generated. This means that in situations where significant quantities of waste heat—but not so much electricity—are needed, Total Energy plants powered by the "inefficient" gas turbine may actually have efficiencies of higher than 65 percent.[64]

The waste heat is collected and applied for a variety of purposes: for hot water; for space heating of buildings in winter; for refrigeration of food; and, if the Total Energy plant is in a factory, the heat can be used for industrial steam or heat processes. The recovered heat can also be used to power a refrigerating or air conditioning system by means of an absorption refrigeration machine that uses steam directly to produce chilled water. The design of absorption refrigeration is based on the principle of the heat pump, which is discussed in some detail below.

As was noted earlier, the second largest category of energy use in the United States is for space heating of buildings and residences. Rather than use electricity for this purpose, decentralized Total Energy plants can accomplish much of the job at cheaper costs and with tremendous savings in energy.

Installations that require less heat and more electricity will find the more efficient Diesel engine more appropriate than the gas turbine for the Total Energy plant. Determining the relationship between heat and electricity is the secret of Total Energy, and understanding the relationship between these energy forms is the first, basic step in solving the country's energy crisis. Because this interrelationship is both important and complex, Total Energy intallations call for the integrated skills and thought of engineers, architects, builders, and specialists in many other fields.

As for the economics of Total Energy plants, here are some case histories:

The builder of a suburban Kansas City garden apartment project of 490 units decided that he wanted to use a Total Energy plant as the power system, but could not afford to install it initially because of the high purchase costs. A modified heating and cooling plant—which did not provide electricity—was purchased from a local utility and installed in the development. After 96 of the units had been occupied, the owner invested in an electrical generator to complete the system. Before the switch to the complete Total Energy plant, the cost of the purchased electricity was $6,000 per month. After the conversion, the costs of power dropped to $1,000 per month.[65]

A smaller Kansas City apartment project, with only 90 units, was equipped with a Total Energy system for a combined cost of $195,000. The Total Energy plant, mortgaged over a 20-year period, was estimated to incur about $15,600 per year in maintenance, fuel, and other expenditures vs. a conventional electrical system, which would cost the owner and tenants $31,000 per year. The calculated net savings of the Total Energy plant, including all mortgage investment costs, was estimated at $5,600 per year.

The installation of a Total Energy plant was recently completed at the Worcester, Massachusetts, Science Center. The center wanted a Total Energy plant installed so that the public could see its actual operations in the "Hall of Energy." In addition, the center required a number of unusual services from the plant, including keeping an indoor tropical rain forest at constant 80° to 85° F temperature. The Total Energy plant, fueled with natural gas, cost over $300,000 to purchase, but compared with a conventional gas-and-electric system from local utilities, a net savings of almost $24,000 per year was computed for the Total Energy facility. The center's director of public affairs explains that "from a conservation point of view, the Hall of Energy illustrates the fact that gas Total Energy is a very efficient system recovering much of the energy which would normally be lost with other systems."[66]

To date, not much interest has been shown in the Total Energy concept by either American industry or government, but at least two pioneering federal programs of the Department of Housing and Urban Development (HUD) are designed to use Total Energy systems. The first HUD program, called Operation Breakthrough, was initiated in mid-1970, and called for construction of 2,800 dwelling units* to encourage the modular, industrialized approach to housing by the nation's builders, who have never been noted as a particularly innovative group.

HUD requested the National Bureau of Standards (NBS), which does research on energy conservation in buildings, to ascertain whether or not a Total Energy plant could be used at one or more of the "Breakthrough" sites. The bureau decided on a small site in Jersey City, New Jersey, that is designed to have a total of 488 dwelling units in

* A "unit" is defined as the housing facility for one family irrespective of the kind of building the unit is in.

6 buildings plus 2 schools, a swimming pool, and a commercial building. NBS conducted a study of the specific type of Total Energy system that would be most appropriate for the project, and recommended the following components:

1. Five Diesel engines, each with an electrical generating capacity of 600 kilowatts. Three of the engines can supply all the energy needs of the community, so that the others can be serviced or repaired without eliminating the self-sufficiency of the community power system.

2. Heat recovery boilers to recover 60 to 70 percent of the energy content of the Diesel fuel, with the recovered heat to be used to provide heating in winter and hot water year-'round. In summer, air conditioning is to be supplied by the production of steam from the engine's waste heat to operate 2 large (546-ton) absorption chilling machines.

3. An automated, pneumatic waste collection system for household garbage and trash. The waste system, powered by the Total Energy plant, consists of horizontal steel pipes 20 inches in diameter, pressurize so that wastes funneled into the pipes from the apartments in the complex can be quickly sucked into a central collection plant. The wastes are moved under pressure at a speed of 60 miles per hour. Pneumatic waste systems are in use in Sweden, France, England, and West Germany; and at least 2 U.S. installations are using the systems: Disney World in Florida (which has several miles of underground tubes) and Roosevelt Island, a new community planned in New York City.†

The Jersey City Total Energy plant was constructed in 1972 and 1973, and the National Bureau of Standards and the Department of Housing and Urban Development will carefully monitor the performance of components and the use of energy to determine the extent to which the agencies will be involved in the future with the Total Energy approach.[67]

The NBS has concluded that the Total Energy system in the Jersey City project would produce cheaper annual energy costs than other alternatives studied, including conventional electric heating and cooling‡ or natural gas or oil-fired conventional heating plants in combination with electricity for summer air conditioning. The NBS also studied other

† In terms of energy savings, the use of automated systems such as these require additional energy to operate, energy that would otherwise be derived from human muscle power, from conventional garbage trucks, etc. The pneumatic systems may require less over-all energy than a system of individualized garbage trucks collecting garbage, but other systems might be developed to enhance collection of garbage by people, eliminating the use of automated, energy-consuming equipment.

‡Using either electric heat pumps or conventional equipment.

Project Breakthrough housing sites in the nation and concluded that Total Energy systems would save from 25 to 60 percent of energy costs in comparison with other systems.[68]

In addition, the bureau concluded that the potential environmental problems from the operation of the Total Energy plant at Jersey City would include emissions from the engine's exhaust as well as noise problems and vibrations from the operation of the machinery. The detailed investigation of the Jersey City site is designed to measure the full impact of the environmental factors.

Excited by the prospects of the Total Energy system program they had initiated, HUD has expanded the concept into a full-fledged combined community waste-disposal, water-treatment, and energy system that the agency calls "MIUS"—for Modular, Integrated Utility System. HUD's MIUS project involves a several-year program, which began in early 1973. The full program is in three phases. The first phase is designed to enlist the cooperation of a number of government agencies, including NBS, the National Aeronautics and Space Administration (NASA), the Environmental Protection Agency (EPA), and the Oak Ridge National Laboratory, which is operated for the Atomic Energy Commission by Union Carbide Corporation.

The ultimate goal of the MIUS program is to fully develop Total Energy technologies in combination with the waste disposal and other community systems so that private utilities will become involved. The greatest obstacles to utility involvement today are political and institutional problems, not technical ones. Private electric utilities, for example, have extensive rights to the transmission and distribution of power across public thoroughfares, and use of this legal weapon is a fundamental reason why there are few Total Energy systems or more advanced MIUS-type systems in use in the United States.

The key to the MIUS program is the development of *modular* components that can be used with Total Energy (and utility) systems. Currently, all Total Energy plants installed in the United States are custom-built and -engineered for specific locations. HUD hopes that the development of modular components—such as engines, control systems, and heat-recovery equipment—will lead to mass production and intensified development of Total Energy and related technologies. With mass production, the agency predicts that the plants can be built and installed in industrial, commercial, and residential sites much faster than any other kinds of power plants. "Because of the smaller size and maximum use of factory assembly," HUD says, "the planning-to-operation time span can be reduced by 75 percent."[69] This means that Total Energy plants could be installed two years after they were ordered, as opposed to large central-station power plants, which take an average of eight years from date of order to completion.

Environmental Implications

According to HUD, the environmental impact of MIUS and Total Energy facilities would be significantly less than any other types of power plants and urban facilities. HUD estimates that thermal pollution would be cut in half, and combustion-generated pollutants would be reduced by 35 percent compared to conventional power facilities. By incorporating solid-waste and water-treatment technologies in the MIUS complex, liquid-waste effluents would be reduced by 80 percent, and 65 percent less land would be needed for waste disposal.*

With the widespread introduction of MIUS facilities on a national scale, HUD foresees the following national gains from the introduction of the plants: †

- Assuming 16 percent of the electricity market were met by Total Energy plants, the amount of electricity saved in eliminating the need for costly and inefficient central facilities would be worth about $3.6 billion.
- The fuel savings represented by use of the efficient Total Energy plants would represent the equivalent of $2.7 billion by 1986 (same conditions as above) in fuel that would otherwise have to be imported.
- By 1986, Total Energy plants would represent an $80 billion annual market in the United States.[71]

Politics, Economics, and Total Energy

If the Total Energy power concept is such a significant improvement over the central electric station, then why hasn't it caught on; why are there only six hundred such facilities in the United States?

The answer to this question has to do with human responses, politics, and economics. One of the stumbling blocks to the development of Total Energy is the fact that each plant is unique—a custom-engineered installation built for the specific needs of the building or small com-

* A description of ways to use solid wastes as fuel for Total Energy and other power plants is discussed in Chapter 6.

† MIUS-type technology has been developed by the natural gas industry at the Southwest Research Institute (SRI), 8500 Culebra Road, San Antonio, Texas 78284. While SRI calls its approach a "Total Utility Plant," it incorporates the same items as the MIUS concept. An excellent service of SRI is computer-programed energy analysis of proposed Total Energy plant needs. The computer service, called the "E Cube Program," is offered to prospective Total Energy customers so that they can determine the energy requirements, equipment selection, and prospective economic of a Total Energy installation.[70]

munity it serves. This fact alone runs counter to the primary trend in energy use, that of convenience. It's easier to let the local electric utility furnish the power supply than to worry about the complexities of designing and engineering one's own power system. The Total Energy plant is also a "new" concept, and—in the minds of many—unproven; whereas the local utility is familiar and, before the recent years of system collapse with ever-more frequent blackouts and brownouts, the local utility was considered fully reliable.

In economic terms, most of the Total Energy plants installed to date have operated successfully, but there have been some failures as the result of poor design, resulting in the replacement of units and a switch to electricity. In a Total Energy installation, the heat and electricity requirements have to be precisely calculated so that the system will perform optimally year-'round. Some systems constructed in the past have not been well thought out, leading to improperly sized engines and heat recovery systems, which resulted in failure of the facility.

Other than the question of poor design, the major economic stumbling block to the development of the independent power plants is the fact that the building owner must decide to invest the extra money for a power system when the building (or housing project or community) is still on the drawing boards. Even though Total Energy installations result in decreased over-all costs, this saving is experienced over the lifetime of the building. In our economy, this concept of "life cycle costing" is in opposition to the accepted method of lowest possible first costs. Not only in power equipment, but in almost every other economic area, the aim seems to be to get the cheapest possible item, not the best possible item, to the consumer. Had builders and investors given greater attention to the lifetime savings that Total Energy plants can bring, far more Total Energy plants would be in operation today.

Another, more pernicious trend has prevented the widespread appearance of Total Energy plants: an open and hostile opposition to the plants by most private electric power companies, which view Total Energy plants as competition that must not be tolerated. At the time the New York 1965 blackout pinpointed the lights of Rochdale Village, showing the nation the merits of Total Energy, there were fewer than 100 of the independent power plants in the country. By 1968, there were more than 450. This situation clearly distressed the private utilities, which saw a clear and present danger to their rule over the power market. In typical fashion, the electric companies—through their trade association, the Edison Electric Institute (EEI)—set up a special "war council" to do battle with the gas companies. The first meeting was held in October 1968, when the EEI set up the "Program to Combat Isolated Generation."[72]

The major weapon in the electric utilities' war chest was, and continues to be, the practice of setting low electric rates. In cooperation with the state and municipal utility rate commissions, which supposedly regulate the utilities, rates traditionally favor bulk users of power. To eliminate the threat of Total Energy, the utilities went a step farther: They set special promotional rates for projects and buildings that might find it advantageous to plan for a Total Energy plant. They also adopted extremely high electrical rates for the owners of Total Energy plants so that, in case of failure of their independent equipment, they would have to pay a premium price for any assistance from the utility in getting over the crisis period. This second measure would have the added effect of convincing other prospective Total Energy customers that the way was not clear for independence from the rule of Reddy Kilowatt.

Perhaps the best case study of the attacks on Total Energy plants is that made by lawyer-consultant C. Girard Davidson, who reported to New York City's Consumer Council in 1968 on the practices and reliability of electrical service to the city by the Consolidated Edison Company (Con Ed). Davidson told the Consumer Council that the giant private electric company, which charges New York City residents the highest electrical rates in the country, was careful not to lose a customer: "For even though it is a supposedly publicly regulated utility, special rates designed to fit the load are promulgated when there is danger of losing a customer."[73]

Davidson's example of this practice on the part of Con Ed was the utility's successful attempt to prevent the installation of an energy-conserving Total Energy plant in the World Trade Center, New York's newest superskyscraper. The World Trade Center, which consumes more electricity (80,000 kilowatts) than cities like Stamford, Connecticut, or Schenectady, New York, was originally designed by the architect/engineers with a Total Energy plant power supply. When they learned this, Con Ed went into action and offered the builders of the World Trade Center a special "promotional" package of electrical rates—at a cost far below what other New York consumers pay for electricity. This offer successfully prevented the installation of the more logical Total Energy plant in the building.

Con Ed President John V. Cleary, defending against charges that electric utilities constitute a monopoly, publicly substantiated Davidson's testimony. "There is no semblance of monopoly," Cleary said. "We compete with the oil industry to heat homes and industry. A striking example of competition is the coming World Trade Center. They could put in their own power supply. The center will use more electricity than Albany, or perhaps, even Schenectady. We have to give them a good price."[74]

During the same period in the 1960s that Con Ed thwarted the installation of the World Trade Center's Total Energy plant, the utility applied to the City's Public Service Commission for special cheap rates—called "Service Classification 13—Bulk Power—Housing Developments," which it could use to get the business of projects like Rochdale Village. Commented Davidson:

> It is generally understood that this rate was filed by Con Ed in order to obtain the electric load of Co-op City, a 15,000 [population] co-operative apartment development in the Bronx. This project had decided to install its own generating plant since the then-existing rate of Con Ed was much higher than the project's costs in generating its own power. This special rate schedule served the purpose of preventing installation of further generating plants as was done by two sister co-ops, Warbasse Houses in Brooklyn and Rochdale Village in Queens.[75]

The consequence of Con Ed's policy is the expansion of its own central station electrical system, which has proved to be the most blackout-prone, expensive, and unreliable electricity generating and distribution system in the United States. At fault here is not only the utility, with its consummate greed, but also the Public Service Commission, which supposedly serves the interests of the people of New York City. The complicity of the Public Service Commission in permitting Con Ed's deliberate war against Total Energy plants takes its toll today with each blackout and brownout and each pound of extra pollutants choking the atmosphere of metropolitan New York City, compounding the miseries of its nearly 12 million residents.

A further footnote to the story is the fact that in 1972 Con Ed installed 348,000 kilowatts of gas turbines mounted on barges at Verrazano Narrows; and announced in 1973 that 880,000 more kilowatts of various peak power plants would be added to the utility's system (presumably mounted on barges or in some location in the city) during the late 1970s. By coincidence, the peak plants eagerly purchased by the utility contain engines virtually identical to engines used in Total Energy plants. Were the Verrazano Narrows gas turbines located inside the skyscrapers of New York, the waste energy from their combustion of fuel could be used for heating and air conditioning, using up to 85 percent of all the energy in the fuel. Instead, the short-sighted utility has them out of reach of the buildings, so the heat—representing three fourths of the energy of the fuel—is simply wasted.

The Heat Pump

The heat pump is a remarkable device that seems to violate the laws of thermodynamics; i.e., it produces more energy as heat than is contained in the fuel used to operate it. In terms of energy conservation, it is one of the most significant developments of our time, and its use can extend from the fully decentralized application of a heat pump for each home to heat pumps for large buildings and communities.

The heat pump does not, of course, actually violate the laws of thermodynamics. The prodigious amount of energy it yields is due to the fact that it employs two energy systems: the energy that is supplied to it in the form of fuel or electricity, and the energy of the natural environment—the heat of the atmosphere, soil, or water—which it "pumps" into the space to be heated.

The heat pump operates on the same principle as the common household refrigerator: the intermittent evaporation and condensation of a refrigerant, or heat transfer agent, in a closed cycle. The heat pump has one additional feature, however: It can reverse its cooling cycle to provide heat instead of removing it.

In a refrigerator, a refrigerant, or heat transfer agent—such as Freon, ammonia, isobutane, etc.—is pumped into a chamber containing a near-vacuum created by a compressor. Under this lower pressure, the heat transfer agent readily vaporizes by evaporation, a process that requires a relatively great infusion of heat. This heat is drawn from the inside of the refrigerator, resulting in the well-known cooling effect that serves to keep food properly chilled. As the heat transfer agent continues through the cycle, it is pumped through a condenser, which returns it to the liquid state. This condensation process releases most of the heat that was absorbed earlier in evaporation. In the case of the refrigerator, this heat is released to the kitchen area.

In the case of a space-conditioning heat pump, the heat required for evaporation may be supplied by the ground, water, or the outside air. Outside air is the most common heat source, and heat pumps that draw their heat of evaporation from this source are called air-to-air heat pumps because they transfer heat from the outside air to the inside air, or vice versa.

The heat transfer agent is circulated through coils, some of which are inside the building and some of which are outside. In the summer, the heat required for evaporation is drawn from the inside of the room or building. The evaporation of the refrigerant in the inside coils cools the coils as well as the immediately surrounding air. The cooled air is then blown by a fan into the room to provide the familiar air-conditioning effect. The heat absorbed from inside the room in the evaporation phase is then released to the outside air when the heat transfer agent is returned to the liquid state in the condensation phase, which takes place in the coils outside the building.

In the winter, this cycle is reversed. The heat is drawn from the outside air which—while cold by human comfort standards—is still "warm" enough to provide the heat required to vaporize the heat transfer agent under the low pressures provided in the system.‡ Thus the outside coils absorb heat for vaporization from the outside air, further cooling that air by a minute fraction of a degree. When released inside the building during the condensation stage of the cycle, that heat serves to supplement the heat provided by standard space-heating equipment —either electric-resistance heating or fossil fuel furnaces.

Thus the utilization of the heat of nature reduces the need for artificial heat. It is this natural booster effect that makes the system inherently more efficient than other space-heating and air-conditioning systems.

The heat pump concept is wedded to the development of thermodynamic theory. Sadi Carnot, Rudolf Clausius, Lord Kelvin, and other founders of the science of energy behavior in engines used the concept of the "reversible engine" to test the applicability of their theories on heat and power.

The first application of the heat pump in the nineteenth century was for refrigeration. Use of the heat pump for both heating and cooling required another century of development, though the development was not strictly of a technological nature. Heat pumps were developed for cooling because of convenience.

Research on heat pumps in the 1920s and 1930s led to the installation of large building heat pump systems in the 1930s. General Electric Company (GE), one of the leaders in the heat pump field, installed large electric heat pump systems in electric utility company business offices in the 1930s. By 1945, GE produced its first three hand-made home-size heat pumps; and by the mid-1950s, a large commercial market was developed for residential heat pumps. Heat pump systems manufactured by GE, as well as by a number of other electrical equipment manufacturers (including Westinghouse, Carrier Corporation, Fedders, and others), were sold by the tens of thousands in the southeastern United States as air conditioning took hold.

‡ In order to reduce initial costs and the size of the heat pump, electric heat pumps available today generally are sized to have a "balance point" of 30° F or less. The balance point is the outdoor temperature at which the heat pump can heat the house without any supplementary heat. Above that point, the heat pump compressor will maintain the temperature set on the thermostat. Below that point, the compressor must be assisted by an auxiliary source of heat.[76]

The heat pump market developed largely in the South, because heat pumps operate better in warm climates, where winter temperatures generally do not go below freezing. Freezing temperatures interfere with the heat pump's operation by forming frost on the outside coils, which markedly reduces efficiency and increases cost. Analysis of electric heat pump operation in one northern city (Canton, Ohio) showed that the heat pump and defrost cycle consumed 80 percent of the energy boost the heat pump was designed to provide.[77] This problem can be alleviated by burying the pump's external coils—which carry the heat transfer fluid—in the ground, or by using water as the source for the heat supplied to the outdoor evaporation coils. Both underground water supplies and the earth itself below the surface remain above the freezing level, and such specially adapted heat pumps could be used in most areas of the United States. Heat pumps using on-site combustion, rather than electricity, as the external energy source can be designed so that part of the heat they generate for the pump operation can be supplied to the heat pump's outdoor air coils to keep it from frosting in freezing weather. This type of heat pump (nonelectric), then, can be easily adapted for northern climate operation. (In a subsequent chapter, the use of solar energy-supplemented heat pumps, which could be used anywhere in the United States, is explored.)

The electric heat pumps of the 1950s were notoriously unreliable; and, as early buyers can attest, the mechanical components of the system required frequent and expensive maintenance. The Air Force purchased thousands of heat pumps for air base housing in Arkansas and other states, but many of the units subsequently had to be replaced because of their unreliability. The problem centered primarily around the faulty operation of the electric-powered compressor—which compresses the heat transfer agent—and in the special reversible valve of the heat pump. The reversible valve allows the flow of the heat transfer fluid to be switched from season to season, so that in winter the outside coils serve as evaporators and the inside coils serve as condensers; and in summer, the inside coils become the evaporators and the outside coils the condensers.

So completely did the Air Force heat pump experience fail—compressor failures ran as high as 30 percent per year—that in 1965 the Department of Defense banned the purchase of heat pumps for Defense Department construction. In the meantime, however, numerous corrections and refinements have been made, and reliable heat pumps are now available for use in all but the northernmost regions of the United States.

General Electric tested its heat pumps in northern markets for ten years, but withdrew them (the "Weathertron" line) from the market in 1962 for extensive retooling. They were returned to the commercial market in 1964. Other companies also redesigned and improved their heat pumps in the mid-1960s, and a wide number and variety of electric heat pumps are commercially available today in the United States.[78] By 1970, an estimated 11 percent of electrically heated households in the United States were equipped with heat pumps.[79]

The reason heat pumps are not in wider use today is twofold: First, the reputation of unreliability has haunted the electric heat pump since its premature introduction on the commercial market in the 1950s. Additionally, since the improved heat pumps have become available, virtually no attempts have been made nationally by electric utilities, manufacturers, or retail outlets to inform the public of their superiority—or even of their existence. This blackout of advertising is astounding in the light of the heat pump's remarkable efficiency and the fact that the initial cost of central home heat pump installations are rarely more than 10 to 20 percent more expensive than those of conventional central air-conditioning system.* Assuming the higher operating expense of home electric heating equipment, it is quite possible that heat pumps are cheaper in most areas than the combined cost of a central air-conditioning and heating system; and the operating costs of the heat pump will save an average of two thirds of the yearly electric power bill—while using one third the energy of the common electrical resistance heating system.

* In Florida, one of the most suitable areas of the nation for the operation of heat pumps, the author surveyed newspaper advertisements on air conditioners and heat pumps over a period of several months. On no occasion were heat pumps found advertised as *energy-saving* devices—not even in a June 1973 Miami *Herald* special supplement containing numerous manufacturers' advertisments, including some describing air conditioners that saved energy by incorporating more efficient equipment than other air conditioners. Heat pumps, called "reverse cycle air-conditioners" and "heat and cool" units, are commonly available in the state in both window units (which in certain cases may cost 30 percent more than conventional air conditioners having an equivalent summer cooling capacity) and for central home installations. The following comment from the manager of an air-conditioning sales firm in Fort Myers, Florida, is typical of the response from commercial outlets: "People don't seem to be buying the reverse cycle units any more. Most of our sales now are central air-conditioning systems with strip heaters." Strip heaters are ordinary electrical resistance heaters. The manager confirmed that the combined cost of central air conditioning with strip heaters was in many cases more expensive than a heat pump installation, but he noted the convenience factor that heat could be provided more quickly with strip heaters. He was not aware that the heat pump saved energy.[80]

The Heat Pump and Energy Conservation

There are two types of heat pumps, the electric and the heat-actuated, although the only heat pump in wide commercial use today is the electric heat pump. The heat-actuated heat pump operates on direct heat provided by on-site combustion of a fossil fuel, such as gas, oil, or coal. However, it is still in the development stage and is not yet available commercially.

The electric heat pump uses electric current to drive a motor, which operates the compressor and condenser. Thus the electric heat pump—relatively efficient though it is in comparison to electric space heating and cooling without it—is still burdened by the inherent 70 percent-plus inefficiencies of the electric power generation and transmission system.

Despite these inefficiencies, a recent study on building insulation indicated that even in the relatively cold Minneapolis, Minnesota, climate, the heat pump significantly reduced electricity use for space heating. If the same number of people use electric heat in Atlanta and New York as use electric heat in Minneapolis, the study concluded that installation of a heat pump to boost the electric heat output in every electrically heated home would reduce electricity consumption for space heating in those cities by 45 percent.[81]

However, the way to gain real efficiencies in heating and cooling is to combine the central station plant and the heat pump at the site of use—the home—in a small-scale Total Energy plant, thereby substantially reducing the central station inefficiencies and transmission losses; in other words, installing a fuel-activated heat pump combined with an electric generator in the home that would generate the electricity required for household operation while channeling the waste heat into a heat pump for space heating and cooling.

One way of measuring the efficiency of machinery is by determining its "coefficient of performance," or COP. In the case of the heat pump, the coefficient of performance expresses the ratio of the energy put into the system by electricity or direct use of fossil fuels, to the work done to accomplish heating or cooling. Because the fuel or electricity that operates the heat pump is augmented by natural energy, many heat pumps have a coefficient of performance greater than 1; i.e., more energy is derived from it than was put into it in the form of electricity or fuel.

The coefficients of performance of electric heat pumps and various types of fossil fuel heat-actuated heat pumps show clearly the superiority of the heat-actuated pumps, as reflected in the following chart:

Air-to-Air Heat Pump	Air-Conditioning COP	Heating Season COP
Electric heat pump using Freon†	.60	.75
Heat-actuated pumps		
External-combustion stirling engine with Rankine energy conversion cycle	1.20	1.60
Conventional internal-combustion engine (Otto) with Rankine cycle	1.20	1.50
Gas turbine engine with Rankine cycle	1.20	2.0

[82]

A recent comparison of the performance of a residential electric heat pump with gas heating for thirty-four cities throughout the United States indicated that notwithstanding the relatively high efficiencies of the electric heat pump, it would be more efficient than a gas furnace—without a heat pump—in only ten of the thirty-four communities. And a test of relative efficiencies in Canton, Ohio, conducted from October 6, 1971 to April 26, 1972, indicated that a late-model, high-quality electric residential heat pump consumed 25 percent more energy than the equivalent gas system—again without a heat pump. These figures suggest that once a commercial fossil-fuel heat-actuated heat pump is developed and marketed, energy savings can be dramatic. Use of the system to generate the electric power needed in the home would result in even more startling savings.[83]

The former director of the American Gas Association's climate control research program wrote in 1973 that, given the "ambitious but conceivable assumption" that heat-actuated heat pumps with a "competitive first cost and a heating coefficient of performance of 2" could be introduced into all segments of the marketplace within the next few years, "it is possible to conserve close to 3 trillion cubic feet of natural gas in the year 1990 in heating applications," which he said would be equivalent "to approximately 75 percent of the gas the Federal Power Commission predicts will be produced by a successful coal gasification program."[84]

Conclusion

The decentralization of America's electrical generating and distribution system will bring tremendous energy savings, and will allow for

† Based on 29 percent over-all efficiency after central station generation and transmission losses.

greater stability in the distribution network. Today's electrical systems are overcentralized, and even minor problems in one part of the distribution grid can bring power loss to the entire system.

The technologies for decentralization are vital to the future development of power systems that can conserve energy and retain stability. What is needed is a national recognition of the importance of such systems, and the development of nationally directed research to bring them into widespread use.

5

Energy Resources in the Future

Within the next thirty years, major energy and materials shortages caused by a decline in the availability of cheap sources of oil, natural gas, and coal will force the advanced technological nations to develop options to avert social crises as economies decline through severely limited productivity. In order to develop decentralized technologies to carry us through this difficult period, it will be necessary to adapt the American society to energy *sources* that today appear novel and unimportant.

This chapter is devoted to an analysis of the energy sources upon which the nation is presently depending to fuel the American system over the coming years. Subsequent chapters are devoted to long-range energy sources that are not immediately available, but that offer the promise of unlimited, renewable energy—from the sun.

Fossil Fuels in the Future

With 3.2 trillion tons of coal deposits in the United States, coal is by far the nation's most extensive fossil fuel resource. The U. S. Geological Survey estimates that it represents 88 percent of proven reserves of all fuels in the United States, and 74 percent of the ultimately recoverable resources.[1]

With the stark realization that the nation is fast running out of petroleum products, coal is thought by many in government and industry to be the one potential fuel that could bail the United States out of its short- and middle-range energy crisis.

It is generally believed, however, that in order to fill the oil and natural gas gaps satisfactorily, coal itself would have to be converted to liquid and gas fuels resembling the petroleum products that have become the mainstay of the nation's fossil fuel-based transportation, space heating, and electric power generation systems.

Besides the convenience offered by gasification and liquefaction in terms of not having to convert or replace existing fuel-burning devices to utilize them, these two related technologies also offer another attractive advantage over coal in its natural state; and that is that the pollutants—the sulfur, the nitrogen oxides, the ash, etc.—are removed in the gasification process, leaving a clean fuel. It is both cheaper and simpler to remove pollutants at this stage than to try to remove them from the smokestack after coal is burned as a fuel.

Coal Gasification

Coal gasification technology involves subjecting coal to intense heat and capturing gases as they are released from the combusting coal. The principle can be observed in a wood-burning fireplace when occasional jets of flame may shoot out of the burning wood with a hissing and crackling sound. The same thing occurs with coal. The heat drives volatile gases out of the fuel that is decomposing in the center of the log or piece of coal.[2]

There are several methods being developed to bring coal gasification to a commercial scale. While it may look simple in the fireplace, to achieve large-scale controlled gasification calls for complex chemical and engineering procedures.

Depending upon the type of process used, the final gas product that emerges can range widely in energy value, as measured by BTU content. Low-BTU (125 to 175 BTUs per cubic foot) gas—also called "producer gas" or "power gas"—may be produced for burning in electric power generators. It is made up of hydrogen, carbon monoxide, some methane, and inert gases. It may also contain some toxic substances.

Other more complicated processes produce a "synthetic natural gas" that is very pure and has a very high BTU content—1,000 BTUs per cubic foot, the same as natural gas. It can be substituted directly for natural gas in the existing commercial pipeline network into homes, industries, electric power plants, etc.

Natural gas is predominantly methane. As the formula of methane

(CH_4) implies, it is a combination of one part carbon to four parts hydrogen. Because coal is largely carbon, it can be caused to react with hydrogen under high-temperature and -pressure conditions to produce methane and small amounts of hydrogen that are suitable for any purpose for which natural gas might be used.

Coal gasification is not new. References to it go back to 1688, when an observant English cleric, the Reverend John Clayton, heated coal in a retort and recorded the reaction. At first, "there came over only phlegm, afterwards a black oil and then, likewise, a spirit arose . . . the spirit which issued out caught fire . . . and continued burning with violence."[3]

By 1807, the first commercial coal-gas plant was furnishing fuel for lighting homes and factories in Manchester, England; and in 1812 the first gas utility, the Chartered Gas Light and Coke Company, was formed in London to furnish gas for the city's street lights.[4] Other experiments and commercial applications continued throughout the nineteenth and twentieth centuries in both Europe and the United States. By 1875 the United States boasted more than 400 operating gas companies.[5]

Devices called "gas producers," which made low-BTU, nitrogen-diluted gas from coal, remained in common use in America through the first several decades of this century. They finally disappeared completely in the years following World War II, when cheap and readily available natural gas eliminated the need for them.[6]

Germany maintained the lead in coal gasification research and development, and in 1933 a company called Lurgi Gesellschaft für Mineralöltechnik developed a new coal gasification process that still remains the only commercially available coal gasifier.

Named for the company that developed it, the Lurgi process produces gas through chemical reactions that take place among coal, oxygen, and water (as steam) at high temperatures and pressures. It yields a "reasonably high" heating value gas—some 475 BTUs per cubic foot—at an efficiency of 70 percent.

It still has been commercially utilized only in Europe, although El Paso Natural Gas Company has announced plans to build a $250 million, 250 million cubic feet per day Lurgi plant in New Mexico, with operation scheduled to begin in 1976. Commonwealth Edison Company of Chicago also plans to install 3 Lurgi units to gain experience with coal gasification.[7]

There are several disadvantages to gas produced by the Lurgi scheme. First, its relatively low BTU content means that it must be substantially enriched to substitute for natural gas. M. W. Kellogg

Company of Houston, Texas, has estimated that pipeline-quality gas could be produced by using a Lurgi gasifier followed by catalytic methanation, a commercially unproven reaction that converts carbon monoxide and hydrogen to methane and water vapor in the presence of a catalyst, releasing about 65 BTUs per cubic foot converted. The temperatures reach levels of 4,000° F and higher. Nickel appears to be the most effective catalyst, but iron also works satisfactorily.[8]

The cost of pipeline gas produced this way, the Kellogg firm has estimated, would be about $1.10 per million BTUs to the consumer, significantly higher than the cost of domestic natural gas but roughly the same as the anticipated price of other high-BTU processes and imported natural gas.

Another disadvantage to the Lurgi process is that the coal must be carefully sized, and many grades of coal cannot be burned in the Lurgi gasifier. Capacity is limited, which means that daily production is low and capital and operating costs are high. In addition, it is necessary to use large amounts of steam to prevent burning out the grades that hold the coals. This adds still more to the costs.[9]

Because of these problems, the United States has been focusing its research and development efforts, minimal though they have been, on more sophisticated—hence more complex and difficult—but potentially less expensive approaches. Four major processes under development include the HYGAS process of the Institute of Gas Technology, Chicago, Illinois; the Carbon Dioxide Acceptor process of Consolidation Coal Company, Pittsburgh, Pennsylvania; the Two-Stage Superpressure Gasification process being tested by Bituminous Coal Research, Inc., Monroeville, Pennsylvania, in affiliation with the coal industry's trade association, the National Coal Association; and the Synthane process, developed by the U. S. Bureau of Mines.[10]

The over-all scheme is the same for each of them. The gasifier in which the various reactions take place maintains pressures ranging from 20 to more than 70 atmospheres, and temperatures up to 1,500° C (2,732° F). Coal enters the gasifier under pressure and at temperatures of 600° to 800° C (1,112° to 1,472° F), under which conditions the volatile ingredients are driven off, and either converted to methane or collected for industrial uses. The devolatized coal then goes to the second stage, containing steam at temperatures of 900° C (1652° F) and above to form synthesis gas, which may be 40 to 65 percent methane. The resulting gas may pass a further step, during which the carbon dioxide, hydrogen sulfide, organic sulfides, and water vapor are separated out. The resulting gas then undergoes catalytic methanation, described above, to bring the heat value up to the necessary 1,000 BTUs necessary to qualify it for pipeline gas.[11]

"Power Gas" and Combined Cycle Plants

While high-BTU pipeline gas has received most of what limited federal research and development funds have been available, extremely low-BTU gas, or "power gas," with 125 to 175 BTUs per cubic foot, could be made from coal for use as a low-sulfur fuel in electric power generation. The type of power plant ideally suited to use power gas is the "combined cycle" in which hot gases are used to turn a turbine before going on to heat steam for a conventional steam cycle system at efficiencies of up to 60 percent—twice that of nuclear reactors and many fossil fuel plants.[12]

Because power gas has such an extremely low BTU content—150 BTUs per cubic foot—building both the gasifier and the power plant at the coal mine site is being proposed, since it would be economically more attractive to transport the energy of power gas by electric power transmission lines than by pipeline.[13] This is only true, however, if the fuel is going to be used for electric power generation only. A greater return from the energy would be achieved by skipping the electric power stage altogether, and upgrading power gas to pipeline quality for direct combustion of the site of ultimate use. Pipeline-quality gas is produced from coal at an average efficiency of 65 percent. A minute fraction of this is lost in transmission from plant to point of use.[14]

Power gas is converted at an average efficiency of 80 percent.[15] When this is again converted to electricity—even in a combined cycle plant at 60 percent efficiency—the resulting energy is only 48 percent of the original coal energy. Ten to 15 percent of that is lost in transmission to the point of use, so that by the time the original coal energy arrives at the point where it is going to be used, only 33 to 48 percent of the original coal energy is left. Further losses are incurred in the actual use of electricity.[16]

Thus, although pipeline gas conversion at 65 percent is less efficient than low-BTU conversion, it loses very little of that 65 percent en route to the point of use. Once again, space heating or cooking by direct flame is a much more efficient process than by use of electric power, so that much more of the original coal energy is eventually put to work if pipeline gas is the transportation medium than if a power gas/electricity combination is utilized.

Some promising future applications of power gas, however, are in fuel cells, which would convert the power gas to electricity at high efficiencies at the point of use; and as a fuel for magnetohydrodynamic (MHD) electric power generation.[17]

Future of Synthetic Gas

The National Petroleum Council (NPC) has projected that by 1985, synthetic gas from coal will supply amounts ranging from 500 billion cubic feet per year under the most pessimistic research and development circumstances, to almost 2.5 trillion cubic feet per year "at the maximum rate physically possible without any restrictions due to environmental problems, economics, etc."[18]

The plant size now being generally discussed would produce 90 billion cubic feet of synthetic natural gas per year from 10 million tons of coal at a plant investment of $200 million.[19]

Under the National Petroleum Council's most pessimistic projection, this would mean a total of 6 coal gasification plants in 1985; and under the most optimistic, about 28 plants.

However, this would not begin to fill the projected gap between the demand for natural gas and the domestic availability. The 1985 demand is predicted to be a little more than 41 trillion cubic feet, according to the NPC. The domestic supply will fall short of that amount by about 10 trillion cubic feet under the most optimistic conditions, and by more than 25 trillion cubic feet if the worst possible supply situation prevails.[20] The most likely 1985 availability lies somewhere in between, at a demand/supply gap between 14 to 20 trillion cubic feet.[21] To fill this gap with synthetic natural gas from coal would require the construction of from 166 to 233 coal gasification plants before 1985, each producing 90 billion cubic feet per year. The capital costs of such a massive engineering and construction undertaking—if it could be physically accomplished in the necessary time period—would range from $33.2 billion to $46.6 billion!

To make up the 1985 gas shortage in even the most optimistic projection of the National Petroleum Council would require some 120 such plants at a capital investment of $24 billion.

Environmental Problems

Some of the environmental implications inherent in the construction and operation of 120 such plants have been summarized in the American Chemical Society journal, *Environmental Science and Technology*:

> From an environmental viewpoint, land, water and air are all critically involved. One hundred twenty coal gasification plants would more than double present coal consumption. Strip mining is expected to be utilized to a large extent. The greatly increased amount of mining will require restoration of mined areas. . . .
>
> Thermal pollution in coal gasification is significant, since the conversion of coal to pipeline gas is only about 65 percent thermally efficient (120 plants would require disposal of about 5×10^{15} BTU per year). Heat disposal to the air is planned by evaporative cooling, which consumes most of the water. . . . Disposal of coal ash from the coal process is also necessary. . . .[22]

The article goes on to state that to synthesize the total 1985 U.S. natural gas demand would require 500 such plants.[23]

While synthetic natural gas is much more desirable as a fuel than coal from the standpoint of pollutants released when the two fuels are burned, the actual gasification process produces the same pollutants. It simply transfers the point of pollution from the electric power plant to the coal gasification plant many miles away. However, as noted earlier, it is cheaper to remove these pollutants by gasification than by trying to recover them from the smoke that goes up the stack when coal is burned at the electric power plant or in an industrial furnace.

A major environmental concern with the coal gasification process is the strip mining inherent in the cycle. Economic viability is based to a large degree on the assumption that the large amounts of coal that will be required (a single plant would require 605 tons of coal per hour, 10 million tons per year) would be stripped from the enormous western deposits of Wyoming, Montana, New Mexico, North Dakota, and Texas.[24] Strip-mined coal's price advantage is reflected in the 1970 price of $4.69 per ton for strip-mined coal compared to $7.40 per ton for deep-mined coal.[25]

Presumably with a conscious awareness of the objections that will be raised about massive increases in strip mining to satisfy coal gasification requirements, the American Gas Association (AGA) has been careful to assure that half of 176 potential gasification sites identified by the organization are deep-mining sites. Also, AGA director of research and engineering Douglas King believes that strip mining doesn't have to be objectionable. The land can be reclaimed, he points out, at a price. But if one plant reclaims land while another doesn't, the negligent plant immediately gains an unfair competitive edge. King calls for national legislation requiring comparable restoration of all stripped areas.[26] However, reclamation costs would likely result in the removal of the price advantage of strip mining vs. deep mining.

Despite reassurances, however, widespread coal gasification could lead to wanton devastation of some of the nation's most beautiful and rugged western wilderness.

Another major limiting factor in the development of coal gasification is water availability. A typical plant will require cooling water at the rate of 100,000 gallons per minute, one fifth of which would be consumed through evaporation.[27] Ironically, the very western lands whose coal reserves offer so much promise for coal gasification are also the same lands whose limited water supplies are already greatly overutilized in most cases. It has been assumed by most observers that the American Gas Association's 176 potential gasification sites were limited as much or more by availability of water as by availability of coal.[28] This was confirmed recently by a National Academy of Sciences (NAS) report that concluded that, while enough water is available for mining and land reclamation at most potential coal mine sites, not enough water exists for large-scale conversion of coal to other energy forms at the western locations. "The potential environmental and social impacts of the use of this water for large-scale energy conversion projects would exceed by far the anticipated impact of mining alone," a National Academy of Sciences official reported in 1973.[29]

Principal NAS staff officer Ralph A. Llewellyn wrote in a letter to the House Interior and Insular Affairs' environment subcommittee that "those areas receiving 10 inches or more of annual rainfall can usually be rehabilitated if the landscapes are properly shaped and if techniques that have been demonstrated successfully in rehabilitating rangeland are applied." Rehabilitation for areas receiving less rainfall than that will be difficult, however, he said, because "artificial revegetation . . . can probably be accomplished only with major, sustained inputs of water, fertilizer and management [and] may occur naturally only on a time scale that is unacceptable to society . . . decades, or even centuries." He said the driest areas are in the Four Corners area formed by New Mexico, Utah, Arizona, and Colorado, where massive strip mining is now under way to support coal-fired electric power plants in the area.*[30]

Thermal pollution is also a significant consideration in coal gasification. The 120 plants that would be required to make up the difference between natural gas demand and supply in 1985 would release 5 quadrillion BTUs of heat per year. Effects of this amount of heat on local climatic and weather conditions have not been studied.[31]

* For more extensive discussions of the Four Corners area, see Chapter 2.

A typical plant would also produce 14.7 tons of sulfur per hour, and 11.4 tons of slagged ash per hour, both of which would require disposal or recycling for commercial use.

Oil from Coal—Liquefaction

Coal could also be converted to a substitute for oil by a somewhat similar process, called liquefaction. Germany utilized it in attempts to overcome oil shortages in World War II. At the present time, only one coal liquefaction plant is in operation. That plant, in South Africa, first gasifies coal and then subjects it to high heat and pressure in the presence of a catalyst to form gasoline, Diesel fuel, and waxes.[32] This method, a variation of another German technique called the Fischer-Tropsch process, is a roundabout method considered economically unattractive in the United States.[33]

More direct processes appear promising, but research has lagged to the point of being almost nonexistent. The threat of petroleum shortages has sparked new interest in the technology, however, and it has begun to appear that liquid gasoline and oil from coal might soon be economically competitive with rising costs of imported petroleum products.[34]

Liquefaction has been pursued both for the production of low-sulfur boiler fuel and for gasoline, although most research has concentrated on boiler fuel. There are four general concepts under investigation for production of oil from coal. All of them involve the same basic process, in which crushed coal would be mixed with an oil solvent, subjected to pressures of at least 1,000 pounds per square inch, and heated. At about 250° C (482° F) the coal begins to dissolve in the solvent, at which point it is admitted into a chamber with hydrogen. The various processes differ slightly in the remaining steps, but they produce substantially the same end product.[35]

Major U.S. processes include:

1. The Pittsburgh and Midway solvent-refined coal process in Pittsburgh, Kansas, supported by the Office of Coal Research and carried out by Pittsburgh and Midway Coal Mining Company.[36]

2. The Consolidation Coal Company process, at Cresap, West Virginia, under the auspices of the U. S. Office of Coal Research. It was originally designed to investigate the production of gasoline from coal, but incompetent management and canceled federal funding put the project in abeyance for several years. Proposals have been made since then to alter it to a low-sulfur heavy-oil process for power plant boiler use.[37]

3. The Hydrocarbon Research, Inc. (HRI) project, operated by HRI Laboratories under Office of Coal Research auspices until funding was discontinued several years ago. A consortium of 6 companies has resumed research on the process, also called "H-oil."[38]

4. Project Seacoke, undertaken by Atlantic Refining Company (now Atlantic Richfield Company). Office of Coal Research support for this project has been discontinued in favor of a similar process, called the COED (Char Oil Energy Development) project, developed by FMC Corporation.[39]

Environmental considerations for liquefied coal are substantially the same as for gas generated from coal, except that there is some question about whether it can be burned with acceptable emissions of nitrogen oxides.[40] Since coal liquefaction is a less efficient conversion process, however, thermal emissions—hence, water requirements—will be greater.

Many are now calling for substantially increased federal funding for coal liquefaction, among them Philip H. Abelson, editor of *Science* magazine, the journal of the American Association for the Advancement of Science. "In comparison with the billions we spend on oil imports," Abelson wrote in a mid-1973 *Science* editorial, "the millions the government is devoting to liquefaction of coal can best be described as a phony commitment—a cosmetic effort whose purpose is to give the appearance, but not the reality, of action. A goal worthy of the world's leader in technology would be to construct in 2 years several plants, each costing about $1 billion and each capable of supplying one percent of the liquid hydrocarbons we consume."[41]

Oil Shale

Oil produced from extensive deposits of oil shale in the western United States is frequently mentioned as one energy resource that might help close the gap between the U.S. demand for petroleum and the amount that can be domestically produced.

Oil shale is a finely textured sedimentary rock that contains a solid tarlike organic material called kerogen, which melts at temperatures of 450° C to 600° C (842° F to 1,112° F). At those temperatures, the kerogen releases vapors that can be converted to shale oil, which can in turn be refined into oil, gasoline, and other petroleum products.

Oil shale deposits, located primarily in Colorado, Utah and Wyoming, are estimated to contain the equivalent of 1.8 trillion to 2 trillion barrels of oil—several times the estimated recoverable oil in the United States. The technology exists to produce oil from shale. The Bureau of

Mines has developed a process that was demonstrated in 1966–67 on a 360-ton-per-day scale by the Colorado School of Mines Research Foundation working with 6 major oil companies.[42] Union Oil Company also demonstrated a somewhat different process at a 1,000-ton-per-day rate in 1958. A consortium of oil companies called The Oil Shale Corporation (TOSCO) in 1967 demonstrated still a third process at a 1,000-ton-per-day rate.[43]

All three of these processes involve what is called above-ground retorting, which is the process of heating the oil shale and collecting and distilling the gaseous products to produce the thick, black, heavy shale oil that can in turn be refined to produce conventional oil products. Another process being studied involves "*in-situ* recovery," which entails breaking up (fracturing) the shale underground with the use of hydraulic pressure, liquid chemical explosives, or nuclear explosives. Instead of removing the fractured shale from the ground for retorting, the process would be carried out in-place underground by injecting air or gas to support the combustion required to heat the shale below the ground. The product shale oil would then be drawn out of the ground for further refining.[44]

The concept of *in-situ* retorting offers several advantages. The main one is that the overriding problems of strip mining and waste disposal associated with above-ground retorting would be eliminated.

Disadvantages are numerous, however, especially if nuclear explosives are employed, as seems most likely under the persistent prodding of the Atomic Energy Commission and the U.S. nuclear power industry. James Rathlesberger, writing in *Not Man Apart,* has enumerated the potential problems resulting from nuclear explosions, including: damage to surface structures by direct vibration or by differential settling of the soil; increased radiological exposure due to radioactive gases and dust emitted into the atmosphere; radioactive contamination of the refined oil; radioactive contamination of ground water passing over or through the rubble; possible sedimentation of underground aquifers and surface waters during the shocks of the blasts; and possible damage to ground waters from mobilization of heavy metals by the heat of the blasts.[45]

Regardless of what explosives are used, reliable methods for fracturing are not developed; and the ability to control the underground combustion is highly uncertain. The pressure at which combustion air must be supplied is still unknown.[46]

Costs of production in *in-situ* recovery are also unknown, although it is speculated that they might be less than in above-ground retorting because mining and waste disposal costs are eliminated. It is questionable whether this assumption takes into account the research and development and continuing support costs incorporated in the nuclear technology involved in *in-situ* retorting with nuclear explosives. It has been estimated that conventional above-ground retorting could produce crude oil at about $4 plus or minus $0.40 per barrel, which could become competitive with imported oil if other oil-producing countries continue to raise their prices.[47]

There are several reasons why production of oil from shale has not been developed into commercial stages. Cost, obviously, is one. Another is that the federal government owns the land on which about 80 percent of the oil shale is located, which has raised questions of several kinds among interested companies, various government agencies, and members of the general public. Some have objected, for instance, to government plans to lease land containing $300 billion worth of oil shale deposits to major oil companies for a token price of $0.50 per acre.[48]

One problem of ongoing concern is the strip mining that would be inherent in an above-ground retorting operation. Rathlesberger has noted that if oil shale development goes beyond the prototype stage, the Department of Interior anticipates the production of 3 million barrels of oil per day, which means the daily mining of 4.5 million tons of oil shale—most of it by stripping.

The greatest problem, and perhaps an insoluble one, is the environmental degradation attached to the disposal of wastes from the above-ground retorting process. Even high-grade oil shale is about 87 percent rock or inert material, and the tailings from spent shale swell up to almost twice the volume of the original shale. The federal government has proposed a prototype leasing program in which 6 sites would be leased to "stimulate development by private enterprise." The idea is for industry to develop shale oil production projects at each of the sites. By 1983 the 6 prototype sites could be producing as much as 1,125,000 tons per day of spent shale, which would require about 1,160 acres a year for surface disposal, with the prototype leases due to run for a period of 20 years. Thus the prototype program alone would cover some 23,200 acres with shale wastes. A full-scale industry producing 3 million barrels of oil per day would require disposal of 4.5 million tons of spent shale per day. This would cover 4,640 acres of land each year.[49]

The oil shale wastes are sterile, and will grow nothing unless prodigious amounts of fertilizer and constant watering are provided. "Even then," Rathlesberger noted, "the re-establishment of the area's natural plant community would be close to impossible." The loss of this vegetation would be a serious threat to the more than 300 species of wildlife found in the area of proposed development.

"Can you see the Rockies," questioned a United Auto Workers publication on oil shale, "gradually reduced to barren rocks as the strip miners cut the trees, tear up the topsoil, and leave the hills of refuse? The roaring rivers would run filthy with industrial wastes and the refineries would foul the once bracing mountain air. Deer, antelope, mountain lions, elk, eagles, trout, and other wild creatures would be seen—if at all—in the confinement of zoos."[50]

There is another major—and perhaps overriding—consideration in oil shale development schemes: They would require enormous quantities of water from the Colorado River and its tributaries. By 1983, a prototype development like that envisioned by the Department of Interior would require 135,000 acre feet of water per year, and a full-scale industry would require 450,000. The impact could be especially severe to Colorado, which only has the potential of 500,000 acre feet more per year under the terms of the Upper Colorado River Compact than it is now using. Thus, since 80 percent of a full-scale oil shale industry would be in Colorado, it would take 360,000 acre feet of the state's water—almost three fourths of all the additional water that is expected to be available to it from the Upper Colorado.[51] As noted earlier, the National Academy of Sciences has already concluded that water will simply not be available in western states for any such massive energy conversion schemes.[52]

The National Economic Research Associates, Inc., concluded in a 1972 report to the electric utilities industry that "the combination of institutional obstacles (governmental leasing policy, title clearance, etc.), possible water supply constraints and environmental problems (especially, disposal of the spent oil shale) make anything more than token shale oil production unlikely in the period through 1985."[53] Oil companies that now hope to reap the bonanza from rich oil shale deposits are on the wrong track.

The massive environmental degradation associated with oil shale exploitation is so severe that it is questionable whether an oil shale industry should ever be developed.

Tar Sands

One other potential source of oil is from "tar sands" in the Athabasca region of Canada, an isolated, undeveloped area in northern Alberta.

Athabasca tar sand deposits contain an estimated 285 billion barrels of recoverable oil.[54] The National Petroleum Council has estimated the sands could produce 275,000 to 500,000 barrels per day by 1980 and 500,000 to 1.25 million per day by 1985. National Economic Research Associates, Inc., estimates that probably all of that would be made exportable to the United States, since Canada's needs would be being met by other sources.[55] Even if this proves to be true—and Canada's increasing reluctance to part with her own energy resources as the "hewer of wood and carrier of water" to the United States suggests it may not—that production level would represent only a minuscule 2 to 4 percent of the U.S. supply needs by 1985, a far cry from the millions of barrels per day the National Petroleum Council says the United States will require by that time.†

Great Canadian Oil Sands Ltd. is already producing about 45,000 barrels of oil per day from the tar sands. Another company, Syncrude Canada, has Alberta government approval to initiate operations on adjoining property, with oil production scheduled to begin by 1976.[56]

Tar sands are made up of lumps of sand particles that are encased in a thin envelope of tarlike substance called bitumen, the same kind of hydrocarbons that make up bituminous coal. As in the case of oil shale, recovery methods include both *in-situ* and above-ground processing, although current operations involve only the latter. The tar sands are now strip-mined and the bitumen is removed by a hot water treatment in which bitumen rises to the top of a separation chamber and is removed. It is upgraded and desulfurized to produce a low-sulfur fuel.

Unlike *in-situ* retorting of oil shale, *in-situ* recovery of tar sands does not involve fracturing. This eliminates objections related to explosives—particularly nuclear explosives. It has been estimated that *in-situ* methods may be the only practical means of recovering the 90 percent of the Athabasca tar sand deposit that is not suited to strip-mining methods. Drawbacks of *in-situ* recovery, however, include the fact that only about half the bitumen at a given site can be recovered using this method. The process also may require great quantities of steam, which means tapping huge amounts of scarce water, some of which may have to be recycled.

Little attention has been given to environmental aspects of recovering oil from the Athabasca tar sands—largely because its remote location has minimized public interest. Great Canadian Oil Sands Ltd., has reported that it replaces the detarred sands in the mine pits, covers them with previously removed topsoil, and plants trees on them.

Because particles are smaller, tar sand waste disposal may be less of

† The National Petroleum Council estimates that oil imports required to meet petroleum demands by 1985 will range from 3.6 million barrels per day under the most favorable circumstances, to 19.2 million barrels per day under the worst possible growth and demand/supply conditions.[57]

a problem than that of oil shale. The same type of air pollution problems occur at the recovery site as at a petroleum refinery.[58]

Conclusion

Though the promise of developing "replacement" fossil fuels for the cheap, readily available fuels of today is held out by government and industry as a significant answer to the fossil fuel energy crisis now at hand, the synthetic fuels would never equal the cheap fuels that up to now have served as the basis of the American economy.

Not only are the coal, oil shale, and tar sand deposits less accessible than the deposits that are mined for fuel today, but much of this energy may be unrecoverable, since the energy required to mine them may exceed the energy derived from them. The *net energy* from the facilities under construction at present has never been measured, but technical warnings that recovery will be more expensive than for today's processes is a key indicator that more energy will be required to make the processes for liquefaction and gasification work. In the event the energy yielded from the processes proved to be less than the energy required to build and maintain the facilities, the nation would almost certainly be facing economic disaster. The country could ill afford to rely on energy from a parasitic technology.

Given the uncertainty of supply of the fossil fuels of dilute nature, such as the oil shales, tar sands, and further coal deposits in the western United States, it would be prudent national policy to carefully explore the energy economics of converting these fossil fuel forms into consumable fuel before irrevocably committing the nation to a disastrous economic and societal dead end. A better policy than raping the West for coal to turn into low-energy gas would be to conserve these rich fossil fuel deposits for possible future use as medicines and other vital petrochemical products.

In addition, strip mining agricultural land in the western United States for shallow fossil fuel energy deposits would bring a curse to future agricultural production. To return the land, once strip-mined, to agricultural production would require the extensive application of artificial irrigation and synthetic fertilizers, both of which would require still further use of limited fossil fuels. The energy trade-offs between food production and strip mining have never been determined; i.e., at what point does the energy gained from strip mining drop below the energy lost from future agriculture productivity?

Other energy sources than the fossil fuels that enrich the economy today are explored in some detail in the following pages—sources of energy such as nuclear fission, nuclear fusion, and tidal and hydroelectric power. However, it should be noted that these energy sources do not yield hydrocarbons, which can be processed into petrochemical products (fertilizers, medicines, plastics, etc.), but are primarily heat or mechanical power sources. Scientists hope that mechanical power and heat sources can someday be used to electrolyze water for the production of hydrogen (a replacement for fossil fuels for motive transportation and heating), but the day for the implementation of such advanced technology is probably well into the next century, and the costs of such schemes are largely unknown.

Electricity from Nuclear Fission

The splitting or fissioning of heavy atoms of uranium to produce heat (for steam) for the operation of conventional steam-electric power plants is the fastest-growing method of electric power generation in the United States. Yet nuclear power plants represented only 3.7 percent of the nation's electrical generating capacity in 1972, according to the Federal Power Commission. The reactors represented about 15,000 megawatts (million kilowatts) of electrical capacity out of a national total of 398,000 megawatts. The bulk of the nation's electricity is still generated by fossil-fueled electric power plants and hydroelectric plants.[59]

The Atomic Energy Commission (AEC), the federal agency formerly charged with promoting and regulating nuclear power development, announced in 1973 that within 27 years, by the year 2000, nuclear power plants would generate 60 percent of the nation's electricity. The AEC predicted that 132,000 megawatts of nuclear power will be "on line" by 1980, although the Federal Power Commission predicts that only 122,000 megawatts will be available by that time. Taking the lower estimate of the Federal Power Commission, this would indicate that one third of the nation's electricity will come from the atom in 1980.[60] The tiny amount of electricity generated by nuclear power in 1973 was approximately 1 percent of the nation's over-all energy supply—about the same amount of energy derived in the United States by burning wood in that year.

This new energy source, scheduled to burst forth from infancy into a pillar of the nation's power supply, has been fraught with technical problems; is associated in the public mind with the atomic bomb; and has triggered a debate over public safety unequaled in the development of any technology in human history. Its proponents in industry and government claim that nuclear energy will usher in an unprecedented age of clean and cheap power. Its detractors argue that its development will bring civilization to a grinding halt in the wake of stifling radioactive pollution of the planet.

The Nature of Nuclear Fission Technology

Nuclear electricity is generated by heat produced by fission, an energy conversion process in which heavy atomic nuclei of elements such as uranium are split by the bombardment of neutrons, subatomic particles that travel at tremendous speed. The colliding neutrons split nuclei of heavy atoms and produce energy in accordance with Einstein's formula: $E=mc^2$. The energy locked in atoms of uranium is prodigious. In one fission event, 50 million times as much energy is released as in the burning of a carbon atom, the primary energy substance of coal.[61]

Harnessing the energy of fission involves controlling the energy release process so that a "chain reaction" is initiated. A chain reaction occurs when the fissioning of one atom releases neutrons that strike the nucleus of another atom, releasing more neutrons, which strike more atoms, etc. The difference between a nuclear reactor in a power plant and an atomic bomb is measured by the number of neutrons that are released in the fission process. If only one neutron at a time is available for triggering the fission event in the chain of atom-splitting, then the chain reaction is under control. The release of more neutrons accelerates the process rapidly, causing an atomic explosion. Nuclear power plants in operation today are designed so that atomic explosions are not possible. Control rods, made of materials that absorb neutrons, can be inserted or withdrawn in the reactor's core to either start or stop the fission process. In most U.S. nuclear reactors, water is circulated between vertical, pencil-thin rods of uranium fuel to control or "moderate" the speed of traveling neutrons. This is done to increase the chance of neutrons colliding with other atoms, since slower-moving neutrons improve the possibilities for fission by 200 or 300 times.[62] For this reason, most U.S. reactors are called "light-water" reactors, to distinguish them from reactors that use deuterium, or "heavy water," as a moderating agent; and high-temperature gas reactors (HTGR), which use helium gas as a moderating agent. Other reactors, discussed later, are called fast breeder reactors, which use sodium (and other agents) as moderating or cooling agents, and produce or "breed" an excess quantity of nuclear fuel—the element plutonium.

Almost all commercial nuclear power plants now operating in the United States are light-water reactors, with the exception of two helium gas-cooled reactors.

The fuel used in nuclear reactors is primarily an isotope (or form) of naturally occurring uranium, called U-235. In nature, only two forms of uranium are present. The other isotope, U-238, cannot be split in a conventional light-water reactor. A key problem with nuclear power is that uranium fuel is very limited, constituting less than two parts per million of all materials in the Earth's crust. Compounding this scarcity, only 0.7 percent of natural uranium is the sought-after U-235.[63]

U-235 must be carefully separated from uranium ore by a succession of chemical processes, and finally enriched to a higher concentration than found in nature before it can undergo fission, or be "burned" in a nuclear power plant. Enrichment of uranium is accomplished in massive facilities called gaseous diffusion plants, which are operated by the Atomic Energy Commission, and were built in the Second World War to make enriched uranium for bombs.

In a modern nuclear power plant, capable of generating 1 million kilowatts (1,000 megawatts) of electricity, about 100 tons of uranium fuel in thin fuel rods are loaded into a half-foot-thick, steel-walled reactor vessel containing coolant water and associated pipes for transferring steam to operate turbines for the generation of electricity. As the control rods alongside the fuel elements are withdrawn, fission is initiated and heat builds up within the reactor core, reaching over 4,000° F in the center of the uranium fuel elements.[64]

Shown below is a schematic diagram of a typical light-water reactor:

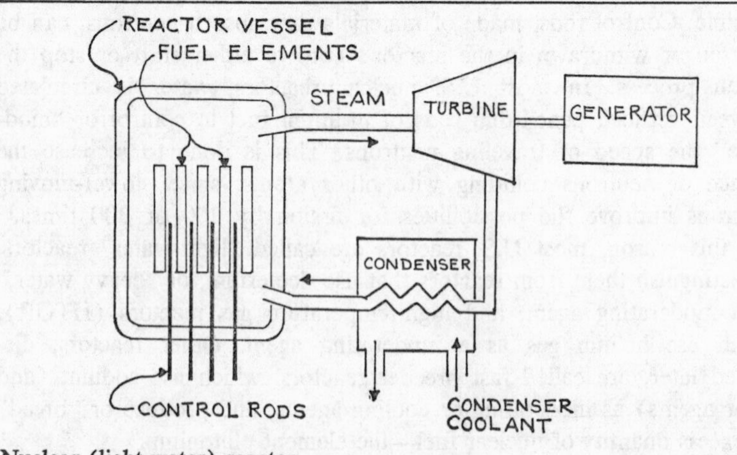

Nuclear (light-water) reactor

The over-all operating temperature is kept lower than the operating temperature of a fossil-fuel power plant to protect the uranium fuel rods. Consequently, the thermodynamic efficiency is lower than for fossil-fuel power plants, and almost twice as much waste heat is produced by the operation of nuclear plants. Therefore, nuclear power plants are heavier thermal polluters to rivers and waterways than their fossil fuel counterparts.

Nuclear power is billed a clean power source by the Atomic Energy Commission and the U.S. nuclear industry, since uranium fuel is not combusted in the presence of oxygen, and plants do not release visible pollutants. However, the fissioning of uranium produces radioactive by-products that are much more lethal than the smoke of fossil fuel combusion, and that—although invisible—do escape from the stacks of nuclear power plants.

The fission process in nuclear reactors that results in the chain reaction, production of heat, and release of neutrons, also involves a second energy process, called radioactive decay or transformation of stable elements into their radioactive forms, or radioactive nuclides. The fissioning of uranium in a nuclear reactor is an intensely accelerated form of a natural process of radioactive decay in which uranium, an unstable element, is transmuted into a stable (nonradioactive) form of lead. The process of atomic decay that starts with uranium and ends with lead is measured by scientists in terms of "half-life." A radioactive substance's half-life is the time required for the substance to decay to one half its initial amount. In the case of U-238, the most abundant form of uranium, the half-life is 4.5 billion years—approximately the age of our planet. Thus, today, the U-238 in the Earth's crust is about half the amount originally present.[65]

Radiation Dangers

Radioactive decay releases intense amounts of ionizing radiation, radiation that can separate or change stable atoms into ions, which are electrically charged atoms.‡

When one form of radiation, called beta radiation, passes through human tissues, it tears negatively charged electrons from atoms that comprise the tissues, leaving positively charged ions. The electrons released ionize other atoms in tissue, until the initial energy of each electron is dissipated. The source of these electrons is the radioactively decaying nucleus of unstable atoms. Radiation impinging on living cells

‡ Ionizing radiation can take several forms, including X rays, gamma rays, beta rays, or alpha particles, depending on the nature of the radiation source, which range from cosmic ray bombardment of the Earth to natural radioactive decay of uranium, medical X rays, and radiation released from nuclear power plants and associated nuclear power sources.

can cause effects ranging from genetic mutation and abnormalities (through chromosomal disorganization) to cancer, to outright death—when tissues are exposed to large doses of radiation, as occurred at Hiroshima and Nagasaki, Japan.

Exposure to ionizing radiation occurs naturally in the environment. Called background radiation, the ionizing rays come from numerous sources, including cosmic rays from space and gamma rays from naturally occurring radioactive materials in the Earth's crust. People living nearer sea level receive smaller natural radiation doses than do people living in mountainous locations, since the atmosphere serves as a filter—blocking much of the radiation from space.

Radiation has been known as a significant cause of disease and cancer since the early years of this century, when many mortalities were reported among early atomic researchers. Public uses of X-ray machines brought another wave of deaths, including that of an assistant of Thomas Edison.[66] A major national scandal erupted in the 1920s, when Dr. Harrison Martland, medical examiner in Essex County, New Jersey, reported that a number of young women employed in a watch factory to paint radium on the dials of luminous watches had succumbed to the effects of radiation poisoning. Their habit had been to wet the tips of the brushes on their tongues, sharpening the brush tips and ingesting daily quantities of radium. Dr. Martland reported blood diseases, bone disorders, and cancer; yet the unrestricted use of radium continued for years thereafter—and it was long listed by the American Medical Association as a *remedy* for many maladies. The U. S. Public Health Service has been active for years in tracking down products containing "miraculous radium water" and other such radioactive palliatives sold by quacks and unscrupulous companies.[67]

Since the development of nuclear reactors and the beginnings of the nuclear industry, the Atomic Energy Commission has pursued an unusual course in regard to the biological effects of radioactivity: It has repeatedly assured the public that radioactive fallout from weapons tests, nuclear accidents, and the like have not produced public injury; and indeed, many official pronouncements of the AEC have asserted that released radiation is "harmless."

An argument advanced by the AEC and the nuclear industry is that radiation releases from the predicted operation of the developing nuclear industry will add only a small fraction of the already present background radiation on Earth. The AEC recently estimated that radiation doses to the population from the nuclear power industry would constitute only a tiny fraction of the background radiation in the year

2000—0.2 millirem from the industry compared to a national average of 130 millirems from background radiation.*[68]

The AEC's standards for radiation protection of the public are based on minimum allowable doses to humans established by the Federal Radiation Council (FRC), an advisory organization to the government established by President Eisenhower in 1959. The FRC determined that the maximum allowable dosage to the U.S. population should be 0.17 rad† per year, and the total amount of radiation received by any single person should be 0.5 rad per year.

The validity of these standards was challenged by several eminent scientists, including Dr. Ernest Sternglass, who reported that levels of radiation lower than the official "safe dose" standard could cause deaths in unborn children, because the fetus is extremely susceptible to the hazards of ionizing rays. Sternglass' own research on the subject of infant mortality indicated that as many as 400,000 unborn children and infants less than one year of age had died as a result of exposures from nuclear weapons fallout in the United States between 1950 and 1969.‡[69]

The AEC was infuriated by Sternglass' theory, and assigned one of its own scientists, Dr. Arthur Tamplin of the Lawrence Livermore Laboratory in California, to refute Sternglass. Tamplin evaluated the calculations and concluded that Sternglass was too high in his estimates, but he concluded that fallout *had* killed infants. In cooperation with Dr. John Gofman, an internationally recognized medical researcher and also an associate at the Lawrence facility, Tamplin began to analyze the weaknesses in the Federal Radiation Council's guidelines for radiation exposure to the public.

Drs. Gofman and Tamplin computed that "if the average exposure of the U.S. population were to reach the allowable 0.17 rad per year average, there would, in time, be an excess of 32,000 cases of fatal cancer plus leukemia per year, and this would occur year after year."[70] The AEC's reaction to this was swift: The two men were threatened with loss of jobs both by the Washington AEC headquarters personnel and by officials of the Lawrence Livermore Laboratory. Dr. Tamplin's

* A millirem is a measurement of radiation absorbed in living tissue. It represents one one-thousandth of a rem, which is an acronym for *R*oentgen *E*quivalent *M*an.

† A rad is a measurement of radiation, and an acronym for "*r*adiation *a*bsorbed *d*ose."

‡ Dr. Sternglass faced considerable difficulty in getting his research papers published in scientific journals, because the AEC waged a campaign against having his material appear in print. The AEC was successful in preventing his publication in *Science,* the journal of the American Association for the Advancement of Science, and attempted to prevent publication in the *Bulletin of the Atomic Scientists,* which published it anyway.[71]

entire staff was reassigned to other jobs, and Dr. Gofman left the facility to devote full time to research at the University of California at Berkeley.

However, Drs. Gofman and Tamplin were joined in their criticism by other leading scientists, including Nobel laureates Linus Pauling and Joshua Lederberg; Pauling estimated that 96,000 cancers would occur from population exposure to the AEC's legal dose, and Lederberg estimated that population exposures to the federal limit would cost the nation an eventual $10 billion yearly in health care. In their book *Poisoned Power,* Drs. Gofman and Tamplin note that Dr. Lederberg's estimate concurs with their additional observation that population exposure to 0.17 rad would result not only in 32,000 yearly cancer deaths, but in from 150,000 to 1.5 million induced genetic fatalities.[72]

The revelations of Drs. Gofman and Tamplin and their subsequent testimony before several congressional committees, including the Joint Committee on Atomic Energy (which scoffed at their findings), led to several blue-ribbon reviews of the FRC data. Responding to mounting pressure, the AEC announced that releases of radioactivity from U.S. nuclear facilities would be kept "as low as practicable," which was later defined as a value about one one-hundredth of the FRC standard.[73] In addition, a panel of the National Academy of Sciences announced that federal standards for radiation exposure were "unnecessarily high" and would lead to 27,000 serious genetic diseases and deaths if the population were exposed to radiation at the federal standard.[74]

What had become a raging national issue simmered down as a result of the AEC's concession in lowering the population exposure standard, and critics of nuclear power conceded that technology was available to control the expected "normal" (i.e., minute) releases of radiation from nuclear reactors. However, the "normal" releases of radiation from nuclear power facilities are an insignificant part of the over-all nuclear picture. What is most significant about the projected development of nuclear power is not the release of radiation from "normal" operations of nuclear power plants, but the *control* of the plants themselves, and the absolute requirement that no significant radiation releases can be permitted of this technology. The issue at stake in the development of nuclear power is primarily safety.

As the Atomic Energy Commission states: "The unique characteristic of nuclear power that imposes an overriding requirement for safety precautions is the generation of large amounts of intensely radioactive materials in the nuclear fuel."[75] The "large amounts" of radioactive materials produced by the yearly operation of one modern (1,000 megawatt) nuclear reactor are equivalent to the fission products (and radiation) produced by 23 megatons (million tons) of nuclear weapons. If the AEC's hopes for nuclear power growth are realized, there will be 500 to 1,000 nuclear plants of this size operating in the United States by the end of this century. The radioactive inventory of 500 of these plants will equal 11,500 megatons of nuclear weapons. According to Dr. John Gofman, "the major long-lived fission products, strontium 90 and cesium 137 [radioactive forms of stable strontium and cesium produced in reactors] have half-lives on the order of 30 years. Therefore, the inventory will necessarily build up, until at a steady state [several times 30 years] the inventory will be . . . 500,000 megaton equivalents of long-lived fission products."*[76]

On the other hand, the combined atmospheric weapons testing by the Soviet Union, the United States, and Great Britain has only amounted to the production of 250 megatons of radioactive products. Put another way, the nuclear plants already in operation, planned, or under construction in the United States will generate the equivalent in radioactive products of 130,000 Hiroshima-size atomic bombs—each year.†[77]

The central issue in the development of nuclear power is whether or not the radioactive inventory of nuclear plants can be sufficiently kept away from all forms of life for periods ranging up to millions of years—the effective half-lives of the myriad radioactive products produced in the reactors. Even a tiny release—on the order of one percent of the fission products produced by a single power plant—would be sufficient to permanently contaminate hundreds of square miles, rendering continued existence impossible for almost all forms of life. Radiation, in the words of the AEC's leading congressional critic, Senator Mike Gravel of Alaska, is unquestionably the "ultimate pollution." Keeping civilian nuclear power under control requires the ultimate in technology: Essentially no errors can be permitted, no major radiation releases allowed. Nuclear technology is a "fail-safe" technology. Once a critical error is made, there can be no turning back; consequently, no errors can be made. The vital question of nuclear power development is whether or not human beings can be responsible for such an awesome task.

The Atomic Energy Commission and the nuclear industry are satisfied that it can be done. One of the founders of nuclear technology and the director of AEC-supported Oak Ridge National Laboratory, Dr. Alvin

* *The Case for a Nuclear Moratorium,* John Gofman, Mike Gravel, and Wilson Clark, Environmental Action Foundation (1973), 30 pp.; Suite 732, Dupont Circle Building, Washington, D.C. 20510.

† The Hiroshima bomb was equal to 20,000 kilotons (a kiloton is 1,000 tons).

Weinberg, says the development of nuclear power offers a reliable and "clean" power source, but poses grave problems.

> Thus, we seem to have struck a Faustian bargain. We are given the miraculous nuclear fire—whose dimensions I saw only dimly 18 years ago—as a means of producing very clean and, with the breeder, inexhaustible energy. The price that we must pay for this great boon is a vigilance that in many ways transcends what we have ever had to maintain: vigilance and care in operating these devices, and creation, and continuation into eternity, of a cadre or priesthood who understand the nuclear systems, and who are prepared to guard the wastes. To those of us whose business it is to supply power here and now, such speculations about 100,000 year-priesthoods must strike an eerie and unreal sound. . . . But the immediate concern for vigilant, intelligent, and responsible operation of nuclear power plants is not theoretical nor remote: It is a heavy responsibility that everyone in the utility industry, public and private, must assume.[78]

Development of Nuclear Plants

Given the extreme nature of the risks, one wonders how and why nuclear power came into being in the first place. Acts of Congress following the successful demonstration of the bomb in 1945 established the U. S. Atomic Energy Commission, which was given an open-ended mandate in the Atomic Energy Act of 1946 (as amended in 1954) to provide: ". . . by national policy that the development, growth and control of atomic energy shall be directed to make the maximum contribution to the general welfare and to the common defense and security, and to promote world peace . . . and strengthen free competition in private enterprise."[79]

From the beginning, the AEC was given total control of nuclear weapons and nuclear reactor development, and a clear license from the Congress to proselytize the atom—to develop a nuclear industry where none existed; where, in fact, there was no serious business interest in the sinister matters of nuclear power.

Under the conditions of the Atomic Energy Act, a special "watchdog" congressional committee was established—the Joint Committee on Atomic Energy (JCAE)—which has seen to it over the years that the promotion of nuclear power has remained the primary goal of the AEC. The JCAE has accomplished its mission by the use of several mechanisms, the most important being the disbursement of almost unlimited funds to the AEC, and the establishment of specific business subsidies for the development of nuclear power. The rewriting of the Atomic Energy Act in 1954 allowed private corporations to share in nuclear technology, but industry was slow to take up the offer. This led to an all-out business recruiting effort on the part of the commission in 1955. In January of that year, the AEC announced the initiation of the "Power Reactor Demonstration Program," which was designed to subsidize the efforts of any electric utility in building a nuclear plant, with a promise of free fuel for five years after startup. A few industries came forward, but the AEC was convinced that "priming the pump" would be necessary.

AEC commissioner Dr. Willard Libby told the JCAE in 1956 that he was worried "that in the atomic power program . . . there is some reluctance of certain companies, I think, to come into this peaceful uses program as quickly and as well as they have come into the weapons thing." The chairman of the joint committee and atom buff Senator Clinton Anderson of New Mexico answered Libby with this suggestion:

> There are a lot of people who would like to build reactors, lots of them. Some of them have good technology. Here is a firm that went down into Savannah River, the du Pont Company, and no attempt is being made to interest du Pont in building a reactor. Du Pont isn't going to bother; its profit and loss statement is pretty good. But if it is given an opportunity under contract, it would be interested in building one of these reactors.[80]

The problem in getting private business interested in nuclear power was an insurance problem: Who could possibly bear the risks for failure of the technology, and the subsequent release of radiation? In 1956, the JCAE asked the AEC to outline the full, potential dangers of radiation release from reactor accidents. A year later, the AEC sent the committee its now-famous report, called, simply, WASH-740 or the "Brookhaven Report," for the AEC-supported Brookhaven National Laboratory on Long Island, New York, where the calculations were made. WASH-740 estimated the accident possibilities for a nuclear reactor considerably smaller than today's 1,000-megawatt giants. The hypothetical reactor dealt with in the report was of 200-megawatt size, and located 30 miles from a major city (in contrast with larger reactors nearer large cities today). The report concluded that a major accident might kill 3,400 people, injure another 43,000, and create up to $7 billion in property damages. Radioactive materials might contaminate up to 150,000 square miles.[81]

The Brookhaven Report served a valuable function to the JCAE: It clearly identified the reasons for industry's reluctance to become involved in nuclear reactors. This problem was solved in the same year, when Senator Anderson and Illinois congressman Melvin Price of the

JCAE introduced legislation in the Congress, which soon became Public Law 88–703, the "Price-Anderson Act."

The Price-Anderson Act amended the Atomic Energy Act to read: "The aggregate liability for a single nuclear incident [reactor accident] . . . shall not exceed the sum of $500,000,000 together with the amount of financial protection required. . . ."[82] "The amount of financial protection required" by the electric utility owners of nuclear plants was left up to the private industry, which formed two insurance pools to provide $60 million in private insurance for potential accidents. What the law did was to assure that the utilities would not be liable for damages above their pooled insurance plus the taxpayers' share of $500 million. Sheldon Novick, in his book *The Careless Atom,* points out that this limits accident liability to one fourteenth of the possible damage outlined in the AEC's study, and insofar as the bill was meant to open the atomic energy field to private enterprise, he notes that "this may sound like an odd sort of free enterprise—the taxpayer assuming the risk and private industry accepting the profits."[83]

The Price-Anderson Act opened the nuclear floodgates, and more than a dozen orders poured in for reactors in the late 1950s and early 1960s. A persistent wish of the JCAE was that the nation's insurance companies would provide substantially more than the original $60 million in private insurance, but this has proved to be a vain hope. Additional hearings on Price-Anderson were held in 1965, when the JCAE sought to have the companies fund at least $100 million in private insurance. They were only able to raise the ante by $14 million, to a total of $74 million. Today, the private insurance pool, at $95 million, still has not yet reached the $100 million sought by the JCAE a decade ago.

The public still bears the major burden for accident liability, and—adding insult to injury—all private homeowners' insurance policies issued in the United States have a "nuclear exclusion" clause, which exempts the insurers from any specific liability for damages caused by radiation, whether that radiation comes from a nuclear weapon or a "peaceful" nuclear reactor. The private insurance companies, almost all of which are members of the nuclear insurance pool, have a captive market. An individual company's liability in case of reactor accident is limited to a small percentage of the $95 million in the private pool, and the public has no recourse at all above the Price-Anderson indemnity ceiling.

When the Price-Anderson Act came up for renewal by the Congress in 1965, the JCAE was told by the AEC that a revised study of the Brookhaven Report was under way, and would be released when available. The purpose of the revised study was to estimate the consequences of accidents involving the five to six times larger reactors going into the power plants of the 1960s. In a June 1965 letter to the JCAE, then AEC chairman Glenn T. Seaborg said of the newer research that calculated damages to the public from a major accident in the bigger reactors "would not be less and under some circumstances would be substantially more than the consequences reported in the earlier study."[84]

This letter was the first and last either the public or the joint committee heard about increased dangers of a reactor accidents for eight years. Between 1965 and 1973, the AEC denied that such a report had been planned or made, and the congressional "watchdog" committee never inquired further about it. Finally, in June 1973, new AEC chairman Dr. Dixy Lee Ray disclosed that the revised Brookhaven Report had been initiated, but squelched in 1965 by the AEC. She released background material, and a draft of the 1965 study, which indeed concluded that accidents in today's larger reactors could result in greater damage than that indicated by the 1957 Brookhaven Report. A clue to the reason for the suppression of the 1965 safety study was given by the AEC's 1973 release of staff documents detailing a series of nearly catastrophic failures in U.S. nuclear power plants between 1957 and 1965, which the AEC said "could have resulted in more serious accidents than any which had thus far been experienced."[85]

Without delving further into the various technical and social questions of nuclear fission technology, the answer to the question, "Why are there nuclear power plants in the United States?" can be delivered simply: An insurance policy. At the 1965 JCAE hearings on extension of the Price-Anderson Act, a nuclear engineer appearing for a major public utility (the Los Angeles Department of Water and Power) put it this way: "Without the protection which presently is provided by the Price-Anderson Act, it is doubtful that any utility would consider it prudent to build nuclear plants."[86]

The ramifications of the national nuclear "insurance policy" go farther than that. The Price-Anderson indemnification of nuclear plants represents the most visible aspect of the largest federal subsidy of any technology, with the possible exception of the space programs of the National Aeronautics and Space Administration. Since the Second World War, more than $40 billion has been spent by the federal government and industry to create the nuclear power business.[87] Spending in the coming decade will more than match that figure. The nation's electric utilities will need to raise $130 billion for capital expansion by 1985, much of which will be nuclear.[88]

Aside from the inherent subsidization of nuclear power by Price-Anderson insurance indemnification, about half of the AEC's yearly $2 billion-plus budget goes for developing nuclear power, pursuing nuclear safety research, and providing for federal waste disposal facilities for use by the nuclear industry. Notwithstanding the euphemistic use of the term "free enterprise" in the Atomic Energy Act, nuclear power development in the United States can hardly be considered private; it has been, and continues to be, a carefully nurtured and subsidized federal program. Virtually all government funds spent for research into *all energy technologies* since World War II have been devoted to the development of nuclear reactors. Adding this to the economic subsidy for the nuclear industry makes it clear why nuclear fission power appears at present to be the only significant alternative energy technology to the burning of fossil fuels in the United States.

Luckily for the public, there have been no major accidents with nuclear reactors of the size being built for electrical production today. However, there have been disastrous accidents in smaller, experimental reactors as well as accidents of varying degrees of severity in other nuclear installations in the United States and other countries. In 1952, technicians operating an experimental power reactor in Canada made several mistakes in manipulating control rods, which triggered a runaway nuclear reaction. The result was a partial meltdown of the uranium fuel, which in turn caused chemical explosions that, in the words of Sheldon Novick, "lifted [the reactor's] four-ton gasholder dome four feet into the air and jammed it among surrounding structures." A number of workers at the plant were exposed to the sudden release of radiation, but no one was seriously injured. The reactor was never rebuilt.[89]

In 1957, the same year that the Price-Anderson Act was passed in the United States, an accident occurred at the English Windscale nuclear plant which released tremendous quantities of radiation into the air, contaminating two hundred square miles of neighboring farmland as well as drifting across the North Sea and exposing several northern European countries to radiation from fission product release.[90] Again, radiation exposures but no deaths were reported among the workers at the facility. Large quantities of milk from cows in the neighboring countryside were found to be contaminated with radioactive iodine and were discarded. The radiation release at Windscale might have taken a great toll in damage to property and human life had the winds blown in a different direction rather than out to sea. The AEC has said that under difference circumstances, "the release of 20,000 curies (a measure of radiation) of iodine 131 could have more severe consequences than those experienced from this incident."[91]

The list of accidents and near-misses in nuclear reactors is long and tedious, but few accidents have involved the loss of human life. One of the most serious occurred in 1961 at the AEC's National Reactor Testing Station in Idaho, when a reactor core exploded, killing three attendants and exposing fourteen workers to radiation. The event is shrouded in mystery, and all that is known is that the accident was caused by the manual withdrawal of the central control rod in the reactor.[92]

Professor T. J. Thompson of the Massachusetts Institute of Technology reported in 1964 that nine serious reactor accidents had occurred from 1949 up to that time in civilian installations, involving the destruction of the fuel cores.[93] In 1973, the AEC reported eleven serious reactor accidents that occurred between 1949 and 1962 in research and civilian plants. The list was included in a public review of reactor safety, and is unusual in that the reporting did not include accidents in the twelve years prior to 1973.[94] A clue to the AEC's policy on public disclosure of unfavorable information is found in the preface to the report, which states that the reader "will, therefore, understand that a certain degree of pro-nuclear bias is probably inherent in the discussion, although [the AEC] staff has tried to make the presentation as simple, factual and objective as possible."[95]

A review of the AEC's draft document on safety in the American Association for the Advancement of Science (AAAS) journal *Science* in 1973 concluded that much of it is taken up with "bland and reassuring discussion of the AEC's regulatory process, the design of nuclear power plants and the elaborate precautions taken to ensure their safety." The *Science* analysis of the report discussed the probabilities of serious accidents occurring at a commercial nuclear power plant, based on the AEC's own calculations of accident probability.

Science reported that the AEC's estimates of the chances of a nuclear plant suffering a serious accident with radiation release might be about one in a thousand for a given reactor per year. "Coupling this estimate with the AEC's projection that about 100 power reactors will be operating in the United States by 1980 and 1,000 by the end of the century, the report indicates that *one such accident each year may become a virtual certainty*" (emphasis supplied).[96] The *Science* article added that draft copies of the safety report had been circulated to about fifty selected reviewers, and that the trade magazine *Nuclear*

Industry praised the report as objective and well-written, but that of two AEC safety researchers contacted (who asked not to be identified), one commented that the report was a "whitewash" and the other said that it was "not worth reading, in its major parts."[97]

Perhaps the most stinging criticism of the AEC's report-WASH-1250, came from the Environmental Protection Agency (EPA), an agency that had never before stringently criticized the AEC's handling of nuclear affairs, or trod on the AEC's turf in matters of radiation safety. As might a college professor react to a student's poorly constructed eleventh-hour term paper, the EPA wrote the AEC that the safety report was "inadequate and consequently it has been given a 'Category 3' rating in accordance with our procedures. . . ."[98] The EPA "Category 3" rebuff was the equivalent of a flunking grade.

What had concerned the Environmental Protection Agency was the AEC's discussion of "cost-benefit analyses" relative to reactors, i.e., the worth of electricity to the public vs. the possibility of a crippling reactor accident that would cause massive injury and property damage. Going farther, the EPA challenged the AEC's description of "Class 9" reactor accidents, accidents of a magnitude sufficient to release large quantities of radiation to the environment. The EPA noted that the AEC's philosophy of providing redundant safety features in reactors was the only substantial safety argument in the report. The EPA added that the AEC staff believed that "because of the extreme precautions taken to forestall and prevent Class 9 events, the probability of their occurrence is so small that the associated environmental risk is extremely low. The draft statement goes on to note that this conclusion is based on the technical judgment of the AEC staff rather than statistical methods."[99]

For years, the AEC's philosophy has been to require reactor manufacturers to install elaborate systems to encase reactors and assure safety by providing back-up safety systems, ranging from thick concrete containment wells around reactors, to back-up cooling water systems, replete with secondary pumps to flush an overheated reactor core in case of emergency uranium meltdowns, such as those that have destroyed more than a dozen other smaller reactors. As the AEC puts it, the safety philosophy is based on practices applied to design, construction, licensing, and operation of nuclear plants that encompass the following points:

1. Plant designs are permitted that are as "inherently safe as practically achievable [e.g., which can tolerate large unanticipated upsets without major damage to the plant or threat to the public health and safety]."

2. Plants are constructed with "multiple barriers against the escape of radioactivity to the environment."

3. Plants are constructed with "a multiplicity of safety devices and systems to protect the public and the plant from human errors, equipment failure or malfunction and from severe accidents or events, including earthquakes and other natural phenomena. . . ."

The AEC's review of its safety philosophy notes that "it is impractical to build and operate large nuclear power plants and fuel reprocessing facilities with an absolute zero-release of radioactivity," adding that "relatively small quantities of radioactive materials do enter the environment from such facilities."[100]

The philosophy of the AEC is that serious accidents, which would release large quantities of radiation, are virtually beyond the realm of possibility. This is *Titanic*-style thinking. The plants are built well, the AEC says, so they're "unsinkable." Not only does the history of accidents at research facilities and small reactors prove this philosophy ill-founded, but a recent history of the AEC's own safety research for large reactors indicates that the official optimism of the nation's largest superscience agency is based more on hopeful thinking than sound analysis.

"Emergency Core Cooling" and Reactor Safety

What the Environmental Protection Agency was concerned about in the AEC's cursory survey of reactor safety was the possibility of an accident called a reactor "Loss-of-Coolant Accident" (LOCA), which might be triggered by the failure of the "Emergency Core Cooling Systems" (ECCS), the back-up cooling systems provided in all U.S. reactors to furnish emergency cooling water in the event of meltdowns in a reactor accident (LOCA). This type of accident is publicly declared by the AEC to be virtually impossible, because of the inherent design of modern reactors, and is referred to as an unthinkable "Class 9" accident. The origin of concern about the possibility of Class 9 accidents in commercial power plants dates back to the 1960s, when the AEC's promotional policies had resulted in more than thirty orders for reactors.

In 1966 the AEC commissioned a blue-ribbon task force of reactor safety specialists to study the manufacturers' designs of emergency cooling systems and report on the effectiveness of proposed safety systems in stopping a potentially disastrous accident. The task force, headed by the late William K. Ergen of Oak Ridge National Laboratory, concluded that numerous technical shortcomings were evident in U.S. reactor designs, which it felt required further research and study before the AEC licensed the larger reactors being considered. (The average size of reactors at the time of the Ergen Report had reached 800 megawatts, four times as large as the reactors studied in the Brookhaven Report of a decade earlier.) ‡[101]

The conditions described in the Ergen Report occur within the span of several hours, but in an accident the die would be cast within minutes. As has been noted, the core of a nuclear reactor contains upward of 100 tons of enriched uranium fuel in the form of some 40,000 tiny pellets enclosed in thin tubes made of special metal alloys (the most common being alloys of steel and zirconium, called Zircaloy). During the nuclear fission reaction, the temperatures inside the fuel pellets may reach over 4,000° F, at times nearing the melting point of uranium (4,982° F). The Zircaloy cladding tubes, however, have a lower melting point, and are kept at a temperature of about 650° F, to prevent melting. The central safety precaution is the circulation of water under pressure in the reactor. If a break occurred in a pipe carrying coolant water to the reactor, a "blowdown" would occur, as pressure loss would cause the superheated water to turn to steam, forcing the original water from the reactor core. The Ergen Report described what would happen to the reactor if emergency cooling water were not provided to the reactor within seconds:

> Core behavior is described in terms of seven periods or events. During the first two periods, blowdown and core heatup by decay heating [radioactive decay heat] in portions of the core could have reached temperatures greater than 2,000° F within 30 to 50 seconds, and localized Zircaloy-steam reaction would become a significant energy source. After the third period, approximately 50 [to] 100 seconds [depending on the type of reactor manufacture] . . . the hottest portions of the cladding could have begun to melt (3,360° F). During the fourth period, core collapse . . . significant accumulation of core debris [following meltdown and metal-water reactions] in the lower pressure-vessel head is expected in from 10–60 minutes after pipe rupture. . . . The time at which the energy from these reactions is added to the containment is past the time of peak pressure due to the blowdown, and the energy would not add directly to this peak load. It is estimated that as much as 50 to 80 percent of the fuel might fall into the lower pres-

‡ The Ergen Report was never officially issued by the commission, and was never given an AEC document number, required for technical material issued by the AEC. It was not even dated, and until recently was virtually impossible to obtain from the AEC. The same occurred with the Brookhaven Report, which the AEC never reprinted. Copies have been available only through select channels and handy Xerox machines.

sure-vessel head. With 50 to 80 percent of the fuel in the lower pressure-vessel head, melt-through of the vessel appears highly probable. Pressure vessel melt-through might be expected to take from 20 to 60 minutes. The time from pipe rupture to melt-through is, therefore, considered to be a minimum of 30 minutes with a likelihood that one hour or more would actually be required.[102]

The report went on to predict that the mass of concrete and steel shielding that protects the reactor vessel would not last more than a few hours, perhaps a few days. The knowledge of consequences of a reactor meltdown was "insufficient to conclude with certainty that integrity of containments of present designs, with their cooling systems, will be maintained," said the select group of safety specialists.[103]

After the dripping, molten mass of highly radioactive fuel broke through the bottom of the reactor vessel, reactor experts half-jokingly refer to the next series of events as the "China syndrome," meaning that the molten mass will continue its course in the general direction of the People's Republic.

In the event of such an accident, the public would undoubtedly be exposed to a large quantity of radiation, either through the collapse of the reactor's containment structure, which would release radiation clouds in the atmosphere; or by means of the "China syndrome," in which case underground aquifers would be contaminated with a deadly assortment of radioactive fission products.

In the years following the release of the Ergen Report, the AEC relied upon various computer models of hypothetical accidents to determine the effects of ECCS operation, and the commission has never tested a full-scale reactor core cooling system to see if it will operate properly in the event of a break in the primary coolant pipes. The only actual tests yet conducted of an emergency cooling system confirmed some fears of Dr. Ergen's safety task force. In late 1970 and early 1971, a series of tests were conducted on tiny nine-inch-high models of reactor pressure vessels at the National Reactor Testing Station in Idaho. The miniature, electrically heated models of reactor pressure vessels were drained of coolant water, and in six experimental efforts to get emergency water into the simulated reactor cores, no successes were reported. Only 10 percent of the emergency cooling water reached the core.[104]

The joint committee was told of the experiments in May 1971. The AEC immediately disputed the Idaho tests, claiming that they were unrealistic, since the nine-inch reactor prototypes could hardly be considered indicative of the tremendous actual two hundred-ton reactors in U.S. nuclear power plants. The AEC's argument pales in light of its own efforts to secure actual tests of emergency cooling systems in real reactors.

In the early 1960s, the AEC planned to build a test reactor at its remote Idaho reactor test facility about one twentieth the size (55 megawatts) of today's large commercial plants. The experimental reactor, called LOFT (for Loss of Fluid Test) was initially designed for an experimental evaluation of a reactor meltdown; then, during the midst of the Ergen study, plans were changed so that the facility would be used to test emergency cooling systems. A series of management failures by the principal private contractor, the Phillips Petroleum Company, and a lack of direction on the part of the AEC failed to get the project off the ground for more than a decade. Tests on the cooling system of the LOFT reactor are not scheduled until after 1974. By that time, more than fifty commercial reactors of various designs will be operating in the United States, all of which dwarf the LOFT reactor's 55-megawatt size.*[105]

In regard to the LOFT tests, one might question the Atomic Energy Commission's own logic regarding criticism of the core experiments on the tiny reactor models. If the tiny reactor models do not produce a valid comparison with the 1,000-megawatt commercial power plants, how will the one-twentieth-size LOFT experiments provide an accurate comparison? To date, all of this has been an academic question, since all reactor safety systems have been designed by computers, with no full-scale field testing.

In 1971, following the Idaho test disclosures, the AEC announced the establishment of new criteria regarding emergency core cooling for the operation of commercial power plants. The criteria established a maximum operating temperature of nuclear reactors that would prevent the nuclear fuel cladding from exceeding 2,300° F. This maximum temperature limit was designed as a safety measure in existing and planned reactors. The AEC reasoned that by prohibiting operation of reactors at higher temperatures, the fuel would not melt and a disastrous accident would be prohibited.

The utilities and reactor manufacturers reacted adversely to this AEC standard, because it meant lowering the operating efficiency of power plants, meaning not only a loss in electricity generated, but a loss

* The world's first actual "destructive safety tests" on an operating reactor were conducted in 1972 and 1973 on the Swedish Marviken reactor by the Swedish national nuclear company, AB Atomenergi. The U. S. AEC was offered an opportunity to participate in the Swedish test series, but declined. The Marviken reactor is three times the size of the AEC's LOFT reactor, and the Swedes planned to conclude the testing with a trial of the emergency core cooling system.[106]

in revenue. On the other hand, a national coalition of environmetal organizations and public interest groups banded together under an umbrella organization, called the National Intervenors, to challenge the AEC's policy as not being conservative enough to provide an adequate margin of safety to protect the public. The AEC held what it called an "experimental" rule-making hearing on the reactor safety criteria, which brought environmentalist lawyers and scientists together with industry's lawyers and scientists to bring their respective cases before a specially established AEC hearing board.

The hearings were conducted in 1972 and 1973, and resulted in an extensive airing of opinions on nuclear safety from the AEC, its national laboratories, and the nuclear industry. The full transcript reached more than 22,000 pages, and the intermittent hearings spanned more than 130 days of testimony.

The hearings were limited to the specific issues of emergency core cooling, and did not cover the wider issues of other accident possibilities from failure of reactor equipment. What emerged from the hearings was the discovery that many scientists at the AEC-supported national laboratories were not contacted for professional opinions of the ECCS safety criteria, which were prepared by the AEC's Washington, D.C., staff. Included in evidence entered in the hearing was the testimony in letters from prominent nuclear safety experts who challenged the AEC standards. Typical of the response (outside the closely knit AEC Washington headquarters) was a letter to the AEC's Director of Regulation, Manning Muntzing, from Dr. William Cottrell, editor of the internationally circulated journal *Nuclear Safety* and head of the nuclear safety division at Oak Ridge National Laboratory (ORNL). The letter concluded that ". . . we are not certain that the Interim Criteria for ECCS adopted by the AEC will . . . 'provide reasonable assurances that such systems will be effective in the unlikely event of a loss-of-coolant accident.'" The AEC's furious Director of Regulation Muntzing returned the letter to Cottrell's superior at ORNL, denouncing it because it allegedly did "not reflect the views of Oak Ridge National Laboratory."[107]

As the hearings proceeded, it became clear that dissension over the AEC's safety policies was not only growing at the relatively independent national laboratories, but also within the cloister of the AEC's Washington headquarters itself.[108] Evidence presented at the hearings by two distinguished scientists and members of the AEC's regulatory staff, Dr. Morris Rosen and engineer Robert Colmar, contained their observation that adequate performance of the emergency core cooling systems "cannot be defined with sufficient assurance to provide a clear basis for licensing."[109] A letter "leaked" to the National Intervenors and presented at the hearings from the AEC's own director of the Division of Reactor Development and Technology stated that "no assurance is yet available that emergency coolant can be delivered. . . ."[110]

Even more damaging to the AEC's position was the release of information by the National Intervenors at the hearing that a major U.S. reactor manufacturer, Westinghouse Corporation, had not informed the AEC of severe fuel rod damage observed at a Swiss reactor Westinghouse had constructed.[111] Later disclosures revealed that reactors built by Westinghouse in the United States suffered from the same defects when hundreds of fuel rods in each reactor were found to be dented, bowed, and crushed by a puzzling phenomenon called "fuel densification." A review of the problem by the AEC regulatory staff concluded that fuel densification posed an additional threat to the performance of reactor safety cooling systems during a potential accident, and recommended lowering the operating temperature levels of reactors as well as finding better methods of enclosing uranium fuel in reactors.[112]

As the hearings on reactor safety criteria ended in 1973, more problems than solutions had emerged. The AEC has not altered its original criteria for operation of the emergency cooling systems, even though overwhelming evidence was presented at the hearings that the criteria were based more on promotional politics than science. With the added weight of the fuel densification problems in U.S. reactors, one certainty appeared: The cost of maintaining and operating U.S. nuclear plants would rise in the future, whether or not adequate margins for public protection were imposed on power plant owners (electric utilities) by the AEC.

The Myth of the Economy of Scale

In the early 1950s, the Atomic Energy Commission optimistically predicted that the development of nuclear power would produce electricity that "would be too cheap to meter," and that the larger the plant, the cheaper the power. At present, a nuclear power plant is the most expensive way to generate electricity, notwithstanding extensive subsidization by the U. S. Government. In 1960, costs for nuclear power were in the neighborhood of $135 per electrical kilowatt of plant capacity; by the late 1960s, costs rose to more than $200.[113] In the mid-1960s, elaborate calculations made by a New Jersey utility indicated that the Oyster Creek, New Jersey, nuclear plant it was building would be cheaper to build and operate than an equivalent fossil fuel plant. Yet the plant was 2 years behind schedule going into operation, and the costs rose substantially from a planned $68 million to $90 million.[114] Indications are that the utility may have learned from this, as it canceled plans for a second nuclear plant in favor of a coal-fired Pennsylvania power plant in the early 1970.[115]

Technological problems and construction difficulties did not abate, but continued in the 1970s. An industry study in late 1972 reported that nuclear plants cost a staggering $555 per kilowatt of installed capacity, 4 times the cost of a decade earlier. The report, by the reputable consulting firm, Nuclear Utilities Services, Inc., also noted that construction time jumped from 4½ years from date of order to plant completion, to 8 years.[116] In its annual survey of electrical power in the United States, the trade magazine *Electrical World* reported in 1973 that orders for nuclear plants were 54 percent behind schedule. The magazine noted that "heavy involvement in nuclear units" substantially contributed to a disastrous situation in availability of U.S. electricity. The survey noted that available power was about 20 percent less than what its experts considered a minimum to meet national peak demands.[117]

The failure of nuclear power to meet its expectations is partially the result of AEC overpromotion and partially the result of its intrinsic characteristics as a new and largely unknown technology. In order to secure the cheapest electricity from the atom, the AEC and the nuclear industry have advocated the development of larger and larger power plants, the argument being that the larger plants have an "economy of scale": the bigger, the better.

The opinion of many power plant operators—the nation's electric utilities—is that this is completely fallacious thinking, based on technological overoptimism. At an annual meeting of the Atomic Industrial Forum, the industrial trade association, Louis Roddis, president of New York's Consolidated Edison Company, declared his pessimism that nuclear power would ever fulfill its "expectation of deliverability." Roddis noted that the AEC and the nuclear reactor builders had long claimed that reactors would easily deliver 80 to 90 percent of their rated capacity, yet the record showed that the average electrical production of U.S. plants was about 60 percent, compared to fossil fuel plants, which averaged 75 percent.

Roddis told the association that the problems of repairing nuclear plants, because of the radioactivity risks involved, added a serious economic handicap to the development of nuclear power. He reported that repairing a pipe failure in Consolidated Edison's Indian Point, New York, nuclear plant had taken 7 months and cost the utility $1 million, not including the cost of replacement power. "A similar repair effort," he said, "if made in a conventional plant, would have required about 2 weeks and probably would not have involved more than 25 men." He added that failures in nuclear plant design might "exhaust our manpower pools when we try to operate and maintain the plants."[118]

A 1973 survey of nuclear plants in *The Wall Street Journal* noted that "their unreliability is becoming one of their most dependable features. The incredibly complex facilities are plagued by breakdowns that experts blame on faulty engineering, defective equipment and operating errors. Failures range from hour-long annoyances to months-long closedowns. Repair costs often run into millions of dollars, and some utilities stoically shell out up to $200,000 a day for replacement electricity to distribute to their customers in the meantime."[119]

Some of the plant problems surveyed by the newspaper included:

• Repairs at a nuclear plant in Connecticut shut the plant down for over 8 months due to extensive salt-water corrosion and various equipment malfunctions. Cost: over $10 million.

• Bolts in a reactor core in a Massachusetts plant failed, necessitating closing the plant for about 6 months for repairs. Cost: $6 million.

• A routine uranium refueling at a Wisconsin plant disclosed turbine and steam generator failures. Costs were not reported, but the plant was closed for 5 months.[120]

Failures of equipment in nuclear plants have rarely led to the loss of human life. The AEC, in fact, until 1973, never fined an electrical utility for improper operation of a nuclear power plant, which might lead to unnecessary exposure of radiation to the public. The exception was the Virginia Electric Power Company, which the AEC fined $40,000 in 1973 for 28 separate violations of safety procedures at its Surry, Virginia, nuclear plant—including falsifying radiation hazard reports to the AEC.[121]

The AEC's interest in the Surry plant had originally come from a mechanical engineer, Carl W. Houston, who reported to the AEC's regulatory division in 1970 that he had spotted evidence that as many as 5,000 seriously deficient welds in vital coolant pipes might be present in the plant. He was dismissed by his employer, the Stone and Webster Engineering Company of Boston, but an AEC review of the pipe welds confirmed Houston's suspicions.[122] The welding errors at the plant were allegedly corrected, but a 1972 steam pipe explosion at the plant killed 2 workers (the plant was under construction and not yet in operation). This may have been the causative factor in the AEC's rekindling of interest in the plant's construction.

Colossal blunders have mounted with the growth of nuclear power. One embarrassing episode was the recent disclosure by the General Electric Company, manufacturer of light-water reactors, that control rod blades in 10 U.S. reactors and 3 foreign reactors made by General Electric had been manufactured *upside down.* The problem had first been reported in the ill-fated Connecticut plant that was already shut down for a host of other reasons, including salt water corrosion of most of its parts. Under prodding from the AEC, General Electric conceded that 5 more *operating* plants had upside-down control rods! The company claimed that the problem was not "technically serious but could be emotionally trying," but the AEC ordered tests on all GE reactors.[123]

A statistical analysis of reactor problems published by the AEC in 1973 indicated the failure of the "economy of scale" in U.S. nuclear power plants. In the following AEC table, "significant events" are defined (euphemistically) as equipment failures and accidents of magnitude requiring immediate notification of the AEC. "Passive components" are condenser tubes, small pipes, and other small-scale equipment that are common to all power plants, not just nuclear plants. "Active components" are defined by the AEC as major pieces of equipment linked to the reactor and steam turbine, such as primary valves and pumps, or control devices. "Human failure" is defined as the failure to follow established procedures. The table lists accident progress ("significant events") and causative factors for the period 1970–72.[124]

Primary and Major Contributing Causes of Significant Events

[Note: Some events may have more than one cause.]

	1970	*1971*	*1972*
1. Failure of passive components	22 (43.2%)	17 (20.5%)	27 (22.5%)
2. Failure or malfunction of active components	21 (41.2%)	51 (61.5%)	83 (69.2%)
3. Human failure	8 (15.6%)	15 (18.0%)	10 (8.3%)
	51 (100%)	83 (100%)	120 (100%)

Analysis of the AEC's data on nuclear power plant failures and incidents shows that reactor operators and personnel are apparently being trained better, because human failure does not seem to be a major contributing factor to plant mishaps. Neither does the failure of small components, which contributed to more than 40 percent of incidents in 1970, but dropped to half that in 1972. Rising failures appear common in the massive new components, specially built for the bulky nuclear reactors being constructed today in the United States. Accidents with large plant components have steadily risen, and contributed to more than two thirds of nuclear plant "significant events" in 1972.†

Engineers often speak of a "learning curve" for a new technology; as progress in developing components and working with a technology continues, the "learning curve" is supposed to register over-all improvements in the technology.

The "learning curve" of nuclear power appears to be in reverse. What has been learned from nuclear power plants is that the more that have been built, the greater the problems that have appeared.

The Breeder Reactor

A primary drawback to the development of nuclear fission power is the availability of uranium to refine into reactor fuel. Currently, uranium reserves in the United States and other countries yield ore suitable for processing at $6 to $8 per pound. The reserves are not infinite. Dr. Ralph Lapp, in a review of AEC estimates, comments that the available amounts of uranum available to the United States at costs up to $10 per pound ($4 higher than current costs) will fuel the nuclear industry only to the year 2000. Then, he says, "the United States would find its uranium fuel bin depleted."[125] According to the U. S. Geological Survey (USGS), the situation may be even worse. USGS says that domestic uranium reserves are only sufficient to meet reactor demands in the 1980s, adding: "Beyond that, however, needs are so great that tremendous efforts in exploration and research in ore-finding techniques will be required to find new resources."[126]

The AEC and the nuclear industry have been cognizant of this essential limiting factor of nuclear growth for years. Their answer to it is the development of a nuclear reactor that will create additional fuel as it produces heat for electricity. The advanced nuclear reactor that will do this job is called the "liquid metal fast-breeder reactor" (LMFBR). The reactor is significantly different from a light-water reactor. It is not moderated with water, as are today's reactors; it is

† A major causative factor in recent electrical blackouts and brownouts in the United States has been the failure of nuclear power plants. Although the public has not yet been injured by exposure to nuclear accidents and consequent release of radiation, the public has encountered the shortcomings of nuclear technology, as evidenced by the blackouts caused by nuclear plants.

cooled with highly volatile liquid sodium. It is called a "fast" reactor because it does not use a moderating agent to slow down neutrons (which improves the chance of fission in a conventional reactor); instead, the breeder uses more compacted fuel, and it has a higher "power density" than conventional reactors. The reactor's fuel is plutonium 239 and uranium 238, the more abundant form of natural uranium.‡ Plutonium is not found in nature, and is produced in small quantities in light-water reactors. The key to operation of the breeder reactor is that plutonium 239 can be fissioned to produce power; and in the process, its neutrons strike the nuclei of otherwise unfissionable U-238 (and some other potential nuclear fuels), producing more fissionable material, including more plutonium 239.[127]

There are two major problems with the breeder:

1. Due to the construction of the reactor's core, an accidental power surge would reduce the density of sodium coolant, which could increase the number of fast neutrons available for fissioning, causing an increase in the power surge. This process could lead to an uncontrolled chain reaction, which could explode the reactor and release radiation into the environment.[128]

2. The second major problem with the breeder reactor is that the fuel it breeds is the most dangerous substance known to man: plutonium 239. Plutonium 239 has a half-life of over 24,000 years, and the quantities generated by the nuclear program in the United States are such that plutonium will be a perpetual hazard, to be guarded by the "nuclear priesthood" for hundreds of thousands of years, longer than previous human habitation of the Earth itself.

Plutonium is fiendish in its toxicity, more poisonous even than the anaerobic bacteria, which cause botulism, and 35,000 times more lethal than cyanide poison by weight. Professor J. B. Neilands of the University of California at Berkeley notes that there are "20 million mortal doses in a weight of the element equivalent to the weight of an American 5-cent piece (5 grams)."[129] Plutonium contamination is nothing new to the AEC. Weapons testing has released enough plutonium in the American Southwest that hundreds of square miles are cordoned off—perpetually. In May 1969, the most expensive single industrial fire in U.S. history took place at a nuclear weapons plant near Denver, Colorado; the fire was triggered by plutonium. Measurements of plutonium scattered by the fire and other incidents at the

‡ Versions of the breeder reactor use U-235 fuel.

plant indicate that plutonium concentration in the nearby soil may be at dangerous levels.

The AEC's plutonium standards for public exposure, according to Dr. Donald Geesaman of the AEC-supported Lawrence Livermore Laboratory, are "acknowledged to be meaningless."[130] Dr. A. B. Long of the AEC's Argonne National Laboratory concurs. He says that "in light of data accumulated over the past 10 years, a serious question exists as to whether the risks associated with the present limits for plutonium oxide inhalation are indeed justified by the benefits derived from activities involving the production or use of plutonium."[131]

Yet in the light of considerable dispute over the dangers of plutonium, the AEC and the Nixon administration have made the development of a "plutonium economy" (a term coined in 1970 by then AEC chairman Glenn Seaborg, codiscoverer of plutonium in the 1940s) via future nuclear breeder reactors the major energy development program in the United States.

Concern over the hazards of breeder reactors has been voiced by many scientists, including the eminent Dr. Edward Teller, "father" of the hydrogen bomb, and by no means considered a critic of the development of nuclear power:

> For the fast breeder to work in its steady-state breeding condition you probably need something like half a ton of plutonium. I do not like the hazard involved . . . if you put together 2 tons of plutonium in a breeder, one tenth of one percent of this material could become critical. . . . Although I believe it is possible to analyze and foresee the immediate consequences of an accident, I do not believe it is possible to analyze and foresee the secondary consequences. In an accident involving a plutonium reactor, a couple of tons of plutonium can melt. I don't think anybody can foresee where one or 2 or 5 percent of this plutonium will find itself and how it will get mixed with some other material. A small fraction of the original charge can become a great hazard.[132]

Dr. Teller wrote these words shortly after a potentially major accident in the only U.S. breeder reactor ever licensed for commercial power production, the Enrico Fermi reactor near Detroit, Michigan. During the embryonic reactor development era of the 1950s, the AEC approved the application of the Power Reactor Development Company (PRDC), a coalition of businesses and utilities led by the Detroit Edison Company, to build the Enrico Fermi plant—which was a sodium-cooled breeder reactor using uranium rather than plutonium as fuel. Opposition to the plant was mounted by several organizations in the Detroit area, including the powerful United Auto Workers' Union, which carried the fight against construction of the plant through the judicial system to the U. S. Supreme Court.* The Supreme Court ruled that the plant should be constructed; but Justices Douglas and Black dissented, saying that "the construction given the [Atomic Energy] Act by the Commission is, with all deference, a light-hearted approach to the most awesome, the most deadly, the most dangerous process that man has ever conceived."[133]

Had the PRDC realized what was to occur only 5 years after the favorable ruling was rendered by the nation's highest court, they would have been pleased to have seen the plant rejected. By 1966, a series of technological problems had occurred in constructing the plant and assembling the reactor, which raised plant costs from an initial investment estimate of $40 million to about $120 million. When the plant was being tested at lower power levels in early October of that year, reactor operators noticed that higher than normal heat levels were being recorded inside the reactor core. In addition, radiation detection devices sounded an alarm within the reactor building, and control rods were inserted within the reactor, shutting down the chain reaction. It proved impossible to get special equipment inside the reactor to examine the core for more than a year, when it was discovered that a sheet of zirconium plate, added as a last-minute device to protect the bottom of the reactor's pressure vessel, had broken free and blocked coolant to the reactor's fuel elements. A substantial meltdown of the reactor's fuel, U-235, had occurred. An emergency meeting was held just after the October incident, when no one knew exactly what had happened. Sheldon Novick reported that those at the meeting "feared that enough uranium had recongealed so that a disturbance of the core—by an attempt to remove the damaged fuel, for example—would jar it into a critical mass too great to be controlled by the control rods, which were already at their maximum."[134]

What may have been on the minds of the PRDC officials was the accident in an AEC experimental breeder reactor in 1955, when a fuel core melted down, destroying the reactor.[135] Of three breeder reactors operated in the United States, two have suffered fuel meltdowns. Although no significant radiation escaped either facility, the record doesn't say much for the promise of the technology.

However, there can be no future in nuclear fission power *without* breeder reactors, regardless of their economics or safety. Understand-

* The battle was joined by the UAW due to the persistence of a leading AEC critic, Leo Goodman, who was at the time a consultant to UAW president Walter Reuther.

ably, the AEC has made development of the breeder its no.-one priority, and with the help of the Joint Committee on Atomic Energy, funds for developing the breeder have flowed freely for years, outpacing research expenditures for other energy technologies by a wide margin.

President Richard Nixon announced in June 1971 that the primary goal in developing new sources of energy would be the fast breeder reactor. He commented that "because of its highly efficient use of nuclear fuel, the breeder reactor could extend the life of our natural uranium fuel supply from decades to centuries, with far less impact on the environment than the power plants which are operating today."[136] Although the President obviously was convinced that the nuclear fuel problem would be solved by swift construction of breeder reactors, one wonders about the presumed minimization of environmental impact.

By order of the President, the AEC swiftly went about preparing an industrial/government coalition reminiscent of the 1950s reactor demonstration program to build one or more demonstration breeders, since industry was not interested in footing the bill alone. The Scientists' Institute for Public Information, Inc., headed by Barry Commoner, Margaret Mead, and other prominent scientists, sued the AEC for violation of the National Environmental Policy Act (NEPA) for not submitting an environmental impact statement (required by NEPA) that would assess the full environmental impact of the "plutonium economy" as well as not listing possible energy alternatives to a network of breeder reactors producing most of the country's electricity. The AEC had only prepared an environmental impact statement for *one* demonstration reactor, a statement that did not consider the consequences of a full breeder program.

The AEC argued before a Washington district court that it was not required to assess the full impact of the breeder economy, and the court agreed. However, a 1973 landmark ruling of Judge Skelly Wright of the U. S. Court of Appeals reversed the earlier court's approval of the AEC's statement. Judge Wright noted that the AEC had taken "an unnecessarily crabbed approach to NEPA in assuming that the impact statement was designed only for particular facilities rather than for analysis of broad agency programs." He added that a government decision to develop a technology such as the breeder reactor "before considering the possible adverse environmental impacts attendant upon ultimate application of the technology will undoubtedly frustrate meaningful consideration and balancing of environmental costs against economic and other benefits. . . ."[137]

The AEC decided to accept the judgment without appeal, since it had backed down only two years previous in another of Judge Wright's decisions over the AEC's "crabbed" interpretation of the National Environmental Policy Act—the 1971 Calvert Cliffs decision. In that case, the AEC's special licensing board had refused to allow certain criticisms of environmental organizations over the planned construction of the Calvert Cliffs nuclear plant (by the Baltimore Gas and Electric Company) on Chesapeake Bay. The environmental organizations were critical of the fact that the plant was planned to release large quantities of heat directly to the bay's water, without provision for cooling towers to decrease the thermal pollution, and sued the AEC to consider a full range of environmental factors. Under the AEC's own rules, consideration of broad environmental effects of nuclear plants licensed by the agency was not permitted, and the commission did not change its rules even after the National Environmental Policy Act became law in January 1970. Judge Wright noted in that decision that the AEC "had continually asserted prior to NEPA that it had no statutory authority to concern itself with the adverse environmental effects of its actions. Now, however, its hands are no longer tied. It is not only permitted but compelled to take environmental values into account."[138]

The AEC was forced in both instances to concern itself about the possible environmental impacts of nuclear facilities before it went ahead and allowed construction of plants, thereby irrevocably committing the nation to its own wishes for the development of nuclear power. In the case of the development of the breeder reactor, the AEC has shown its talents for promotion.

Hearings before the Joint Committee on Atomic Energy revealed details of the AEC's plan for development of the breeder. The AEC's plan originally was to develop breeder reactor research, which it then could use to excite the interest of industry. When the AEC's subsidy amounting to hundreds of millions of dollars per year failed to provide a sufficient inducement, the commission originated a cooperative arrangement whereby it would provide a partnership with the reactor manufacturers and electric utilities in building breeder demonstration plants. The joint venture for the first breeder plant consists of the AEC, the Tennessee Valley Authority, Commonwealth Edison Company (the private electric utility that serves Chicago), and two corporate structures representing manufacturers and utilities, Breeder Management Corporation and Breeder Reactor Corporation.

The AEC reported at the hearings that the demonstration breeder would be built on the Clinch River in Tennessee, near the site of the

Oak Ridge National Laboratory; the site was chosen because of favorable local interest in the AEC, which had provided an economic boost to the local economy since ORNL's origin in the Second World War. The plant's estimated cost, including fuel and operations during a five-year test period, was said to be between $696 million and $716 million. Following the AEC presentation, Chairman John Pastore of the JCAE noted that "this project is going to cost close to $700 million, which means that the public contribution will be better than a ratio of 2 to 1." The utilities (including the public power utilities) had only agreed to contribute $254 million.[139]

Physicist Amory Lovins summed up the argument by environmentalists in his analysis of the breeder program for Friends of the Earth. Lovins said that the scheme represented a "very bad bargain for the taxpayer," with the contract provisions that the AEC pay $422 million outright, pay *all cost overruns,* buy the plant back if the TVA decided it didn't want it after it was completed, and indemnify the private industry for *all claims and liabilities.*[140] The lengths to which the AEC went to accommodate the nuclear industry was more than the subsidies offered the fledgling nuclear industry in the 1950s, and the breeder contract provisions made the federal government's scandalous subsidization of the aerospace industries appear responsible in comparison.

The JCAE was pleased by the contract, and after more criticism by environmental organizations and Ralph Nader in 1973, made minor changes in the contractual arrangements of the AEC, and finally the contract for construction of the Tennessee plant was signed by the AEC and industry. As Lovins points out, the substance of the AEC's arrangement is an acceptance on behalf of the (unknowing) taxpayers of "an open-ended commitment to escalating costs and unlimited indemnities, while the utilities are buttering their bread on all six sides."[141]

At the joint committee's hearings on the breeder reactor, Lovins and others pointed out the risks involved in undertaking the development of this technology at a time when little is known of its safety. Even the AEC's optimistic environmental impact statement on the sodium-cooled breeder managed to sandwich a few comments on the risks involved. For example:

> Sodium is generally nonreactive with a wide range of common structural materials used in system containment and component fabrication. It does react with most nonmetals, and precautions must be taken to prevent exposure of sodium to air and water. Sodium readily oxidizes in air. When exposed to water, sodium reacts vigorously to form

hydroxides and hydrides and releases hydrogen gas which in turn can react explosively with oxygen. Special design features will be employed in the demonstration plant *to preclude sodium-water reactions and to mitigate their effects* [emphasis supplied].[142]

One wonders why it would be necessary to mitigate the effects of something that is already precluded from taking place!

Breeder Economics

The Atomic Energy Commission has attempted to justify development of the liquid metal fast-breeder reactor (LMFBR), or "breeder," on the basis that (1) it offers the best available means of meeting the nation's future electrical energy requirements; that (2) it will "conserve our natural resources of uranium"; and that (3) "significant environmental advantages will accrue from its use." To make its point, the AEC in 1971 produced a cost vs. benefit study concluding that if it is developed, the breeder will save $21.5 billion in energy costs by the year 2020.[143]

Dr. Thomas B. Cochran, a physicist working with Resources for the Future, Inc. (RFF), in Washington, D.C., made a thorough study of the AEC's economic claims for the breeder. He concluded that, in his judgment, "the LMFBR program cannot be justified on the basis of economics; significant environmental *disadvantages* will accrue from its use and these will outweigh the advantages; and an early introduction of the LMFBR will not conserve our natural resources of uranium."[144]

The AEC has acknowledged that the first few breeders will not be economically competitive. To bring them to competitive status, the AEC is relying on achieving nuclear fuel cycle costs of 1.5 to 1 mill per kilowatt hour, considerably lower than commercially available light-water reactors and high-temperature gas reactors (HTGR)—the two nonbreeding nuclear fission technologies with which the breeder must compete to prove economically attractive.

The lower fuel cycle costs would be necessary to offset the higher initial capital costs of the breeder reactor. The AEC has said capital costs of the breeder will be about $20 per kilowatt higher than for the light-water reactor. However, the AEC's estimates are much too low. They are less than one third the average estimate of a reactor assessment panel of the Edison Electric Institute (EEI), a trade organization of investor-owned electric utilities.[145]

The breeder's advantage is dependent on introduction into the market by 1990 of a General Electric breeder reactor design that has serious shortcomings in terms of accident prevention or minimization. The

U. S. AEC nuclear reactor safety program up to this time has incorporated a three-step technology: first, building the installation as safe as possible; second, extensive instrumentation to monitor any possible developing accidents; and third, a construction philosophy that calls for attempting to contain the accident if one should occur, or at least to shield the surroundings as much as possible from its effects. The General Electric design on which the AEC has calculated its cost-benefit studies *does not contain the third-step safety provisions*. In the long-range breeder program, as it has been presented, the AEC has clearly jettisoned plant and public safety in attempts to achieve economic advantages.

The AEC's ground rules for estimating fuel cycle costs also did not include estimates of the effect of possibly more stringent regulations on siting, safeguards, or effluent control.[146] A key factor in the AEC's favorable economic portrayal of the breeder is the assumption that spent fuel from the reactor can be shipped to a reprocessing plant after only thirty days of cooling. The spent fuel rods would still be extremely radioactive and at very high temperatures after only thirty days out of the reactor. Shipping would require them to be sealed in steel canisters filled with liquid sodium as a cooling medium. By the year 2020, the chance of a transportation accident that would release hazardous levels of radioactive iodine, xenon, and krypton to the atmosphere and the surrounding land area at least once each year would be reasonably good.[147] Doses of radiation to the population might be quite high in local areas. The decontamination and associated costs of such an accident could run into the tens of millions of dollars.[148] According to Dr. Cochran, "delaying the spent fuel shipments could reduce the severity of such an accident by at least an order of magnitude. But then," he added, "this would further increase the LMFBR fuel cycle costs."

The AEC is relying heavily on economies of scale to bring breeder fuel cycle costs down to a point where the breeder would be competitive with light-water reactors. "The assumption, or perhaps hope," Dr. Cochran explained, "is that the LMFBR will look so attractive when it becomes available that utilities will 'go breeder' just as they 'went nuclear' in the 1960s."

Nuclear industry spokesmen don't share the eager assumptions voiced by the AEC. Herman M. Dieckamp, president of Atomics International Division of North American Rockwell, said during an Atomic Industrial Forum annual meeting that "the breeder does not have the optimism about a nuclear panacea that aided the introduction of the light-water reactor. On the contrary, its request for support comes at a time when the industry is painfully aware of the costs of introducing nuclear power and also gruelingly aware of the public acceptance problem."[149]

Other major factors in the AEC's bright projections for the economic attractiveness of the breeder are its predictions about electric power demand by the year 2000, which may be overestimated by 25 to 50 percent. If the demand is 25 percent below the AEC projection in the year 2000, then the AEC's projected benefits of the LMFBR program are cut in half. If the actual demand is only 50 percent of the AEC projection, the alleged benefits would be cut to zero.[150]

While the AEC's economic arguments in behalf of the breeder are based in large measure on eliminating safety features, as noted above, Dr. Cochran points out that many of the breeder's characteristics make it potentially more hazardous, so that "greater care *must* be taken in order to insure plant and public safety."

The features that Dr. Cochran believes render the breeder more hazardous than today's fission reactors include higher power density; higher operating temperatures; higher fuel enrichment; greater sensitivity to control; and chemical problems with the sodium coolant, including fire and radioactive hazards associated with sodium.[151]

Another major factor that the AEC has ignored in making its cost-benefit analysis is that, with the large inventory of plutonium in the LMFBR fuel cycle, greater care must be given to safeguarding the breeder fuel than fuel for today's reactors or for the high-temperature gas reactor. These costs of protecting plutonium from theft should be reflected in the breeder cost-benefit analysis.

Radioactive Waste

A major problem of nuclear power that escalates with development of the breeder reactor is the transportation reprocessing of nuclear wastes after they leave the reactor. As has been pointed out, the problem of water and air pollution from nuclear power plants can be minimized to low levels—assuming, of course, the near-perfect operation of the plants. In the case of fuel reprocessing plants, which treat the massive volume of radioactive wastes built up in reactors before their perpetual storage by the "nuclear priesthood," the same analysis is not applicable. The AEC freely admits that reprocessing plants will become a major source of radioactive pollution in the future.†

† Releases by current reprocessing plants, of which three are in operation in the United States, have been recorded. The West Valley, New York, plant of Nuclear Fuel Services, Inc., released in 1970 radioactive liquids—primarily radioactive hy-

The primary pollutants released by the reprocessing plants are liquid radioactive wastes such as tritium (H_3), the radioactive isotope of hydrogen, and strontium 90. The primary waste released in air is the radioactive gas krypton 85. The AEC says that operation of the two newest fuel reprocessing plants in the United States, one of which is in South Carolina and the other in Illinois, will release a yearly total of 18.2 million curies of radioactive krypton per year and 700,000 curies of radioactive hydrogen per year, plus small quantities of other radioactive substances, such as iodine 131.[153] Without adequate controls at the reprocessing plants, which are yet to be technologically perfected, the radiation doses to the surrounding population might reach levels higher than the radiation output of all the nuclear power plants serviced by the reprocessing plants. Tritium is a special problem at the reprocessing plants, and Dr. Edward Radford at The Johns Hopkins University in Baltimore estimates that the maximum permitted tritium release levels at the New York reprocessing plant, "if continuously ingested, would lead to 3,300 new mutations per million human births, or about 12,000 mutations per year in the United States."[154]

As in nuclear power plants, accidents may occur at reprocessing plants, where a variety of chemicals and acids are used to extract radioactive materials from spent nuclear reactor fuel rods. In a survey of environmental problems in the nuclear reprocessing plants, the AEC lists a number of possible accidents that might occur, including zirconium and plutonium fires, tank leaks, acid spills, filter failures, and the like. The AEC says that none of the postulated accidents would result in "significant" population exposures, although plant workers might suffer high radiation doses and might be killed.

As for the problem of high-level radioactive waste storage at reprocessing plants, the AEC says that the wastes are so well contained in stainless steel tanks with coolant systems, etc., that they are "essentially accident-proof." "For these reasons," the AEC adds, "the probability of an accidental release of radioactivity is considered to be extremely low and thus there would not be any significant release to the environs."[155] While it is hardly necessary to note that all the AEC assurances are highly qualified, it is clear that the primary accidental control mechanism is optimism that all systems will work as planned.

---

drogen and strontium 90—into a local creek, which were about 20 percent of the AEC's maximum permissible population dose standards, which have been successfully challenged as adequate for population protection. The plant also released 180,000 curies of radioactive krypton gas in the atmosphere that year, down from the 300,000 curies of krypton 85 it released in 1969.[152]

As in nuclear reactors, the technology is fail-safe. Accidents of a serious nature are presumed impossible.

One of the AEC's admitted fears about the future development of large-scale nuclear power, particularly about the "plutonium economy," is the possibility that shipments of nuclear fuel containing quantities of plutonium might be hijacked and used by saboteurs to fabricate bombs. Either plutonium 239 or U-235 can be fabricated into a nuclear bomb by a knowledgeable person, and a crude bomb might be made by a competent college physics student with access to publicly available illustrations of bomb configurations (published by the AEC and others), say the experts.

Former military weapons specialist Dr. Theodore Taylor, president of the Washington, D.C., thinktank International Research and Technology Corporation, believes the world has little time to bring the problem of transporting nuclear materials under control. Taylor finds the situation especially alarming since "the knowledge, materials and facilities required to make nuclear explosives with yields at least as high as the equivalent of a few tens of tons of high explosive—given the required fissionable materials—are distributed worldwide."[156]

The amount of fissionable material in the form of uranium or plutonium required to make a workable bomb is termed the critical mass—i.e., about 5 kilograms of plutonium and 20 kilograms of uranium. Rand Corporation physicist Victor Gilinsky believes that the production of large quantities of plutonium by breeder reactors (as well as plutonium production to a lesser degree in light-water reactors) is the chief danger associated with the development of civilian nuclear power.[157] He points out that 10,000 to 25,000 kilograms of plutonium alone will be produced in conventional U.S. reactors between 1975 and 1980, but the useful weapons-grade material is plutonium 239, and U.S. reactors will produce plutonium 239 that is "contaminated" with a sister isotope, plutonium 240, making it generally unsuitable for weapons use.[158] Likewise, the U-235 used in conventional reactors is not high-grade uranium suitable for weapons use. Two nuclear developments may erase this safety margin: the development of the breeders, which will produce a purer plutonium, suitable for weapons; and the sale of high-temperature gas-cooled reactors, which use higher-quality U-235. A subsidiary of Gulf Oil Corporation has built one gas reactor in Colorado, and has sold several more in the United States.[159]

One aspect of the problem is the likelihood that many nations will soon have access to high-grade weapons material through the construction and operation of reactors and fuel reprocessing plants.

Gilinsky notes that natural uranium (heavy-water-moderated) reactors produce high-grade U-235 suitable for weapons; countries having these reactors include Canada, Argentina, Czechoslovakia, India, Italy, Japan, Pakistan, and Spain. Both India and Japan are constructing reprocessing plants, which would allow them to build weapons if they desired.[160] In the war-torn Middle East, Israel has developed a sizable nuclear program and, according to some sources, has already developed nuclear weapons capability.‡ With the development of more "peaceful" nuclear power, the nuclear weapons club will be expanded to even the smallest states, which may be more inclined to use the weapons than the giants of international politics.

Considerable quantities of nuclear materials suitable for weapons manufacture have already been lost, either in normal operations of nuclear industries, or in transporting nuclear materials. One U.S fuel fabrication plant is reported to have lost more than 220 pounds of U-235 in a 6-year period, enough to make 5 bombs of the Hiroshima size.[161] Shipments of nuclear materials of weapons grade have even reportedly been aboard airliners hijacked to Cuba. An AEC official at the time of the reported incidents refused to tell the *Washington Monthly* more than the fact that he was at AEC's safeguards division at the time, but couldn't comment in the interests of "national security."[162]

Perhaps as frightening a prospect as a group of saboteurs actually making a bomb to dispose of New York or Washington is the possibility that someone might acquire as much as a few grams of the estimated 720,000 kilograms of plutonium 239 that will be in circulation in the United States by the end of this century. A 5-gram thumbnail-size pellet of plutonium contains 20 million lethal doses, Professor Neilands indicates, and a tiny, almost invisible fraction of this would constitute a quantity large enough to snuff out the lives of occupants of a skyscraper or the U. S. Capitol, were the material properly released into the ventilation system.

One way the material could find its way into general circulation is through the proliferation of plutonium-powered pacemakers and artificial hearts, for which the AEC is trying to find a market. Nuclear pacemakers are already in use in the world, and the AEC is developing a fully nuclear-powered heart. The principle of operation of the

‡ A sign that Israel may be holding out the nuclear threat as a *coup de grâce* to the Arab world is the fact that the country never signed the 1968 Nonproliferation Treaty, which bans nuclear arms from nonnuclear countries. Israel is, then, not subject to international safeguards inspection by the International Atomic Energy Agency.

devices is simple: Radioactive decay heat of unstable elements such as plutonium provides the heat source for a miniaturized engine. Since the radioactive materials used have exceptionally long half-lives, the heat given off lasts years, providing a longer life than conventional batteries, making the nuclear units superficially more attractive to doctors, who would not have to undertake frequent transplants.

Unfortunately, even the tiny pacemaker nuclear batteries contain enough plutonium* to disperse the equivalent of three million to thirty million doses of the element, sufficient to cause individual human lung cancers.[163] The newer nuclear-powered aritificial heart contains many times this amount. A clever saboteur, blackmailer, or deranged person would not have to go to the trouble of stealing plutonium from the AEC; he would simply have to locate a person with a nuclear-powered artificial heart or pacemaker to get the necessary material to mount a considerable threat. A more mundane and realistic consideration is the fate of a nuclear pacemaker-wearer in an auto accident, a plane crash, or a fire. Would the plutonium be released? The AEC's answers to this question are unconvincing.

Ultimate "Disposal"

The laws of thermodynamics show that there is no real "disposal" of energy wastes on Earth; energy is transformed, not destroyed. A small fraction of an energy resource is converted to useful work, the rest wasted. In the case of nuclear fission power, the waste problem is exceptional.

In both today's light-water reactors and the projected breeder reactors, only a small amount of fissionable energy materials are converted into heat to be used for production of electricity. The overwhelming product of fissioning uranium and and other radioactive elements is radioactive waste. The guiding principle of nuclear science is to develop techniques that will minimize the introduction of this material into the biosphere. Unlike many chemical wastes and products of fossil-fuel energy conversion, there is no known way to treat radioactive wastes to remove their deadly biological hazard.[164]

This characteristic of nuclear wastes leaves only one option: to separate the radioactive products of fission reactions from environmental pathways (such as the atmosphere and ground water supplies) leading to living things. The traditional method of disposing of nuclear wastes has been to package radioactive material in drums for burial on land or at sea; the Atomic Energy Commission maintains federal

* The isotope used is plutonium 238.

repositories where wastes are stored in enormous concrete and steel tanks.

Reactor wastes are divided into three categories by the AEC: high-level, intermediate-level, and low-level wastes. High-level wastes contain more than 10,000 curies of radioactivity per gallon and are the deadliest products of reactors. High-level wastes are not only extremely hazardous for long periods, but wastes kept in liquid form are "self-boiling" from the heat of radioactive decay and are chemically corrosive. This feature makes it close to impossible to store the wastes in conventional tanks, even of the most sophisticated metal alloys. The hot wastes corrode and destroy the tanks.

Coupled with the long life of the wastes, measured by the presence of large quantities of plutonium (with a half-life of more than 24,000 years) and other long lived toxic radionuclides, nuclear garbage storage must be permanent. AEC scientists talk of waste storage in terms of over 100,000 years.

In the meantime, the waste-storage policies of the AEC have hardly been able to contain wastes for periods of even a few years.

In 1955, a special scientific advisory committee of the National Academy of Sciences' National Research Council was set up by request of the AEC to study the agency's waste-disposal practices, report to the AEC on problems, and make recommendations for future policy. The committee, which was disbanded in 1967, reported in 1966 to the AEC that waste-disposal practices were inadequate at the AEC's facilities. It recommended several principles for the AEC to follow for the disposal of high-level wastes. Temporary waste-disposal practices, the committee members said, should not be considered "safe" even at the current low level of development of nuclear power. Safety should be defined, they continued, by the AEC's capability of handling the massive quantity of nuclear wastes that would be present in a full-fledged nuclear power economy, when hundreds of reactors will be producing wastes continuously. The committee told the AEC that no funds should be held back in the pursuit of safety in the development of nuclear waste storage.[165]

The committee specifically recommended to the AEC that investigations of waste storage by injecting radioactive waste slurries into underground bedrock formations at the AEC's Savannah River plant on the Georgia–South Carolina border be discontinued, because of the possibility that wastes might leak into the ground water, and possibly reach the Savannah River in a hundred years. This would lead to the permanent contamination of the water supply for that part of the southeastern United States.[166]

The NAS report was essentially ignored by the AEC, and plans are still continuing for underground waste disposal at the Savannah River facility.‡ A second less critical panel was established later by the National Academy of Sciences (NAS) at the AEC's request, and it cautiously recommended that storage in the Savannah River bedrock be considered, but only if guarantees could be made that no radioactivity would leak out for a thousand years. The second panel noted that eighty million gallons of lethal wastes (containing strontium 90, cesium 137, and plutonium 239) would have to be disposed of, and noted that little was known of the possibilities of the radioactive products "migrating" from bedrock into the Savannah River water.[167]

The AEC believes that the best method of long-term waste storage for the concentrated wastes would be to calcine the wastes into a solid material, and then bury the solid wastes in an underground bedded salt formation, since the presence of salt deposits indicates that water has not been present for thousands of years.

A program to bury wastes in an abandoned Kansas salt mine, coded "Project Salt Vault" by the AEC, was initiated in the mid-1960s by the Oak Ridge National Laboratory. Initial tests at the Lyons, Kansas, site were conducted by the Oak Ridge National Laboratory, and the AEC became convinced that the idea was not only a perfect solution to the problems of waste disposal, but indicated that the Lyons site had been chosen as the first pilot facility for a national high-level waste depository.[168]

The AEC announced to the nation in 1970 that the decision to store wastes in the Lyons salt formation was the turning point of the nuclear age.‡ However, many scientists in the AEC and in Kansas were unconvinced that enough research had been done to assure that radioactive wastes might not "migrate" from this site and contaminate ground water.

The principal skeptic of the scheme was the director of the Kansas State Geological Survey, Dr. William Hambleton. Dr. Hambleton's original concern was that another ice age might occur, as the last one had ten thousand years ago in Kansas, exposing the still-active wastes

† The AEC concealed the existence of the report for four years, until a copy was finally given to Senator Frank Church of Idaho in 1970 by then AEC chairman Glenn Seaborg. Church had heard of the report, and had repeatedly asked the AEC for it before finally receiving a copy.

‡ The AEC spent thousands of dollars in public-relations funds just to convince the residents of Lyons that the idea was sound and would provide jobs for the small town. A newspaper poll in Lyons showed that sixty-seven people out of one hundred were for the idea of the town's becoming the nation's nuclear waste dump; twenty-two were against it; and eleven were undecided.[169]

to some future generation of unsuspecting humans. He also suspected the validity of AEC data collected to predict the ability of the salt formation to contain the hot wastes. On a limited budget, Hambleton conducted his own investigation of the Lyons site, and appeared before the Joint Committee on Atomic Energy in the spring of 1971 to warn the AEC's "watchdog" that the commission's research was not adequate to commit the nation to the Lyons site as a storage facility. However, the JCAE was not impressed by Dr. Hambleton, and Congressman Craig Hosmer of California, a leading member of the committee, said of Dr. Hambleton's cautious approach that "I get the impression we should never have invented the wheel if we had thought about it beforehand."[170] The JCAE approved an AEC request to buy the site and after prolonged congressional debate, and over the bitter opposition of Kansas legislators, the bill was passed, authorizing purchase of the site.

Dr. Hambleton's perseverance paid off, however. Later research at the Salt Vault site disclosed the presence of twenty-nine old oil and gas wells nearby, which could release wastes in water seepage. In addition, mining operations at a salt mine only a few thousand feet from the proposed waste facility were found to use water—which could lead to further waste migration. Dr. Hambleton summarized the situation when he later told a national science meeting that the Lyons site "was like a piece of Swiss cheese."[171]

Finally, the AEC bowed out, admitting the uselessness of pursuing further work at the Lyons site. The commission announced that the long-term solution to nuclear wastes would now be "engineered surface facilities" designed so that wastes could be stored above-ground, as in the current practice of "interim storage." The surface facilities would be designed to hold wastes for a minimum of a hundred years.[172] The AEC reasons that this will provide time to find another "ultimate" solution.

Current Storage

Time is running out for the AEC at its present waste facilities, where the mismanagement of wastes has led to a series of accidents and inadvertent spills of liquid wastes kept in tanks for only a few years. Adding to the problem of routine spills at AEC waste facilities is the fact that advisory panels have repeatedly warned the AEC that its facilities are not sited in suitable areas to protect the public from the migration of wastes once accidents occur. The National Academy of Sciences committee, in fact, told the commission in 1965 that "none of the major sites . . . is geologically suited for safe disposal of any manner of radioactive wastes other than very dilute, very low-level liquids."[173] Two studies of the AEC waste sites made by the congressional General Accounting Office found that AEC waste-disposal practices were ill-managed and underfunded, and noted that wastes containing plutonium had been buried at various sites prior to 1970, and that "provision for retrieval was not a primary consideration at the time of burial."[174] What this means is that this plutonium will almost inevitably reach water supplies, perpetually contaminating areas outside the waste facilities and endangering life for centuries to come.

The General Accounting Office uncovered evidence that considerable quantities of wastes had leaked out of tanks and storage drums at AEC facilities over a period of years. The GAO first reported tank leaks at the AEC's Hanford, Washington, waste facility, built during the Second World War for production of the atomic bomb and used in recent years for weapons production and waste disposal. In 1958, a tank made of concrete and lined with carbon steel leaked 35,000 gallons of concentrated waste liquids. In ensuing years, 9 more of Hanford's 150 tanks sprang leaks, releasing up to 55,000 gallons of wastes. Fortunately, the wastes were trapped by layers of clay and did not migrate into the underground water supply.

In addition to leaks at Hanford, tanks leaked at the AEC's Savannah River plant. The GAO noted that reserve tanks were unavailable at the waste facilities, and that in the event of a major accident involving structural failure of any given tank, there would be no chance of pumping wastes into a reserve unit. Particularly acute was the problem at the Hanford facility: The GAO said that "some of the tanks at Richland [the Richland, Washington, AEC reservation] had been in service 10 years or more and that a contractor had estimated that the expected life of those tanks was probably no more than 20 years."[175]

By 1973, when 42 million gallons of the highly radioactive wastes were in storage in the outdated Hanford tanks, the first major disaster warnings became apparent. About one third of the capacity of one of the old tanks leaked out, 115,000 gallons in all. How long Hanford's other tanks will last appears uncertain. The radioactive waste leakage at Hanford includes large quantities of plutonium, and as is true with breeder reactors, the potential for an uncontrolled chain reaction is a distant but very real possibility. One of the former waste-disposal practices at Hanford was simply dumping wastes into trenches—wastes containing plutonium. The AEC reported that so much plutonium had accumulated in one waste trench at Hanford that the agency concluded that a nuclear chain reaction was conceivable.[176]

In that event, one might conceive of an explosion that would not only liberate tremendous quantities of plutonium into the atmosphere, but would rupture other waste tanks and release floods of waste into ground water leading to the Columbia River. Unlike the bedrock project at the AEC's Savannah River plant, the potential for disaster in Washington seems on the threshold, not a hundred years away.

As these leaks and disasters at the AEC's facilities have continued, the official reaction of the commission has been one of measured silence and publicly expressed optimism over the future of waste disposal.

Perhaps the most chilling confirmation that the AEC really has no answer for the problem of nuclear waste disposal is that it has seriously considered rocketing nuclear waste cargoes into space. Aerospace Corporation recommended to the AEC that wastes be compacted into copper spheres 2½ feet in diameter and weighing 11,000 pounds, and rocketed into "heliocentric" orbit, circling the sun like other planets. The flights would cost $9 million apiece, and 40 would be needed in the year 1999 alone.[177] A failure of even a single flight could wreak havoc on an exposed population. When former AEC chairman James Schlesinger seriously proposed the idea, Anthony Tucker of the British newspaper, the Manchester *Guardian,* commented that, "without being too much of a pessimist about space technology, I would suggest that this would be permissible only if all the fall-out from failed shots landed on Mr. Schlesinger."[178]

Citizen Involvement in the Licensing of Nuclear Power Plants

Much public debate has occurred about the advantages and disadvantages of nuclear power. In recent years, a number of citizens groups have attempted to stop or delay the construction of nuclear power plants, but these efforts have been for the most part unsuccessful. The Atomic Energy Commission's regulatory staff reviews all electric utilities applications for nuclear power plant construction, and after reviewing them, prepares a plant Environmental Impact Statement, in conformance with requirements of the National Environmental Policy Act (NEPA).

Public hearings on nuclear power plants are conducted by a special AEC-appointed panel, the Atomic Safety and Licensing Board (ASLB). The ASLB is empowered to determine whether or not the proposed nuclear plant should be licensed (on grounds of safety and environmental effects); and during its hearings, electric utility applicants for plant licenses are pitted against members of the public (intervenors), who may challenge the proposed construction of the power plant on environmental grounds. The hearings are both costly and time-consuming for public intervenors; and since the ASLB is appointed by the AEC, the outcome (i.e., licensing the plant) is usually predetermined. Only in rare instances have proposed plants not been licensed, and then only because of exceptional findings, such as proposed siting of plants on known earthquake faults.[179]

The previously mentioned national citizens coordinating group, the National Intervenors,* disseminates information on the nuclear issue to citizens groups, and has called for a moratorium on the further construction of nuclear fission power plants. Legislation to enact a moratorium on the further construction of nuclear plants "until the electric utilities have demonstrated to Congress and to their respective customers that the use of nuclear fission to generate electricity will not constitute a significant hazard to the health and safety of the public, or render this country vulnerable to blackmail, terrorism, or economic chaos" was introduced in the Ninety-third Congress by Senator Mike Gravel of Alaska. The bill, S. 1217, would allow for the operation of some existing power plants until 1980 (in time to find other, nonnuclear substitutes).[180]

The most ambitious attempt to curtail the operations of nuclear power plants was a recent legal challenge to the AEC posed by the environmental organization, Friends of the Earth, in conjunction with Ralph Nader. Their lawsuit sought to shut down twenty of the nation's thirty-one operating nuclear power plants on the grounds that the hearings on Emergency Core Cooling Systems (ECCS) had indicated that the systems would fail to operate in the event of a plant emergency, and that the AEC was allowing certain plants to operate illegally. The U. S. District Court for the District of Columbia dismissed the suit in 1973, and it is at present under appeal.[181]

Conclusion

Both the detractors and the advocates of nuclear fission power agree that the full costs to society of developing this power source are largely unknown. The tremendous power of the atom beckons an energy-craving civilization, but the major question is whether the civilization can safely use the power. Nuclear power plants have been promoted by an overzealous agency that has consistently argued that the benefits to society—electricity—far outweigh the potential hazards. Economist Allen Kneese of the Resources for the Future, Inc., a think-

* 153 E Street, S.E., Washington, D.C. 20003.

tank in Washington, D.C., has summarized the future significance of making decisions on the basis of the AEC's "cost-benefit" approach:

> . . . The advantages of fission are much more readily quantified in the format of a cost-benefit analysis than are the associated hazards. Therefore there exists the danger that the benefits may seem more real. Furthermore, the conceptual basis of cost-benefit analysis requires that the redistributional effects of the action be, for one or another reason, inconsequential. Here we are speaking of hazards which may afflict humanity many generations hence and distributional questions which can neither be neglected as inconsequential nor evaluated on any known theoretical or empirical basis. This means that technical people, be they physicists or economists, cannot legitimately make the decision to generate such hazards based on technical analysis. Society confronts a *moral* problem of a great profundity—in my opinion one of the most consequential that has ever faced mankind. In a democratic society the only legitimate means for making such a choice is through the mechanisms of representative government.
>
> For this reason, during the short interval ahead while dependence on fission energy could still be kept within some bounds, I believe the Congress should make an open and explicit decision about this Faustian bargain. This would best be done after full national discussion at a level of seriousness and detail that the nature of the issue demands. An appropriate starting point could be hearings before a committee of Congress with a broad national policy responsibility.[182]

The continuation of America's nuclear fission program in the light of such major social and environmental problems is not a bright prospect for the future, particularly in light of the increasing shortages of fossil fuel energy and vital materials in the economy. The development of nuclear fission power plants may actually exacerbate the energy crisis rather than alleviating it. Few studies have examined the net energy output from nuclear power plants, that is, the electricity they deliver after subtracting the energy costs associated with the nuclear fuel cycle. Large quantities of energy (mostly from fossil fuels) must be expended to search for, mine, and transport uranium ore; mine, refine, and fabricate metals used for nuclear power facilities; construct, operate, and maintain (indefinitely) nuclear fuel reprocessing and storage facilities; and other elements of the nuclear fuel cycle.

A particularly energy-intensive aspect of the fuel cycle is the processing of uranium fuel at special government-owned facilities called gaseous diffusion plants. Energy conversion engineer E. J. Hoffman of Laramie, Wyoming, argues that approximately "half of the gross electrical output of a nuclear plant would have to be recycled to supply input for fuel processing."

Hoffman believes that nuclear power plants yield very little net energy to the society at present, and he suggests that in the future, the plants will become a drain on the other energy sources available. "The cumulative energy expenditure of the entire atomic energy program may not be recouped from nuclear fission power plants by the time the reserves of economically recoverable U-235 are used up," he says.† An energy-short United States might not be able to mobilize the necessary technological safeguards to keep the nuclear industry fed with the necessary support mechanisms, such as fossil fuels to enrich uranium (which constitutes a major national energy use at present), or the fossil fuel energy necessary to maintain fuel transportation to reactors and reprocessing plants. Shortages of conventional fuels could prove to be the key that unlocks the Pandora's box of nuclear fission.

A prudent future energy policy for the nation would be to devote its talent and economic resources to developing more benign alternatives to nuclear fission, an energy source plagued with massive problems and one that offers only electricity in return for social investment in the lethal technical machinery of the nuclear establishment. Each new reactor brought to the nation's electrical system adds a grave and disturbing burden to unsuspecting new generations.

Electricity from Nuclear Fusion

As opposed to nuclear fission, which is the process of energy release by the collision of heavy atoms, nuclear fusion processes release energy by the combining of very light atoms.

Scientists have known for a little more than forty years that the energy of the sun and stars is a product of fusion reactions. The first earthbound fusion experiments were conducted in the 1920s by using high-energy particle accelerators to shoot protons, the nuclei of hydrogen atoms, fast enough to force the protons to fuse with other nuclei. Uncontrolled fusion was graphically demonstrated in the hydrogen bomb in the early 1950s.[183]

The concept of attempting to control the thermonuclear reaction, or fusion, as a means of providing heat for the generation of electricity is a relatively new one. Formal work on the concept began in

† Hoffman's arguments on nuclear power are detailed in Jeff Stansbury and Edward Flattau's article, "It Takes Energy to Make Energy: The Net's the Thing," *Washington Monthly*, March 1974, pp. 20–26.

the United States in 1952, when the government initiated a highly classified research effort under the code name "Project Sherwood," a half-serious pun on the response, "It sure would," to the question, "Wouldn't it be nice if we could achieve nuclear fusion?"[184]

There are several approaches now under study in the attempt to develop the fusion process. They all fall into two basic categories: (1) the magnetically confined thermonuclear reactors (MCTR), based on the use of magnetic forces to compress and heat the thermonuclear material to fusion temperatures and then confine it during the fusion reaction; and (2) the inertially confined thermonuclear reactors (ICTR) which utilize high-energy laser beams to heat the thermonuclear material to the intense heats required for fusion.[185]

Magnetic Confinement Fusion Reactors

The magnetic confinement approach is the one initially pursued under Project Sherwood, and is consequently the one on which the most data and experience have been accumulated. It involves subjecting amounts of thermonuclear fuels whose atoms are compressed to densities of 100 trillion to 1,000 trillion per cubic centimeter, to temperatures of 60 to 100 million degrees Centigrade for periods of time ranging from 0.1 second to 1 second.[186] The thermonuclear fuels under consideration include deuterium, tritium, lithium, and helium, all of which are very light elements, which fuse more easily than the heavier ones.[187] There are several combinations of deuterium, tritium, lithium, and helium that are considered likely candidates for the fusion reaction. The one most commonly envisioned as the fuel for the first fusion reactors is a combination of deuterium and tritium. Since the tritium does not appear in nature, and is derived from interaction of deuterium with lithium, the presence of lithium in the reactor is necessary, even though it is not used directly as a fuel.[188] An advanced fusion technology, and the one that is expected to follow in the second generation of fusion reactors, is the deuterium-deuterium reaction, which involves the use of pure deuterium alone.[189] Deuterium is present in virtually endless quantities in seawater, from which it can be cheaply and easily extracted. Hence, it is considered a fuel that is in limitless supply. Supplies of lithium, from which the tritium is extracted, are less extensive, but there is enough available to sustain at least several thousand years of fusion operation.[190]

A roadblock to achieving thermonuclear fusion has been the problem of containment. Obviously, it is not possible to contain materials at temperatures of millions of degrees in any kind of known metal. The solution that fusion research scientists have pursued is magnetic containment; that is, containment in invisible "bottles" whose walls are formed by very powerful magnetic fields.[191]

Matter is usually thought of as having three states: solid, liquid, and gas. There is also, however, what is considered to be a fourth state that substances enter when they are brought to extremely high temperatures. That is the plasma state, which somewhat resembles the gaseous state except that all the electrons have been dislodged from their atoms by the force of colliding at greatly accelerated speeds brought about by the intense heat. The dislodged electrons have a negative charge, and the nuclei from which they were separated are left with a positive charge. It is because these charges react to magnetic forces that the plasma is theoretically able to be contained by magnetic fields.[192]

However, success in achieving these theoretically possible temperatures and containment conditions has been elusive, and the scientific feasibility of this concept still remains to be proven. In fact, an equation derived in the early 1940s by Dr. David Bohm, an American physicist conducting nonfusion plasma and magnetic field research, indicated that reaching this theoretically possible goal was in fact impossible. He found that the ability of magnetic fields to contain the thermonuclear reaction seemed to decrease as the temperatures increased.[193]

Dr. Bohm's equation seemed to hold true in subsequent fusion experiments, and continued to frustrate attempts to achieve thermonuclear reaction until the mid-1960s, when Soviet researchers at the Kurtchatov Institute announced that one of their fusion research devices, called the Tokamak (meaning "large current" in Russian), had succeeded in defying the Bohm equation by instituting a different arrangement of the magnetic systems.[194] The Russian findings have since been verified by U.S. experiments. Recently, an Adiabatic Toroidal Compressor, a Tokamak-like machine at the Atomic Energy Commission-supported Princeton University Plasma Physics Laboratory at Princeton, New Jersey,[195] was used to demonstrate a new plasma-heating technique that produced electron temperatures of about 25 million degrees Centigrade with plasma densities of 100 trillion particles per cubic centimeter.[196] A much larger machine is under construction at Princeton, which Dr. Harold P. Furth, codirector of the AEC-supported Plasma Physics Laboratory at Princeton predicted would "come close" to proving scientific feasibility.[197] Government research on fusion is also being conducted at other AEC laboratories, including

Lawrence Livermore Laboratory in California, Los Alamos Scientific Laboratory in New Mexico, and Oak Ridge National Laboratory in Tennessee. Several universities are also conducting their own fusion research programs, with the University of Texas' Center for Plasma Physics and Thermonuclear Research one of the largest.[198]

Laser Fusion Reactors

The second major fusion technique, inertially confined thermonuclear reaction—more commonly referred to as laser fusion—is a still more recent development. Soviet reseachers first demonstrated in 1968[199] that a laser beam could initiate a fusion reaction, and since then several nations—including the United States—have stepped up their own laser research aimed at achieving laser fusion.[200]

A laser is a device that amplifies light by "pumping up" atoms to high-energy levels. The name is an acronym derived from "*l*ight *a*mplification by *s*timulated *e*mission of *r*adiation." An external source of energy is used to operate a laser, which produces coherent monochromatic light—"pure" and consisting of only one wavelength, contrasted with ordinary daylight, which is incoherent, or consisting of many wavelengths. The coherent laser light can be concentrated at high energy in a very narrow beam.[201]

The laser technique involves the use of laser energy to raise the temperature of small pellets of deuterium and tritium to 100 million degrees Centigrade before the pellets have time to expand, in accordance with Newton's law of inertia. The resulting heat of fusion would heat water to produce steam for a conventional steam turbine electric generator.[202]

Operation of the laser itself requires an enormous energy input. Fusion scientists are optimistic that this energy will be recovered by the potentially greater release of energy in the nuclear fusion reaction.

The most promising laser technique at this time appears to be one developed by KMS Industries, Inc., in Ann Arbor, Michigan. It involves simultaneously focusing several laser beams on the pellet, which then implodes under this high-energy pressure, including more complete burning and greater heat. By using this process, according to John Nuckolls and Lowell Wood at Lawrence Livermore Laboratory, the laser energy required for a practical power reactor could be reduced by a factor of 1,000 below previous estimates—to a value of 100,000 to 1 million joules.‡[203]

‡ A joule is a measurement of power equivalent to 0.239 calorie or 0.738 foot-pound.

The KMS technique still requires many times the maximum energy delivered by the largest laser available today—a neodymium laser at the Lebedev Institute in Moscow that delivers 600 joules in two billionths of a second (2 nanoseconds). However, if computer simulations of the KMS process are borne out in experiments planned over the next several months, large lasers now under construction in the Soviet Union and scheduled for construction in the United States could prove the scientific feasibility of the principle in the near future.[204]

The fuel cycle and environmental problems of laser fusion would be similar to those of magnetic fusion, with the principal environmental problem being the confinement of the highly volatile radioactive tritium. However, laser fusion does pose some unique engineering problems not shared by magnetic fusion. Besides the most important problem of developing a very durable laser that can deliver the very high energy required at one-second intervals, a strong restraining wall will have to be developed to contain the explosive reaction. Methods must also be developed to contain the lithium required to provide the tritium particles and to protect the structural parts of the reactor vessel against radiation damage. Hoped-for solutions to these problems are now under study at several laboratories.[205]

Laser fusion technology might have a size advantage over magnetic fusion. Laser fusion reactors of 50-to-200-megawatt size are hoped for, and much smaller units could be used to provide motive power for such purposes as space propulsion.

The laser fusion process has been presented as a simpler process without as many formidable obstacles to fusion power as its magnetic counterpart, but as William D. Metz observed in *Science* magazine, this may be "only because the concept is newer." In fact, Metz noted, "the technological problems of the laser alone could prove to be as difficult as the development of an entire Tokamak reactor."[206]

Both the magnetic fusion and laser fusion techniques have thus far been discussed in terms of providing heat energy for the generation of electricity through conventional steam turbines. However, what appears to be a much more promising concept in terms of efficient energy conversion is the use of the magnetic containment process to generate electricity directly at efficiencies of 90 percent and higher. The deuterium-tritium$_3$ reaction is considered the most promising for this concept, since the energy of these elements is released as kinetic energy carried entirely by charged reaction products. These charged particles, traveling at very high speeds, could in effect be lined up and sent directly into the electric power line.[207]

Richard Post, a leading fusion scientist at Lawrence Livermore Laboratory, has proposed such a scheme, which would use only the charged particles that "leaked" from the "magnetic bottle." The "leaked" particles would emerge into a one-hundred-yard-long, fan-shaped vacuum chamber surrounded by magnets that would guide the particles to electrodes, charging the electrodes to high voltages in the process.[208] According to Post, small-scale tests have proved many of the principles involved and have achieved measured conversion efficiencies greater than 80 percent. "One of the nice features of the system is that it would generate high voltage direct current directly and that its cost per kilowatt converted might be substantially lower than that of steam turbine-alternator conversion systems," Post said.[209] The promise of direct conversion fusion is that it might significantly reduce or eliminate the problems associated with generation of radioactive material, since some proposed fusion fuels will be nonradioactive.

Policymakers frequently allude to fusion power as "the ultimate energy source," which is at best a calculated gamble. Despite increasing successes in developing the densities, heat, and containment capabilities required to support the fusion reaction, there are some who still seriously doubt that fusion will ever prove scientifically feasible.

Even if scientific feasibility of fusion is proven, some enormous problems remain. Dr. Herman Postma, director of fusion research at the AEC's Oak Ridge National Laboratory, has cited engineering feasibility as one of them. Postma has noted that the serious question of whether a fusion reactor could withstand the radioactivity within it has not yet been explored. In the deuterium-tritium reaction, for instance, 80 percent of the energy produced emerges in the form of very-high-speed neutron particles. The impact of such particles will radioactively damage almost any material over a period of time, reducing most metals, for instance, to the strength of chalk. Neutron damage also causes material to swell as a result of the transmutation of atoms deep within it. These and other materials problems must be overcome before a fusion reactor can be seriously considered, even if the fusion process proves to be possible. Massachusetts Institute of Technology engineering professor David Rose has pointed out that these problems may prove to be as difficult as the plasma confinement itself.[210]

The other major unknown that Postma noted is economic feasibility. It is simply an unknown at this point, although some who doubt the economics of fusion as the heat source for the steam turbine electric power cycle are more sanguine about the economic feasibility of the direct conversion process.[211]

A 1971 report, *Energy Research Needs,* prepared for the National

Science Foundation by Resources for the Future, Inc. (RFF), in cooperation with Massachusetts Institute of Technology Environmental Laboratory, noted that, while cheap fuel will assure low operating cost for fusion power plants, the bulk of the costs will be in the capital investment, "which initially will be high due to the complex nature of the technology. Since the fusion process replaces only the boiler portion of the power plant and half the price of a power plant is involved with conventional technology of heat exchangers, turbines, and sites which are still needed for fusion, large reductions in price seem elusive."

This situation could change, the report added, if direct energy conversion is achieved in the fusion process. The report continued:

> Actual refinements to these studies must await more progress in order to remove the large uncertainties. The judgment on the economics will eventually depend strongly upon the complications or simplifications brought about by the improved understanding in scientific feasibility and the engineering solutions to the problems of reactor design. Economic values to be attached to the short doubling times, the lessened environmental insults, the inherent safety of such plants. Flexibility in location will certainly enter the eventual evaluation, but again true assessment can be made only after the uncertainties dominating the confinement and engineering tasks have been determined by the building of several demonstration fusion reactors.[212]

Under the most optimistic circumstances, fusion researchers state that it is virtually impossible—short of a multibillion-dollar national scientific effort—for electric power from a fusion reaction to become available before the year 2000. Given the unsolved and very complex problems cited above, some scientists talk about the first fusion in the mid-21st century. Fusion scientists contend that fusion reactor development is directly proportional to the amount of money available for research and development, and contend that U. S. Government funding for these programs is inadequate.[213]

AEC fusion scientist Paul W. McDaniel told a recent congressional hearing that "at the present rate of funding we are not going to achieve feasibility in this century," but that with a "wartime urgency" funding level, feasibility could be demonstrated in the 1970s and commercial plants could be on line by 1993.[214] Dr. T. K. Fowler, associate director of the AEC's Lawrence Livermore Laboratory, says that predictions about the fusion time scale are "beyond the credible"; but assuming achievement of scientific credibility by 1980, "it is popular to estimate commercial fusion power by the year 2000. Any number of technical and socio-economic variables can either accelerate or slow this development," he cautioned.[215]

Dr. Herman Postma noted that to demonstrate scientific feasibility will call for "large engineering feats that seem at this time much more difficult to achieve than was the case of the demonstration of fission" in Chicago in 1942.*[216]

Since fusion reactors would use the conventional steam cycle in producing electricity, with its attendant low conversion efficiencies of about 40 to 50 percent,[217] the fusion reactor electric power generating plant would be accompanied by all the same thermal air and water pollution problems as today's conventional steam plants.

In addition, several environmental problems involving radioactive materials would be added, although they would be of a magnitude considerably smaller than the radioactive materials and emission problems inherent in today's fission reactors and the breeder reactor being heralded as the successor to today's generation of nuclear reactors.

Radioactivity in the fusion reactor would result from two of its built-in features: (1) the fact that the radioactive isotope of hydrogen—tritium (H_3)—would be a fuel source, at least in the first generation of fusion reactors; and (2) the high-speed bombardment of neutrons against the walls of the reactor structure, inducing radioactivity in the structure itself.[218] Thus the radioactive structure poses a waste-disposal and handling problem. Every 20 years—the projected life cycle of a plant—150,000 kilograms of the structure would have to be safely buried, transported, or recycled. However, the problems of radioactive fuel reprocessing inherent in the light-water and breeder reactors would be eliminated by fusion.[219]

The amount of tritium that would have to be handled—100 million curies per day—would pose a significant potential radioactive hazard. Emission of 15 curies per day is permitted under present federal radiological health standards. During normal operation, very small amounts of tritium are released to the steam cycles and to the water cooling systems.[220] While the dangers of tritium are much smaller than for other nuclear fuels, they still are considerable, and proper care must be maintained to prevent tritium contamination of the environment.

There are other problems inherent in fusion technology that are less readily apparent. For instance, the technology calls for large amounts of exotic, energy-intensive materials such as niobium and vanadium. The superconductor magnet required for the reactor in each 1,000-megawatt plant would require almost 2.8 million pounds of copper

* The demonstration of scientific feasibility of the nuclear fission process is described in Chapter 1.

(1,261,000 kilograms). Each plant would also require an inventory of almost 1,000,000 pounds (454,000 kilograms) of lithium out of a total world supply of 13 billion pounds (6 billion kilograms), and an inventory of 308,644 pounds (140,000 kilograms) of vanadium or niobium out of a total world supply of 22 million pounds (10 million kilograms).[221]

While the lithium supply at the projected level of use in fusion reactors would last "thousands of years," according to Herman Postma, it would still be necessary to develop the technology for deuterium-deuterium fusion reaction: ". . . it is not possible to satisfy forever the energy demands" with the deuterium-tritium cycle."[222] Mining of lithium will also have some environmental impact. Approximately 80 percent of the U.S. reserves are in the heavy brine deposits in Nevada and California, and the remainder is in the pegmatite ores in North Carolina and South Dakota.[223]

The fusion reactor would initially occupy approximately the same area (1,000 acres) and require the same facilities for cooling water, cooling towers, electrical substations, and exclusion areas as other future central power plants, Postma pointed out in testimony prepared for hearings before the Committee on Science and Astronautics of the House of Representatives. "It would also initially be located remote from cities because radioactivity would be involved and precautions would be taken for very careful monitoring." This remoteness, he noted, would also require transmission facilities and corridors. However, in the longer range, he said, if fusion reactors proved safe it would be possible to site them close to or within urban centers, "thus allowing far more flexibility in location and permitting greater energy utilization, since waste heat, in addition to electricity, would be available."[224]

The hope of nuclear fusion is indeed bright, although there is no assurance at all that this potential power source will prove workable. Nuclear fusion is not appreciably closer to a reality today than in the early experimental days of the "Sherwood" program. Although recent experiments have been promising, fusion scientists are apparently in agreement that fusion can offer no real alternative to this time of decreasing fossil fuel stocks. Undoubtedly, funds for fusion research—which are only a fraction of the funds expended for nuclear fission—should be increased, but the hopes of society should not rest on a power source that may never be proven.

Even though fusion is seen as an alternative to nuclear fission, the talent and funds for developing new sources of energy should be shared among other promising power sources.

Geothermal Energy—Heat from the Earth

Since the beginning of the 1970s, increasing national attention has been focused on the possibility of using geothermal resources for electrical power generation.

As its name implies, geothermal energy is the natural heat of the Earth. The source of this heat is the Earth's molten center, where temperatures in excess of 1,000° C (1,832° F) are reached as the result of natural decay of radioactive core materials and frictional forces resulting from solar and lunar tides as well as the relative motion of crustal "plates," which form the bases of the continents.[225]

The core heat migrates outward toward the surface at the average worldwide rate of 54 calories per square meter per hour.[226] Starting from the opposite direction, the heat of the Earth normally increases 30° C (54° F) per kilometer as one proceeds from the surface toward the Earth's center. However, there are countless pockets where special geological conditions have created pressures and heat much higher than those that would normally occur at that depth. These concentrated heat pockets, called geothermal reservoirs,[227] most often occur along areas of past volcanic activity at relatively shallow depths of a few thousand feet.[228] Where ground water comes into contact with the hot molten rock in these geothermal anomalies, natural deposits of geothermal steam or hot water are formed.[229]

World Geothermal Resources

Geothermal energy takes three forms: dry steam; wet steam and water, called a hot water system; and hot, dry rocks. United Nations geothermal consultant John Banwell has estimated that total energy from these three geothermal resources is virtually limitless—the equivalent of 21 *billion billion* tons of coal, many times more energy than the world could ever use without destroying itself with the waste heat of the energy conversion process.[230]

Banwell's UN colleague, Dr. Tsvi Meidav, has estimated that the geothermal energy stored in the upper 7.5 kilometers (24,600 feet) of the Earth's crust is equivalent to 21 million tons of oil per square kilometer of the Earth's surface.[231]

"It's like offering a hungry man an enormous meal," Banwell explained. "You simply can't eat it. And we, likewise, as a species will never be able to use that much energy."[232]

The problem, Banwell noted, is not *whether* geothermal energy resources can be tapped to provide energy for the world's needs, but *how* they can be tapped when and where they're most needed—efficiently and economically and with the least environmental impact.

Several countries, including the United States, have already undertaken limited development of their geothermal resources. Italy was the first to use geothermal steam for power generation, at Larderello in 1904.[233] The Soviet Union, Hungary, and Iceland use geothermally heated water and wet steam to heat homes and buildings and for some industrial purposes; New Zealand has had a network of geothermal wet steam electric power plants since the early 1950s;[234] Boise, Idaho, and Klamath Falls, Oregon, have used geothermal energy for heating buildings since the 1890s and 1930s, respectively;[235] Kyushu Electric Power Company in Japan has been generating power from geothermal steam for the past several years; and the Mexican government recently completed a 75,000-kilowatt geothermal generating station at Cerro Prieto, 30 miles south of the California border.[236]

Altogether, by 1972, some 790 megawatts (million kilowatts) of geothermal steam electric power was being generated in the United States, Italy, New Zealand, Iceland, Japan, Mexico, and the Soviet Union, at costs ranging between one half and two thirds the cost of other alternatives.[237]

U. S. Geothermal Resources

Commercial geothermal development in the United States has been centered up to now on the Geysers, an area in northern California that has been known for centuries for the escaping steam that marked it as a geothermal resource. A three-way combine of Magma Power Company, Union Oil Company, and Thermal Power Company has been drilling for and producing steam at the site since 1960. Pacific Gas & Electric Company (PG&E) buys the steam for electricity generation. Present total capacity of the Geysers is about 200 megawatts. However, PG&E has announced plans to continue to increase capacity, until by the end of the 1970s, geothermal steam would be generating several thousand megawatts of geothermal power at predicted costs of less than conventional coal or nuclear power plants.[238]

The second area now receiving major attention in the United States is the Imperial Valley in southern California, a wet steam and water geothermal resource area that is being closely scrutinized by virtually every major oil company for possible geothermal development. San Diego Gas & Electric Company, in conjunction with oil companies, has proposed construction of a $3 million geothermal pilot power plant in El Centro, and in early 1973 had already budgeted $8 million to $10 million to acquire right-of-way for power lines to San Diego, 80 miles away.[239] The total power potential of the Imperial Valley has been estimated at 30,000 megawatts.[240]

Geothermal heat deposits have also been noted in the Mono Lake/Long Valley/Casa Diablo area of California's eastern Sierra Nevada regions, where shallow wells have produced hot water.

A less publicized but perhaps even more promising area is the Gulf of Mexico geopressure zone, an area covering about 150,000 square miles in Texas and Louisiana where normal temperature gradients occur to depths of 10,000 to 12,000 feet, below which geopressure zones occur with much higher temperatures to a level of about 15,000 feet. Since it is an area in which countless oil wells have already been drilled, some consideration is being given to the possibility of utilizing existing unused wells for tapping into the geopressure zones.[241]

There are several other areas in western states in which known geothermal resources exist, but many of these are located on federal lands, which until 1970 were not available for geothermal development, and still have not been leased.

Dry Steam

Of the three types of geothermal heat, dry steam is the most desirable for purposes of electric power generation because it is very clean and requires little more than drilling a well into the steam pocket, filtering the steam as it issues from the well, and channeling it through conventional steam turbines to turn electric generators.

However, while dry steam is the most economically attractive, it is also the least abundant, and occurs only under unusual geological conditions, which create deep-level temperatures approximately ten times higher than the normal heat gradient for that depth.[242]

Banwell has estimated that the total heat output that could be derived worldwide from geothermal dry steam is equal to six billion tons of oil per year for several hundred years.[243] However, Banwell and Meidav have pointed out that probably only 10 to 20 percent of this heat could be effectively utilized for electric power because of the low efficiencies of geothermal heat for power generation in accordance with the first law of thermodynamics and because of the low pressures under which the steam enters the plant.[244] Furthermore, they add, these high-temperature geothermal reservoirs are often located in island areas and other remote regions where their development may not be feasible in the near future.[245]

Several dry-steam fields are under development, however, notably the Geysers field in northern California, and fields in Japan and Italy.

One estimate for the total U.S. dry-steam electric power potential capacity is 100,000 megawatts, with a lifetime of at least 20 years—probably much more.[246]

Wet Steam and Water

A much more abundant form of geothermal energy is wet steam and water. There are about twenty times more of these geothermal hot water systems than dry-steam systems, and they are thought to be larger.[247] Their temperatures are about three times higher than for the normal Earth heat gradient for that depth.[248] The combination of wet steam and water that these systems produce[249] can be utilized either as direct heat for various uses such as heating buildings, or to generate electricity. In New Zealand and Cerro Prieto, Mexico, it is used to generate electricity. In Hungary, Iceland, and the Soviet Union, it is used for direct heat.[250] Banwell and Meidav have estimated that the volume of stored hot water in one large geothermal basin of this type in Hungary alone has a heat reserve equal to one half the heat value of the known 1963 petroleum reserves of the entire world.[251]

It is estimated that the Soviets are saving 125,000 tons of oil per year by using geothermal wet steam and water for direct heat applications.[252]

The Imperial Valley is one of approximately 1,000 geothermal hot water systems identified so far in the United States. The total power potential in the United States from geothermal hot water systems is not known, but it is estimated to be between 1 million and 10 million megawatts, with a resource life estimated between 100 and 300 years.[253]

There are two ways in which geothermal wet steam and water can be used to generate electricity. One is by using the steam (about 20 percent of the total in a hot water reservoir will be wet steam) to turn a steam electric generator, much as in the dry-steam process. This method is in use in New Zealand. Another way, and one that is being demonstrated in the Cerro Prieto plant in Mexico, is to use the hot water to heat another liquid with a lower boiling point—such as isobutane—and use this secondary steam to turn the electric generator.

Hot, Dry Rocks

The third class of geothermal resource, and the one that seems to promise the ultimate in limitless energy, is the normal heat gradient of the Earth itself, available in the hot, dry rocks at depths of several thousand feet.

Banwell and Meidav have estimated that the total energy reserve to a depth of 7.5 kilometers (24,600 feet) is equivalent to three quadrillion tons of oil, which they noted is "much more, by about two orders of magnitude than the estimated maximum world total fossil fuel and minable uranium ores."[254]

Since the heat of the Earth's crust increases at the rate of 30° C (54° F) per kilometer of depth, the 90° C to 150° C (194° F to 302° F) temperatures occurring normally at 3 to 5 kilometers beneath the surface would be adequate for domestic heating, water desalination, chemical processing and many other industrial uses. Temperatures at levels of 5.5 to 6 kilometers would be adequate for electric power generation using a secondary system such as isobutane.

However, it would be necessary to go to depths of 9 kilometers (almost 30,000 feet—nearing the limits of present drilling technology) to extract heat that could be used to generate electric power directly.[255]

The most commonly considered method of tapping normal heat gradients is by combining drilling and hydrofracturing, a technique proposed by the Atomic Energy Commission's Los Alamos Scientific Laboratory at Los Alamos, New Mexico. By this method, one hole would first be drilled—perhaps to a depth of 15,000 feet. Water would then be pumped into it at a pressure of about 7,000 pounds per square inch. This pressure would be sufficient to create great cracks, or hydrofractures, in the rock. Once the crack was formed, a second hole would be drilled to tap it at its uppermost level, thus completing a fluid loop. Water could then be pumped into the first hole, where it would be heated by the hot rocks at that level before being drawn back up to the Earth's surface through the second well. The system would create a natural pumping effect; i.e., the cool water in the "down" well would be denser and heavier than the rising hot water. The difference in weight between the two columns of water would force the hot water to rise without artificial assistance from expensive pumps. Once at the surface, the heat captured by the water would be released to a heat exchanger for electricity generation, after which it would be reinjected to go through the cycle again.[256]

This hydrofracturing method is commonly used in oil drilling technology. Another method of fracturing rocks for the hot rock geothermal process has been proposed by the Atomic Energy Commission. It calls for detonating a nuclear explosive at the desired level to provide the necessary channel between the first and second wells. Inherent in this approach, however, are all the hazards of radioactivity and seismic damage of any nuclear explosion. Nonetheless, the AEC claims it would be "technically feasible," and that it promises "economic potential" under favorable conditions.[257]

Federal Geothermal Leasing Program

Up to now, all geothermal development in the United States has been on private lands, but the Geothermal Steam Act of 1970 authorized the Interior Department to facilitate and supervise the exploration, development, and production of federally-owned geothermal resources.[258] Involved in the federal leasing program are fifty-nine million acres in fourteen western states. In addition, several states, including California, Idaho, Utah, and New Mexico, have established regulations to administer the leasing of state lands for geothermal development. Indian lands are also available for leasing through direct negotiations with tribal officers and the Bureau of Indian Affairs.[259]

The federal leasing program has been the target of considerable criticism. Geothermal industry interests have complained that the Interior Department has been deliberately slow in developing guidelines and initiating the leases, and protest the requirement to file environmental impact statements, which they contend is delaying development of geothermal resources in the West. Environmentalists, on the other hand, have objected that the Interior Department's leasing regulations fail to provide public and environmental protection, and don't satisfy the requirements of the National Environmental Policy Act. They also have argued that leasing regulations fail to provide for public participation in deciding when and where geothermal operations might be allowed on public lands.[260] Still another group, publicly owned utilities represented by the American Public Power Association (APPA), has protested that the Interior Department in its geothermal leasing regulations has failed to provide safeguards against monopolistic practices by oil companies, private investor-owned electric utilities, and other major private interests considering the development of federally-owned geothermal resources.[261]

Environmental Effects

The quality of geothermal resources varies widely. Iceland's geothermal waters are pure enough to drink. However, U.S. geothermal resources discovered so far are unfortunately laden with impurities that pose some serious potential pollution problems that must be resolved before geothermal development can take place on a large scale.

The water drawn from U.S. geothermal reservoirs contains very high levels of salts and dissolved and suspended minerals. If discharged into streams, it could pose enormous environmental problems. This major problem could be turned into economic—though not environmental—advantage. The water can be desalted economically (because relatively small amounts of heat must be added to the existing natural temperatures to achieve distillation) to help solve another major problem—water shortages—for the American Southwest. The problem of brine disposal from the desalination process still remains, however.

Another possibility is the commercial extraction of minerals from the geothermal fluid. Sulfur and high-purity silica, as well as valuable minerals such as silver, lithium, cesium, etc., are known to exist in some geothermal fluids, and their extraction for commercial markets might prove economically attractive.

The briny, high-mineral water could be reinjected into the Earth. In some cases this may even prove necessary to prevent the sinking of the Earth's surface, which has occurred in New Zealand and some other areas where excessive amounts of geothermal fluids have been withdrawn.[262] Dr. T. V. Grose of the Colorado School of Mines, on the other hand, has warned that reinjection could produce earthquakes.[263]

Air pollution is a potential problem with geothermal power plants. A number of noxious gases are released into the air, including hydrogen sulfide, with its pervading, rotten egg smell, which can make life near a geothermal plant very unpleasant. Dr. Martin Goldsmith of the California Institute of Technology Environmental Quality Laboratory has estimated that on the basis of experience with the Geysers geothermal plants, a 1,000-megawatt geothermal power generator would release 97,000 pounds of hydrogen sulfide per day—about the same as a fossil-fueled power plant of the same size, burning low-sulfur oil.[264] A 1,000-megawatt plant using hot water instead of dry steam would release 1.18 million pounds of sulfur per day—more than the same size fossil-fueled plant burning high-sulfur fuel. Technology is available to prevent the release of these gases, however, if this is found to be necessary as further development occurs.[265]

Dr. Goldsmith has estimated that 3,630 megawatts of heat are rejected by a water tower-cooled 1,000-megawatt geothermal electric plant—almost twice as much heat as the 2,000 megawatts of heat rejected by a nuclear plant of the same size. If 10 such plants were installed in the Imperial Valley, as is presently being anticipated in some quarters, the total heat added to the 1,000-square-mile area of the valley would be 5 percent of the total summer solar heat. The effect of this heat on the local weather is unknown.[266]

Heat emissions from an air-cooled plant would be even greater.

One significant potential danger is posed by the possibility of a well blowout, which has already occurred in the early production wells at Cerro Prieto. Days were required to bring the blowout under control while the well spewed steam and salt water. Goldsmith has noted that such a blowout might produce as much as 10 acre-feet per day of steam and salt water, which would clearly pose a major environmental problem in an agricultural area.

An aesthetic intrusion on the environment threatens in the possible use of above-ground insulated pipes for transporting steam and water from the well to plants over distances of up to a mile. Noise also presents a significant problem. The blast emanating from the plant as the steam is periodically vented from the pipes is deafening, even with the installation of gigantic mufflers.[267] These factors, plus the possibility of noxious odors, would render a geothermal plant almost totally incompatible with a residential environment.

In addition to environmental and sociopolitical problems, some technological obstacles also remain to be overcome before geothermal resources can be developed on a wide scale. Further research and development of deep-drilling methods is perhaps the most pressing need. Present costs of geothermal drilling have proven to be in the range of $43 to $50 per foot, compared with average gas and oil well costs of $19 per foot.[268] In addition, drilling technology does not exist for depths much beyond 30,000 feet, and is also often inadequate for conditions at shallower levels.

A very promising drilling technique is under study at the Atomic Energy Commission's Los Alamos laboratory, where a molybdenum drill has been developed which, when heated to 1,600° C (2,912° F), melts through any material—including granite—with relative ease. The intense heat also forms a glasslike, hard lining for the hole as it is drilled, so that casing appears to be unnecessary. Costs, feasibility, and environmental implications have not been studied; and while this approach appears to solve many problems, extensive additional research and development must be undertaken before its usefulness can be determined.

Turbines operating with geothermal wet steam and water have proved generally unsatisfactory. While several new turbine approaches appear promising, much more research and development is required here also. A geothermal resources research conference in 1972, chaired by former Secretary of the Interior Walter J. Hickel, cited a number of other areas where further research is also necessary, including resource exploration and assessment, utilization technologies and economics, and effects of geothermal development on the general public, private industry; and local, state, and federal governments.[269]

At this point, geothermal energy appears to offer the hope of increased energy, but much remains to be done before it can begin to play a significant role in the total U.S. energy picture. The Hickel study concluded that:

> Geothermal resources can be developed to supply power in less time than more conventional power supplies. Measured from the time when the selection of power supply type is final, geothermal power plants should be operative in about two years, whereas fossil fuel power plants may require up to five years and nuclear plants up to ten.
>
> In summary, geothermal resources, by approximately 1985, can have a potentially enormous impact in supplying the nation's need for energy and augmenting the supply of water in regions with insufficient natural water. . . . The development of geothermal resources could substantially increase national energy self-sufficiency and provide a dramatic improvement in the U.S. balance of payments posture.[270]

Before this can become a reality, however, the federal government will have to move much faster with research and development and the carefully regulated leasing of federal geothermal resources than it has done thus far.

At the same time, geothermal power should be carefully examined for possible earthquake effects, since the removal of large quantities of Earth heat may cause geological shifts that could prove disastrous.

Geothermal power may prove to be a great boon to civilization; but this source, like nuclear fusion, should not be viewed as the definitive alternative to this society's energy crisis. Geothermal heat may be far too costly in both economic and energy terms to bring to the surface from great depths. As the costs of conventional fossil fuels (and other scarce materials) rise, so will the cost of the technologies needed to make widespread geothermal power a reality. Geothermal power has received very little funding in the past, however, and deserves far more in the future.

Tidal and Hydroelectric Power

Among methods for using the natural, renewable energy of the sun to produce power are the forces of the tides and the energy of falling water. In both cases, kinetic energy is intercepted and turned into mechanical power, which can be used to generate electricity.

Tidal power is electricity produced by the movement of water in the incoming and outgoing tides of the sea turning some type of water turbine. Falling water must move against the blades of the turbine, which turns a generator and produces electricity. The difference in levels of water between which the turbine wheels work is called the "head of water"; in the case of tidal power it is the difference between the high-tide level and the low-tide level.[271]

The use of tidal power to perform work goes back many centuries, and the prospects of capturing power from the sea have continued to fascinate man. In the American colonies, Slades Mill in Chelsea, Massachusetts, was grinding spices in 1734 with a four-waterwheel tidal installation that generated some 50 horsepower under optimum conditions.[272] And an eighteenth-century Rhode Island tide mill had reached considerable sophistication with its 20-ton wheels 11 feet in diameter and 26 feet in width.[273] Tidal mills were also in use during this period on Passamaquoddy Bay, the only site in the coterminous United States that is considered a serious candidate for tidal electric power generation today.[274] Some 280 patents relating to utilization of tidal energy were registered between 1856 and 1939.[275]

The ultimate demise of eighteenth- and nineteenth-century tidal power, like wind power, can be attributed to the introduction of cheap electricity. F. L. Lawton, former study director of the Atlantic Tidal Power Programming Board, in Halifax, Nova Scotia, told a 1970 conference on tidal power:

> With the advent of power generation on an industrial scale, utilizing for the most part the hydroelectric possibilities of rivers and the thermal energy of fossil-fuel-fired power plants, the price of energy began to decline. Today it is roughly one twentieth, in terms of man hours, what it was at the beginning of the eighteenth century. Small tide mills of the past could not meet this fall in price.[276]

This remains essentially true, except for some 15 unique coastal areas around the world where tidal extremes in partially enclosed coastal basins present the kind of water forces considered essential for economically feasible power generation. Only one of these, the Bay of Fundy, is in North America, in which a total of 9 separate tidal power projects are possible, including the only one in the United States, on the Maine side of Passamaquoddy Bay.[277]

The French have pioneered in the realm of commercial tidal power with the construction in 1965 of a 240,000-kilowatt generating plant across the mouth of the La Rance estuary in Brittany.[278] To harness the 44-foot tidal range required the design of a completely new kind of turbine generator unit whose blades have adjustable pitch and are reversible. This permits the turbine to not only adjust to the tide's varying flow speeds, but also to continue to generate electricity no matter which direction the tide is moving.

Edward P. Clancy, professor of physics and chairman of the Physics Department at Mount Holyoke College in South Hadley, Massachusetts, in his book *The Tides,* explained the operation of the unique system:

> As the ocean starts to rise from its minimum level and the head decreases to the point where efficient generation is no longer possible, the turbines are shut down, but the sluice gates open to allow continuing outflow from the basin. When basin and sea are at the same level, all gates are closed and there is a waiting period until sufficient head builds up. Then power generation begins, and continues until the combination of decreasing sea level and increasing level in the basin results in too little head. The turbines are stopped, but the sluice gates allow further filling of the basin. When levels within and without are equal the gates are closed and there is a waiting period until sea level falls sufficiently to create a new head—this time in the opposite direction. Meanwhile the blades of the turbines have been reversed and a new cycle of power generation begins, continuing until the head is again no longer adequate.[279]

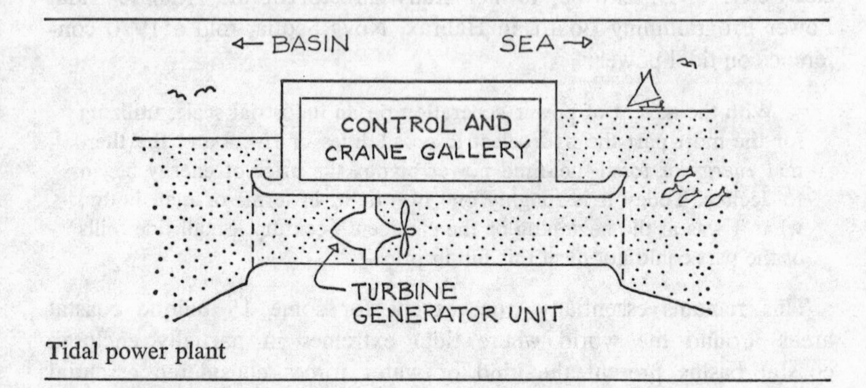

Tidal power plant

The Soviet Union followed with another unique design, in constructing the Kislaya Guba tidal power plant, a 400-kilowatt experimental generator located in a narrow neck connecting the White Sea with Ura-Guba Bay. The plant was designed to determine the feasibility of tapping the Soviet Union's 210 billion kilowatt-hour per year tidal power potential.[280] While cost and operational figures have not been published, the chief engineer of the project said that the project's objectives were reached and that as a consequence a new look at the economic feasibility of several tidal-electric power plant sites—including the controversial Passamaquoddy Bay project in the United States—is warranted on the basis of the Russian technology.[281]

The maximum world potential of electric power from harnessing the tides has been estimated at 63.8 million kilowatt capacity, with potential production of a little less than 559.5 billion kilowatt-hours per year[282] —about 6.5 percent of the total world generating capacity in 1968, and some 13 to 14 percent of its 1969 electricity consumption.[283]

However, areas considered suitable for development of tidal power are limited to San Jose Gulf in Argentina; Severn estuary in England; several coastal sites in France; four locations in the Soviet Union; Cook Inlet, Alaska; and the Bay of Fundy, in North America.

The Bay of Fundy, which is made up of nine smaller bays lying mostly in Canadian territory between land areas in Maine in the United States and New Brunswick and Nova Scotia in Canada, has a total potential capacity of 29,027 megawatts,[284] although probably not more than about one third of that could be practically developed.[285]

The three most promising Bay of Fundy sites have an estimated potential capacity of 9,000 megawatts, which is the equivalent of about 3 percent of the total current U.S. electrical capacity. However, the amount of power they would be capable of producing in a year—some 13 billion to 18 billion kilowatt-hours—could have supplied 20 to 27.5 percent of the 1970 electricity consumption of New England and the Canadian Maritime Provinces.[286]

The only one of the nine Bay of Fundy sites that the United States can lay even partial proprietary claim to, Passamaquoddy Bay, has a potential generating capacity of 1,800 megawatts, with potential production of 15.8 billion kilowatt-hours per year,[287] according to some authorities. Others have estimated potential output much lower—at only 1.8 billion kilowatt-hours per year.[288] In any case, harnessing the Passamaquoddy Bay tides has been an issue of controversy since the 1930s, and appears no nearer resolution now than 40 years ago.[289]

Disadvantages of Tidal Power

There are some disadvantages inherent in building gigantic structures across the mouths of estuaries and inlets for the purpose of tapping tidal movement for generating electricity. There would inevitably be some damage to the marine environment, although available studies indicate that hazards to the biogeochemical cycle would probably be minimal, and certainly less than for fossil fuel and nuclear alternatives.[290]

Aesthetically, these damlike structures with their turbine housing and networks of high-voltage power lines stretching across the bays and rivers would likely constitute something of an eyesore, in addition to creating inconvenience and perhaps some hazard to shipping and navigation.

Only two tidal power electricity-generating projects have been undertaken to date in the entire world. However, as available conventional fuel supplies drop, environmental considerations are taken into account, and fuel costs and energy demands rise, tidal power will no doubt become more feasible economically and will probably be developed in the limited areas where it appears practical. It cannot, however, be viewed as a significant solution to the energy and fuels crisis—and certainly not in the United States, where almost no potential sites exist.

Hydroelectric Power

Hydroelectric power works on somewhat the same principle as tidal power—electricity produced by the force of falling water intercepted by the blades of a turbine attached to an electric generator. The "head of water" so important in the engineering and economics of tidal power is also a key factor here. In this case, it is the amount by which the water behind the dam stands higher than at the foot of the dam. Obviously, the greater the drop, the more force the falling water can gather and the more power can be generated.

The total conventional hydroelectric power potential of the 48 coterminous states of the United States is estimated at about 146,000 megawatts (million kilowatts) of capacity, which can produce an average of 530 million megawatt-hours of electric energy each year.

While hydroelectric power installations are generally smaller than thermal plants, some are quite large. There are 5 plants of more than 700 megawatts on the Columbia River, for instance, and one of 2,000 megawatts on the Niagara.[291]

About 94,000 megawatts—representing 274 million megawatt-hours of production per year—are still undeveloped in the United States. The 52,000 megawatts of hydroelectric power that have been developed—mostly in the Pacific Northwest—represent some 16 percent of the total demand for electric power today.[292]

Economics, environmental opposition, and many other factors will prevent many of these sites from being developed. Some will be prohibited by legislation, such as the Colorado River Basin Project Act, which prohibits the Federal Power Commission from issuing licenses for projects on the Colorado River between the Glen Canyon and Hoover Dam projects—a stretch of river that represents a potential of 3,500 megawatts of hydroelectric capacity; and the Wild and Scenic Rivers Act, which prohibits the FPC from licensing the construction of any power facility affecting a river included in the 37 national wild and scenic rivers system established by the law. Sites offering a potential of 9,000 megawatts of hydroelectric capacity are included along these 37 river reaches.[293]

The cost of building a hydroelectric project per kilowatt of installed capacity depends on the type of project, its location and size, the water-head capacity, the cost of lands involved, and the cost of relocating facilities within or adjacent to the reservoir area, such as railroads, bridges, roads, buildings, etc. While costs vary widely, initial capital cost per kilowatt is substantially higher than for thermal-electric plants or pumped storage projects. However, because the installations use naturally flowing water instead of fossil or nuclear fuels, the operating expenses are much lower. Thus, as fuel costs increase, hydroelectric power sites have become economically attractive.[294]

Specific cost figures for making economic analyses of most potential hydroelectric power sites are not available. However, the Federal Power Commission has estimated that 22,000 megawatts of new hydroelectric capacity could be added by 1990 at costs that would make the plants economically feasible.[295]

There is also some possibility of importing electricity generated by Canadian hydroelectric power, but long-range prospects of this appear increasingly remote as Canada's own energy requirements continue to escalate. In eastern Canada, the Churchill Falls project in Labrador and other large projects in Quebec and New Brunswick are under construction, from which possibly up to 300 to 350 megawatts might be available to the United States on a short-term basis. Quebec has large undeveloped hydroelectric resources along the rivers flowing into James Bay, Ungava Bay, and the north slope of the St. Lawrence. If this power were developed, possibly as much as 5,000 megawatts might be available to the United States over a 20-year period. In Manitoba, the first phase of major development of the Nelson River is under way, with construction of the Kettle Rapids project of 5,000 to 6,000 megawatts of hydroelectric capacity, possibly as much as 800 megawatts of which might be available for export to the United States.[296]

Environmental Aspects

Although hydroelectric power generators are free of chemical pollutants and emit no noxious gases, many objections have nevertheless been raised to building new plants on environmental grounds. As John Holdren, a physicist at Lawrence Livermore Laboratory in Livermore, California, observed in the thoughtful and indispensable book *Energy,*

> for those who enjoy power-boating, water-skiing, and warm-water fishing, [damming water] may be considered a benefit. Probably the most powerful conservation argument on this point is that there are already a great many lakes where such activities can be pursued, but very few remaining gorges and wild rivers. In some instances, hydroelectric

> projects cover prime farmland, which eventually proves to be a poor trade. For those whose homes and livelihoods have been submerged, the personal cost of hydroelectric power is high.[297]

Holdren goes on to note that dams and reservoirs involved in the hydroelectric system sometimes also destroy the spawning grounds of migratory fish like the salmon. Seepage from reservoirs may raise the water table, he notes, bringing with it subsurface salts and minerals and reducing the fertility of the soil. And in some instances, filling large reservoirs has set off earthquakes because the sudden weight of the added water has caused the balance of stresses on the Earth's crust to shift.[298]

Some of the environmental effects of the proposed James Bay project in Canada, from which the United States stands the possibility of being able to purchase 5,000 megawatts, have been reviewed by author Boyce Richardson in the *Sierra Club Bulletin.* The project would flood 133,000 square miles—an area twice the size of England—and destroy the homes and the culture of several thousand Indians whose lives have up to now been remarkably unaffected by the encroachment of the white man. The Indians have sadly observed that "if you flood the land, you will destroy the animals. If you destroy the animals, you destroy the Indians."[299]

The animals on which the Indians' lives depend live near the shorelines of lakes and rivers. If these shorelines fluctuate constantly, as they will with a hydroelectric project, the animals will have to move away. They will be forced to eat in summer the hilltop browse on which they usually depend in winter. In Richardson's words, "the Indians are haunted by the prospect of their glorious wilderness turned into a wasteland of decayed stumps and mud-flats. The best scientific estimate supports their fears, though no one in a position of authority will admit it."[300] The entire project has been conceived as if the Indians did not exist, Richardson contends. Taking its cue from 300 years of precedent in the United States to the south, the Quebec Government plans to shunt the Indians off the land they have always roamed and take it for the use of the white man—in this case, to turn the wheels of his high-energy civilization.

Richardson reports that "not one second of biological research was done in James Bay before the decision to build the project," although a few months later, in September 1971, a team of officials from the Canadian and Quebec governments was given two months to come up with a report on the environmental consequences of the project. The team concluded that the proposed human intervention would improve, rather than harm, the ecological system, with perhaps the exception of the Indians. That conclusions could be reached so quickly is surprising, Richardson points out, in view of the fact that a scientific conference had agreed that "if research on northern ecosystems could be increased ten times and development delayed ten years, Canada might be in a position to develop its northern lands on the basis of real knowledge."[301]

Another disruptive effect the James Bay project might have is on the millions of geese that use part of the area as a resting place during their annual migration. It also threatens to wipe out the salmon run in the river, one of the world's great salmon-spawning grounds.[302]

In some parts of the world, the lives and health of millions are adversely affected by the damming of the rivers in a hydroelectric development. One of the most recent examples is in Egypt, where the flow of the Nile River en route to the Mediterranean has been slowed by gigantic Lake Nasser, part of the Aswan Dam project. One result has been the increase to epidemic proportions of a slow-killing disease called schistosomiasis, or bilharziasis, which is carried by a certain type of snail found in the Nile and other rivers of Africa. In years past, the annual Nile floods have washed enough of the snails out to sea to keep the snail population to manageable proportions. The calm, gentle waters of Lake Nasser, however, provided the environment the snails need to establish a real foothold and multiply without restraint. The snails serve as an interim host to the larvae of a blood fluke which, upon entering the human body, attacks the liver, stomach, heart, and lungs. The recorded death rate from the disease is high, and no one is certain how many additional deaths from heart attacks, kidney failures, and other organic disorders may actually be caused by it.[303] An increase in dam construction in Rhodesia has also mirrored the increase in bilharziasis, and construction of the Sennar Dam in Sudan brought the same deteriorating health conditions there.[304]

As M. Taghi Farvar, of the Center for the Biology of Natural Systems at Washington University in St. Louis, Missouri, has observed:

> The tremendous, continuing increase in the incidence of bilharziasis is one more manifestation of a biological dilemma: The basic vulnerability of an artificial ecosystem. Disease and suffering for millions of people are a direct outcome of the attempt to control the processes of nature with the simplistic solutions that modern technology offers in the form of simple, managed ecosystems in place of the intrinsically complex natural systems.[305]

It is for these and other reasons that Dr. M. King Hubbert, of the U. S. Geological Survey, has expressed doubt about the full development of the world's remaining hydroelectric potential. Hubbert cites two considerations in expressing this view:

> The first is the aesthetic one of whether the people of the world wish to sacrifice some of their most beautiful natural scenery in order to develop fully the associated water power. The second concerns the fact that in all reservoirs formed by dammed streams, the streams are continuously depositing their loads of sediments, so that in periods of a century or two most man-made reservoirs are due to become completely filled by such sediments. This problem has not been satisfactorily solved, and may never be. Hence, although the stream rates of discharge may remain relatively stable for millennia, most water-power sites may have periods of maximum usefulness measured by a century or two. It is accordingly questionable to what extent the world may be able to depend upon water power as a substitute for the depleted fossil fuels.[306]

Pumped Storage

A rapidly growing hydroelectric power storage method that is based on the same principles of falling water as the conventional hydroelectric dams and reservoirs is pumped storage. This process involves two reservoirs, a high one and a low one, connected by pipes containing dual-purpose pump turbines. As with conventional hydroelectric power facilities, electricity is produced by water flowing from the upper level to the lower. There is one major difference, however: Unlike the conventional hydroelectric station, after the water has flowed downward through the turbines, generating electricity, it is pumped back up again for later use.

Pumped storage may be used in conjunction with either hydroelectric or thermal power generators. Its purpose is to provide greater supplies of electricity during peak electric demand periods. At such times, water is released from the upper reservoir to generate added electricity. Then, during the late-night and early-morning hours, when there is a surplus of electric power, the water is pumped back up again for peak periods the following day. Electric utilities find it economically attractive because power used during peak periods is sold at a rate that is about three times higher than during the low-use periods, when the water is pumped back up.[307]

Pumping the water back up again requires about one third more energy than it generates flowing down; but this added use of energy is more than offset if pumped storage eliminates the need for building an additional power plant to meet the peak demands.

The pumped-storage method is being adopted with increasing frequency. Prior to 1961, only 4 pumped storage projects existed in the United States; today, several hundred are being implemented or designed.[308] The Federal Power Commission has also identified 55 that could be built and brought on line by 1990, to add 58,798 megawatts of electrical storage capacity to the nation's power plant grid at an estimated cost of $100 to $140 per kilowatt of installed pumped storage facility.[309]

The environmental impact of a pumped storage facility is usually less than for a hydroelectric power plant because the upper storage reservoir can be relatively small, while the lower reservoir is usually an already existing lake or stream.[310] However, pumped storage may have a greater negative aesthetic and recreational impact because the water level would fluctuate wildly—depending on the energy storage needs at any given time—which would create unsightly mud flats and prevent any recreational uses, despite the fact that electric utilities claim increased recreational facilities as a major "advantage" of the projects.

The use of pumped storage reservoirs is expected to increase dramatically in the future, largely due to government and industry projections of the growth of nuclear power plants, which are operated as "base load" plants; i.e., always producing a constant output of electricity in order to maximize the use of uranium fuel in the plants' reactor cores. Nuclear plants cannot be operated at "full steam" to meet peak demands, and then cut back in power during periods of minimum electrical demand, as can other thermal generating plants. The development of nuclear power would entail the virtual necessity of cementing over much of the nation's hill and mountain country for water storage reservoirs.

As is the case with any method of storing energy, the entropy law requires that energy will be lost in the process of converting the electricity into the potential energy of water and then back to electricity. Pumped-storage facilities (aside from taking considerable energy to build) will use more energy than would a power plant running alone without being connected to a pumped-storage facility. The sole justification of the technology is that human use of energy (in the form of electricity) is not a constant demand, but varies over the twenty-four-hour day. Efficient operation of the electrical system to meet this demand requires some form of storage.

The difficulties in getting power for public consumption will surely be greater in the future than they have been in the past. If the hoped-for sources of nuclear fusion, solar and geothermal power do not arrive in time to forestall the burgeoning nuclear fission era during the general decline in availability of fossil fuels, chronic energy shortages and social disruption will be commonplace in the advanced technological societies of the world.

As fossil fuel availability declines, a wise technological shift within the technology-rich world would be to examine ways for transformation of the liquid-fossil-fuel economy into either an electrical economy (with attendant automobile battery stations, etc.), or into an economy that could utilize a transportable fuel substitute for fossil fuels.

The exciting prospect of a hydrogen-fuel economy is examined in the following pages as a technical alternative to the fossil-fuel age in which we now live.

Hydrogen

Hydrogen, the lightest chemical element, was discovered in 1766, when the English chemist Henry Cavendish observed what he called "phlogiston" or "inflammable air" rising from a zinc-sulfuric acid mixture. It was identified and named by the eighteenth-century father of chemistry, Antoine Lavoisier, who showed that Cavendish's "inflammable air" in fact did burn in air to form water.

Thus it was a true element, he concluded, and called it hydrogen, the Greek term for water-former. Its lightness (it has an atomic weight of 1.008) also endows it with a unique readiness to join other elements, which has rendered it a basic component of many of the most common chemicals.[311]

Jules Verne, the famed nineteenth-century science fiction writer, first predicted the use of hydrogen as fuel in his 1847 novel *The Mysterious Island*. Asked what men will burn when coal and other fuels are gone, an engineer in the book replies, "water. Yes, my friends, I believe that water will one day be employed as fuel, that hydrogen and oxygen which constitute it, used singly or together, will furnish an inexhaustible source of heat and light. . . ."

The concept was extended by British biochemist J. B. S. Haldane in a Cambridge, England, lecture in 1923. He predicted that the power problem England would face four centuries from then would be solved by "rows of metallic windmills working electric motors which supply current at a very high voltage to great electric mains." He continued:

> At suitable distances, there will be great power stations where during windy weather the surplus power will be used for the electrolytic decomposition of water into oxygen and hydrogen. These gases will be liquefied, and stored in vast vacuum jacketed reservoirs, probably sunk

in the ground. . . . In times of calm, the gases will be recombined in explosion motors working dynamos which produce electrical energy once more, or more probably in oxidation cells.[312]

A small-scale version of Haldane's proposal had already been demonstrated more than a quarter of a century before by Danish professor Poul La Cour, who used wind-generated electric power to produce hydrogen and oxygen, which he burned in the presence of the element zirconium to illuminate the Askov, Denmark, high school where he taught.*

The same basic concepts embodied in Verne's, La Cour's, and Haldane's visionary pipe dreams and experiments are now being intently discussed—not, however, as a fuel alternative four hundred years away, as Haldane supposed, but for use within the next fifty years.

There are several light, synthetic fuels that might be considered as substitutes for petroleum products in the future. In addition to hydrogen, they include acetylene, ammonia, hydrazine, methane, and methanol. Dr. Lawrence W. Jones of the University of Michigan has pointed out that any of them—and many more—could be synthesized as a fossil fuel substitute, but the two reasons for selecting hydrogen, he noted, are that (1) it contains the greatest energy per unit weight of all the alternatives considered and (2) it is completely cyclic—i.e., water drawn from lakes, streams, or oceans is broken down into hydrogen and oxygen; and when the resulting hydrogen is subsequently burned, water is again created as a by-product and is returned to the biosphere.[313]

Hydrogen offers many attractive advantages as a fossil fuel substitute. Because it can be made from water, the potential supply is theoretically endless. It might be substituted for natural gas in the natural gas pipeline system, or for gasoline in automobiles and other vehicles—thus making possible a smooth transition from present to future fuels. It can provide the fuel for devices like the fuel cell, which could decentralize production of electricity by generating electricity at the site where it is needed. Because its only combustion by-products are water vapor and minor amounts of nitrogen oxide, it would contribute minimal pollution to the environment. The technology for storing and handling hydrogen in its concentrated form—as a liquid—is at hand, developed by the National Aeronautics and Space Administration (NASA) for the U.S. Apollo and other space programs, for which liquid hydrogen has constituted the basic fuel.

As a natural gas substitute, hydrogen is proposed for use in the home for cooking, cooling, and heating; and in industry, for heating and cool-

‡ See Chapter 8 on wind power.

ing—with only some minor adjustments or redesign of burners. It could also serve as a chemical raw material in many industries, including the fertilizer, foodstuffs, petrochemical, and metallurgical industries. Hydrogen could also be used to generate electricity in local power plants, either through conventional steam turbine or with gas-turbine or magnetohydrodynamic (MHD) generators or—most efficiently—in large fuel cells.[314]

Hydrogen as a fuel may be viewed in terms of three basic market components: production, transmission, and storage.

Hydrogen Production

Hydrogen is already being produced for many industrial applications. Its production in the United States has more than tripled since 1960. In 1968 alone, total production of hydrogen for industrial use was twelve billion pounds. Its uses include the hydrogenation of fats, oil, margarine, and soap; the production of ammonia for fertilizers; the production of metal powders; annealing stainless steel; inflation of weather balloons; cooling electrical generators at power stations; uranium extraction and processing; synthesis of chemicals for products such as nylon and polyurethane; glass manufacture; corrosion control in graphite nuclear reactor cores, and others.

Most of today's hydrogen is produced by reforming hydrocarbon "feed" materials, such as natural gas, propane, butane, naphtha, and unused petroleum refinery gases. However, due to the increasing costs of these feedstocks, and the growing unavailability of natural gas due to shrinking reserves, industries have turned to other manufacturing processes for hydrogen.

One commercially viable process, electrolysis, uses electricity to break water down to produce hydrogen and oxygen. Electrolysis was initially demonstrated by Sir Humphrey Davy in 1806. Any source of electricity can be used, which means that the original fuel for the electrical generator may be coal, natural gas, oil, nuclear power—or solar or wind power.

Unfortunately, most electrolytic processes are highly inefficient. In the case of coal, for example, the original energy of the fuel is converted first to electrical energy at a power plant, and the electrical energy is then utilized to electrolyze water and produce hydrogen. Each conversion process entails significant energy losses.

Electrolytic cells for producing hydrogen from water now operate at efficiencies of only 60 to 70 percent, although experimental electrolysis cells manufactured in 1966 by Allis-Chalmers have achieved almost 100 percent efficiency in laboratory operation. Dr. Derek P. Gregory of the Institute of Gas Technology (IGT) points out that, *in theory,* the maximum electrical efficiency of electrolyzers is close to 120 percent, because an ideal unit would absorb heat from its surroundings and convert this into hydrogen also.[315] If such experimental processes prove commercially feasible, they offer promise for manufacturing hydrogen at much lower levels of cost and fuel waste. In the meantime, however, contemporary electrolytic hydrogen production techniques waste dwindling fossil fuel supplies and are prohibitively expensive.

Institute of Gas Technology scientists recently estimated that with further improvements in electrolyzer technology, hydrogen may soon become available at prices competitive with imported natural gas. They have calculated, for instance, that if a cheap source of electricity were available to fuel an advanced water electrolyzer, the cost of hydrogen would range between $0.78 and $1.03 per million BTUs—and conceivably lower. Wholesale cost of imported Algerian natural gas in Boston is currently above $0.70 per million BTUs. Natural gas and other traditional fuel sources are expected to continue to rise sharply in price. As they do so, hydrogen offers more and more promise as an alternative fuel, particularly as more efficient production techniques are developed.[316]

Dr. Gregory points out that the theoretical power necessary to generate hydrogen by water electrolysis is 79 kilowatt-hours per 1,000 cubic feet of hydrogen; however, since present-day plants are only about 60 percent efficient, a more likely power-consumption figure is 150 kilowatt-hours per 1,000 cubic feet.

The amounts of hydrogen that would be required in a hydrogen economy are enormous. For instance, according to Dr. Gregory, to produce enough hydrogen to fully substitute for the natural gas produced in the United States at the present time—i.e., 70 trillion cubic feet of hydrogen—would require more than 1 million megawatts of electric power to produce. Total electrical generating capacity in the United States is only 360,000 megawatts. *To meet the projected hydrogen requirements for natural gas alone would call for a fourfold increase in generating capacity, which would mean building 1,000 additional 1,000-megawatt power stations!* This does not provide for increased electric power demand for other purposes, nor does it take into account the generation of hydrogen for transport fuel or as an additive in chemical and industrial processes.[317]

On a less ambitious scale, some are proposing the use of electricity in off-peak hours to electrolyze water for hydrogen production. Since the consumer demand for electricity varies widely both seasonally and during the day, the electric power generating rate must be adjusted continuously. This would not be necessary if there were some means of storing electricity. The only practical way available today to store large quantities of electrical energy is the pumped-storage plant, a reversible hydroelectric station. However, there is a limited number of sites that are geographically suitable for such systems.

Hydrogen could be utilized as an alternative form of electric power storage. The electric industry frequently operates at as little as 50 percent of installed capacity during off-peak hours. If this surplus electricity were converted to hydrogen and stored, the industry could operate at near 100 percent of its capacity at all times, and the stored hydrogen could be used to meet peak demands for power.[318]

Cryogenic storage of natural gas is already a rapidly growing technique; at 76 locations in the United States "peak shaving" operations involving liquefied natural gas (LNG) are in use or under construction. There is no technical reason why a similar peak-shaving technique cannot be employed with liquid hydrogen, according to Dr. Derek Gregory.

It is estimated that more than 22 million people will live in the Greater New York City area in 1985. They will probably consume 2.38 quadrillion BTUs of energy a year. If hydrogen were produced using off-peak power, it could supply more than half the energy requirements for transportation of nearly 90 percent of the household and commercial requirements within the city limits.[319]

An extension of this would be to substitute hydrogen completely for electric power transmission and storage. According to Dr. Gregory, it is already cheaper to send hydrogen through pipelines than it is to send energy in the form of electricity through overhead transmission lines for distances of more than 250 miles—and about 20 miles in the case of underground transmission lines. Dr. Gregory and others would like to see the energy system of the future consist of very large nuclear power plants—isolated from population centers for health and safety considerations—that would utilize the electric power they produced to generate hydrogen by electrolysis. The hydrogen, rather than electricity, would be distributed for consumption by consumers.

Such an arrangement, proponents point out, would ease the design demands on the nuclear electric power stations. They would not have to operate within very tight constraints imposed by variations of electric consumer demands; they would only have to supply continuous power to the electrolyzer. They would no longer have to maintain their present high electric power reliability because they would have enough hydrogen storage in the pipelines and elsewhere to allow the electrical supply to be periodically interrupted. They would also no longer have to accurately synchronize the alternating-current generator with all other generators on the network, because such precise control of rotor speeds or voltages would be unnecessary.[320]

Another way than electrolysis to generate hydrogen is to thermally decompose water using the heat from nuclear reactors. Euratom scientists in Ispra, Italy, have devised a thermal dissociation process that would enable hydrogen to be produced at temperatures below 730° C (1,346° F). Dr. Cesare Marchetti, director of Euratom's Materials Division, reports that the experimental method, called the Mark I, promises to be half as expensive and twice as efficient as producing hydrogen by electrolysis of water.

Other researchers are pinning their hopes on the nuclear fusion reactor. Dr. Bernard Eastlund, of the Atomic Energy Commission's Oak Ridge National Laboratory, has proposed that instead of using the fusion reactor to produce electricity, which would then electrolyze water to produce hydrogen, the superheated (100,000,000° C) fusion "plasma" might be used to produce photons, the energy of the sun's rays. After being synthesized from the hot plasma of the fusion reactor's core, the photons would be channeled to a "water vapor cell," where this synthetic solar energy would directly convert water to its constituent parts—hydrogen and oxygen—as it does in the Earth's outer atmosphere.[321]

Storage of Hydrogen

There are three possible approaches to hydrogen storage: as a gas under pressure, like today's propane; as a cryogenic liquid under pressure, like today's liquefied natural gas (LNG) discussed above; and in metal hydrides. Which approach is used depends to a great extent on the purpose to which the hydrogen is ultimately going to be put.

The simplest and most common method of hydrogen storage is as a gas under pressure, either in a cylinder or in underground vaults or caves. However, gas under pressure in a cylinder is bulky and impractical for hydrogen-powered vehicles, which many consider one of the largest potential hydrogen applications. For instance, a normal 20-gallon tank full of gasoline weighs about 120 pounds. To store the equivalent amount of energy in the form of hydrogen gas would require an enormous container holding 66 cubic feet of gaseous hydrogen and weighing a little more than 1 ton (including the fuel weight of 45 pounds).

A lighter and more compact means of storing the hydrogen fuel—and one that is more suitable for vehicle application—is as a supercold liquid. Liquefied hydrogen must be stored at —434.6° F (at 1 atmosphere of pressure), compared to present-day storage of liquefied natural gas at —260° F. A liquid hydrogen storage tank, equivalent to the normal gasoline tank, weighs 353 pounds and stores the hydrogen in 10.2-cubic-feet of space. When compressed and cooled to liquid form, hydrogen has about 2½ times the energy by unit weight of gasoline.[322] This is the form in which hydrogen comes nearest to being competitive with a fossil fuel. Today it is only 50 percent more expensive than gasoline on an energy-per-unit weight basis because of hydrogen's higher energy content.[323] Even so, liquid hydrogen would require a 50-gallon tank to get the same mileage as a 20-gallon tank of gasoline, which poses obvious problems of automobile design, refueling, service station storage, etc.

It also poses problems of cost. A dewar vessel (the container resembling a Thermos bottle required to hold liquefied hydrogen) that would hold hydrogen with an energy content of a 15-gallon gasoline tank would cost about $1,800.[324]

For these and other reasons, Dr. Lawrence Jones, writing in *Science* magazine,[325] observed that logical first uses for liquid hydrogen fuel might be in airplanes, long-haul trucks, and city buses, which would carry liquid hydrogen in replaceable tanks, or dewars, that could be quickly exchanged. Many believe that aircraft make the strongest immediate case for liquid hydrogen because of their high energy-to-weight ratio. NASA research has shown that hydrogen burns with high efficiency in modified gas-turbine and ramjet engines, and that it could also be used efficiently in today's jumbo jets, perhaps permitting lighter, more efficient aircraft to be built because of its lighter weight.[326]

The Hydrogen Automobile

In any case, hydrogen economy enthusiasts see automobiles as one of the most promising applications of hydrogen as fuel. Forty years ago, when England was concerned—as is the United States now—with growing pollution from automobiles and with national security aspects of fuel imports, inventor Rudolf A. Erren, who had escaped from Nazi Germany to do research in Britain, was proposing large-scale manufacture of hydrogen from off-peak electricity. In anticipation of substituting it for gasoline in automobiles, he also undertook extensive work to modify the internal-combustion engine to operate on hydrogen.†

† Reports have persisted that Erren was unable to solve difficulties that arose with hydrogen detonation in his modified internal-combustion engines. Former colleague Kurt Weil, now professor emeritus at Stevens Institute of Technology, says this is not true, that "scores of 'errenized' trucks" were traveling the highways between Ber-

Erren's early interest in the hydrogen automobile is being revived and expanded in an increasing number of university and industrial research centers. As Dr. Jones noted, "if a future energy system can build on the engineering experience of our current automobile and aircraft industries we will be able to make any transition more expeditiously and efficiently."[327]

Many hydrogen efforts to date have been aimed at reducing auto pollution to meet automobile emission restrictions. In the 1972 clean-air automobile race, the only two hydrogen-powered cars in the competition came in first and second in a field of sixty-three entries. First place went to a converted Volkswagen modified by a group of students at Brigham Young University, Ogden, Utah. Its emissions were so pure that it would actually exhaust air cleaner than it received in many cities. Second place went to an American Motors Gremlin, converted to hydrogen by a group of students at the University of California at Los Angeles.[329]

Dr. Roger Schoeppel of Oklahoma State University has built an internal-combustion engine that employs direct injection of gaseous hydrogen into the cylinder, similar to the fuel injection process in a standard diesel engine. The gas jet flame configuration during the injection period eliminates engine "pre-ignition" and "detonation," which have proved to be problems in many designs.[330]

Dr. Schoeppel has found that his hydrogen engines operate smoothly using different types of ignition, as opposed to the single conventional method of spark plug ignition in today's standard automobile engine. The hydrogen engine, he has found, is "easy to start, responds rapidly to different rates of fuel injection, and runs cooler under low-to-intermediate loads than its gasoline equivalent."

A University of Miami engineering team has converted a 1971 standard Toyota station wagon to operate on compressed gaseous hydrogen. Tests of the automobile indicate that at an average cruising speed of 40 miles per hour (2,400 revolutions per minute engine speed), the car's engine produced no atmospheric pollutants except nitrogen oxides (NOx), and these amounted to only 18 parts per million (ppm).

Federal law will limit NOx emissions to 3.1 grams per mile, or about 984 ppm. With its 18 ppm, the Miami car produces only a tiny fraction of this strict federal limit. Average NOx emissions from today's conventional gasoline-powered autos are about 5 grams per mile.

---

lin and the industrial Ruhr Valley when Weil left for the United States in 1938. Erren's hydrogen work in Germany was well documented, he added, but the documentation was destroyed when Allied bombs destroyed Erren's Berlin office.[328]

Even the most sophisticated technology for modifying the gasoline-fueled combustion engine is estimated to reduce NOx emissions only to 0.4 gram per mile, about 7 times the level of the Miami hydrogen engine today.[331]

Using yet another method of modifying the internal-combustion engine, the Perris, California Smogless Automobile Association has converted two automobiles to hydrogen operation. At the Energy Conversion Engineers' Conference in Boston in 1971, Patrick Underwood and Paul Dieges described a fuel system that is based on liquid hydrogen and liquid oxygen storage.

The fuels are stored in special vacuum "Thermos" bottles, or dewars, which keep them supercool at —400° F temperatures. The engine is modified to burn a rich hydrogen-to-oxygen mixture which, according to the inventors, eliminates the traditional hydrogen engine bugaboos of combustion knocks and crankcase explosions. The engine's only exhaust "emission," water vapor, is collected and recycled as engine coolant. Thus, the engine is not only pollution-free, but it also requires no exhaust system.

The Perris group has converted two vehicles to the system, a 1930 Ford Model A and a 1960 Ford pickup truck.

A unique concept in the development of the hydrogen-fueled internal-combustion engine is that developed by Marc Newkirk, president of a small Boston firm, International Materials Corporation. Newkirk's system is based on the use of gasoline as a primary fuel, with a slightly modified internal-combustion engine. The gasoline in the vehicle is "reformed" with steam to produce hydrogen, which is the operating fuel in the vehicle.

This final mixture is about 75 percent hydrogen and 25 percent carbon dioxide. Even though this system differs from the other hydrogen cars in that pure hydrogen is not burned, the resulting pollutants are extremely low. While Newkirk's experimental car was aimed at demonstrating the use of gasoline as a hydrogen storage medium, his basic concept of burning hydrogen with carbon dioxide remains valid regardless of the source of the gases. Thus, while it hardly makes sense to use gasoline as a hydrogen fuel source when much of the energy of the gasoline is lost in conversion to hydrogen, the hydrogen and carbon dioxide might be obtained from other sources in a hydrogen economy.[332]

An Illinois Institute of Technology research group is working on a project to build an automobile in which hydrogen would be burned to drive an electric generator to power electric motors on each wheel.[333] Another group at Brookhaven National Laboratory has operated a

Mazda Wankel engine successfully on hydrogen, and reports that its operation is, if anything, smoother with hydrogen than with gasoline.[334]

The potential for the use of hydrogen as fuel for the automobile seems generally promising. As Dr. Lawrence Jones of the University of Michigan observed: "I have been frankly surprised to note that apparently every serious effort to run an engine or an automobile on hydrogen fuel has been successful."[335]

Hydrogen Distribution

The development of safe storage and distribution systems for hydrogen fuel will probably be based on the accumulated experience of owners of currently operating natural gas and liquefied petroleum gas fuel systems.

Natural gas tanks were tested in a crash test by Dual Fuels Systems Company in a vehicle having over twenty-five bottles of natural gas attached in many places including the front seat, rear seat, trunk, and underneath the car. The vehicle was dropped from a height of forty-six feet, equivalent to a thirty-mile-per-hour head-on collision. According to members of the Clean Air Car Project at the California Institute of Technology, damage was almost nonexistent. "Three tanks in the front seat slid forward three fourths inch. None of the other tanks slipped in their mountings, nor did any of the mountings break loose despite the fact that the chassis of the car buckled," according to the Cal Tech observers.

Charles Gray of the Environmental Protection Agency says that it is conceivable that hydrogen fuel tanks could be designed "so that on impact it might rupture and quickly disperse. Even if the hydrogen ignited, it would probably be rising so rapidly that you wouldn't get continuous burning." However, such tanks have not been built, even experimentally, and detailed laboratory evaluation would be necessary to assess the construction possibilities.

Gray adds that a hydrogen fire is less dangerous in many respects than a gasoline fire, since "there's radiant heat transfer of energy to a much greater degree with a hydrocarbon (gasoline) flame than with a hydrogen flame." This means that a person would have to be physically closer to a hydrogen fire than to a gasoline fire[336] to be burned.

A team of scientists at Brookhaven National Laboratory in New York have developed a promising storage tank for hydrogen fuel. Their process uses a fuel "tank" composed of powdered magnesium metal that can be chemically charged with gaseous hydrogen. Hydrogen is stored in the tank as magnesium hydride. To release gaseous hydrogen to the engine, engine exhaust heat is pumped into the tank under pressure to release the pure hydrogen from the metallic chemical bond. A 1968 staff study of the metal storage system called it more convenient than other hydrogen storage systems. The tank would take up to six times less space than compressed hydrogen gas tanks, and would weigh only one third as much.[337]

Storage and transmission requirements for hydrogen as a natural gas substitute pose much less of a problem. Hydrogen might be distributed by underground pipelines that already serve most industries and more than 80 percent of the nation's homes.[338] Hydrogen might enter the fuels economy first as a dilutant to help "stretch" natural gas during a phase-in period. George Long, research director for Northern Illinois Gas Company, cites studies that show that as much as 37 percent hydrogen can be used in pipeline gas without significantly altering the distribution system, and "town gas" systems in the past have delivered gas with a hydrogen content of up to 50 percent.[339]

Some technological difficulties would have to be resolved, however. Because of hydrogen's lower heating value (325 BTUs per cubic foot compared with about 1,000 BTUs per cubic foot for natural gas), 3 times as much hydrogen would have to be moved through the pipeline to deliver the same energy. But since hydrogen's lower density enables it to move more than twice as fast as natural gas in a pipeline, this would almost compensate for its lower unit heat value, some scientists believe.[340]

Dr. Marchetti of Euratom's Ispra research facility has contended, however, that a methane or natural gas duct adapted to carry hydrogen would need 3 times more powerful pumps, and that the existing pumps already account for about one third of methane transport costs.[341]

Another problem might be posed by "embrittlement," or brittle hardening of the pipeline through which pure hydrogen would be distributed. Because of its small molecular size and highly diffusive nature, hydrogen is able to penetrate solid metals or alloys and cause embrittling and structural fatigue, which would weaken the pipeline.[342]

However, Dr. Marchetti has pointed out that a 127-mile network of hydrogen pipelines connects various firms in West Germany, and that similar networks exist in Texas as a part of more complex systems for distributing chemicals between refineries and petrochemical works. A town in Italy, Basilea, distributes a gas to its residents that is 80 percent hydrogen.[343] Embrittlement has not been reported as a problem with any of these systems.

Dr. Derek P. Gregory believes that converting to a hydrogen economy would require no major breakthrough in science or technology. It would be, he contends, a straightforward engineering problem—a complex one, but one for which the solutions appear to be at hand.

While hydrogen appears to offer many advantages as the dominant fuel of the future, there are major potential drawbacks. One of them is safety—i.e., overcoming the dangers, real or exaggerated, of storing, transporting, and burning hydrogen under pressure.

The biggest safety obstacle is what has been termed the *"Hindenburg* Syndrome," the label applied to the distrust and fear of hydrogen as a result of the explosion and burning of the zeppelin *Hindenburg* over Lakehurst, New Jersey, in 1937. In an effort to counteract this fear, a group of hydrogen-oriented scientists have formed the *"H_2indenburg* Society (H_2 being the symbol for hydrogen). It is an ad hoc organization formed at a 1972 meeting of the American Chemical Society on the thirty-fifth anniversary of the *Hindenburg* disaster, and is "dedicated to the safe utilization of hydrogen as a fuel."[344]

Hydrogen does pose legitimate hazards. In a confined place, it can form an explosive mixture with air. It has a low ignition point (it is believed that the *Hindenburg* explosion was sparked by static electricity) and burns with an intensely hot, colorless flame that can make its way out of the most minute openings.

On the other hand, in open air or in well-ventilated areas, hydrogen leaks and spills diffuse so rapidly because of its very light weight that it actually poses less of an explosion and burning hazard than gasoline. For this reason, proponents point out, hydrogen explosions are very rare.[345]

Dr. Lawrence Jones agrees that "the safety problems with hydrogen are different than with gasoline or with methane; they are not necessarily worse."[346] Dr. Derek Gregory argues that "just as we have designed apparatus and procedures to enable us to fill our automobile tanks with gasoline and carry the resulting twenty-gallon 'fire bomb' at speeds of up to seventy miles per hour along a crowded highway and park it overnight right inside our home, we can surely devise safe practices for handling hydrogen."[347]

While hydrogen cannot be detected by the senses, so that a leak of pure hydrogen might be particularly dangerous, chemical odorants could be routinely added to it, as has been done with natural gas for the same reason. An illuminant could also be added to make the flame visible.[348]

The Economics of Hydrogen as Fuel

Whether a hydrogen economy could be developed to substitute for the fossil fuel economy of today is largely dependent on the relative cost of building the necessary technologies for extracting power from other energy sources (nuclear power, geothermal heat, solar power, wind power, etc.). A hydrogen energy economy would surely be different than the fossil fuel age we live in at present, since the cost of making hydrogen will be far higher than the cost of drilling oil and refining gasoline today. In all probability, a hydrogen economy might be possible only for industrial uses of hydrogen and for some large-scale social uses of hydrogen fuel, such as mass transit in urban areas, or for fueling some fuel-cell installations.

The high future costs of hydrogen will prohibit the use of this synthetic fuel for an auto economy.

Research and Development

To bring any of the new energy sources into actual use will require a substantial commitment of the advanced, technologically rich nations in the future. No such commitment has been seen in the past. Far more money has been spent on consuming fossil fuels than in planning for the coming age of limited (or no) fossil fuels. Whereas the United States has seen fit to expend nearly $60 billion on interstate highways and support facilities, only a few billion dollars has been spent on energy research and development. The following analysis of the U.S. federal energy budget for fiscal 1974, prepared in an age of "concern" for the massive energy shortages of the future, still only allots $1 billion for *all energy research.*

Having devoted almost all energy research efforts into developing nuclear fission power through the Atomic Energy Commission for more than two decades, the federal government reversed its policies in 1972 and 1973, and began to realize that the energy shortages long predicted were not years away, but were actually occurring in America.

As a result, the fiscal 1974 federal energy budget of $1 billion was realigned to devote more money to other energy research areas. A breakdown of the budget shows the following general amounts in these areas:[349]

Area	Amount
Coal research (including liquefaction, gasification, etc., and oil shale)	524 million 99 million
Nuclear fission technologies (including safety program and breeder reactor development)	11 million
Nuclear fusion	7 million
Geothermal power	
MHD (magnetohydrodynamics) and combined cycle power plants	4 million
Energy storage technology (including hydrogen research)	13 million
Solar energy technologies (including wind power)	$170 million
Energy conservation methods	15 million

What is astonishing about the federal budget is the lack of commitment to the promising technologies of the future, with the possible exception of nuclear fusion. The government has devoted almost all its effort to the exploration of fossil fuels from dilute deposits and to harnessing electricity from dangerous nuclear fission processes—both areas posing significant problems to society. In the case of the dilute fossil fuels, the possibilities of economic recovery of energy are laden with physical limitations (the lack of water, for instance); the likelihood that agriculture in many states will be impaired by the rush to strip mine coal (and oil shale); and the distinct possibility that the processes will be completely uneconomic.

Little or no money is devoted to the development of decentralized power technologies and energy storage methods that might conserve electricity on a broad scale. The only federal programs designed to implement energy conservation are hasty attempts to allocate fuels in the midst of distinct crises, such as the winter of 1973–74, but no significant funding is devoted to long-range plans to salvage the American citizen from the hazards of energy-short years ahead.

The following chapters of this book are devoted to an analysis of the least-funded areas of national energy research: the potential for continuing civilization—as we know it—on renewable and dependable sources of energy derived from the sun. The federal commitment to solar (and wind) energy research, like its commitment to energy conservation, is negligible, and could well prove to be the major failure of American civilization. Without the abundant fossil fuel energy to which we have become accustomed, and without adequately developed natural energy sources to fall back on, the way of life our nation has long taken for granted will cease to exist, and the years of realization and readjustment will be bitter and long.

The current struggle to develop safe, clean, and dependable sources of energy constitutes the most crucial battle for survival man has yet fought. Without solutions to today's energy shortages, and the even greater ones that await us tomorrow, there will be little chance of preserving the legacy of this civilization for future generations. Some of the important answers do not lie in the paths currently pursued by the great institutions of our society now racing to strip mine the nation for "cheap" coal and line our rivers and estuaries with "safe" nuclear plants. The answer to the energy crisis of our age lies in taking up once more man's relentless pursuit of the sun.

6

Energy from the Sun

Human life—as well as all other forms of life on this planet—is completely dependent on the daily flow of solar energy. Not only is our food production linked to the sun's rays, but all the life-support systems of the natural environment are dependent on the sun. A great hope of man for ages has been to somehow harness the sun as a mechanical source of power. This chapter describes attempts—both ancient and modern—to realize this dream.

The sun, our nearest star, bathes the Earth each day in a sea of radiant energy—energy that travels in small particles called photons, reaching the Earth from a distance of 93 million miles in only 8 minutes. This constant stream of electromagnetic energy—radiating from the sun in all directions—strikes the Earth with a force that is only one 200-millionth the total energy of the sun. The Earth's share is equal to about 180 trillion kilowatts of electricity—far more energy than could conceivably be harnessed or utilized for human power generation.[1]

The sun's great energy release is the result of an elaborate chemical process in the sun's core—a process of thermonuclear fusion like the reaction in a hydrogen bomb. At the tremendous heat of more than 45 million degrees Fahrenheit in the sun's interior, hydrogen atoms fuse with helium atoms to produce the energy that powers our planet. The sun's furnace can release this energy in only one way. Since the star is in the vacuum of space, its heat cannot be transferred by conduction or convection in the way heat is dissipated on Earth. So the energy can only be released in the form of radiation—rays of heat energy. Scientists describe this radiation in terms of quanta, or photons of sunlight. The energy of a single quantum unit (or photon) is always proportional to its frequency. Rays of energy that are short in wavelength (such as cosmic rays) have a high frequency and greater energy content than longer wavelength rays, such as those in ordinary sunlight.[2]

This concept of radiation—or more precisely, of electromagnetic radiation—is the basic means of describing and comparing all the sources of radiation with which we are daily bombarded. The wide range of radiant energy in the universe makes up the total electromagnetic spectrum. One portion of it, which includes visible light rays, is referred to as the solar spectrum.

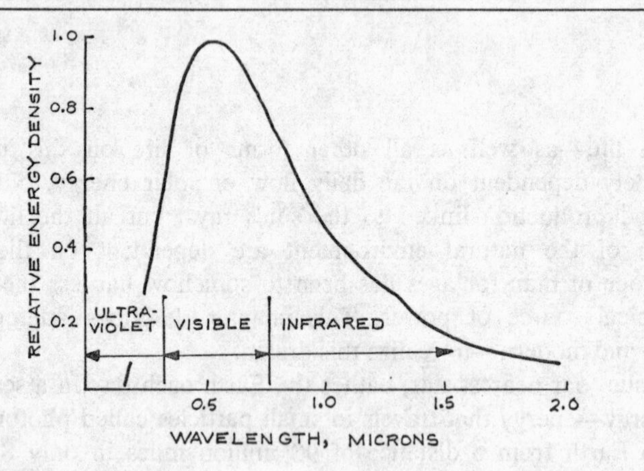

Solar spectrum

We cannot normally "see" even the visible part of the solar spectrum; only by the use of a prism or by observing the natural prismatic effect of an afternoon rainbow can we discern the multicolored bands of ordinary "white" light.

The most intense, short-wave length, high-energy radiation in the solar spectrum is found in the ultraviolet band, which is invisible. The visible "windows" or bands of the spectrum are comprised of blue, green, yellow, and red light. The bands of infrared radiation, like ultraviolet radiation, are invisible.

What we see is not light, but the reflection of objects in the visible part of the solar spectrum.

Scientist/author Lyall Watson, in his book *Supernature,* oberves that life on this planet is tuned to the narrow frequencies of radiation emanating largely from the sun, and that life forms filter most of sound waves by the use of sense organs. He adds:

> Light waves carry both energy and information. It is no accident that the amount of energy contained in visible light is perfectly matched to the energy needed to carry out most chemical reactions. Electromagnetic radiation covers a vast range of possible frequencies, but both sunlight and life are confined to the same minute section of this spectrum, and it is difficult to avoid the conclusion that one is directly dependent on the other.[3]

The most important chemical reaction on Earth is the reaction of sunlight with the green plants—the reaction in which the radiant energy of the sun is absorbed by the chlorophyll in the plant, which produces chemical energy to sustain the Earth's biological cycles. This process, called photosynthesis, breaks down the water and carbon dioxide gathered by the plant into sugar—(carbohydrate)—a composite of carbon, hydrogen, and oxygen—and pure oxygen. The chemical formula of the photosynthesis-energy reaction is written as follows:

CO_2	+	H_2O	+chlorophyll+light→	(H_2CO) +	O_2	+chlorophyll
(carbon dioxide)		(water)		(sugar) (carbohydrate)	(oxygen)	

A single photon of ultraviolet light contains the energy necessary to break molecular bonds. Since in photosynthesis ultraviolet light does not play a major role and the red and blue light does, several photons of the low-energy intensity of the red and blue light are required to break the chemical bond in the photosynthetic reaction.

A beautiful description of the photosynthetic process is that offered by Donald Culross Peattie in his book *Flowering Earth:*

> How does the chlorophyll, green old alchemist that it is, transmute the dross of earth into living tissue? Its hand is swifter than the chemists most sensitive analyses. . . . Light, in the latest theory, is not waves in a sea of ether, or a jet from a nozzle; it could be compared rather to machine gun fire, every photo-electric bullet of energy travelling in regular rhythm, at a speed that bridges the astronomical gap in eight minutes. As each bullet hits an electron of chlorophyll it sets it to vibrating, at its own rate, just as one tuning fork, when struck, will cause another to hum in the same pitch. A bullet strikes—and one

electron is knocked galley west into a dervish dance like the madness of the atoms in the sun. The energy splits open chlorophyll molecules, recombines their atoms, and lies there, dormant, in foods.[4]

A more technical explanation is this one by David M. Gates, professor of botany at the University of Michigan:

It turns out that photosynthesis is a complicated stepwise process. Light is absorbed by the chlorophyll molecule (and by other pigments in the plant) and is transferred to electrons in such a way as to create strong oxidants and reductants; that is, molecules that readily remove electrons from other molecules (oxidize them) or readily supply electrons to other molecules (reduce them). In photosynthesis the oxidants and reductants assist with the storage of energy in chemical bonds, notably those of carbohydrate and of adenosine triphosphate (ATP), the basic energy currency of all living cells.

Animals, by eating plants, are able to release the energy stored in them by means of the various oxidative reactions of metabolic processes. ATP interacts with the carbohydrate glucose to prepare it, through glycolysis for a long series of complex reactions in the metabolic sequence known as the citric acid cycle. The energy released is employed to do muscular work, to generate nerve impulses and to synthesize proteins and other molecules for the building of new cells. The entire chain of life proceeds in this way as energy cascades through the communities of plants and animals.[5]

The photosynthesis reaction actually appeared some billions of years after the planet's earliest sun-triggered chemical revolution, the photodissociation reaction, in which high-energy ultraviolet radiation breaks down water (H_20) into oxygen and hydrogen. According to the theory of the Soviet biochemist Oparin, the early atmosphere of the Earth was primarily methane (CH_4) and hydrogen (H_2), with trace amounts of water and nitrogen compounds, such as ammonia (NH_3). The addition of oxygen to the Earth's atmosphere was brought about by photodissociation (or photolysis: *photo,* light; and *lysis,* loosening), a reaction expressed as follows:

$$2H_2O + \text{radiation (light)} \rightarrow 2H_2 + O_2$$

The liberation of atmospheric oxygen some four billion years ago was followed by the presence of the first living cells, which are not believed to have had the capability of photosynthesis, and thus did not produce oxygen. The first cells that manifested the photosynthetic reaction were those of the blue-green algae, appearing on Earth some 2 to 3 billion years ago. During this seminal period of the Earth's development, the processes of photodissociation and photosynthesis generated much oxygen, some of which found its way into the upper atmosphere, where the ultraviolet rays of sunlight transformed the oxygen (O_2) into ozone (O_3). This upper atmospheric layer of oxygen and ozone shields the surface of the Earth from the high-energy ultraviolet radiation, making life as we know it possible.

By the time the first oxygen-*consuming* cellular organisms appeared on Earth some 1 to 1.5 billion years ago, 2 very important building blocks were present on Earth: (1) the abundance of oxygen, to fuel the coming ages of multicellular organisms—fishes, plants, reptiles, mammals, birds, and humans; and (2) the stored legacy of the Earth's great mineral and energy resources, including the basic metals, ores, and the fossil fuels, which would go untouched for millions of years, until humans found out how to use them.[6]

The role of solar energy in powering the Earth's energy cycles—the conversion of sunlight into chemical energy in the plants—and the role of solar energy in powering the Earth's atmospheric cycles have been explored in great detail, but scientists still know precious little about the exact interactions of either plants and trees or hurricanes and waves. What *is known* is that the sun is the power source that makes the conditions of life possible through a myriad of complex energy flows in nature. The investigation of attempts to maximize our use of the sun's energy begins with a precise understanding of the quantity and quality of sunlight available to us.

An aim of solar scientists for years has been to establish the exact value of the intensity of solar radiation reaching the Earth's atmosphere. This energy value is defined as the *solar constant*. The solar constant now accepted is the equivalent of 1.94 calories of energy striking each square centimeter per minute, *or* 430 BTUs of energy striking each square foot per hour. In more easily understood electrical terms, this amount of energy is equivalent to 1.36 kilowatts of electricity irradiating each square meter. If this amount of solar energy were accessible, our current yearly energy uses of *all kinds* could be supplied by the solar radiation striking an area the size of this country in only 32 minutes.

This is not possible for 2 reasons: First, the solar constant does not take into consideration the absorption of much of the sunlight by the atmosphere of the Earth; second, we cannot convert sunlight to other more useful forms of energy at 100 percent efficiency.*

* Even the green plants on Earth operate at about 1 percent efficiency. In certain applications, we can convert sunlight to heat at very high efficiencies—above 60 percent—but the heat is too diffuse for conversion to electricity, a high-quality

If we could convert solar energy to useful power at 100 percent efficiency, the intensity of the sun's rays that do pass through the atmospheric filter would be equivalent to 700 times the U.S. energy consumption. The average intensity of solar energy that reaches the surface of the United States is about 1,400 BTUs per square foot per day. This means that at known efficiencies of conversion of solar energy to other forms of energy (heat and electricity), the continental United States receives about 160 times as much energy from the sun as our current consumption from other energy sources.[7]

Solar intensity increases from the northerly to the southerly latitudes, and varies from a peak in the summer to low values in the winter. Some comparative values of the amount of solar radiation available at various places in the United States show the range of energy available.[8]

Average Solar Energy
in Megawatts of Electricity Per Square Mile Per Day

	December	*March*	*June*
New York and Chicago	130	330	575
Southern California and Arizona	260	420	730
Florida	260	420	575
Nevada	180	420	675

The power generated in modern electrical power plants ranges from a few hundred megawatts (a megawatt is equal to 1,000 kilowatts) to 1,000 megawatts. From this chart, it is evident that many areas of the country receive over-all solar energy equivalent—in just 1 square mile—to the electrical production of a large electrical power plant. Since power plants utilize fuel and cooling water in large quantities, they may take up more land than would be needed by a hypothetical solar electrical power plant. In Florida, for example, a large power plant complex near Miami (Turkey Point) contains several thousand megawatts of electrical generating capacity. To cool the condensers of the steam electric generators, the local utility (Florida Power & Light Company) has constructed a network of cooling ponds, which cover more than 11 square miles. Assuming an average yearly solar intensity just on the area covered by the cooling ponds of 400 megawatts of solar energy per square mile, the total energy intercepted by

form of energy. Efficiencies for converting sunlight to electricity have historically been very low—less than 10 percent—although claims indicate that higher than 20 percent may be possible.

the Earth is more than 4,400 megawatts (4.4 million kilowatts) of power—more than the output of the generating station.

History of the Direct Use of Solar Energy for Power

The use of solar energy in antiquity formed the basis for the development of civilization through the harnessing of plants and domestic animals in agriculture. Through its indirect forms—water power and wind power—solar energy was also harnessed for the production of mechanical power. The adaptation of houses and cities to climate, and thus to the sun, was a widely practiced art, lost in the current era of high-energy civilization.†

The following history describes various attempts to harness the energy of sunlight *directly* for the production of useful power, without utilizing the sun through intermediate forms of conversion, such as chemical energy (food and wood), kinetic energy (tides and wind), and orientation of the structures of civilization in relation to the sun.

The most stunning historic example of the direct use of solar energy is attributed to the famed Greek mathematician and engineer Archimedes. According to the second-century A.D. Greek historian Galen, in his *De Temperamentis,* Archimedes used solar reflecting mirrors to set fire to a Roman fleet besieging Syracuse in 212 B.C. The legend has been disputed because two other historians of the period, the Roman Livy and the Greek Plutarch, do not mention the event in their works.‡[9]

Livy and Plutarch do, however, mention other innovative weapons designed by Archimedes, who was unsuccessful in driving off the Roman attack and was killed following the Roman invasion of the city.

In his *Chronicon,* the twelfth-century writer Ioanne Zoaras claimed that Proclus used solar mirrors similar to those of Archimedes to drive off the fleet of Vitellius during the siege of Constantinople in A.D. 626.[11]

The American astronomers and solar scientists Aden and Marjorie Meinel devoted an entire chapter of their 1970 book *Power for the People* to a hypothetical reconstruction of Archimedes' feat, concluding that he probably used 4,300 gold and bronze shields on wooden towers. The shields, with a total reflective surface area of 2,330 square yards, could have concentrated such intense solar power on an 8-square-yard target that the temperature would have reached 1,550° F!

† See Afterword on climate and architecture.

‡ Plutarch does mention that solar concentration was used in the seventh century B.C. by the vestal virgins, who lit their sacred fires with sunlight focused by goblets of reflective metal.[10]

This would have been more than enough heat to ignite the dark-colored sails of Marcellus' fleet in a few seconds.[12]

Centuries passed before history again records the harnessing of solar power by man. In the seventeenth century, following the development of high-quality optical glass and Galileo's telescope in 1609, a few inventors duplicated Archimedes' legendary feat—though on a much-reduced scale. Using 5 mirrors to concentrate sunlight, Athanasius Kircher (1601–80) set fire to a pile of wood. A century later, the French scientist Georges Buffon used 168 6-inch-square mirrors to set fire to a woodpile 200 feet away. Buffon also used his mirrors to melt lead and silver at distances up to 100 feet. From these experiments, he concluded that Archimedes must have set up his solar weapon at a distance of 100 to 140 feet.[13]

Such rudimentary experiments set the stage for the practical and peaceful use of solar power.

Swiss scientist Nicholas de Saussure (1740–99) constructed the first solar "hot box" or oven, and used it for cooking. His box was made of wood, painted black on the bottom, and covered with plates of glass to trap the sun's heat. He found he could produce 191° F in full sunlight, and by chemically coating the outside surfaces of the glass, he was able to raise the temperature in the box to 320° F.[14]

This simple application of trapped solar heat has remained the largest and most widespread use of solar energy. Since the sun's energy is diffuse, and elaborate equipment is required to concentrate it for high-temperature application, simple boxes such as De Saussure's remain the cheapest and most reliable means of extracting power in the form of heat from the sun. Variations of the solar "hot box" technique have enabled it to be used for heating and air conditioning homes and buildings, cooking food, desalting water, and a variety of other purposes, which will be described in greater detail in the following pages.

During the eighteenth and nineteenth centuries, other experimenters devised more sophisticated methods and machines for harnessing solar energy. The father of the science of chemistry, Frenchman Antoine Lavoisier (1743–94), used a 51-inch-diameter glass lens filled with alcohol, which increased the lens' refractive power to concentrate sunlight to obtain high temperatures for melting metals. He succeeded in melting platinum—at 3,190° F—with his solar concentrator, and relied on it for a substantial part of his pioneering work in chemistry.

Beginning in 1860, French physics professor Augustin Mouchot conducted experiments with solar ovens in which sunlight was focused with metal reflectors, an exercise that led him to invent a solar-powered steam engine. Emperor Napoleon III was impressed with Mouchot's solar engine when he viewed it in Paris in 1866, which led to the French Government's subsequently paying Mouchot to continue his work. M. L. Simonin described a later version of Mouchot's engine at work in the *Revue des deux Mondes* in 1876:

> The traveler who visits the library of Tours sees in the courtyard in front a strange-looking apparatus. Imagine an immense truncated cone, a mammoth lamp shade, with its concavity directed skyward. This apparatus is of copper, coated on the inside with very thin silver leaf. On the small base of the truncated cone rests a copper cylinder, blackened on the outside, its vertical axis being identical with that of the cone. This cylinder, surrounded as it were by a great collar, terminates above in a hemispherical cap, so that it looks like an enormous thimble, and is covered with a bell glass of the same shape.
>
> This curious apparatus is nothing else but a solar receiver—or, in other words, a boiler—in which water is made to boil by the heat rays of the sun.[15]

Simonin also described Mouchot's calculations of the power output of the solar steam engine:

> Finally, on July 22, toward 1 P.M., an exceptionally hot day, the apparatus vaporized 5 liters of water per hour, which is equal to the consumption of 140 liters of steam per minute, and one-half horsepower. For these experiments the inventor used an engine which made 80 strokes per minute under a continued pressure of one atmosphere. Later on it was changed for a rotative engine—that is to say, an engine with a revolving cylinder—which worked admirably, putting in motion a pump to raise water, until the pump, which was too weak, was broken.[16]

In collaboration with another French solar pioneer, Abel Pifre, Mouchot developed a solar steam engine, which operated a printing press in Paris in 1882. The newspaper printed was appropriately titled *Le Soleil.*[17]

The Swedish-American inventor John Ericsson, best known as the designer of the Union Navy's much-chronicled ironclad ship of the Civil War, the *Monitor,* experimented with a number of solar-powered hot-air engines. Between the 1860s and 1883, Ericsson built 9 solar hot-air engines, which were also adaptable to steam production. In New York in 1883 he demonstrated a solar steam engine that used a cylindrical metal reflector, as opposed to the conical reflectors employed by Mouchot and Pifre.

Ericsson's machines were the most efficient solar devices built up to that time. He used a rectangular collector made of silvered window glass to concentrate the solar heat, and designed the device to "track" the sun across the sky in order to maintain a constant power output. The boiler's efficiency of heat conversion was claimed to be 72.5 percent, compared to 49 percent for Mouchot's engine.

Ericsson described his work in the British journal *Nature* in 1884, citing the traditional opposition to solar power:

> Practical engineers, as well as scientists, have demonstrated that solar energy cannot be rendered available for producing motive power, in consequence of the feebleness of solar radiation. The great cost of large reflectors and the difficulty of producing accurate curvature on a large scale, besides the great amount of labor called for in preventing the polished surface from becoming tarnished, are objections which have been supposed to render direct solar energy practically useless for producing mechanical power.

In an 1878 letter, Ericsson reluctantly concluded that "the fact is . . . that although the heat is obtained for nothing, so extensive, costly, and complex is the concentration apparatus that solar steam is many times more costly than steam produced by burning coal."[18] Ericsson's hopes for supplying his engines to benefit what he had termed the "sun-burnt regions of our planet" were never realized, and he ingeniously converted his engines to run on coal and gas, selling more than 50,000 of them in the international market to turn the sun-inspired failure into a commercial success.[19]

One of the most successful solar-powered installations of the nineteenth century was built in the desert near Salinas, Chile, for desalting water. It produced fresh water for use in a nitrate mine. The plant, which contained 51,200 square feet of solar collectors, cost $50,000, including windmills, which were presumably used for pumping the brine to the solar plant. An 1883 description of the plant reported that:

> The sun evaporated the water, and the resulting vapor condensed on the glass, for the temperature in the box was far higher than that of the atmosphere and hence of the glass. The pure water trickled down the sloping glass and dripped from its lower edge into a small channel on the top of each side of the box. These channels delivered into large ones, and thus the distilled water was collected. The plant yielded 5,000 gallons of pure water per day in summer, i.e., one pound of water per square foot of glass.[20]

A solar engine using a cone-shaped reflector was built in 1902 in Olney, Illinois, by H. E. Willsie and John Boyle, Jr. Their engine was designed to allow solar energy to heat water, which was then used to vaporize volatile fluids—such as ammonium hydrate, ether, or sulfur dioxide—which were in turn utilized to drive an engine. This machine was the first to demonstrate the use of two fluids (water and ammonia) in connection with solar power. Later, the two men built several plants in the American Southwest, including a 20-horsepower engine at Needles, California.

An ingenious method for generating electricity with solar energy and storing it for later use was developed by mine owner J. A. Harrington in the early years of the twentieth century in New Mexico. Harrington used solar concentrating mirrors focused on a boiler to produce steam, used for pumping water into a 5,000-gallon tank. The water was released to a water turbine for producing electricity (for lights) and operating a small electric motor in his mine.[21]

Frank Shuman, a Tacony, Pennsylvania, engineer, began experimenting with solar engines in 1906. He concluded that the demise of Ericsson's machines was caused by the tremendous expense of the "tracking" collectors, as Ericsson himself had guessed.

Using a variation of De Saussure's "hot box" design, he connected a vertical single-cylinder engine to a 1,200-square-foot solar collector constructed of glass plates. The solar heat was trapped in the box in pipes containing ether. The ether boiled, providing steam for the 3½-horepower engine. The exhausted vapor of the ether was condensed and returned to the collector's pipes, providing a continuous cycling of the system. Shuman's power plant used 2 plates of glass to trap the solar heat, and the flat-plate collector did not have to be tracked with the sun. It was a simpler and less costly device than solar engines of the nineteenth century.

In 1908, Shuman formed the Sun Power Company and convinced English financiers to back his efforts to build larger plants using the flat-plate collectors. In 1911, he demonstrated a plant in Philadelphia with more than 10,000 feet of collector surface. It produced 816 pounds of steam per hour and was used to operate a steam-driven water pump.[22]

The following year, he built a plant of slightly modified design in Meadi near Cairo, Egypt, with the technical cooperation of British physicist C. V. Boys. Boys suggested a change in the flat-plate collector to a parabolic design, which would increase the amount of solar energy received on the black-painted pipes. The collector area was increased to 13,269 square feet to furnish the heat for powering a 100-horsepower engine. The plant's boiler, made of thin zinc, failed in the first tests; but a second boiler, made of cast iron, was successfully erected at the site in 1913.[23]

The Egyptian plant, which was designed to pump irrigation water from the Nile River, was significantly different from Shuman's previous machines. The addition of Boys's parabolic collector meant that the plant was converted into a suntracker, as opposed to the stationary collectors originally designed by Shuman. The collectors were automatically oriented along the sun's path by some of the power from the plant's engine. The plant operated successfully, and the costs of construction could have been amortized in three years.

A. S. E. Ackermann, a consulting engineer of the British Division of the Sun Power Company, described the power production of the plant:

> The best hour's run gave 1,442 pounds of steam at atmospheric pressure; hence, allowing the 22 pounds of steam per brake-horsepower-hour, the maximum output for an hour was 55.5 brake horsepower—a result about 10 times as large as anything previously attained, and equal to 63 brake horsepower per acre of land occupied by the plant.[24]

Like many solar experiments before it, Shuman's plant was abandoned. Technically, the plant was a success, but World War I intervened and forced the plant to shut down.

In 1901, a large solar reflector was built by A. G. Eneas to provide heat for powering a water pump on a Pasadena, California, ostrich farm. The reflector weighed 8,300 pounds, and the steam boiler was very efficient—about 75 percent. It was rated at 4 horsepower, and could pump 1,400 gallons of water 12 feet per minute.

The March 16, 1901, *Scientific American* noted that:

> The motor is a result of a number of experiments by a band of Boston capitalists. . . . Dwellers in the East, where rain falls every few days throughout the year, cannot realize what such a perfected motor means to the West, where arid land awaits but the flow of water to blossom as the rose. . . . In such locations the solar motor is a boon. The skies are comparatively free of clouds and the machine can begin work an hour after sunrise, possibly earlier, and continue until one-half an hour before sunset.[25]

The Arizona *Republican* of Phoenix, Arizona, commenting on another plant constructed by Eneas in 1904 near Tempe, Arizona, predicted that the facility demonstrated the coming era of solar energy in the western United States:

Any attempt to predict possibilities in this field would be mere guesswork, but it is safe to say that solar power will be one of the great influences of the new century and will make the arid regions of the West and other parts of the world the theater of the greatest industrial revolution of the future.[26]

The venerable dean of America's solar scientists, Dr. Charles Greeley Abbot of the Smithsonian Institution, began experimental work in the nineteenth century dealing with scientific aspects of solar radiation as well as the application of the sun's energy to the production of power. Dr. Abbot's contributions continued until 1973, when he celebrated his 101st birthday; he died later that year. The oldest known holder of any U.S. patent, he was recognized internationally as a pioneer in several scientific fields.*

Dr. Abbot developed measuring instruments to analyze the solar spectrum in collaboration with fellow scientist Samuel Langley in the nineteenth century. This early work was followed in 1916 by experiments with solar cookers on Mount Wilson, California. Abbot's cookers not only employed parabolic mirrors to reach high temperatures, but also used a heat storage device that enabled food to be cooked long after the sun had set. Instead of water as the heating fluid, he substituted engine cylinder oil, which did not boil at the high temperatures (approaching 200° C to 382° F) that the concentrators produced, making possible greater heat storage capacity.

In the 1930s, he built and demonstrated other solar machines using parabolic concentrators, including water distillers and steam engines. In his book *The Sun and the Welfare of Man,* he described the operation of the solar water distiller:

> The arrangement of the mirror is similar to that [of one used] . . . for cooking purposes. In this case . . . the elongated vacuum jacket like a thermos tube except that it is not silvered within, is supported in the focus of the mirror with its open end at the bottom, and its closed end extending a foot or more above the top of the mirror, which rotates on rollers bearing the hollow trunnions of the mirror. In the case of the cooker, and also of the flash boiler . . . the absorber of rays is made as small in diameter as possible in order to reduce heat losses, so that the temperature may run high. . . .

* In the late 1960s, a group of Soviet scientists suggested that a moon crater be named for him, an honor reserved thus far—with the exception of the astronauts—for the dead. When they learned he was still alive, they withdrew the request, and Abbot tersely commented that he would prefer a few more years of life to perfect his solar inventions.

The efficiency of the device is very high. The steam being condensed by flowing through the entering water, that water reaches the lower end of the boiler tube at almost boiling temperature. Thus it is only the latent heat of steam that must be provided by solar radiation, and not the heat required to raise water to boiling. In experiments made in Florida in March, 1938, the stinking water of Arcadia was distilled to perfect purity and odorlessness. Distillation commenced within 5 minutes after the sun came out of a cloud. A mirror of 11 square feet of surface distilled between 2 and 3 gallons of water, entirely automatically, in one cloudless day.[27]

In the 1930s Dr. Abbot also worked on solar "flash boilers," which were operated by reflecting parabolic mirrors that tracked the sun. The boiler tube was made of black-painted copper connected to a vacuum jacket of Pyrex glass. The device was able to pick up solar energy very quickly. On cloudy days, for example, it could generate full steam pressure within 5 minutes after the sun emerged from behind a cloud.

Dr. Abbot estimated that such a device large enough to produce 3 to 5 horsepower could convert into actual mechanical work 15.5 percent of the sun's energy intercepted by the mirror. He recommended development of such boilers in combination with heat storage materials for retaining the heat captured by the sun for later use at night or the next day.

Another innovation consisted of encasing the blackened metal boiler tube in vacuum to conserve heat. "With proper design," Dr. Abbot noted, "there would be sufficient solar heat to keep the engine liquid far above its boiling point—in short, in a super-heated condition. Thus, super-heated steam or vapor would burst forth to the engine whenever the communicating cock was opened for that purpose, either by day or by night."[28]

Dr. Abbot secured several patents for using the flash boiler to produce both electricity and mechanical power. He exhibited a 0.5-horsepower solar-operated steam engine at the International Power Conference in Washington, D.C., in 1936, and later demonstrated a model of an improved version of the engine using a flash boiler, which generated 0.2 horsepower.[29]

Dr. Robert Goddard, the American inventor who paved the way to rocket propulsion with his 1919 study *A Method of Reaching High Altitudes* (published by the Smithsonian Institution), was also a leading pioneer of solar-powered engines.

Goddard's interest in solar energy was kindled by his friendship with Dr. Abbot, who encouraged him to pursue work in both rocket propulsion and solar energy. In the 1920s and '30s Goddard acquired several patents on solar engines that used parabolic mirror reflectors.

Goddard's primary innovation in the construction of the engines was a design approach aimed at minimizing energy losses at all stages in the systems. His first patent, for an "accumulator for radiant energy," was dated June 10, 1924. His studies of earlier equipment (including the Shuman-Boys plant in Egypt) revealed a common flaw: The collectors, steam engines, and electrical dynamos were all extremely inefficient machines, so that only a small fraction of the original solar energy was retrieved by the systems.

The concentrator of his engine boiled various liquids—including isobutyl alcohol—to produce a steam vapor that was then passed through a cone-shaped vaporizer to turn a turbine to produce power. The vaporizer closely resembled the nozzle of a rocket engine, and was in fact almost identical in shape to one used in Goddard's 1914 patent for a "rocket apparatus."†

Goddard was issued 5 patents on various aspects of the solar turbines; and, while he built only laboratory-scale models, his ideas for development of the machines were clearly commercial. He wrote in the October 1929 issue of *Popular Science Monthly* that:

> There is solid scientific backing for the belief that a solar motor of the size illustrated (20 feet in diameter) would produce upwards of 30 useful horsepower when operating under a clear sky between 10 in the morning and 3 in the afternoon . . . when sunlight is at its maximum . . . in the United States. That amount of power, converted into electrical energy, would far exceed the requirements of a large farm; the unused current could be employed to charge batteries. These, in turn, could maintain the normal current supply on cloudy days and nights.[30]

Goddard's solar and space work overlapped somewhat, and he wrote of the future possibility of powering planetary space travel with solar engines. His actual work in the solar field, however, was limited by his overriding dedication to the field of rocketry. In the late 1920s, Charles A. Lindbergh convinced the Guggenheim family to give Goddard $100,000 to pursue his work in rockets. With this grant, Goddard was able to move to New Mexico, open a laboratory, and devote his full time to this embryonic space science—leaving solar energy research to Dr. Abbot and other former colleagues.[31]

† However, Goddard was not the first to propose a cone-shaped vaporizer. The original was designed by Dr. Carl Gustav de Laval in 1889.

The History of Solar Water Heaters

Between the turn of the century and the 1930s in the United States, the first widespread commercial use of solar energy came into being with the installation of solar water heaters in California and Florida. The collectors were of flat-plate design, like De Saussure's "hot box," and used usually a single pane of glass to cover a metal box containing blackened copper tubes. Tens of thousands of these heaters were sold in both states until the middle 1950s.

The method of heat transfer and storage of these water heaters is called thermosiphon circulation. This means that a water tank was attached physically above the solar collector—as in the usual false "chimney" position on the roof—and pipes circulated cold water from the bottom of the tank to the collector pipes. As water is heated by the sun, the hot water collected flows automatically (by convection) via a second pipe to the top of the water tank, forcing the colder water to flow through the collector.

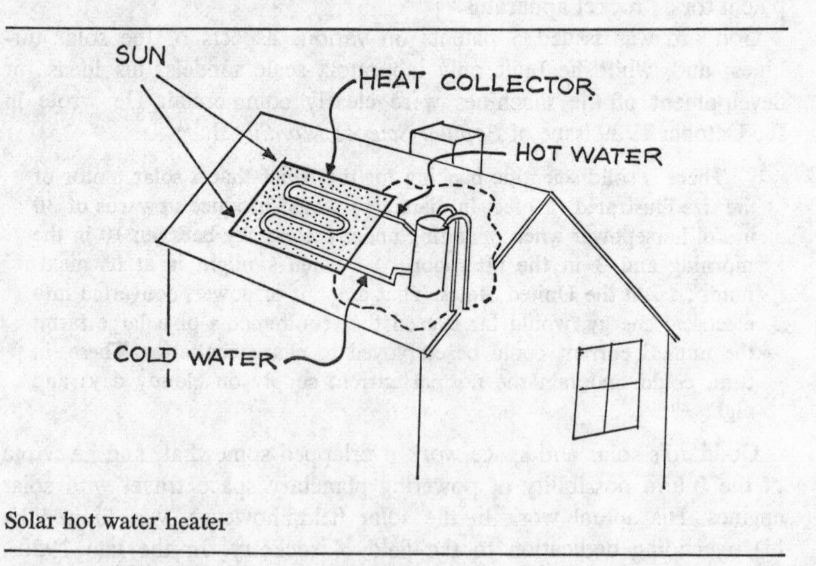

Solar hot water heater

The construction and installation of the water heaters was performed either by the homeowner himself or by one of a number of small companies in local areas in the sunny states. In either case, they were largely hand-crafted and unsophisticated.

The solar water heater industry remained a home-grown phenomenon, and knowledge of the design and construction of these devices has not been at all widespread in the United States. In 1935, Harold

L. Alt, an engineer with the Gibbs & Hill engineering firm of New York, presented a paper to the American Society of Heating and Ventilating Engineers on the potential of solar water heating for the nation. His study of solar water heaters convinced him that, with modifications such as the addition of an auxiliary conventional heater, or "booster," the heaters could find extensive use in the United States.

"To date," he noted, "the design of solar heaters seems to have been one of hit-and-miss, modified by the original ideas of the designer, who may have copied one he has seen somewhere else, or gone ahead blindly to the best of his ability with no guidance at all." The consequences of such a haphazard approach were "just what might be expected." He added:

> Some of the solar heaters give good service, and others do not. In recent years, the advance in the art of heating due to research on heat transfer and sun effect problems, makes it possible to apply some of the data which have been developed to the design of solar heaters. . . .
>
> It is rather surprising to notice the number of textbooks and reference books on heating that do not even mention solar heating. It was thought that with the presentation of this paper and discussion by the members, that the guide publication committee would have enough material to prepare a short chapter for the [engineering] guide [of the Society in] 1936.[32]

In 1936, Professor F. A. Brooks of the Agricultural Experiment Station of the University of California at Berkeley published a comprehensive booklet listing the advantages and characteristics of solar water heaters, commercial and home-made. He noted that the 2 most common heaters in California were (1) simple, dark-painted water tanks (or boilers) mounted in sunny places; and (2) glass-enclosed heaters with storage tanks in which solar-heated water was kept during the daytime for hot showers. While heaters of the first type were adequate for hot showers at the close of a sunny day, they would not keep water hot overnight.

Professor Brooks found that the most economical solar water heating system—at least for a farmhouse—was a solar glass collector with black pipes connected to a water tank and an auxiliary water heater. The solar unit would be used from early spring to late fall, at which time it would be disconnected for the winter, with the auxiliary unit taking over to furnish the hot water needs until springtime again.‡

‡ Solar water heating systems for year-'round operation sold for about $5 per gallon of installed water tank capacity, a price that included insulation around the

A number of companies prospered in the solar water heater market until cheap natural gas and petroleum became available to consumers, some of whom by that time had probably tired of the necessity of periodically draining, cleaning, and otherwise maintaining the solar units. The Day/Night Company in California, for example, manufactured solar water heaters in the early years of the century. Later versions of their heater used a supplemental gas boiler. With the onslaught of cheap gas prices, the company discontinued the solar heater, and is now a major manufacturer of gas water heaters.[34]

Solar water heaters survived longer in Florida than in California. It was estimated that more than 50,000 such heaters were in operation in the city of Miami alone in 1951.[35] By 1973, the industry had almost vanished. Only 5 U.S. companies still manufacture the units, and then only on a custom basis.

Materials used for the collector surface in Florida ranged from glass panes to durable plastics. In a brochure distributed in the 1950s by the Solar Products Company of Opa Locka, Florida, the following characteristics, indicated on the basis of tests, were listed for a solar water heater panel made of contoured plastic.*

- The plastic cover would lose only 2 percent of its light-transmission qualities over a 10-year period.
- Its high efficiency would deliver as heat energy more than 90 percent of the solar radiation it intercepted.
- A standard 3-foot-by-6-foot panel weighed less than 40 pounds, and could be easily installed by the purchaser.
- Three of the panels would deliver sufficient hot water for the yearly requirements for a 4-bedroom home.

These panels, made under the trade name SOLARTEK, are no longer produced. It is interesting to note, however, that some two decades ago, the company foresaw the use of such collectors for other

---

collecting box and water tank; pipe materials; and labor. An average system with a 60-gallon storage tank cost about $300 at that time. Professor Brooks found that the cost of solar water heating with these commercial units—assuming an 18-year lifetime—was equal to the cost of heating water electrically at 6 mills per kilowatt-hour. A less expensive system designed for use only during sunny months cost about $3 per gallon of water tank capacity, installed. Amortized over a lifetime of 15 years, heating water with solar energy cost about the same as heating with natural gas at the time—one ninth cent per thousand BTUs of heat.[33]

* The panel fitted over the absorber surface like a bubble, enabling some concentration of solar heat, compared to the flat-plate panel, which cannot concentrate the available energy.

purposes. Their brochure describes how the small panels could be easily interconnected to supply house heating, swimming pool heating, and heat for industrial plants.[36]

Solar water heaters have been produced in great quantities in such sun-rich countries as Israel, Japan, and Australia. America solar engineer John Yellott reported in 1960, after a visit to Japan to examine uses of solar energy, that more than 200,000 solar water heaters were in use—ranging from simple plastic "pillows" filled with water to sophisticated flat-plate collectors with aluminum pipes. He noted the sophistication of some Japanese designs:

> Simplicity in construction and operation is the keynote. . . . Control equipment is kept to an absolute minimum, since man-power is usually available to perform such chores as opening and closing valves, and turning pumps on and off. Most solar heaters are designed to drain automatically when the water supply is shut off, to minimize the danger of freezing. A supply of water is solar-heated during the day, and available for evening hot baths. Since automatic dishwashers and laundry machines are not widely used, the demand for hot water during daylight hours is usually small.† [37]

In both Australia and Israel, the production of solar water heaters is widespread. Australian water heaters are produced primarily for use in nonurban locations, but Israeli water heaters are in use in both cities and the countyside. The father of Israel's solar sciences, Dr. Harry Tabor, now director of the National Physical Laboratory at Hebrew University in Jerusalem, described the growth—and decline—of the Israeli solar water heater industry in a 1972 memorandum:

> Solar water heaters were widespread in use—over 100,000 domestic installations in Israel. This represented near saturation, as the housing program moved in the direction of high-rise buildings for which the solar collector is less attractive due to higher plumbing costs and increased transmission losses. Centralized piped gas and low-cost night rate electricity reduced the competitive status of domestic solar water heaters and sales are now low.[38]

Solar water heaters, as we have seen, were in widespread use in a number of countries—not only less-developed countries, but also the

† Undoubtedly, the reason for success of Japanese solar heaters is twofold: (1) The Japanese are chronically short of conventional energy sources, and the solar boost is a welcome free energy supply; and (2) they are also willing to adapt their lifestyle to the supply of solar energy—i.e., showering at night instead of in the morning. The American consumer, long dependent on a multitude of energy-expending machines, is not as adaptable to the natural environment and its supply of sunpower.

technologically advanced nations—until the advent of cheaper forms of energy, including electricity. The few remaining locations where the heaters are in use are either isolated from sources of cheap fuels or reflect individuals' personal preference for solar energy to other fuels.

Solar Cookers

Internationally, the most widespread historic uses of direct solar radiation—excluding basic agricultural practices—have been for cooking, drying, and desalting water. It is of related importance that, since earliest history, man has been forced by the natural climate to seek means of sheltering himself from the elements. As will be demonstrated in a later chapter on natural architecture, the greatest application of solar energy has been in the judicious use of solar-oriented building design.‡

Solar cookers and ovens have been built by a number of scientists and inventors over the past few centuries, primarily as demonstration devices. The French solar engine pioneer Mouchot demonstrated the art of cooking beef in a solar oven at the World Exhibition in Paris in 1878.[39] However, this and other demonstrations were intended only to illustrate the incidental potential of solar energy, while the real scientific attention was concentrated on converting sunlight to power.

In the United States in this century, Dr. Abbot and others have devised and built a number of cookers, but never with a commercial market in mind because of the inconstancy of the devices in comparison to conventionally fueled stoves and ovens.

Foreseeing a potential need for cheap cooking devices in underdeveloped countries, however, a number of universities as well as the federal government researched the potential of manufacturing solar stoves in the 1950s. A wide assortment of small, relatively inexpensive cookers were developed, ranging in cost from less than $10 to more than $100. Ovens in the cheaper categories consisted of simple sheets of reflective materials (such as aluminum) that focused heat on a pan.

At the United Nations' 1961 Rome Conference on New Sources of

‡ The adaptation of solar energy in the design of shelters and buildings underwent a revival in the United States between the mid-1930s and early 1950s. A number of "solar oriented" houses were built—i.e., oriented in a southerly direction to catch winter heat in large windows and minimize summer heat. However, the advent of home air conditioners in the United States in the 1940s and 1950s halted this common-sense architectural development.

Energy, international solar scientists discussed the development and potential of these devices. Dr. George Löf, a U.S. expert in solar energy, reported that the cheap ovens could not deliver energy greater than 100 watts, raising a question of usefulness.

More expensive devices discussed at the meeting utilized sophisticated parabolic reflectors, plastic focusing lenses, and heat storage systems to deliver the greater amounts of heat required for cooking. These cookers could reach temperatures higher than 400° F in bright sun. However, the high cost prohibited their widespread development and use in poor countries. In 1960–61, field tests were conducted to determine the acceptance of 2 dozen 4-foot-diameter cookers among the 266 families of a Mexican village. Through Rockefeller Foundation grants, anthropologists were employed to live with the villagers and report on their use of the devices. Even though some of the cookers developed mechanical problems (such as collapsing in the wind), the villagers appeared to like them, and used them frequently. They were not intended to be used as the primary cooking device, but as a supplement to the conventional wood and oil stoves in wide use. The villagers adapted to the fact that the cookers were useless at night and in cloudy weather.

The one hitch was that each cooker was estimated to cost $25, the equivalent of 2 weeks' salary for the average villager. Costs were predicted to drop to $8.66 apiece for cookers produced in quantities of 1 million per year. But even this relatively low figure would be more than people in poor countries could afford—even though the cooker's cost could be amortized in a year at prevailing fuel rates in the area.

Other papers presented at the conference described cookers developed in the United States, in India, and by the United Nations; but all suffered from the same problem: high capital costs, the bane of solar projects throughout history.

Solar Distillation

The direct use of solar energy to distill water was first commercially proven in the nineteenth-century Chilean nitrate mine discussed earlier. Since then, solar stills have been used on a small-scale basis in isolated communities and houses for purifying brackish and salt water. The essential technique of the solar still is to admit solar radiation through a transparent plastic or glass cover to a shallow basin of salt water. Pure water quickly evaporates from the brine when light strikes it, and the vapor is condensed on the transparent cover, which is sloped to direct the condensed water into a collection trough.

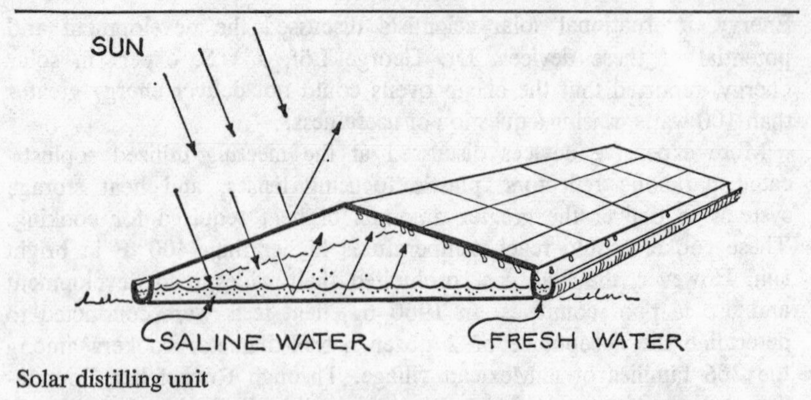

Solar distilling unit

Small solar stills have been produced in the United States, Spain, France, and Australia that require little maintenance and provide pure water at low costs in areas where cheap fuels are not available to operate sophisticated high-temperature distillation equipment. The capital costs of large solar stills may range as low as $1 per square foot of basin area, which would make possible economic water production.

Both plastic and glass covers pose maintenance problems for solar stills. The glass can be broken by hailstones, rocks, or other flying objects, and the plastic is subject to cuts and tears from sharp instruments or the claws of animals.

Researchers at the University of Arizona Institute of Atmospheric Physics have developed a multiple-effect solar still under the direction of Carl Hodges and Dr. Richard Kassander. Seawater, preheated under a plastic dome, is pumped to the top of a small tower and released in droplets that serve to saturate air in the tower. The saturated air then flows to a separate "condenser tower," where the water condenses and is collected for use.

The system requires various pumps and blowers, power for which was produced by a diesel-electric generating plant in the Arizona experiments. The waste heat produced by the diesel can be utilized to supplement the solar heat, making it possible to operate the system at night. The diesel exhaust can also be used to furnish additional carbon dioxide to plants in plastic-covered greenhouses, which are irrigated by a modification of the solar still concept. According to John I. Yellott, director of the Yellott Solar Energy Laboratory in Tucson, Arizona, "this combined system has real promise for arid areas that have plentiful solar radiation as virtually their only energy resource."[40]

Use of a solar still to replace or supplement the radiator for cooling the diesel engine offers several advantages, according to an Australian mechanical engineer with considerable experience in solar distillation. For instance, water costs are reduced, much less area is required, production is less seasonally dependent, and solar still efficiency is markedly improved.[41] The first multiple-effect still of this type was built in the late 1960s at the University of Arizona Energy Laboratory. The experience led to the construction of a 9,000-square-foot pilot plant at Rocky Point, Sonora, Mexico, operated by the Mexican Government.

While solar stills produce drinking water in various areas of the world, the solar distillation process is still too expensive to be applied to agricultural irrigation.

Under sponsorship of the U. S. Office of Saline Water, Battelle Memorial Institute, Columbus, Ohio, has prepared a manual on solar distillation. In the manual, solar scientists J. A. Eibling, S. G. Talbert, and G. O. G. Löf conclude that ". . . large durable-type, glass-covered solar stills will produce water on a consistent, dependable basis for a cost between $3 and $4 per 1,000 gallons in most situations." Furthermore, they predicted that "improved designs and cheaper materials can reduce the costs somewhat below this range."[42]

Water distilled in solar stills has been more expensive than water distilled in processes using cheap fossil fuels. The cost range decribed by the Battelle Memorial Institute may bring solar distillation processes into commercial viability in the 1980s, however. The construction of solar stills would be a major means of conserving energy that otherwise would go for fueling conventional distillation processes, which use fossil fuels and electricity generated by other, limited energy resources. Both on a small, community scale and in large installations in favorable areas with plenty of sun and access to the ocean, the processes of solar distillation can be expected to play a major role in the future development of needed fresh water resources.

Applications of Concentrated Solar Energy

Although the most successful applications of solar energy have historically been low-temperature uses, sunlight has been concentrated with various devices for such sophisticated purposes as the operation of scientific furnaces (for materials testing, etc.) and for the conversion of sunlight into electricity.

Solar Furnaces

Since World War II, much international interest in solar energy applications has been devoted to the construction of solar furnaces, similar in principle to those built by Archimedes and Lavoisier. The furnaces are used for concentrating intense heat on metals and other materials for evaluating their performance in high-temperature applications such as aeronautics and space testing.

French scientist Felix Trombe has built a number of solar furnaces at Mont Louis in the Pyrenees Mountains. Beginning with a furnace using concentrating mirrors to develop 50 kilowatts of power (as heat) in the early 1950s, Dr. Trombe continued to refine his designs, and in 1970 he completed the construction of the world's largest solar furnace, which develops more than 1,000 kilowatts of power. The furnace is comprised of a central parabolic mirror, which receives concentrated solar rays directed onto it from 63 mirror panels, each about 3 times a man's height. The furnace reaches temperatures over 6,000° F and can melt a three-eighths inch steel plate in less than a minute. It is used for materials testing and for chemical vapor depositing of exotic materials such as molybdenum and tungsten borides.[43]

As a result of Dr. Trombe's research efforts at Mont Louis, the uses of solar furnaces by industry have been discussed often, but because processes using conventional fuels are comparatively inexpensive, no serious attempts have been made to manufacture solar furnaces for widespread applications. Solar furnaces might also be utilized on a small scale by small glass or metal industries for material reduction. Small solar furnaces might be used to melt and recycle glass, for example.

Solar Electricity

As a result of research activities associated with the electronics and aeronautics industries, techniques for producing electricity directly from solar energy began to be investigated in the 1950s. In 1954, researchers at Bell Telephone Laboratories announced the discovery of the photovoltaic method of producing electricity from silicon photocells.[44] The silicon cells are tiny wafers of electrically conductive material that generate direct-current electricity when exposed to solar radiation. Uses for the silicon cells have been confined to space applications—such as powering satellites—and in a few instances for supplying electricity for remote earthbound applications, such as powering lighthouses and for isolated military applications. The main drawback for the cells is their extremely high cost. The electricity they produce costs thousands of times more than that from conventional processes.

Other experiments conducted in conjunction with America's aeronautics and space program resulted in the development of thermoelectric and thermionic methods of generating electricity from solar radiation. Thermoelectric devices are based on what is called the "Seebeck effect," which describes a flow of electrical current when a temperature difference is maintained between the junction of two dissimilar metal junctions in a closed circuit. Government experiments conducted in the late 1950s used solar flat-plate absorbers as well but the systems were costly and sophisticated and yielded relatively little power.[45]

Thermionic power conversion is a process originally observed by Thomas Edison in 1883, and is often called the "Edison effect." Edison found that metals emitted electrons when exposed to high temperatures. In a conventional vacuum tube, heat applied to a cathode (the emitter) causes the release—or "boiling"—of electrons from its surface. A positive voltage applied to the anode (the collector) draws electrons across the vacuum gap. In a thermionic converter, the emitter is exposed to high temperatures (1,300° C to 2,000° C—i.e., 2,340° F to 3,632° F), and electrons flow freely to the collector without the necessity of applying additional voltage. Solar concentrating mirrors were used as the heat sources, but the devices—as in the case of thermoelectric converters—proved quite costly for a return of very little electric power.[46]

The potential for producing electricity from the "direct conversion" solar devices appears to be limited to photovoltaic cells, which—though costly—have made the exploration of space and the development of satellites for long missions practical by using lightweight solar panels to harness the sun's energy. Research is currently under way to bring the costs of the direct conversion devices to levels suitable for widespread development of solar energy on Earth as well.

Solar Politics

Although much interest in the scientific community has been focused on solar energy at various times in history, widespread development of solar power equipment has never been achieved—primarily because of the high cost of developing solar power compared to that of technologies utilizing cheap fossil fuels. Many scientific projects have been undertaken in attempts to realize the enormous energy potential of solar energy for man's use, but scientists have found that a public already well supplied with cheap energy sources is not willing to dabble in solar collectors and power devices.

Only with society's very recent realization that the precious stocks of fossil fuels are becoming depleted, and that other hoped-for energy sources may not be available in time to stem the waning tide of the fossil fuel era, has serious interest been focused on widespread application of solar power. When similar fears about energy shortages were expressed in the United States in the late 1940s and early 1950s, solar scientists also suggested then that—as a national policy—the collection of solar energy on house rooftops should be explored. In fact, "solar houses," using an expanded version of the solar hot water heater technology, were built in several areas of the nation in the 1940s.

So much excitement had been drummed up in the 1940s and early 1950s at Massachusetts Institute of Technology (MIT), Colorado State University, and by individual engineers and architects across the country by the initial development of solar-heated houses that the 1952 report of the President's Materials Policy Commission predicted the widespread adoption of solar-heated dwellings. The famous "Paley Commission," named for its chairman, William S. Paley, was created by President Truman to appraise the nation's natural resources position in the year 1975 and to formulate policies appropriate for meeting future resource demands at lowest cost. Written during a period of public concern over America's postwar dependence on foreign lands for energy and mineral resources, the report predicted widespread sale of solar heating units in the United States during the 1950s, '60s, and '70s. By 1975, the report predicted, 13 million solar heating systems—at a cost of more than $2,000 each—would be installlled on U.S. homes and commercial buidings and would account for 10 percent of the nation's over-all energy needs of all kinds.[47]

The anticipated solar power era did not occur for a number of technological, economic, and political reasons.

Technologically, the solar heating experiments on which the commission based its judgment were largely doomed to failure. Almost all the structures using the initial prototype solar heating units were later converted to conventional heating systems due to a series of errors in design and installation of the solar equipment. Equipment for collection, storage, and distribution of solar heat was not nearly so adequate as that available today.

In terms of economics, the cost of the solar heating system has remained many times the cost of cheap fossil fuels.

In political terms, the government never developed an interest in furthering research on solar energy because of continuing pressures from the fossil fuel industry—both coal and oil—which saw solar energy as a competitive resource that would reduce demand for—hence, profits from—the fossil fuels. For similar reasons, the only energy agency with any clout in Washington—the Atomic Energy

Commission (AEC)—has been steadfastly opposed to spending funds for any application of solar energy research. That resistance has remained active and effective up to the present.

The political climate of the 1950s was inimical to the development of solar energy. The growing power of the Atomic Energy Commission and the fear of scientists in opposing its technological and political might were widespread. Then, as today, many scientists and technicians were afraid to publicly announce their opposition to nuclear power for fear of recriminations by the AEC or the nuclear industry. Then, as today, solar power advocates were for the most part critical of nuclear power because of its disastrous environmental effects.

In March 1959, the University of Maryland's Bureau of Business and Economic Research published a prescient document, *Solar and Atomic Energy—A Survey,* which assessed the future effects of the development of nuclear reactors vis-à-vis the development of solar power. It concluded:

> Modern industrialized society is now in possession of a concentrated atomic power which could destroy it. In the history of science, development of nuclear fission was a notable accomplishment. The thrill of its potentials and the challenge of its application have influenced the planning of the military and the allocation of funds in the Nation's budget. . . . Application of atomic energy to peaceful uses is under test, it is true; but the emphasis is minor, and the basic dangers of contamination of the human race through radioactive waste, medical overdose of isotope therapy, and accidents, particularly in transportation, are terrifying. . . .
>
> Fortunately, there is an alternative source of energy waiting to be developed, and an international community interested in its application to the welfare of mankind. This is solar energy. . . . Now, before our conventional fuels are in short supply and we and our environment are more contaminated by radioactive substances, is the time for solar research.[48]

John H. Cover, an economist who at that time was serving as director of the Bureau of Business and Economic Research, recalls clearly the hostility with which the report was met. "I had the advantage of a number of scientific and technological consultants," he explained recently, "but was dismayed that only two would permit reference to their participation, although each read and all approved of the manuscript."*[49]

* The two men who signed the final report were scientist/author Ralph Lapp and engineer Ellery Fosdick.

Solar Energy Today

Interest in large-scale applications of solar energy has been tremendously accelerated in the United States in the early 1970s. Although it was in 1968 that Dr. Peter Glaser suggested the use of enormous orbiting satellites to collect solar energy and transmit it to Earth below via microwave rays (described in subsequent pages), it took the increased public awareness of environmental deterioration and the growing energy shortages of the early 1970s to focus significant new attention on more expansive use of the sun's power.

In 1970, Dr. Aden Meinel and Marjorie Meinel privately published their book *Power for the People,* which describes the potential development of a national solar power facility in the southwestern United States capable of providing electric power for the entire nation. In 1971, the International Solar Energy Society held its annual meeting at the National Aeronautics and Space Administration (NASA) Goddard Space Flight Center in Greenbelt, Maryland. The meeting was surprisingly well attended, and the society—a normally dormant group of professionals—was clearly alive with the sound and fury of the coming solar era. Papers were presented on all aspects of solar energy conversion at the two-day meeting; and afterward, tours were made of several solar-heated houses in nearby Washington, D.C.

A few weeks after the meeting, President Richard Nixon mentioned solar energy in a summer energy message to the American people—even though he didn't accompany the allusion with a suggestion for increased federal funding for solar projects. However, through a special program of the National Science Foundation—the RANN (Research Applied to National Needs) program—solar research money increased from virtually zero in the 1960s (except, of course, for the expensive solar cells for use in the space program) to 13 million dollars in the 1974 federal budget. This figure seems minuscule in comparison to the hundreds of millions that the Atomic Energy Commission is given each year to spend on development of nuclear reactors; but compared to the lack of funding they were accustomed to, it was a godsend to advocates of solar power.

In the early 1970s several major corporations—principally aerospace industries—began investigating solar power for earthbound applications, including electric power plants. The aerospace industries, smarting from the demise of the Vietnam War technology profits, the winding down of the space program, and the death of the supersonic transport (SST), were prime candidates for new sources of federal support. Both McDonnell Douglas and the Aerospace Corporation of

Los Angeles applied to the National Science Foundation (NSF) for grants to investigate the construction of solar power plants.

Non-aerospace industries however showed a singular lack of interest. While some oil companies—including Texaco and Standard Oil of New Jersey—have begun devoting funds to investigate the potential of solar power, these research programs have remained small efforts of no great significance.

Interest in solar power in the academic-scientific community has been much more visible than in the business community. Meeting after technical meeting in 1972 and 1973 drew record crowds of engineers, scientists, and scholars to hear of the progress in solar research, minimal though it was. A special committee of solar scientists, organized by NASA and the National Science Foundation, reported in 1973 that by the year 2020 solar power could provide 35 percent of the nation's heating and cooling needs for buildings, 20 percent of the country's electricity, and 30 percent of its gaseous fuels. The total federally financed research program would cost $3.5 billion, but would include constructing a number of demonstration plants, including electric power plants; organic conversion plants (algae to fuel); and wind generators.†

Solar Electricity

The potential for supplying electricity for Earth uses through the application of solar-electric cells is largely undeveloped, even though modern solar cells—which were developed by Bell Research Laboratories in the 1950s—are produced in quantities for specialized use.

The tiny 2-centimeter-by-2-centimeter solar cells, made of various materials, are linked together in panels, complex mosaics of cells that can supply thousands of watts or kilowatts of electric power. The most common material used in the cells is silicon, the Earth's second most abundant element—found in sandy beaches and deserts everywhere. Today, the silicon cells are made for powering space vehicles and satellites. More than 600 American and 400 Soviet space machines had been orbited by 1973—all powered by solar energy.‡ NASA's 1973

† These are described in Chapter 8.

‡ As the National Aeronautics and Space Administration learned with the failure of 2 huge solar panels in the launch of the Skylab orbiting space station experiment in May 1973, solar panels are the most important elements of space power. The solar panels were not the cause of the failure. Within 63 seconds after blastoff of the $294 million research station, a critical heat shield designed to insulate the interior of the craft from the sun's heat and to protect the space labora-

Skylab space station boasted the largest solar panels ever launched into space. Had all the panels been properly deployed, more than 500,000 silicon cells, covering 2,500 square feet of panel area, would have provided electricity for the mission.[50]

With the exception of fuel cells, the only other alternative power source for spacecraft is heat-emitting radioactive isotopes contained in small power packages. These nuclear power units use the heat from radioactive decay to operate miniature electric power plants. Unlike nuclear reactors on Earth, which have some mechanisms to mitigate the effects of deadly accidents, nuclear power sources on satellites and space vehicles have no such contingency provisions, and rely solely on the ability of the capsule itself to contain the radioactive materials—including plutonium, the most lethal poison known.*

The pervasive influence of the nuclear power establishment is reflected in the government funding for solar research and development vs. that for nuclear-powered spacecraft. Even though nuclear power plants have seen service in only 10 American unmanned spacecraft (compared to solar power systems in more than 600), and delivered only one sixtieth the power, the federal government has spent more than 3 times the research funds for the dangerous nuclear systems as it has for solar power in space.

Between 1966 and 1973, NASA spent about $41 million to develop solar power for space applications. In that same period, the agency spent more than $120 million for space nuclear systems. This reflects only a partial figure, since Atomic Energy Commission costs of developing the small nuclear systems frequently matched NASA's share, but are not included in this sum.[51]

Several companies were building the tiny silicon cells for American spacecraft during the 1960s, but by 1973, only 2 American manufacturers were left to serve the dwindling market of only a few million dollars a year: Heliotek-Spectrolab Division of Textron Inc., and Cen-

---

tory from micrometeor bombardment was inexplicably ripped off the craft's exterior. This triggered a number of other failures, including the loss of 2 solar panels designed to supply 1,900 watts of electricity because they were unable to unfold from an accordionlike array attached to the space station. Four other solar panels were isolated from the heat shield accident, and opened properly. These remaining solar cell arrays supplied 9,700 watts of electricity to the Skylab station.

* Several incidents have occurred in which space nuclear power systems have been lost in re-entering Earth's atmosphere. Two notorious incidents involving American spacecraft include one satellite in the 1960s, which burned up during re-entry over Italy, scattering plutonium in the upper atmosphere; and an Apollo re-entry vehicle in 1970, which aborted, carrying a plutonium power package to a Pacific grave. Fortunately, the package did not strike land, spilling its contents.

tralab Division of Globe-Union, Inc., both located in southern California.

From the 3 to 4 percent efficiencies of the embryonic period of the 1950s, the efficiency of converting sunlight to electricity in the cells increased to a level of 10 to 11 percent in the mid-1960s and early 1970s.† That level remained essentially unchanged until 1972, when a significantly improved silicon cell was produced at the Communications Satellite (COMSAT) Corporation laboratories in Clarksburg, Maryland (described in detail in the following pages).

For space applications, the 2 remaining American manufacturers utilize what has been described as precision "jeweler's techniques" in fabricating the solar cells. Using a special pure grade of single-crystal silicon made especially for electric cell production, the 2 companies turn out thousands of tiny cells per year. The cells are thin wafers of silicon only a few thousandths of an inch thick and 1½ inches square. The cottage industry nature of solar cell manufacture and the costs of pure silicon combine to boost the cost of electricity from the cells to astronomical levels.

The silicon cells alone cost about $100,000 for every kilowatt of power they produce in space—and this does not include costs of the special satellite power system or launching, which together raise the cost of space electric power to $200,000 per kilowatt and upward (some space solar cell systems have cost as much as $800,000 per kilowatt). Assuming that panels of silicon solar cells made for NASA's satellites could be arrayed in sunny locations in the United States, the costs would be enormous compared with conventional power plants. Just the cost of the cells—at $100,000 per kilowatt of the plant's electrical capacity—is far more than the cost of an entire fossil-fueled electric power plant today (around $300 per kilowatt of installed capacity).

This apparently crippling factor still has not daunted scientists and commercial interests who are persuaded that technical discoveries of the 1970s combined with mass-production techniques could lower the costs of solar power from electric cells to a level competitive with con-

† All the energy conversion efficiencies listed in this chapter are based on the efficiency of conversion of sunlight to electricity in a space environment. In the space environment (air mass: 0), more light falls on the cell, and the average conversion efficiency is 10 to 11 percent. On Earth (air mass: one), the same cell will convert the same amount of sunlight to the same amount of electricity—but at a seemingly higher efficiency of about 13 percent. This is due to the fact that, since the energy content of the sunlight at the Earth's surface is less than in space, the cell's conversion efficiency appears to be greater.

ventional electric power—and without the environmental costs of conventional plants.

As noted above, silicon cell energy conversion efficiencies remained in the 10 to 11 percent range from the 1960s to 1972—despite technical calculations suggesting potential efficiencies of up to 22 percent. The explanation for this lack of progress in developing better cells lies partially in technology; but politics must share an equal responsibility.

Technology

It is useful to examine the way in which the silicon cell converts sunlight to electricity, in order to evaluate the possibilities of producing more efficient and less expensive silicon cells. The silicon solar cell is able to convert electricity by the unique characteristics of silicon itself, which is a semiconductor material; that is, it is both an electrical insulator and a conductor. Semiconductor materials are in widespread use outside the solar cell industry (which is one of the smallest users of silicon and other semiconductor substances), and their availability has made possible the familiar transistor, which replaced cumbersome vacuum tubes in electronic equipment and revolutionized modern electronic science.

Silicon produced for solar cells is "grown" in large single crystals, and the waferlike thin strips of silicon that form the basis for the cell are painstakingly cut from the crystal ingots with precision diamond saws. The silicon strips are then coated with other materials, such as boron, to produce a positive electrical layer, which interacts with the underlying negatively charged silicon layer. This critical sandwich layer, or "p-n (positive-negative) junction," is the key to the production of electricity, and gives the silicon cell its "photovoltaic" property.

The word "photovoltaic" is derived from "photo," meaning light, and "voltaic," from Alessandro Volta, a nineteenth-century pioneer in the discovery of electricity—voltaic indicating, then, electricity. When the elementary energy particles of sunlight, the photons, strike the silicon cell, they are converted to electrons in the p-n junction. The "p," or positive, layer accepts the electrons, and the "n," or negative, layer rejects them, thus setting into motion a flow of direct-current (DC) electricity. The current is diverted to electrical wires by an electrical conductor imbedded in the surface layer of the cell. In space applications, a grid of silver, which is an excellent conductor of electricity, is used for this purpose. In a typical 2-centimeter-by-2-centimeter-square cell in use on a contemporary spacecraft, 6 tiny silver fingers embedded in the cell pick up the cell's electricity. It is then channeled to storage batteries and to power equipment.

The conventional 10 to 11 percent efficient silicon cells have a p-n junction relatively deep (even though it is only 4,000 angstroms, or about 1.5 hundred thousandths of an inch), which is considered necessary for conducting the electrical current from the cell. This deep junction requires the use of a large quantity of the silicon crystal in the manufacture of each cell—a vital factor in the cell's high cost.

There are also other factors: highly skilled labor involved in individually cutting ingots of silicon; the time-consuming and somewhat primitive practice of individually "growing" and slicing the ingots; the considerable waste—as much as one third of the silicon—inherent in cutting the round silicon ingot slices to square cells; and the high base price of the silicon raw material.

Silicon raw material—ordinary sand—is abundant, but to purify crystalline silicon requires an enormous amount of energy to produce. Dr. Martin Wolf of the University of Pennsylvania noted the impact of these costs at a meeting of solar cell specialists in 1972:

> With the additional experience factor of approximately 1 gramm of silicon being consumed for every solar cell completed, 7,500 kilowatt-hours of electrical energy will have to be delivered from each square meter of a newly installed [power] system before the energy required for the fabrication of its solar array has been recovered.[52]

This means, Wolf explained, that if one were to install a solar panel of silicon cells on the roof of a house in the United States to power the house needs with solar-generated electricity, the electrical power used to manufacture the cells would not be recovered from the collectors for 40 years.

The Violet Cell

This fact, combined with the extreme costs of space-grade silicon cells, motivated Dr. Joseph Lindmayer, director of the Communications Satellite Corporation's (COMSAT's) physics laboratory, to pursue development of a silicon solar cell that would produce more power per gram of silicon while using less silicon to produce—the combination of factors needed to achieve greater power output at less cost.

Lindmayer was in a unique position at the COMSAT laboratories, since the satellite corporation was one of the few users of silicon space cells outside of NASA. COMSAT's satellites all used (and continue to do so) silicon solar cells for supplying power for transmission of information from space to Earth.

Dr. Lindmayer has explained that for years it puzzled him that the silicon solar cell industry had stagnated in the 1960s and 1970s despite the fact that tremendous strides were continuously being made throughout this time in the development of semiconductor materials—particularly silicon—for use in a number of other applications.‡ Lindmayer concluded that the silicon solar cell industry must take advantage of the revolution in its sister technology of silicon semiconductors.

He determined that by using manufacturing techniques developed in the general semiconductor industry, he would probably be able to develop improved solar cells cheaper than those currently available from NASA's suppliers. By May 1972, Lindmayer's research team (which initially included no scientists with previous experience in silicon solar cells) announced that the research goal had been met: A remarkable new solar cell had been produced in their laboratory. The new cell was called the "violet cell," because it converted more sunlight to electricity from the violet/ultraviolet range of the light spectrum than did conventional silicon cells.

The violet cell was designed using the modernized processes of the semiconductor industry, including (1) the use of special coating techniques, which add layers of silicon by vapor deposition in a vacuum tank; (2) the use of a shallow (1,000 angstroms, or about 4 millionths of an inch, deep—compared to four times that depth in the conventional cell) p-n electrical junction, reducing by half the amount of silicon needed in the cell; and (3) the use of a new electrical conductor circuit for the cell. The conventional cells, as noted above, have 6 silver fingers, which conduct electricity from the cell. The violet cell uses a microscopic network of frail silver threads to conduct electricity—a technique utilizing photographically etched microcircuits of silver, as opposed to the bulky fingers of silver on conventional solar cells.

At COMSAT's Clarksburg, Maryland, labs, the new violet cells developed more than 15 percent efficiency in converting sunlight to

‡ An example is the development of sophisticated semiconductor technologies using silicon "chips"—tiny blocks of silicon with etched conductor layers—which make possible miniature computer-calculating machines. By 1973, these electronic calculators sold in the United States for prices ranging from $50 to about $500. The inexpensive calculators provided, for the first time, a desktop minicomputer for use by ordinary individuals, as opposed to previous limitations of the use of such equipment to businesses, laboratories, and individuals who could afford steep prices. One calculator, selling for $400, combined computer sophistication with desktop size—features not possible only 2 decades earlier, when costs were thousands of times higher and equipment bulk necessitated whole rooms for installation rather than the corner of one desk. It should be noted, however, that silicon for solar cells is 100 times more pure than the "metallurgical grade" used in most semiconductor applications, such as the electronic calculators—leading to a present cost differential between the two applications.

electricity, compared to the 11 percent efficiency of conventional cells—an increase of more than one third. Thus, for a small investment (probably not more than $300,000), Dr. Lindmayer's innovative research team succeeded in only 1½ years in halting more than a decade of stagnation in the silicon solar cell industry.*

He is confident that a new era is in store for the solar cell. He points out that the new manufacturing techniques used in the violet cell can lead to tremendous decreases in the manufacturing costs of solar cells especially produced for Earth applications.

Additional new techniques of solar cell manufacturing may provide the economic impetus to break the current cost stranglehold on the technology. The fledgling but vital efforts of Lindmayer's group proved that the way is open for further breakthroughs.

The Promise of Mass Production

Despite the fact that cost considerations are not the only—perhaps not even the major—factor in deciding whether solar or fossil or nuclear fuels should provide future electrical power, there are many who are convinced that, with further research, solar power can nonetheless become as cheap as or cheaper than either fossil or nuclear power. Certainly a key determinant in bringing this into being will be the development of mass-production techniques.

William Cherry, an engineer who has worked with solar cells since the 1950s and is now a NASA official at the Goddard Space Flight Center in Greenbelt, Maryland, has advocated a combination of new techniques to mass-produce solar cells for use in power plants in the United States. He foresees an efficient, highly automated assembly line for the production of the silicon cells, in which one end of the machine is being fed a roll of silicon raw material while from the other emerges a finished solar cell "blanket."[53] Using sophisticated mass-production techniques, solar cells would cost about $0.50 per square foot, or $50 per kilowatt, he has said.[54]

A machine very similar to that proposed by Cherry has been developed for mass-producing "ribbons" of high-quality sapphire for various technological applications. The developers of the sapphire-ribbon machine are engineer Harry E. Labell, Jr., the inventor, and Dr. A. I. "Ed" Mlavsky, director of the Corporate Technology Center of Tyco, Inc., in Waltham, Massachusetts.

* The violet cell technology was sold by COMSAT in 1973 to the Centralab Division of Globe-Union, Inc., one of the two American silicon space solar cell companies. COMSAT and NASA have announced that violet cells will be used on future satellites.

The Tyco mass-production process is based on the principle of capillary action. At a blistering 2,620° C, molten sapphire is drawn through a capillary tube. When it reaches the top of the tube, a specially designed "die" made of sapphire-molybdenum causes the sapphire to conform to its shape—i.e., square, circular, rectangular, etc. Another solid piece of sapphire is attached to the molten die, and then pulled away, bringing with it the molten sapphire from the capillary tube in the shape of the die. Additional sapphire rises in the capillary tube, replacing the sapphire drawn through the die and continuing the process. A sapphire ribbon is produced by drawing the molten sapphire through a specially designed gap, or opening, in the die. The gap can be almost any shape, so that the sapphire ribbon may emerge cylindrical, as a rectangular strip, or in some other desired configuration.

In a 1972 interview, Dr. Mlavsky outlined preliminary research and development work aimed at producing silicon ribbons in the same way that Tyco now produces the sapphire ribbons. The temperature of molten silicon is lower than that of sapphire, which could eliminate some of the problems attached to sapphire mass production. (At 1,420° C, silicon's melting point is more than 1,000° F lower than that of sapphire.)

However, silicon poses other additional difficulties, the principal one being its reactive chemical nature, which causes the dissolution of many elements and the formation of new chemical compounds with others. This creates substantial problems in the choice of a suitable die material that could withstand the rigors of continuous mass production of silicon strips. However, Dr. Mlavsky said that preliminary efforts were quite encouraging, and predicted that the problem could be solved—particularly through the application of an advanced development program and several hundred thousand dollars in financing.

By 1973, prototype machines were producing crystalline silicon ribbons on a small scale, but—as feared—some reaction problems with the die materials were preventing the silicon ribbon from achieving a quality high enough for use in solar cells.

Nonetheless, the Tyco process has raised the hopes of solar power advocates that the automation problems can be solved so that the costs of finished silicon cells will drop dramatically.

Dr. Mlavsky has estimated that, on the basis of future raw silicon costs of $10 per pound, a small factory of only 12 machines could continuously produce 2,400 1-inch-wide ribbons of silicon pure enough for solar cells at a cost of $17.50 per pound of solar cell ribbon. This would bring the cost of silicon per kilowatt of solar electrical power produced to $90. Of course, this represents only about half the cost of the finished solar cell, which Mlavsky conservatively anticipates could be built for $180 per kilowatt utilizing these new techniques.[55]

Dr. Mlavsky believes that the successful development of the Tyco mass production process would make possible the future use of silicon as an energy "storage" material—i.e., as a transportable fuel. Silicon ribbons—suitable for fabrication into solar electric cells—would be manufactured in a location having available sand and electric power (which could come from a solar power plant). After production, the ribbons, or perhaps the completed silicon solar cells, would be shipped to their ultimate destinations for use in converting sunlight to electricity. In a sense, this would be similar to the present-day shipment of gasoline, oil, gas, or the transportation of electricity in power lines. Dr. Mlavsky believes that silicon shipment would be more advantageous economically and environmentally than conventional energy transport.†

A coalition of solar cell scientists from the solar cell industry (with one representative from the government-funded Jet Propulsion Laboratory, Pasadena, California) presented a technical paper on the economics of solar cells in 1972, arguing that the Tyco process could provide solar cell silicon "blanks" (the pure silicon slices on which the additional electric conductor layers are deposited)‡ at a cost of $250 per kilowatt, more than a hundred times cheaper than the costs of cutting silicon from furnace ingots—$30,000 per kilowatt. "The development of a continuous ribbon growth technique is the key necessary for more economical silicon solar cells," they said. From this, they extrapolated that a production line for finished solar cells (as William Cherry suggested) could become a reality, in which all the tasks of making the cell's electric junctions, conductors, and coatings would be automated. They concluded that at 10 percent cell efficiency, the costs of electricity would be about $375 per kilowatt of plant capacity at a sunny location.[57] This would bring the silicon solar cell to commercial realiza-

† "Take Africa, for example," Dr. Mlavsky said. "There is the Sahara Desert as well as an abundant supply of cheap hydroelectric power. The ribbon would be made from sand on site, using hydroelectric power to heat the furnaces (to make the ribbon). . . . It's a completed loop system where silicon is used as an energy storage medium. Silicon is portable. One truckload of silicon (in completed solar cells from ribbons) would generate many, many thousands of kilowatts."[55] Dr. Mlavksy added that this sort of energy storage concept should be compared to the production of hydrogen fuel from solar energy, referred to at length in a subsequent section. His argument is that silicon has more energy content capability and is cheaper to transport than hydrogen fuel.[56]

‡ The Tyco process would produce a continuous ribbon, which would be automatically cut into blanks for the cells—a much simpler process than cutting silicon ingot slices, and one not requiring precision diamond saws.

tion, since today's conventional power plants cost $300 to $600 per installed kilowatt. The additional costs of storing solar power for nighttime use would have to be taken into account, but this cost would probably not double the installed power costs. With mass production of silicon solar cells, solar power could become competitive with nuclear power plants.

Representatives of the Heliotek-Spectrolab Division of Textron, Inc., have attempted for several years to encourage federal and private funding of advanced technologies to improve the performances and lower the costs of silicon solar cells. Heliotek-Spectrolab engineer Eugene Ralph in 1970 and 1971 advocated a solar cell development program that would lower the cost of solar cells for Earth power from $100,000 to less than $15,000 per kilowatt.

Ralph's comprehensive plan would initially consist of redesigning the shape and composition of NASA-quality silicon cells. The first step would be producing the cells—not in squares and rectangles required for satellites, but round, in the shape of the silicon ingots from which they are cut. This would eliminate the waste of energy-intensive silicon, which is discarded as the square cells are cut from the round ingots. This one achievement, coupled with mass production, he said, would be sufficient to produce the projected cost cut from $100,000 to $15,000 per kilowatt of electricity.[58]

Additional savings could be achieved by using mirrors or optical lenses to concentrate more photons of sunlight on the solar cells, making possible more power output from them.* Under Ralph's direction, the Heliotek-Spectrolab laboratory produced special concentrator solar cell arrays, called "egg carton" concentrators, because the solar cells are placed at the bottom of mirror-lined cavities that look like the inner walls of egg cartons. This concentrating device produced 2½ times the electric power output of similar cells without concentrating mirrors.

Ralph believes that much higher concentration ratios will increase the power output, while lowering costs still further. One experimental approach he has developed would increase the power output of a cell equipped with mirrors 125 times above that of the naked solar cell.

"The commercial solar cell cost of $15,000 per kilowatt would then decrease to about $150 per kilowatt," he predicted, "which would not only far exceed the . . . cost goals, but . . . would bring solar energy conversion into a competitive status with conventional fossil fuel power generation stations."[59]

* Some of the problems with mirror concentration are discussed in this chapter in the section on a space solar power system.

Other Types of Solar Cells

There are a number of other solar conversion devices that can be made to convert sunlight directly to electricity. Most of these materials have proven to be quite costly, are made of very rare materials, or are very inefficient. A panel of scientists chaired by Dr. Paul Rappaport, of the Radio Corporation of America (RCA) laboratories in New Jersey, reported to the National Academy of Sciences (NAS) in 1972 that the most promising solar cell technology was that of silicon. The report listed a number of other technologies that showed some interest.[60]

One technological area discussed in the report is the use of semiconductor materials of rare origin (found in Columns III, IV, and V of the Periodic Chart of the Elements). These include gallium arsenide, indium phosphide, gallium phosphide, and aluminum-gallium arsenide. A great deal of research work in the late 1950s and early 1960s was performed to develop efficient solar cells using these materials. In the 1970s, IBM Corporation investigated gallium arsenide technology and produced a few solar cells developing higher than 15 percent energy conversion efficiency.

The NAS panel concluded that the materials showed some promise, justifying about $20 million in government research funding to fully develop the potential of solar cell technology in this area; but it said that the cell would probably be more expensive than silicon in addition to having only limited advantages over silicon. One might add that for large-scale use on Earth, the required raw materials would be exceptionally difficult to provide at all, notwithstanding economics.

Other materials that produce the photovoltaic effect are organic substances and natural materials (products of photosynthesis). Assuming that photosynthesis-produced "natural" photovoltaic materials could be found and developed into solar cells, the outlook still would not be optimistic, the panel concluded. "The maximum efficiency of conversion of solar energy into stored chemical energy via photosynthesis is about 3 percent," they noted, "and to produce heat and electricity would cause further degradation so that less than 1 percent efficiency can be expected."[61] This amount of energy is so minimal compared to other types of solar cells that it can be discounted as a major source of future electric power.

The most promising candidate material for solar cell development outside of abundant silicon are materials such as cadmium sulfide and cadmium telluride (materials found in Columns II and VI of the Periodic Chart of the Elements). Although cadmium is a rare material, a number of solar scientists are hopeful that supplies will be sufficient to afford a lasting future availability for solar cell applications.

The major problem with cadmium sulfide solar cells is low efficiency (5 to 6 percent, compared to 11 percent and higher for silicon cells) and an unfortunate tendency to degrade, or lose power, over a period of time. On the plus side, the cadmium sulfide cells are much lighter than silicon cells, and are made on thin sheets useful for space applications.

The cadmium sulfide solar cell concept is being investigated intensively at the University of Delaware's Institute of Energy Conversion, where, under the direction of Dr. Karl W. Böer, a former German semiconductor scientist with extensive expertise in cadmium technology, an automated assembly line for production of cadmium sulfide cells has been developed for use in power applications, including panels that would utilize the material in solar electric-powered houses.†

Dr. Böer's program was funded in 1972 with a $291,000 grant from the National Science Foundation; almost $40,000 from the Delmarva Power & Light Company; grants from the University of Delaware; and over $150,000 in equipment donated by various business firms in the area. The Institute for Energy Conversion purchased patents for cadmium sulfide solar cell technology from Gould Inc., a Cleveland, Ohio, battery manufacturer, and hired some of the staff from Gould's project, including Frank S. Shirland, a project leader and expert in cadmium sulfide cell fabrication.

In a 1972 interview, Dr. Böer predicted that the current low efficiency—about 5 percent—of the cadmium sulfide cells could be boosted to 8 percent within 5 years, assuming a $5 million development program could be completed. Such cells would have an unspecified lifetime, which he suggested would be more than 10 years.‡ Using mass-production techniques being developed at the Institute of Energy Conversion, he said that the costs of cadmium sulfide cells could be brought down below $5,000 per kilowatt. As for the problem of obtaining sufficient cadmium, a by-product of zinc mines, Dr. Böer said one newly discovered zinc mine in Tennessee would supply most of the material needed to cover all single-family house roofs in the United States. In addition, most of the cadmium would be eventually recycled and used again in solar collectors.

† The Delaware solar houses are discussed in Chapter 7.

‡ This lifetime should be compared with that of silicon, a material not expected to suffer the loss of power output noted in cadmium sulfide. Silicon cells probably have a useful lifetime 3 times or more than that of cadmium sulfide cells.

As for the problem of deterioration—or loss of power output—experienced in previous work on the cadmium sulfide cells, a sophisticated design used in production at the Institute of Energy Conversion seals the cadmium—and other solar cell layers, including a copper layer—inside a plastic sandwich, in which is circulated an inert gas to prevent degradation of the sensitive metals.

Technological Outlook

Even though relatively minuscule amounts of research funds have been devoted to the development of efficient, low-cost photovoltaic solar cells since their initial development in the mid-1950s, tremendous strides have been made on shoestring budgets by researchers in the past few years. Now that national attention is focused on energy problems, however, specialists in the field of photovoltaic conversion are hopeful that sufficient government and private industry funds will now be forthcoming to permit further development of some of the processes outlined in these pages so that better and less costly cells can be mass-produced to make practical and economic solar plants on Earth a reality.

Large Solar Power Plants

The promising achievements made by scientists pursuing the mass-production aspects of solar cell technology have motivated a number of individuals to speculate on the utilization of solar cell power plants for supplying bulk electrical power to United States consumers. Though still some distance in the future, several proposals have captured the attention of the scientific community and the public.

These center around two basic approaches: A NASA technology-based orbiting solar cell power plant beaming power back to Earth; and the less spectacular but more easily achievable approach based on building a solar cell plant on terra firma—probably somewhere in the sunny southwestern United States.

A Solar Power Plant in Space

In 1968, Dr. Peter Glaser, vice president of Arthur D. Little, Inc.—the Cambridge, Massachusetts, thinktank—and head of its Engineering Sciences Division, published an article in *Science* magazine that unleashed something of a furor in the U.S. scientific community.[62]

In the article, Dr. Glaser suggested that not only was solar energy an excellent source of energy for the future production of electricity, but that it could be collected in space with silicon electrical cells and transmitted back to Earth in a microwave beam. The space station would be of enormous size, and the microwave beam transmitting power to Earth would span 22,300 miles. It would supply 10 million kilowatts of electrical power, or the equivalent of 10 large, new electrical generating plants—enough to supply the voracious electrical needs sequently recalled that “what led me to say this was that I realized

Looking back over his decision to publish the paper, Dr. Glaser subsequently recalled that “what led me to say this was that I realized solar energy was dying. Most of the interest was in the small-scale specialized uses, and what I thought was necessary was to be really outrageous. I said to myself, in order to be outrageous, we have to reach far”;[63] and a solar satellite station 22,000 miles in space was, he calculated, pretty far. It was, in any case, far enough to cause considerable consternation among his peers of the science establishment.

Since then, Dr. Glaser has unwaveringly pursued the topic of a satellite solar power station. Whether it ever actually comes into being or not, it has already succeeded in achieving Dr. Glaser’s initial goal: reawakening widespread interest in large-scale solar power for generating electricity.

He has spent the interim years refining various aspects of his proposal—size, the nature of materials to be used, type of rocket transportation necessary to transport the giant solar collector to “geostationary” orbit, etc.—and trying to generate interest in the business community.

The latter he has achieved with remarkable success. In Washington, D.C., in November 1971, he announced before a small congressional seminar on energy problems (a panel of the House Committee on Science and Astronautics) chaired by Representative Mike McCormack (D-Washington), that his company, Arthur D. Little, Inc., had assembled a high-powered industry coalition to begin work on the project.

Appearing with him were representatives of the partner companies: Grumman Aerospace Corporation, which would be responsible for the rocket transportation of the station; Heliotek-Spectrolab Division of Textron, Inc., which would make the solar collector; and Raytheon Corporation, which would manufacture the facilities for converting the solar electricity to microwave power and for transmitting it and collecting it on Earth. By the time they appeared before this small seminar—held appropriately in an unused conference room of the Congressional Joint Committee on Atomic Energy—the 4 companies had spent more than $1 million of their own funds to research the possibility of space solar power.[64]

The major reason for the Washington presentation was to begin acquainting legislators with the feasibility of the project, which would undoubtedly require government funding higher than that of the manned space program.

Why Solar Power in Space?

Dr. Glaser reasoned that to place a solar power plant on Earth would require locating a site where weather conditions were almost perfect, to enable the plant to be exploited to its maximum potential. Even so, the sun can only be counted on during daytime hours, so special energy-storage facilities would be needed at night. He found that one of the best spots in the United States for a solar electric plant would be Arizona. Even Arizona's excellent "sun-fall," however, isn't perfect. In January, for instance, there are an average of 9 days with cloud cover over at least 80 percent of the state, and only 15 days when the cloud cover is 30 percent or less. This atmospheric problem would not preclude the possibility of placing a solar plant in operation, but it would mean that in addition to energy-storage facilities, larger collectors would have to be built, which could cover more land, in order to compensate for the lack of constant sunshine.

The great advantage of the space station, on the other hand, is that sunlight can be captured 24 hours a day during most of the year, so that, without the Earth station's limitations of only about 8 hours a day of solar power input, the satellite would collect more than 10 times the solar energy of its earthbound counterpart (since there is no atmospheric block to solar radiation). In a synchronous orbit, the satellite power station would hover at one point above the Earth's surface. During the summer and winter months, the giant solar collectors would be illuminated around the clock, but during the fall and spring equinoxes, sunlight striking the satellite would be eclipsed by the Earth for just over an hour at midnight. During these twice-yearly equinox periods, temperatures would drop severely on the space solar cells, possibly lowering their performance. For this reason, the station will have to be designed to withstand the temperature fluctuations, as well as other obstacles of the space environment, such as degradation of the sensitive silicon cells by bombardments of micrometeors and the deleterious effects of prolonged exposure to gamma rays and the intense ultraviolet radiation from the sun.

There are ways to alleviate the problems posed by the equinoxes, however, either by putting two satellite power stations in orbit at roughly twice the cost, or by changing the orbit altogether. For example, a satellite station might be placed in a "nonequatorial" orbit

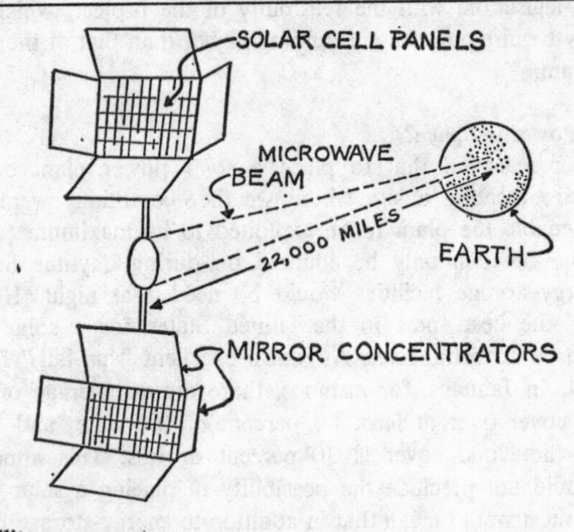

Solar satellite power system

with a seasonal rhythm pattern. The satellite in this orbit would appear to move up and down above the southern horizon, the swing increasing during the equinox periods.

However, due to added problems in collecting the energy transmitted by a satellite in this orbit, this alternative has been shelved, and Dr. Glaser's work has proceeded toward developing the single stationary satellite.

What the System Would Look Like

The basic components of the proposed satellite power system include:

1. A pair of solar collectors—2 4-kilometer square wings made of a thin sheet of thousands of interconnected silicon solar cells, which directly convert sunlight to electricity.

2. A microwave power converter for converting the DC electricity from the solar cells to microwaves, and a transmitter. The one-kilometer square converter-transmitter would be located between the giant solar panel wings.

3. At the other end of the system, on Earth, a microwave "rectenna" with associated equipment for converting the microwaves from space to electricity. The microwave receiver, or rectenna, would in itself take up more than 7 square kilometers.

The silicon solar cell "blanket" would be built by Heliotek-Spectrolab. According to Eugene Ralph, director of research for Heliotek-Spectrolab, the satellite power station presents a unique challenge to his industry, which is now engaged in manufacturing solar cells for NASA's satellites. The satellite itself is at least 10,000 times larger than the largest space power system now designed, including the jinxed Skylab station orbited in May 1973. Ralph points out that a totally new method of manufacturing solar cells must be developed to cope with the number of cells the satellite would require. Such methods of manufacture as Tyco Laboratories' process of constant silicon production would be mandatory to supply the cells at low cost.

The solar cells would have to be of high quality, since the object of the satellite is to catch as much of the available solar energy as possible at high solar cell efficiency. On Earth, the silicon cells might be mass-produced with impure silicon; but for space, the silicon quality must be more than 99 percent pure, as well as being lightweight to trim the costs of getting the satellite into orbit.[65]

The success of the satellite power station depends on two additional technical advances: the development of a much more efficient solar cell (similar to the "violet cell" made by the COMSAT laboratories), and the development of special mirror concentrators to focus additional sunlight on the cells to further upgrade performance.

Heliotek-Spectrolab scientists are confident that solar cells delivering 19 percent energy conversion efficiency (sunlight to electricity) will be available by 1985. These cells would be attached to "blankets" of heavy-duty plastic, which would be rolled up for their space shuttle journey, then unrolled into structural sections in space for attachment to the power station.

Huge mirrors affixed to the sides of the solar cell wings (see page 398) would concentrate sunlight on the cells, requiring the use of less solar cell area to achieve the same power output. Unfortunately, the power output of the solar cell goes down as temperature increases, and the mirrors would tend to increase temperature with the increase in solar heat, so a prime research goal is to try to minimize temperature effects on the cells.

Calculations made by Heliotek-Spectrolab have shown that concentrated sunlight on 18 percent efficient solar cells lowers their output to 11.7 percent efficiency. However, the concentrating mirrors can focus more energy on the cell than is lost in the cell's efficiency. With concentration, the size of the solar blanket can be cut by two thirds.[66]

The second major component in the space satellite solar power proposal is the microwave space converter station. Under the direction of Dr. William Brown, Raytheon Corporation—which first demonstrated microwave power transmission in 1963—has conducted tests of prototype equipment that might be used in the satellite station for microwave conversion to power, and for transmission and recovery. Microwaves are short-wavelength radio waves, which have a wide variety of applications, ranging from power transmission (even helicopters have been powered by microwave beams) to microwave ovens. In the space station, microwave power transmission is about the only known technological means by which such enormous amounts of power can be transmitted across the 22,300-mile distance.

According to Dr. Brown, a special microwave power amplifier-transmitter, called an Amplitron, would be used. The device has been tested extensively by Raytheon, and models exist that can convert and transmit electrical power to microwave beams at 76 percent efficiency. Dr. Brown is confident that with additional technological advances, this figure can be boosted to more than 90 percent transmission efficiency in the space environment.

Transmitting microwave power from space to Earth has two advantages: Since space is a vacuum, no appreciable transmission losses would occur in the system, and short-wavelength microwave energy can pass through the Earth's atmospheric mask without significant loss (long-wavelength electromagnetic waves cannot).

The microwave transmitter—or antenna—capable of handling the 10 million kilowatts of direct-current (DC) electrical power generated by the twin solar blankets might look something like an enormous parabolic mirror, covered with thousands of individual microwave amplifier-transmitters (Amplitrons) aimed at the microwave receiving station on Earth, the rectenna.

As the powerful microwave beam enters the Earth's atmosphere, clouds and rainfall could divert the energy of the beam, diffusing and scattering it outside its projected course. Dr. Vincent Falcone, Jr., of the Air Force Cambridge Research Laboratories, reported in the *Journal of Microwave Power* that microwave beam scattering is important in light of the fact that on average, clouds are present in the path of the beam 20 percent of the time, and rain 6 percent of the time. "When moderate (10 millimeters per hour) and heavy (25 millimeters per hour) rainfall . . . [occur] . . . wavelengths less than 9 centimeters are highly attenuated [scattered]. In fact, 15 to 65 percent of the transmitted energy is lost," he said.[67] Obviously, this aspect of power loss is important in selecting the right frequency to transmit microwave power. The project leaders do not believe that this atmospheric attenuation would be significant, however, and that even in heavy rainfall, no more than a few percent of the energy would be lost by scattering.*

Perhaps a potentially more serious problem with microwave transmission from a space power satellite to Earth is the possibility of "pollution" of the microwave bands of the electromagnetic spectrum. Microwaves are used extensively for transmission of information from point to point on Earth as well as to Earth from existing satellites. The injection of yet more microwaves in the large quantities necessitated by space-to-Earth power transmission might seriously overload existing wavelengths.

The third major component of the system is the actual collecting device—called a rectenna, a contraction of rectifying antenna—located on Earth to receive the microwave power. It would be about 7 kilometers square and, according to project leaders, constructed mainly of "chicken wire." In fact, the rectenna would be the least expensive element in the satellite power system. The rectenna would, they say, be suitable for mounting above an industrial park, where it would collect the microwave power and convert it to direct-current electricity, then to alternating-current electricity for direct use in power grids and the industrial park beneath.

Efficiency of the System

The biggest stumbling block to the successful technological development of the system is the question of weight. Orbiting a satellite station 10,000 times larger than anything in past history would require thousands of individual rocket flights carrying people and materials to a fabrication point in space. At every stage, weight of the essential components (the solar blankets, the microwave transmitters, etc.) would have to be reduced by using ultralight, flimsy plastic materials as well as exotic substances (such as platinum for the power system and samarium cobalt for the microwave transmitters). The heaviest element in the system is the solar blanket, which alone would weigh 114 million pounds (based on current space technology). The project sponsors believe that this could be trimmed to a seventh of this weight by the time the satellite could be orbited—about 30 years from now. Nonetheless, the expenditure of effort, national resources, and money would be enormous.

How much would it cost? The panel of solar scientists established by NASA and the National Science Foundation referred to earlier

* Environmentally, this could be significant, since the microwave radiation would go *somewhere*, possibly having detrimental effects on people or wildlife.

estimates that installation of a single station would cost $20 billion, contingent on the development of the space shuttle. Development of the space shuttle is itself a hotly contested federal program being urged by NASA. The shuttle, a sort of space ferry, would be reusable, unlike today's conventional "throw-away" rockets. To develop the space shuttle would cost $10 billion, and the NASA/NSF report suggested that 30 shuttles would be required to build the first few satellite power stations. If each station cost $20 billion, electricity on Earth would cost 38 mills per kilowatt hour, about 10 times the cost of electricity today.[68]

Dr. Glaser, however, points out that this cost figure (about $2,000 per kilowatt of installed electrical capacity) is about the same as that of the nation's first prototype nuclear power plant, the Shippingport reactor built in Pennsylvania in the 1950s. He says that this cost would be somewhat high, but that it would be reduced substantially with further technological developments.

In addition to the question of costs, another related criticism of the satellite solar power project concerns the amount of energy that would be required to fabricate the satellite station, to build the rockets and the space shuttle, and to put them into space vs. the amount of energy that would ultimately be transmitted back to Earth. Estimates have ranged anywhere from 1 to 20 or more years for energy amortization.

However, in a 1973 interview, the 3 leading members of the satellite solar power station team, Dr. Peter Glaser, John Mockovciak of Grumman Corporation, and John Yerkes of Heliotek-Spectrolab, discounted this as a serious problem. They claimed that less energy would be required in electrical terms to produce and orbit the station than the city of Los Angeles uses in 6 months (Los Angeles' present consumption of electricity is about 1 billion kilowatt-hours per week). They contend that the energy required to manufacture the 30,000 tons of aluminum for the station would be recovered by 100 hours of operation. The amount of energy required to manufacture the silicon solar cells, which require huge inputs of power to break the silicon molecular bond, would be recouped in 3 months of operation of the station. To supply the rocket propellant for the station, probably oxygen/hydrogen, would, they say, require 6 months of operation to break even.†

† The prototype station now under consideration by the team is not as large as the original facility proposed by Dr. Glaser. The current concept for the station, which could be orbited within 30 years, is a 5,000-megawatt, or 5000,000-kilowatt, facility, which consists of 2 solar collectors about 5 kilometers square in space, and a 7-kilometer-square land-based microwave receiver, or rectenna.

In addition, they do not believe that rare resources would be used in great quantity for the station; e.g., only 2 percent of the U.S. supply of platinum would be tied up in future development of space power stations.[69]

Dr. Martin Wolf, of the University of Pennsylvania, estimated in 1972 that the energy costs of the space station would be amortized in four years.[70]

Environmental Effects of the Solar Satellite Power Station

Persistent criticism of the satellite station has focused on possible dangers of microwave radiation being beamed to the Earth receiving station. Microwave radiation is a relatively unknown environmental pollutant. It was developed during the Second World War for radar, and its use since then has been largely limited to radar and communications and microwave ovens.

Research conducted in the Soviet Union and the United States has shown that there are significant physical dangers to human beings from extended microwave exposure. As microwaves penetrate the body, radiation causes overheating in bodily tissues and an increase in the metabolic rate, which can cause damage to the heart, the respiratory and digestive systems, and the brain.

Research in the Soviet Union on microwave radiation has led to much more stringent standards for exposure than the U.S. government allows.‡ In both Sweden and the Soviet Union, microwave standards for exposure to the general public are set at 1 milliwatt per square centimeter—10 times more conservative than the corresponding U.S. standard. According to Terri Aaronson, writing in the May 1970 issue of *Environment* magazine,

> The Soviets have demonstrated a firm belief in the validity and gravity of their own research on microwave radiation effects be setting a limit of .01 milliwatt per square centimeter. If a person is to be in a microwave field for an entire day, this level is 1,000 times less than current U.S. recommendations.
>
> Because the general public as well as numerous industrial workers are being exposed to a rapidly increasing number of microwave sources, it seems quite clear that more research in the field of microwave effects is indicated.[71]

One of the most persistent critics of Dr. Glaser's orbiting solar power satellite has been another American solar scientist, Dr. Aden

‡ The U. S. Government will engage in a $63 million research program beginning in 1974 to examine microwave effects.

Meinel of the University of Arizona.* Dr. Meinel, in his book *Power for the People,* said in 1970 of Dr. Glaser's satellite that

> The real problem faced by employing a microwave beam carrying currents like 25,000 megawatts is that it would be the ultimate in a death ray should it stray accidentally or be deliberately diverted from the receiving antenna. The possibility of an accident is so high that in our opinion it jeopardizes the entire concept. Any failure in the attitude control could cause the beam to sweep over inhabited areas doing harm that would make the explosion of a nuclear power station look trivial in comparison. One would be faced with a giant microwave oven cooking all people, animals and plants caught by the wandering beam![72]

However, the three project leaders—Dr. Glaser, Mockovciak, and Yerkes—discounted this as a serious problem. In a recent interview, they noted, first of all, that the microwave levels would meet federal microwave radiation standards. But more to the point, they contended that the microwave station could not stray off course, since, if it did attempt to stray, controls would be activated cutting off power altogether.

Solar Plants on Earth

The 1973 report of solar scientists commissioned by NASA and NSF recommended that the government devote a total of $780 million to a long-range research and development program for the commercialization of photovoltaic technology. Their report recommended the construction of a prototype power plant on Earth using photovoltaic cells; the development of photovoltaic technology for the eventual development of small solar cell arrays for house and building rooftops; and the development of Dr. Glaser's solar power stations in space.

The potential of solar energy for conversion to electricity through the photovoltaic process cannot be determined until major research programs aimed at reducing costs are initiated. The NASA/NSF panel report pointed out that

> A large central solar powered plant with energy storage that is capable of delivering electricity at a constant rate, or on demand, must compete with other sources of electricity costing about 9 mills (nine tenths of a cent) per kilowatt hour. Smaller plants, located closer to [consumers of electricity] may compete with power worth up to three times the value indicated due to savings in distribution costs and the higher fuel cost of conventional "intermediate load" plants.[73]

* Dr. Meinel's proposal for a land-based solar power facility is described in the following pages.

A Solar Power Plant Using Photovoltaic Cells

Assuming a suitable development program that would provide inexpensive solar cells for use on Earth, a large solar power plant using the direct photovoltaic energy conversion would include the following components:

1. Long rows of solar cell panels oriented toward the south, for maximum exposure to the sun. The panels might be a few feet wide and several hundred feet long, arranged on their east-to-west axis.
2. Near the rows of cells, a building housing electrical conversion equipment would convert the direct-current (DC) electricity produced by the cells to alternating-current (AC) electricity for transmission on conventional electric power lines.
3. Another subsidiary building would house equipment for storage of the solar-generated electricity for use at night, or during periods of inclement weather, when the plant is not producing electricity.†

NASA engineer William Cherry has described a large plant of this nature, which would utilize ordinary lead-acid batteries for energy storage. To build a solar plant that would serve Washington, D.C., and the surrounding 545 square miles of suburban area, 73 square miles of silicon solar cell panels would be needed—assuming the panels were 7 percent efficient in converting sunlight to electricity.‡ This would be about 1 square mile of solar cell collectors for every 7.5 square miles of electric service area.

The same low-efficiency (7 percent) solar photovoltaic power plant, if placed in a sunnier location than Washington, D.C., would need somewhat less land for more power production. Cherry calculates that the same plant in the southwestern United States would cover 53 square miles.

A one-square-mile power plant of Cherry's design would produce over 210 million kilowatt-hours per year in electricity and would be rated at 60,000 kilowatts of capacity—a small plant compared with today's nuclear and fossil-fueled plants of 1 million kilowatt capacity.

† A solar plant producing electricity by the photovoltaic process would look quite different from conventional power plants, since power is converted directly from sunlight without the need for turbine generators to use heat-generated steam for electricity production. The solar plant would not need to be located near a lake, river, or ocean for cooling water.

‡ Of course, modern cells having twice this efficiency would need half that much land, and the "violet cells" available today would require only about 35 square miles of land to serve the same electric service area.

It would cost about $100 million to build—not including the cost of land, which would be considerably less expensive in the southwestern United States than near any urban area.

If mass-produced, Cherry says, the solar panels might be manufactured for about $0.50 per square foot, or $50 per kilowatt of electrical capacity, for a total of $14 million. Assuming that the solar panels would be replaced twice in 20 years, another $28 million is added to the capital costs. Energy storage in the form of chemical energy would be in an enormous bank of lead-acid batteries similar to those used in telephone exchanges. While lead-acid batteries are not ideal for this purpose, no other advanced battery systems are available for use at the present time. The initial cost of lead-acid batteries for the solar power plant would be about $10 million, including the electricity conversion equipment to transform the cells' direct current to alternating current. Adding 2 changes of batteries over the plant's projected 20 years of operation, the total cost for storage adds up to $30 million. Finally, the costs of preparing the selected 1-square-mile site for the plant would represent another $28 million, Cherry estimates.

It is clear that many of these costs can be eliminated in the future by application of advanced technologies. The development of inexpensive solar cells designed to last more than 20 or 30 years could eliminate the need to replace the solar cell arrays twice during the plant's lifetime.

The same is true of batteries. Current industry and government research programs investigating numerous ways to store energy less expensively than the lead-acid batteries are discussed in a later chapter. The use of better energy storage techniques could sharply reduce the $30 million lifetime storage costs of the plant by eliminating the necessity to replace batteries.

Nonetheless, even with today's technology and 7 percent efficient solar cells, Dr. Cherry observes that—on the basis of future electricity costs of 30 mills ($0.03) per kilowatt hour—the plant would produce $26 million in revenues over a 20-year period in a sunny southwestern United States location. "Subtracting the installation, maintenance and operating costs of $100 million for the square mile leaves about $26 million net income per square mile. . . . This land is then producing a 'crop' which yields about $2,000 per acre per year. Farm land yielding such a net return is considered a premium," he adds.[74]

A solar power plant would not necessarily have a capacity of 60,000 kilowatts. It could be many times larger, assuming land was available in an appropriately sunny area, and assuming the existence of relatively inexpensive methods to transport electricity or possibly an electricity-generated fuel such as hydrogen. On the other hand, given the existence of inexpensive solar cells, small power plants using solar cells might be utilized for community power in various areas. There is no reason why solar cells could not be used in certain locations where raw land is not available, but where land already in use for other purposes could also be covered with solar collecting panels. For example, there are more than 15,000 square miles of paved roads in the United States. Perhaps special solar panel "roofs" could be built above some highways, providing a dual use of the land.

Environmental Aspects

The 1973 NASA/NSF solar power report reflected serious concern over the ability of solar power plants to match the costs of conventional power plants. It must be kept in mind, however, that strict dollar-for-dollar cost competitiveness is not necessarily the overriding objective of large-scale solar technology, since there are many other economic comparisons between solar and conventional power plants that are not readily apparent.

For instance, environmental effects from power plants using solar cells would be near zero, since the plant would create no air or water pollution; and thermal pollution would likewise be nonexistent at the site, since there would be no steam turbine generators in the plant to require cooling water. The photovoltaic solar power plant is not subject to the same thermodynamic laws as conventional steam plants. It uses a process of direct energy conversion that does not involve the exchange of heat from one source to another in the solar power plant, as occurs in a conventional steam-electric plant. However, the thermodynamics of the Earth's heat engine must be understood in regard to the solar power plant. Energy is neither created nor destroyed, and the photovoltaic power plant cannot create energy out of the ether. The plant would convert some of the available sunlight to electricity; and waste heat, while not appearing at the site of the solar power plant, would reappear at the point of ultimate use—the house, building, or electric vehicle (in addition to energy losses in transmission, assuming electricity is the power source). If the solar plant were large enough to cover an enormous area of the Earth's surface, it might affect the albedo—or reflectance—of the Earth's surface. In other words, the solar cells would absorb more heat than the surface of the Earth had absorbed before the plant was built. This could trigger local changes in the weather, perhaps bringing more precipitation to the area, hence more clouds—which might upset the operation of the plant, since less sunlight would then be available for the solar electrical cells to convert to electricity.*

The major environmental costs occur during the processing of solar cells. At today's rate of power consumption for the production of silicon cells, 7,500 kilowatt-hours of electricity would be required to produce the solar cells to cover 1 square meter of a solar panel. Thus, the electricity needed to produce the cells might require the operation of a great number of electrical power plants just to produce the power needed to purify the silicon for a single solar power plant. Clearly, research must be quickly initiated to determine whether or not cells can be produced that use only a fraction of the silicon in use in the space-quality silicon cells manufactured today.

Another environmental cost frequently cited is the use of land for a solar power plant. However, a comparison of land needs of solar plants and conventional power plants shows that the solar facilities still have the lesser land impact of the two. Over the lifetime of a conventional plant, fuel needs have to be met by the extraction of either fossil fuels or uranium from the earth, requiring mining and the subsequent despoliation of large land areas. In a solar facility, however, all land that will eventually be used is part of the initial construction.

For example, at the 10 percent conversion efficiency of today's silicon solar cells, a solar electric plant would need 40 square miles to equal the eventual several thousand megawatts of power output of the abominably filthy Four Corners plant in New Mexico—less land than will be strip mined to furnish the coal for the plant, and without the tens of thousands of tons of air pollutants that will be released to the environment over the coal plant's lifetime.

The Meinels' Proposal for Desert Solar Power

In 1970, 2 University of Arizona scientists attracted national attention with an ambitious proposal to turn more than 5,000 square miles of the southwestern United States desert into what they called a "national solar power farm" capable of supplying all U.S. electricity needs in the twenty-first century.

The two scientists, Dr. Aden Meinel, director of the university's Optical Sciences Center, and his wife, Marjorie, an astronomer, devoted several years of study to historic approaches to the use of solar energy for large-scale power production. They summarized their conclusions—

* This is discussed in greater detail in the section on the Meinels' proposal for desert solar power.

along with their own plan for electricity production—in their privately published book *Power for the People*.[75]

The "national solar power farm" proposed by the Meinels is based on the concept of concentrating solar energy with mirrors to heat a fluid, which would in turn provide the heat to produce the steam to power a conventional turbine electric power generator. Aside from its use of solar energy as a heat source (and related heat storage equipment), the power plant itself would resemble a conventional unit powered with fossil or nuclear fuels.

Their research into the history of solar power efforts convinced them that the primary obstacle in earlier experiments had been the inability to reach operating temperatures high enough to produce the heat needed to generate electric power.

The Meinels' research into relative costs persuaded them that using heat concentrators to produce steam (having a temperature of 1,000° F) was not only more efficient (reaching 30 percent efficiency) than converting solar energy to electricity with solar cells, but that it would also be more expensive. Reaching the unheard-of level of 30 percent operating efficiency—3 times that of solar electric cells—seems audacious in comparison to the fledgling efforts of previous pioneers in this sector of the solar field. The nineteenth- and early twentieth-century solar engines probably never exceeded more than a few percent efficiency. Not only were the boilers and engines inefficient, but the actual collection of solar energy had never been high enough to fuel the boiler with heat approaching a fraction of the Meinels' anticipated 1,000° F —a temperature that seems more designed for a solar furnace than a power plant!

To achieve the twin goals of concentrating sufficient solar heat *and* delivering it to the steam turbine of an electrical power plant, the Meinels suggested the adaptation of 2 modern technologies to the problem. To insure a high collection of sunlight, they proposed the use of special coatings on the fluid-carrying pipes of their solar collectors, which are called selective surfaces. In addition, special antireflection coatings would be applied to the glass tubes surrounding the solar absorber pipes. A similar application of such materials is the coating of lenses (binoculars, cameras, etc.) with special chemical films to decrease light reflection. The principles of such coatings have been known for more than a century, but their first successful development occurred during the Second World War, when Great Britain, Germany, and the United States coated military optical equipment with single antireflection films of chemicals such as magnesium fluoride.

Thin films of this sort are also common in nature. Peacock feathers and oyster shells (as well as oil slicks) have an iridescent beauty due to the presence of thin optical interference layers, which affect the reflection of light.

The importance of the optical thin films in solar energy applications is found in the fact that chemical depositing techniques have been developed to increase the penetration of short-wavelength, high-energy solar radiation on a collecting surface, while preventing the escape of long-wavelength infrared heat energy from the collector surface.

Since the pioneering days of the 1940s, optical thin-film coatings have progressed significantly. While the early films were in single layers, as many as 100 layers capable of performing specialized optical tasks are possible today. As Dr. Aden Meinel told an energy panel of the U. S. House of Representatives' Committee on Science and Astronautics, "we want a surface that is 'black' to sunlight but looks like a perfect mirror in the infrared. . . . These coatings are only one one-hundred thousandths inch thick and are deposited within a vacuum tank." In his Tucson laboratory, multilayer coatings have been produced from special silicon materials—the same abundant raw material used in the fabrication of silicon solar electric cells.†

This coating "is ordinary silicon," he stressed, adding that:

> Silicon is opaque to sunlight but quite transparent in the infrared. If one places a highly reflecting metal coating between the steel substrate (the plate on which the coating is applied) and the silicon, sunlight never penetrates to it because the sunlight is absorbed in the silicon. In the infrared both the silicon and the nonreflecting layers added to the silicon to improve collection efficiency are invisible. As a result, the surface attempts to emit as though it were a highly reflective metal, and since bright metallic coatings have very low infrared emissivities, the surface traps the heat.‡

Such specialized coatings as Dr. Meinel describes must be inexpensive in order to be useful in solar applications. Recently developed techniques have in some cases brought costs down from hundreds of dollars for coating a single square inch of material to only a few dollars for coating an entire square yard!

In 1972 and 1973, the Meinels' research team at the University of Arizona—working with Itek Corporation and Helio Associates, a small business they created to exploit the solar power concept—refined

† Some of his coatings are only 25 atoms thick!

‡ An example of this may be found when we enter an automobile in the summertime and touch the shiny metal surface of a seatbelt clasp that has been lying exposed to the sun's rays. This specular metal surface traps the solar heat without emitting it—leaving the buckle almost too hot to touch.

special silicon coatings and used them to cover the collecting surfaces of a small prototype model of the proposed solar power plant. A thin silicon layer only one fourth of one-thousandth inch thick is used to absorb the solar energy, and several other optical coatings, made of silicon nitride and silicon dioxide (quartz), were used to improve the solar energy absorption. This increased the normal absorption of solar rays by silicon from 70 percent to 90 percent with the additional antireflection layers. To prevent the reradiation of collected energy (in the form of infrared energy), a metallic mirror surface was placed under the silicon optical layers.

The mirror layer traps enough heat in the intense Arizona sunlight to raise the temperature of the solar collector to 1,600° F!

Problems

The optical coatings produced at the University of Arizona* have demonstrated their capability of providing high enough temperatures to operate the solar power plant, but one potential stumbling block remains: How long will the coatings last? No one seems to know, although it is certain that they must withstand years of desert sun and heat without significant deterioration. However, Dr. Meinel's group seems confident that the coatings will hold up for their projected 30 or 40 years of operation.

The second major technological problem that the Meinel team has been seeking to conquer is how the intense solar heat captured in the collector can be transported, stored, and used to make steam for an electrical turbine generator.

Their research program has investigated the use of various fluids and gases contained in the optically coated pipes to move and store the solar heat for use in a power plant. Leading candidates are chemicals called molten (sodium) salts, chosen because there is a wide temperature range between their melting and boiling points. Such salts are called "eutectic," which means that they melt when absorbing heat (in this instance, from the solar collector pipe) and release heat as they solidify. Unlike water, which turns to steam at 212° F, the molten salts can be used to transport and store solar heat at the much higher temperatures needed in the solar power plant.

A potential problem might be excessive corrosion caused by certain molten salts and metals. These materials react with most forms of steel that might be used in the storage tanks, and research work is

* The University of Arizona scientist in charge of the selective surface work is the developer of the coatings, Dr. Bernard A. Seraphin.

under way to identify metal alloys that do not react and will allow long-term usage without deterioration.

Interestingly, some of the molten salt materials being considered for heat transfer and storage are identical to molten salts being considered for use in the liquid metal nuclear breeder reactor. In a memorandum to the National Science Foundation in 1972, Dr. Aden Meinel pointed out that "liquid metals have a bad reputation because of their contribution to the total LMFBR (breeder reactor) problem. In the case of solar energy systems we need not consider nuclear cross-sections of contaminants, radioactive handling, or catastrophic risks associated . . . with . . . the coolant metal."[76] The problems experienced in breeder reactors with this heat transfer material would not be as severe in the Meinel power plant, although the explosive interaction between liquid sodium and water must still be taken into account. In any case, Dr. Meinel's team has been able to draw on much of the AEC-financed research on breeder reactors without concomitant concern for the deadly aspects of radioactivity.

In addition to molten salts and liquid metals, Dr. Meinel's group is investigating the possibility of using high-pressure gases such as nitrogen in the pipes of the solar collectors to transfer the heat to tanks of molten salt, to be stored for later use in turbine gnerators.

A Closer Look at the System

The Meinels' book *Power for the People* foresaw a great power facility consisting of 1,000 1-million-kilowatt solar power plants covering an area of the Southwest about 70 miles square (assuming all the collectors were wedged together in this area, which would probably not be possible). All great projects begin small, however, and the prototype model solar collector built on the roof of the Optical Sciences Center in Tucson provides a starting point for examining the whole system.

The collector pictured here is the prototype "credibility model" built by the Meinel team. It is similar in many aspects to those used by Shuman and Boys in the Egyptian solar irrigation plant of 1913 (described in preceding pages). In fact, the Meinel team drew on the history of the Shuman and Boys design, as well as on the experiences of other solar pioneers of the twentieth century, including Israel's Dr. Harry Tabor, who built small-scale power plants for agricultural uses along the lines of what the Meinels are suggesting.

The prototype model in the above illustration is 10 feet long and consists of a shiny metal parabolic mirror that focuses sunlight onto a glass-enclosed steel pipe coated with the optical selective films. The 1½-inch-diameter steel pipe, which carries the hot gas (air in this

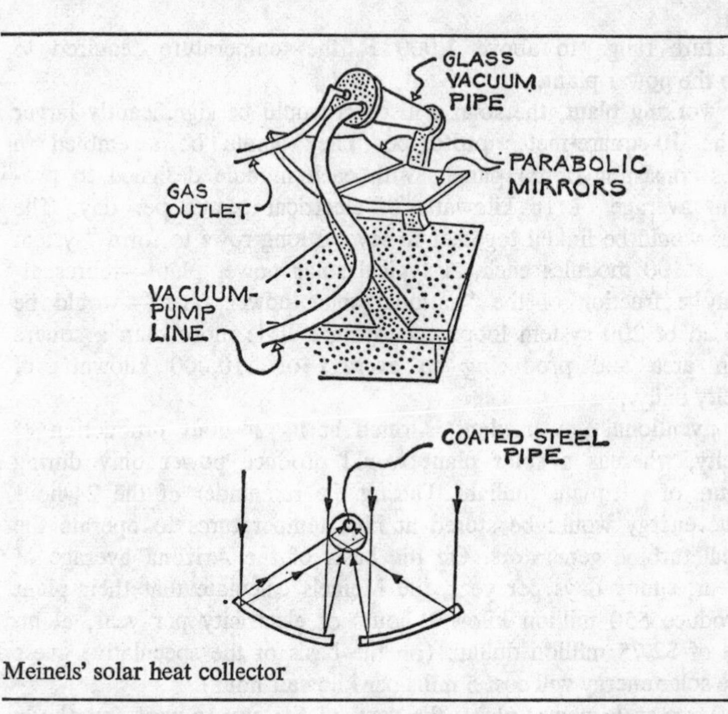

Meinels' solar heat collector

case), is in the middle of a sort of Thermos bottle. The 6-inch-diameter glass tube is coated with an interior thin layer of silver to further reflect the light focused by the parabolic concentrator. Only a small "window," a narrow horizontal clearing in the glass, admits the focused light, which is 10 to 15 times brighter than the sunlight striking the underlying concentrator.

The collector is arranged on an east-to-west axis, and is equipped with a tracking motor to follow the sun's path across the sky. The collector's elaborate design—the optical layers, polished metal, and silver-tinted glass—is calculated to minimize all heat losses by re-radiated infrared energy; in other words, to trap all possible solar heat once it enters the glass tube.

This prototype model was not designed to actually produce heat for running an electrical generator, but to permit scientific measurements of its heat output. As air is circulated through its inner steel-coated pipe, precise temperature measurements are recorded. The prototype collector produced superheated air at temperatures higher than 600° F in early tests conducted in 1972 and 1973. More sophisticated design refinements, including the use of special heat-transfer gases and better selective surface coatings, are expected to raise the temperature range to above 1,000° F, the temperature required to operate the power plant.

In a working plant, the solar collectors would be significantly larger than the 10-square-meter prototype. They would be assembled in modules consisting of 24 panels, with each module designed to produce an average of 16 kilowatts of electrical power per day. The modules would be linked together in several long rows to form "system loops" of 100 modules each. A typical solar power plant—representing only a fraction of the "national solar power farm"—would be comprised of 200 system loops occupying a little more than a square mile in area and producing an average of 310,000 kilowatts of electricity daily.†

A conventional power plant is rated at its 24-hour production of electricity, whereas a solar plant would produce power only during the hours of adequate sunlight. During the remainder of the 24-hour day, the energy would be stored at high temperatures to operate the electrical turbine generators. On the basis of the Arizona average of 300 clear, sunny days per year, the Meinels calculate that their plant will produce 550 million kilowatt-hours of electricity per year, at an income of $2.75 million dollars (on the basis of the speculative guess that the solar energy will cost 5 mills per kilowatt-hour).

In a large-scale power plant, the rows of the east-to-west (southerly oriented) collecting panels—carrying nitrogen gas in their steel pipes—would be linked to centralized massive and well-insulated storage tanks where the heat in the nitrogen would be transferred to liquid sodium in which the heat energy would be stored for nighttime use and for periods of cloudiness or inclement weather. Special heat exchangers would transfer the 1,000° F heat from the storage tanks to the power plant boiler, where it would produce steam to drive the electrical turbine generators.

The solar power farm, located near the Colorado River, would require conventional power plant water-cooling facilities. While this may be viewed as a potential liability, the Meinels believe it could be turned into a great asset in the southwestern United States, where an even greater need than energy is water. In their national solar power facility, they conceive of applying the turbine's waste heat to desalt ocean water, which would be brought to the facility from the Gulf of California through many miles of aqueducts. The aqueducts

† Of course, both daily and seasonal variations would affect the power output of the farm, and the 310,000 kilowatts of averaged solar power would in actuality equal the 120,000 kilowatts of electrical power produced by a conventional power plant burning a fossil fuel.

would pump enough ocean water to the power facility to desalt 50 billion gallons of water per day—enough to supply the needs of 120 million people.

This venture would require the cooperation of the Mexican Government (with which the Meinels have already established some rapport), a step that should be facilitated by the fact that the fresh water would be used to supply parts of Mexico as well as the American Southwest and water-deprived southern California.

Although 1,000 such solar-thermal plants could theoretically be built on an area only 70 miles square, the support facilities, desalination plants, power transmission lines, and connecting roads might entail the use of up to 13,000 square miles of desert. This would represent more than 10 percent of the desert land in the southwestern United States, a fact that is of significant environmental concern.

The Meinels foresee the blossoming of entire new cities.

> The facility of 40 generating stations would have 6,200 employees and dependents and an additional satff of 100 employees at the group headquarters, bringing the total community to 6,700 persons. Any small city with this many people . . . will require . . . secondary persons—to run the schools, markets, shops, gas stations, laundries, movies, restaurants, and the host of other services that constitute a city. We are not certain what the multiplier is to arrive at the net size of the community, but we feel that it will be between 1.5 and 4 persons, so the total population of each of the new cities will be between 17,000 and 35,000.[77]

Interestingly, they also point out that such new accumulations of people might adversely affect the solar power plants by attracting new industries, with the likelihood of accompanying air-polluting activities that would reduce the amount of solar energy reaching the collectors, thus lowering the electrical output of the plant.

Other adverse effects might arise from the operation of the plant, though the Meinels seem confident that they would not occur. One obvious problem is the change in the albedo, or reflectance of the Earth's surface. The addition of thousands of solar thermal collectors to the desert's surface would mean that energy normally reflected from the Earth would be retained in the collectors, changing the heat balance. This change could trigger local climatic effects, perhaps bringing cooler temperatures and more rain to the area. The Meinels tackled this criticism in a February 1972 article in *Physics Today:*

> The solar collectors do not noticeably darken the color of the desert. This is good: Otherwise the extraction of energy in the form of electri-

cal power and the delivery of it elsewhere would cool the desert. One finds that the solar power farm is so close to an exact local thermal balance that painting the iron-work of the collectors black or white makes a difference. . . . Collection of solar energy does, however, increase the net thermal balance of the earth. The only way to balance this term is to "paint" other areas near the power-consuming areas white, to offset the blackness of the solar collectors. Then we could truly have "alabaster cities gleam. . . ."![78]

The exact effects of placing highly efficient solar collectors in the deserts, however, are unknown. Presumably, tests could be conducted on small-scale plants before committing the development of large land areas to giant solar farms.

Another potential major environmental problem recognized by the Meinels would be the by-product of water desalination, the production of brine. If this highly concentrated salt water were simply piped back to the Gulf of California, it would alter the natural chemical balance of the gulf, probably killing many species of aquatic plants and animals. Each gallon of fresh water produced by the desalination plant would also generate a gallon of brine. The Meinels suggest that if the concentrated brine were pumped far enough out in the Gulf of California, it would become sufficiently diluted. This, however, may be too rosy a prospect. Other approaches might include using the brine in a further solar process —pumping it into large shallow ponds where the sun would evaporate the remaining water, leaving a salt by-product, which could be concentrated and sold.‡

Economics

The Meinels suggest that their power plants will eventually produce energy at an estimated cost range of 5 to 10 mills (one-half to 1 cent) per kilowatt hour; i.e., the cost of power averaged over the lifetime of the solar plant would be twice as expensive as the costs of conventional power from fossil fuel plants. Furthermore, they note that this is the cost of wholesale power, not power delivered to the ultimate consumers of electricity, although "doubling the cost of generating power would only add 10 to 20 percent to the cost to the public for power. In the second place," they noted, "[conventional] power costs will rise due to increasing fuel costs."[79]

Since virtually all the costs of producing electricity from solar power

‡ This is already being done in California by several manufacturers of natural sea salt. Consumer demands are growing for this natural product, and the sale of sea salt concentrated by the solar power plants might add to the economics of the plant.

reflect the initial cost of the plant (the fuel is free), it is necessary to consider the capital intensity of a project on the order of the Meinels'.

For a 1,000-megawatt (1-million-kilowatt) plant—one one-thousandth of their projected power facility—the costs of the solar collectors and the energy storage system would be more than $2.2 billion.* The Meinels point out that the cost of the Navajo coal-powered plant in northern Arizona is estimated at $616 million. In addition, they compute the value of conventional fuel saved by uses of the solar plant. In 35 years of use, the solar plant would generate electricity equivalent to the energy content of 333 million barrels of oil, worth $1.7 billion at 1970 market prices. Nonetheless, they believe that federal subsidies will be required to meet the initial costs of building the solar plants, much as they were necessary for the development of nuclear power.[80]

Other Meinel-Type Concepts

Through the National Science Foundation's RANN (Research Applied to National Needs) program, significant federal interest has been generated in the Meinel concept. In 1971 and 1972, the NSF supported the Meinels' work at the University of Arizona with grants totaling about $180,000. With this money to supplement their other resources, the Meinels were able to demonstrate the feasibility of the solar-thermal plant concept.

Largely as a result of the Meinels' active promotion of their ideas, interest in the concept spread to other companies, research foundations, and universities. Several proposals were forwarded to the National Science Foundation for financing of Meinel-type schemes. The three big winners in this new federal grant race emerged as Aerospace Corporation of Los Angeles; Honeywell, Inc. (working with the University of Minnesota); and Westinghouse Electric Corporation (in collaboration with Colorado State University). The research proposal submitted by another aerospace giant, McDonnell Douglas Corporation, was rejected.

The Aerospace Corporation of Los Angeles was granted $125,000 by the NSF in 1972 to evaluate the economic, technical, and operating features of a Meinel-type system, which they labeled with the acronym STEC (*S*olar-to-*T*hermal *E*nergy *C*onversion system)† the Aerospace

* This cost includes $825 million for solar "boosting panels" originally designed by Dr. Harry Tabor.

† The Aerospace Corporation requested to evaluate their own concept, the Meinels', and a design proposed by Honeywell. Given the nature of the request, it came as no surprise that their initial report to the NSF concluded that their own concept "appears more promising now than at the beginning of the contract period. Although current cost projections appear high, the economics of improved design

study focused on the technical and economic problems anticipated in building a 1-million-kilowatt solar-electrical plant in the American Southwest. They used research data developed by themselves and by the Honeywell/University of Minnesota team of solar scientists. While the preliminary study submitted to the NSF in the fall of 1972 showed no major differences in their projected power plants vis-à-vis the Meinels', there were some interesting changes proposed.

For one thing, the study did not consider using cooling water for ocean water desalination, as did the Meinels'. The Aerospace report presumed that water would be available from rivers or aquifers, not necessarily the Gulf of California. To cool the condensers of the electrical turbine generators, cooling towers such as those found on conventional power plants would be used. A potential problem here ruled out the consideration of conventional "wet" towers, which are actually gargantuan open-ended water tanks that discharge hot steam directly to the atmosphere. The hot steam would produce moisture-laden air in the vicinity of the plant, possibly interfering with the amount of sunlight striking the collectors. Another problem would be the use of "dry" cooling towers, which are more expensive than "wet" towers, but discharge hot air—not water—to the atmosphere.

The Aerospace study cautioned that the dry towers might add ecological problems that are unknown at this time and would require further research. The study also pointed to other problem areas, such as the construction and maintenance of the solar collectors in the plant. The collectors would have to be specially fabricated to withstand years of potentially damaging winds. The study attempted a comprehensive analysis of the available sources of equipment and chemicals for the exotic solar facility—including such things as liquid sodium (and other heat transfer materials), optical coatings for the glass tubes, aluminum parabolic concentrators, and an assortment of vacuum pumps and electrical power equipment.

On the basis of their conferences with manufacturers and estimates of the cost of future technologies required for mass-producing the special thin film optical coatings, the Aerospace report concluded that the 1-million-kilowatt STEC plant designed for high-temperature operation (like the Meinels') would cost about $1.6 billion. On the basis of the plant's peak power—its power output in maximum sunlight, about 1,415,000 kilowatts of electrical capacity—the power plant would cost

---

and indirect benefits of STEC power are sufficiently attractive to warrant continued investment in this concept. . . . It is recommended that both Aerospace and Minnesota/Honeywell, and others, continue to pursue the types of system studies and research programs outlined in the Aerospace Task 2 Report."[81]

just under $1,200 per kilowatt of capacity to build. This is about 3 times the cost of a conventional coal plant. This conceptual plant envisioned 176,000 rows of solar collectors costing $875 million.

The Aerospace Corporation found the solar plant concept advocated by Honeywell/the University of Minnesota (working under a $446,000 grant from the NSF) to be a potentially cheaper system than their own—probably costing about $1 billion for the 1-million-kilowatt facility.

The Honeywell/University of Minnesota plant is similar to the Meinel concept, but uses a different means of transferring heat from the solar collectors to the power plant. Instead of heat exchangers, the Honeywell/Minnesota scientists would use a new device called a "heat pipe" to transfer the sun's heat from the collectors. Heat pipes are closed tubes containing a liquid that can "change phase" from liquid to vapor and vice versa. It has no moving parts, and can transfer heat more than 1,000 times as quickly as does copper.

The secret of its operation is capillary action. The liquid within the pipe is in contact with a "wick" of cloth or metal. As heat is added at one end of the heat pipe, the heated liquid inside condenses on the wick and quickly moves—by capillary action—to the other end.‡

In the Honeywell/Minnesota solar power plant, heat pipes are used—instead of steel tubes carrying hot gases—in the center of the solar collectors. The heat pipes transfer the boosted solar heat to other pipes, containing liquid sodium or other molten salt.

The Honeywell/Minnesota solar plant would probably operate at lower temperatures than the Meinels' plant or the Aerospace facility. A 1972 statement of the Minnesota/Honeywell research team explains this:

> The temperature level at which the heat pipes operate will have a strong influence on the operating temperatures in the power plant. If water is employed as the heat pipe medium, then the maximum temperature is limited by the critical temperature of water (374.1° C). . . . The present-day technology of water heat pipes is farther advanced that is the technology of liquid metal heat pipes. For a demonstration system that is to operate in the near term, the water heat pipe appears to be more attractive.[82]

Researchers at Honeywell have also developed a number of promising optical thin films, which would be used for coating the solar

‡ Heat pipes are manufactured by two companies in the United States: Q-Dot Corporation in Texas, and Isothermics Corporation in Clifton, New Jersey. A commercial use of the heat pipe is a commonly available $10 instrument that is sold in supermarkets to help cook meat faster.

absorbers (using ultrathin layers of molybdenum and other metals as the reflecting substances). Coupling this expertise in optics with promising developments in solar heat exchange work under way at the University of Minnesota,* the joint effort has produced exciting promise for the concept of solar power plants.

In 1973, the third major industry-university team entered the solar power plant sweepstakes—a coalition of scientists at Colorado State University and Westinghouse Electric Corporation. The $500,000 federally financed project, announced in April 1973, is under the direction of one of the country's best-known solar engineers, Dr. George Löf of Denver. Dr. Löf, active in the solar field for more than 20 years, lives in a home heated by solar energy and has published some of the most significant research available on the economics of solar energy applications.

Under the NSF grant, the Colorado State/Westinghouse team will analyze work performed by the Meinels,† Aerospace Corporation of Los Angeles, and Honeywell, and will apply computer modeling techniques to calculate the technical and economic feasibility of solar-thermal power plants. According to the NSF, the project will include specialists in solar energy conversion, optical physics, the design of turbines and heat engines, heat transfer, properties of materials, economics, computer sciences, and environmental analysis.

In addition to the $180,000 from the NSF, the Meinels' pioneering efforts have been supported by the University of Arizona, a few electric power companies, and the Itek Corporation—at a level totaling probably $300,000. With this funding, they succeeded in 2 years in kindling nationwide interest—of both the scientific community and the general public—in the conversion of solar energy by concentrating thermal collectors.

On the other hand, the NSF has awarded over 3 times this amount—more than $1 million—to other university and industry research teams to study the Meinels' idea.

Since the publication of the Meinels' book in 1970, the concept of

* The director of the University of Minnesota Department of Mechanical Engineering and head of the Honeywell/Minnesota solar project, Dr. Richard C. Jordan, has long been active in solar energy research. The university team includes several leading scientists in the field of heat transfer, including the internationally known specialist in this field, Dr. Ernst Eckert.

† Dr. Löf has been a persistent critic of the Meinel project. At the 1972 winter meeting of the American Society of Mechanical Engineers, he confronted Dr. Aden Meinel, who had just delivered a paper on the Arizona project, and challenged the validity of Dr. Meinel's claims on the efficiency of the solar collector that he built on the rooftop of the University of Arizona's Optical Sciences Building.

energy "farming" in the sun-drenched southwestern United States has gained a wide following. Within these few years, the interest of many observers in government, the science establishment, and industry has been aroused, and solar power on a large scale has found its first major —if precarious—footing. The powerful chairman of the Committee on Science and Astronautics of the U. S. House of Representatives, Republican George P. Miller of California, summed up the awakened awareness in a 1972 congressional speech. "At the risk of seeming quixotic," he said, "I suspect that the time has come for civilized man to stop taking the sun for granted. . . . What I assert is the spiraling need for the human race to tap the sun's enormous energy and put it to work as a major power supply—soon!"[83]

So far, however, the funds necessary to make large-scale solar power a reality have been slow in coming. The combined spending of industry and government for the desert solar plant concept has been under $3 million—less than 1 percent of the yearly research budget for nuclear energy!

To make the Meinel concept work, more will be required than a few million dollars' worth of computer studies and the construction of a few demonstrator collectors. A major new industry would have to be built, almost from scratch, just to manufacture the solar collectors required for a few 1-million-kilowatt power plants.

The development of a desert power facility might become reality by 1985, about the same time frame proposed for the liquid metal fast breeder reactor (LMFBR) being developed by the Atomic Energy Commission and the American nuclear industry. As we have seen, many of the elements of a solar-thermal power plant use a similar technology —the transfer of heat by molten salts—yet without the pollution by radioactive materials and threats to public safety of the breeder reactor.

Other exciting possibilities for utilizing solar heat to operate large power plants visualize the production of fuel, rather than electricity.

"Storing" the power produced by solar plants as thermal energy in tanks of molten salt to produce electricity on a continual basis is not the only way to approach the problem of a reliable energy supply. Another possibility is the production of the element hydrogen—one of the best-known fuels.

Hydrogen Fuel Production

Hydrogen has long been known as an excellent fuel. It can be readily stored and transported to power homes, aircraft, and other vehicles. In fact, the natural gas in use in the United States during the first few decades of this century—known then as "coal gas," because it was extracted from coal—contained 50 percent hydrogen.

Hydrogen contains more than twice as much energy per pound as gasoline, with virtually no polluting emissions. It is used at the present time in its liquid state—along with oxygen—as fuel for U.S. space rockets. The familiar clouds enveloping the oxygen-hydrogen-fueled rockets at Cape Canaveral are not air pollution in the ordinary sense. They're clouds of steam, or water vapor—hydrogen's only pollutant.

The Meinels have rejected the concept of producing hydrogen for fuel storage in their projected solar plants because it requires 2 energy conversion stages in the solar plant process. The first stage is the production of electric power by the solar plant itself at efficiencies of about 30 percent. In the second step, the electricity produced by the solar plant would be used to "electrolyze" water—i.e., reduce water to its 2 constituent elements, hydrogen and oxygen. Commercially available machines called electrolyzers utilize electricity to break down the water, liberating its hydrogen and oxygen.‡ The oxygen is released, and the hydrogen is stored for later use as a direct fuel.

"Reconversion of a chemical fuel [hydrogen] by burning is not particularly attractive," the Meinels asserted in *Physics Today,* "because it entails a second Carnot loss of thermodynamic efficiency and reduces the total system efficiency by 60 percent. If cheap and long-lived fuel cells were available, this problem would be greatly reduced, but such devices are also still 'around the corner.' "[84]

However, a number of U.S. scientists have proposed concepts using solar energy concentrated with collectors to produce heat or electricity that would in turn be used to make hydrogen. A Michigan consulting engineer, William J. D. Escher, has proposed an ambitious plan to create a vast network of solar concentrating-mirror collectors to produce steam for electricity, then hydrogen fuel from electrolysis of water.* Escher's solar-electricity-hydrogen power plant—called "Helios-Poseidon" for the ancient gods of sun and sea—would be built on floating platforms in the Pacific Ocean. This conceptual approach, unveiled by Dr. Escher at a 1972 meeting of the American Chemical Society, involves the use of solar mirror collectors similar to those built by the venerable solar pioneer Dr. Charles Greeley Abbot in the 1930s, and by Dr. Harry Tabor in the 1950s.

The "Helios-Poseidon" plant would occupy a square area of ocean

‡ Electrolyzers can convert water to hydrogen at 60 to 70 percent efficiency, although laboratory tests indicate that 100 percent efficiency is attainable.

* Escher's ocean plant would be equipped with solar distillation units to feed distilled water to the electrolyzers, which cannot operate with the chemically impure ocean water.

about 4.3 miles on a side. Forty-eight individual "modules"—floating components of the plant mounted on special semisubmerged buoys—would contain the solar collectors, electrolyzers, and underwater hydrogen fuel-storage tanks. The hydrogen storage tanks would be linked together by pipes to feed a central pumping station. There, the gaseous hydrogen would be cooled to the supercold (cryogenic) temperature of —423° F, at which it would become liquid hydrogen, to be shipped in tankers to the United States or to other nations that might purchase it.

Escher believes that a "hydrogen fuel economy" in the United States will follow the current natural gas economy, which now faces extinction unless new gas fuel supplies are developed. He argues that the technology for ocean transport of liquefied natural gas at —260° F in current cryogenic tankers has already demonstrated the technological feasibility of his system. He contends that all that is needed for his cryogenic transport system is a "modification of present technology."[85]

The Helios-Poseidon facility would cost about $1.5 billion to construct, and would produce 670 tons of liquid hydrogen and 5,360 tons of liquid oxygen per day. The power output of the system would be roughly equivalent to a 1-million-kilowatt electric power plant. As Escher observed in his initial 1972 proposal, the cost is comparable to that proposed by the Meinels for the desert facility. "Despite zero 'fuel' and land costs, this translates to a cost-of-energy of the order of 5 times that of today's conventional electrical and cryogenic production means," he said.

In evaluating the system, a number of energy costs should be taken into consideration. Energy losses would occur in the collection of solar energy through parabolic mirrors; in the conversion of solar heat to steam in boilers; in the water electrolysis process; and in cryogenic fuel conversion and transportation.

Despite these losses and the consequent drop in over-all plant efficiency, Escher believes that the $1.5-billion plant would produce $62 million in revenue per year from the sale of hydrogen fuel, which would easily justify the expenses of the project.

Another version of the solar concentrator concept applied to a hydrogen energy storage system has been proposed by a coalition of physicists, architects, and engineers at the University of Houston.

This unique solar thermal power proposal calls for rows of solar-concentrating fresnel lenses—or concentrating mirrors—surrounding an enormous boiler mounted on a 1,500-foot-high tower (higher than the tallest skyscrapers in the United States, including the 1,454-foot-high

Sears Building in Chicago). The boiler might either be part of a conventional steam turbine electrical generator system, or it might be an advanced magnetohydrodynamic (MHD) generator system.‡ Heated by solar rays, the boiler would supply heat for the production of electricity, which would then provide the power for the electrolyzers to generate the hydrogen.

Chief advocates of the project, Dr. Alvin F. Hildebrandt, chairman of the University of Houston Physics Department, and Dr. Gregory M. Haas, a physicist with extensive experience in the Atomic Energy Commission's fusion power program, are particularly interested in developing MHD for the solar project.

They note in their project summary that the key to successful development of any large-scale solar energy conversion project is the minimization of energy losses at all stages of the system. MHD, an advanced technology that produces electricity directly from ionized gases passed through a magnetic field, is much more efficient than conventional steam-turbine electric power systems. MHD can extract 50 percent or more electrical power from the heated gas provided by the solar boiler, whereas the best conventional power plants are able to convert at most 40 percent of the fuel's energy content to electricity.

Additionally, the MHD generator has no moving parts and can be made smaller than a conventional steam-turbine system. Since the electrical generating equipment and hydrogen-oxygen electrolyzer equipment would have to be emplaced atop the 1,500-foot tower, size and weight are important priorities.

The concept envisions not only the use of a massive tower, but also a considerable quantity of land. One plant would require more than 250,000 individual mirror lenses covering a square mile of area. The plant's power output would range from 40,000 kilowatts in the winter to 80,000 kilowatts in the summer, when more sunlight is available.

Drs. Hildebrandt and Haas believe that the over-all efficiency of the plant in converting sunlight to electricity would be about 30 percent, assuming all the heat losses from the solar collectors, the boiler, and the MHD system. The over-all efficiency of the solar-electric-hydrogen storage system, then, might be less than 20 percent.

One of the inspirations for this conceptual system is the solar furnace built in the 1950s and 1960s in the French Pyrenees by Dr. Felix Trombe. "We assume that we can build the mirror system for one tenth of what Trombe built his—$2 a square foot. . . . That would be $25 million for this mile-diameter system," Dr. Haas noted. Other costs would include $15 million for the tower and $40 million for interest

‡ See MHD, in Chapter 4 on fossil fuels.

paid back over the 30-year amortization of the plant. Thus the amortization costs alone would equal the cost of the system's hardware, and bring the total outlay to $80 million—not including the costs of the MHD generator and the electrolysis-hydrogen storage system. Dr. Haas believes that the cost of power from this system—computed as an average of all costs of materials for the plant over an operating lifetime of 30 years—would be about 6 mills per kilowatt-hour of electricity. Of course, if the electricity were used to produce hydrogen, the additional expense would have to be incorporated, which would double the costs. Still, the economics of this unusual solar furnace system seem compatible with economic estimates of other large-scale solar thermal and solar electric power plants.

From this cost summary, Drs. Haas and Hildebrandt conclude that, "consistent with the assumptions made, solar energy will become competitive with the costs of conventional energy production in the near future. The required technology is sufficiently understood to make development of large-scale solar energy conversion possible."[86]

Physicists Norman Ford and Joseph Kane of the University of Massachusetts proposed a related but still different approach in 1971. It called for focusing solar mirrors and lenses on a boiler to produce temperatures high enough to break water down thermally into hydrogen and oxygen. Again, the oxygen would be released and the hydrogen stored as fuel.

They proposed reaching the necessary temperatures by erecting large arrays of inexpensive plastic Fresnel lenses. For a power plant that during sunlight hours would produce 1 million kilowatts of electricity, they calculated that 2 square miles of Fresnel lenses would be needed—arranged on the familiar east-to-west axis common to all solar power plants. On a tower in the center of this array would be mounted the boiler on which the lenses would be focused to produce the required 1,500° C heat. "At this temperature," they said, "some of the water vapor thermally dissociates; in equilibrium about 0.07 percent of the vapor is dissociated into hydrogen and oxygen. If we can selectively pump the hydrogen out as fast as it is produced we will have accomplished our goal."

As for economics, they suggested that if hydrogen were to become a major fuel in the United States, their projected power plant could become a significant factor in producing cheap, marketable hydrogen. They based this projection on estimates of the cost of the power plant, which they put at $33 per square meter of area (with the bulk of this in the boiler and the lenses at $10 each). If hydrogen were sold at gasoline prices—i.e., an equivalent energy content at an equivalent price—the fuel collected in the facility would be worth $10 a year for each square meter of plant area. "Assuming a 10 percent return per year on investment," they concluded, "this means that one might reasonably spend between $30 and $90 per square meter. . . . This compares . . . very favorably with our estimates ($33 per square meter).[87]

The Costs

One of the major problems in the development of a large-scale solar power plant—whether it produces electricity or hydrogen—is the tremendous capital cost, since the initial outlay corresponds to the purchase of both the plant and the lifetime fuel needs of a conventional fossil fuel or nuclear plant. Financing such a facility over the legally prescribed 30-year period of amortization required for utilities means that the interest costs on the initial capital investment will be tremendous—far higher than for cheaper nuclear or fossil-fueled power plants. The Meinels told the House Committee on Science and Astronautics Energy task force in March 1972 that their suggested energy cost for solar power of 5 mills per kilowatt-hour would be doubled if their conceptual power plant were to be amortized over 30 years instead of over a shorter period. "When one studies the relationship between average power cost and amortization time," they said, "one finds that the shorter the time, the lower is the average cost to the consumer. Fifteen years looks much more reasonable than 30 years."[88]

The Aerospace Corporation study of solar thermal power plants also noted that inflation of 8 percent per year would add significantly to the cost of a capital-intensive solar project. They computed all the known costs—based on existing economics and technology—of building such a power plant, and concluded that the solar power would cost 8.4 cents per kilowatt-hour in 1990 (in inflated 1990 dollars).[89]

Increased research and development efforts on the part of government and industry should be devoted to this concept of solar-thermal power, which offers the potential of more power and safer power than the conventional sources. None of the solar thermal power schemes have been funded, with the exception of the desert-based Meinel-type plants. Some interest in these technologies has been expressed by the National Science Foundation, but no money has been allocated for further investigation.

The January 1973 NASA/NSF solar panel suggested the expenditure of $1.13 billion to develop and build land-based solar thermal plants. This is only a fraction of the budget suggested by the powerful Atomic Energy Commission for various forms of nuclear power (not including the past subsidy of billions to give birth to this deadly technology). Even so, as *Science* magazine recently observed, Aerospace Corporation's estimates of the cost of solar thermal power plants "put the cost of electricity . . . in 1990 closer to economic competitiveness with nuclear power."[90] Again, it should be noted that these cost estimates do not take full advantage of the lower costs brought about by mass production of components of the solar systems, or the already internalized (almost zero) environmental costs of the solar plants.

Sea Thermal Power

The basic economic problem with the large solar power concepts is the cost of the collectors required to concentrate the diffuse solar energy. Only through expensive concentrating collectors can a Meinel-type power plant achieve temperatures high enough to generate steam to operate an electrical turbine generator. Even the direct energy conversion solar cells are usually designed with mirrors to further concentrate available sunlight on them, thus boosting both the output of electricity and the costs.

One unique solar power scheme would circumvent the need for expensive collecting mirrors or lenses; instead, this solar power plant would utilize not only the free energy of the sun, but a free solar collector as well—in fact, the largest solar collector on Earth: the sea.

More than 70 percent of the Earth's surface is covered by the oceans; between the tropics of Cancer and Capricorn, more than 90 percent of the area is ocean. If we visualize the ocean as a great solar collector, we find that the surface of the ocean—particularly in the tropics—is daily subjected to intense solar radiation, which raises the temperature of the surface water. At the same time, the sun also cools the ocean; i.e., solar rays melt the ice caps at the North and South Poles, causing great, deep currents of cold water to flow toward the tropics from the polar regions.

Not only does the sun heat and cool the ocean's water through the interplay of natural cycles, but the enormous mass of water in the ocean serves as a storage vessel for the daily solar radiation. In the tropics, a temperature differential of some 40° F exists between the 80° F surface and the 40° F depths a few thousand feet below. These ocean thermal water layers—or thermoclines—remain at remarkably constant temperatures throughout the year.

For almost a century, engineers and scientists have known that a specialized power plant could produce power from this natural oceanic heat engine. As was noted in the first chapter, the laws of thermodynamics codified in the nineteenth century show that a heat engine can be placed between a high-temperature source and a low-temperature source. Some of the heat—which naturally flows from the high-temperature region to the low-temperature region—can be converted to work for power production. The principle of power production from the ocean heat gradients was first discussed by the French physicist J. D'Arsonval, writing in the *Revue Scientifique* in September 1881. The concept became reality when another French scientist, Georges Claude, operated a steam turbine in the 1920s in Cuba using the solar-supplied heat of the tropical ocean.[91]

In the 1920s and 1930s, Claude built and tested sea thermal power plants both on his ship, the *Tunisia,* and on land in Cuba and Africa. In the Matanzas, Cuba, plant, a vacuum pump was employed to bring warm surface water far below atmospheric pressure, causing it to boil. The steam released was passed through a turbine for conversion to power. The remaining steam was cooled in a condenser supplied with cold water brought up from an 1,800-foot depth in a rigidly anchored pipe. Using the Matanzas plant as a prototype, Claude built a larger plant at Abidjan, an Ivory Coast port in West Africa, but he experienced severe equipment costs, since the size of the plant's components was huge (compared to other conventional plants) in order to compensate for the use of low-temperature heat in the system, which requires a large heat-exchange equipment area. The early plants were rated at less than 50 kilowatts, and they did not prove economically competitive.

The French Government was fascinated by the concept of sea thermal plants, and formed a state corporation—*Energie des Mers*—after the Second World War to carry on research in the area. In 1947 a 7,500-kilowatt sea thermal power plant was built in the Mediterranean, shipped to the Gold Coast of Africa, and placed in operation for a few years. This small plant did not prove economic and never developed its full power rating, which prompted the French to abandon the research work—along with their African colonies.[92]

The Andersons' Proposal

A revival of interest in sea thermal power in the United States in recent years has not resulted in the construction of any power plants, but a number of university research teams have applied for federal grants to develop the concept on a large scale, and a father-son engineering team has made extensive calculations of the costs of building a sea thermal plant near the southern coastline of the United States.

The engineers are James H. Anderson and his son, James H. Anderson, Jr., consulting engineers in York, Pennsylvania. The senior Anderson is a noted engineer who designed a revolutionary process for extracting power from geothermal steam reserves. His power system, called "Magmamax," uses a vapor turbine—not entirely unlike the vapor turbine used in the sea thermal power plant—for geothermal power. A prototype unit is in use at Brady Hot Springs, Nevada.

The Andersons propose building a sea thermal power plant rated at 100,000 kilowatts' capacity, which they say would cost about $166 per kilowatt to build, a cost below that of almost any conventional power plant.

In a conventional plant, a highly concentrated source of energy is burned to supply intense heat to a boiler. This use of intense heat—about 1,100° F in a modern coal plant—requires only relatively small surfaces to exchange the heat from the source to the boiler. Since sea thermal power plants—as well as many other solar plants—operate in a temperature range about one third to one fourth that of a conventional plant, a much greater surface area is needed for heat exchange. Even with these much larger components, the plant's over-all efficiency would be limited to 6 to 7 percent—about one sixth that of a modern fossil-fueled electric plant.

Technology

The Andersons' proposed plant would be similar in principle to the plant built by Georges Claude in the 1920s, but it would take advantage of many improvements in technology that have reduced costs and increased efficiency. The principle of the plant's operation may be compared to the operating principles of a refrigerator. In each case, a fluid is circulated to transfer heat in a closed system. In a refrigerator, a refrigerant fluid having a low boiling point is used to absorb the heat energy inside the refrigerated space and the heat is dissipated—or condensed—outside the refrigerator in condenser coils behind the machine.

The mechanics of sea thermal power involve the use of a fluid (which may be the same fluid used as a refrigerant), which is boiled by the ocean's warm surface water. This vaporizes the fluid, which expands and is passed through the power plant's vapor turbine. After work is extracted from the vapor passing through the turbine to produce electrical power, the vapor is condensed back to a liquid in a condenser cooled by the colder ocean water hundreds of feet below the surface.

In the Andersons' conceptual plant, the power-generating equipment—the vapor turbine generator—would be incorporated within an ocean-

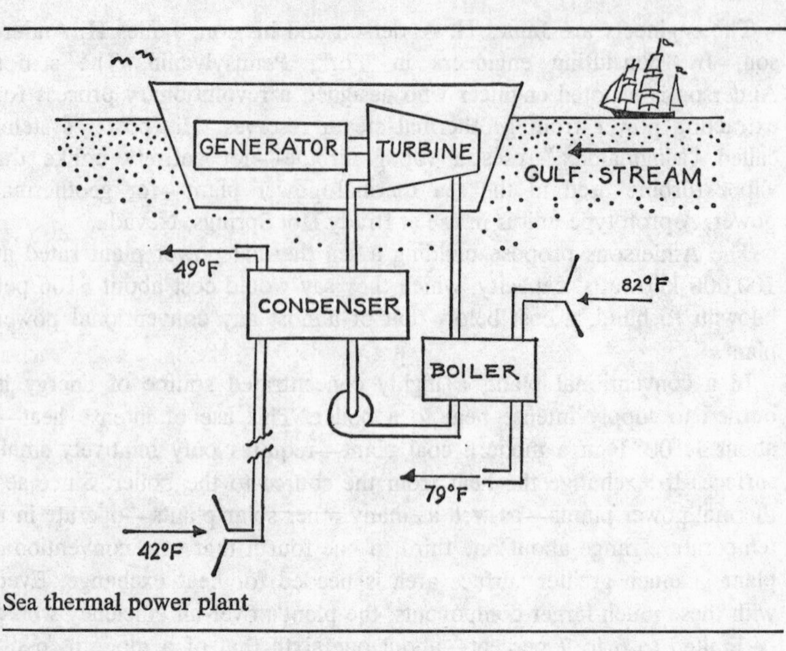

Sea thermal power plant

going barge of advanced design. This combination ship/power plant would be towed to a suitable location and then almost completely submerged. Similar vessels have been built. In fact, recent tests of a similar U.S. naval vessel—appropriately called *Flip*—provide the basis for their own calculations.

Suspended beneath the semisubmerged power plant would be a complex arrangement of pipes to supply hot and cold water to other components: the underwater boiler and condenser.

The Andersons studied a number of refrigerant fluids for use in the plant, and selected propane, the well-known fossil fuel derivative, as an ideal working, or heat transfer, fluid. Propane has several qualities that make it well-suited for use in the plant. First, its pressure at the 78° F temperature of the plant's boiler and the 55° F temperature of the plant's condenser, is well above atmospheric pressure. Second, the use of propane allows for the placement of the underwater boiler and condenser within diving range of a maintenance crew. James Anderson, Jr., told a 1972 meeting of the American Society of Mechanical Engineers that "other fluids [than propane] could be chosen. It is important to note that the depth of the heat exchanger is determined by hydrostatic pressure. Then there is no pressure differential across the heat exchanger surface. Since men will have to inspect the heat exchangers, it is best to place them within working depth of divers."[93]

Using propane as the working fluid of the system—to transfer the heated vapor to the power turbine—the Andersons calculate that the boiler would be situated 278 feet under the ocean's surface, and the condenser would be 154 feet deep. (As noted above, these depths are determined by the water pressure acting on the boiler and condenser.) At these depths, the pressure of the propane approximates water pressure on the surfaces of the heat exchangers of the boiler and condenser. This is largely done for economic reasons, because by leveling the pressure in the system, thin metals can be used for the heat exchangers—limiting size and initial capital costs. They suggest the use of Alclad aluminum, a durable aluminum alloy that does not react to corrosive salt water—an absolute prerequisite for long operating life.

Another advantage of the undersea design is that the plant would be buffered from the effects of storms and the buffeting of waves. Assuming that the plant were located near Florida to tap the heat resources off its coastline, the effects of hurricanes would probably not be felt on the submerged plant, excepting some movement of the submerged upper structure. This places a great premium on the value of a sea thermal plant vis-à-vis proposed barge-mounted nuclear (fission) power plants off the U.S. coastline, which would be subject to years of battering by natural forces—including the lethal hurricanes.

Even though the vapor turbine power system, including the boiler and condenser, can be built using less costly materials than a conventional power plant, significant costs in the sea thermal plant occur in the actual use of water to supply heat for power production. Very large and very lengthy water-carrying pipes must be employed to supply warm surface water to the 278-foot-deep boiler, and cold water from depths of more than 1,000 feet to the plant condenser, located 154 feet beneath the surface.

Nevertheless, the Andersons point out that the total length of the pipes is less than 3,000 feet, which is considerably less than that required in a conventional hydroelectric plant. The high costs occur because—since the plant operates on the diffuse energy of the heated surface water and cold deep water—tremendous quantities must be pumped to the heat exchangers of the boiler and condenser. The Andersons have calculated that the cold-water pump would require 8,500 horsepower, on the basis of a water flow of 10,000 cubic feet of water per second to the plant's condenser. The smaller pump supplying the boiler would require 4,650 horsepower to "fuel" the plant with warm water. Translated into electrical power, this means that almost 9,800 kilowatts of electrical power—about 10 percent of the over-all power capability of the plant—would be required as "parasitic" power to operate the water pumps. Additional power would be required to operate a much smaller pump that circulates the propane in a pipe loop from the boiler to the condenser and back.[94]

How Much Would It Cost?

The Andersons have prepared a number of detailed analyses of the cost of the 100,000-kilowatt plant, suitable for operating in tropic waters. The largest costs in their plant are represented by the boiler and condenser, which would comprise about one third of the over-all $16.6 million price tag on the plant (in 1972 dollars). Other major items are the turbine generator, water pumps, and plant assembly. Structure and assembly of the plant alone would cost $4.2 million. In the basis of a minimum yearly electrical power output of 656 million killowatt-hours, they conclude that the cost of electricity would be 2.85 mills at the plant boundary—not including distribution of electricity to land via underwater cables. They suggest that a sea thermal plant might be designed for dual uses—producing desalted water and electricity, for example. A modified plant of their design would produce 50,000 kilowatts of electrical power and would desalt 800 million gallons of salt water per day at less than $0.04 per gallon of fresh water.

The cost of the electricity alone from the plant is significantly less than the prevailing rates being charged for any forms of conventional power (nuclear, coal, natural gas, or oil-generated electricity), with the exception of some small-scale uses of geothermal energy in the American West. The added benefits of fresh water production could prove a godsend to water-starved southern Florida, for example, where water supply is as pressing a problem as energy.

According to James Anderson, Jr., at this $0.04 per gallon cost, "fresh water could be barged to most major coastal cities cheaper than municipal water systems gather water from surrounding lakes and reservoirs."[95]

Other Proposals

Sea thermal power plant concepts have been proposed by a number of other American scientists and engineers. Smaller versions of a sea thermal plant have been proposed by H. E. Karig, power plant consultant to the Naval Undersea Center in San Diego, California, and engineers P. G. Wybro and C. S. Chen at the University of Hawaii and the University of Virginia, respectively. These concepts were presented to a meeting of the American Society of Mechanical Engineers in 1972, and are designed for such applications as supplying power to undersea research facilities, where scientists might live for months at a time off the natural power of sun and ocean. Karig noted that sea thermal power plants might be placed on the ocean bottom to produce both electrical power and desalted water.

Additionally, such a plant need not be confined to tropical regions. On the basis of thermodynamic theory, a plant could be designed for use near the North or South Pole, using the chilly ocean as a "heat" source, and the still-colder air or polar ice as a heat sink for the plant's condenser.[96]

Engineers Wybro and Chen conceive of a 75-kilowatt sea thermal power plant (less than 1 percent of the Anderson plant's capacity), which could also be used specifically for underwater research habitats (submerged minicities, for example). They point out that no present available power systems are adequate for power in the depths for such applications, and that "the dependence on surface support for power supply and logistics during extreme environmental conditions may prove hazardous for habitats operating at great depths and duration." They concluded from a conceptual study of sea thermal power that a small plant operating on Freon (instead of the Andersons' propane) working fluid would require only 4.7 kilowatts of power to operate the various pumps for the parasitic power of the plant, freeing 75 kilowatts for undersea power supply. In this case, only 6 percent of the system's over-all power production would be lost for internal power. They say that although the system would be costly to install, this "should be offset by the lower operating costs, which makes STE (Sea Thermal Energy) systems appear to be quite attractive."[97]

Similar interest in the potential output and low cost of sea thermal power has been expressed by several university research groups interested in the Anderson-type concept for large-scale power production.

In a 1971 grant request to the National Science Foundation, a team of scientists from various departments of the University of Massachusetts (at Amherst) proposed the development of a large-scale sea thermal power plant based, in part, on the work of the Andersons. Led by engineering professor William E. Heronemus, the Massachusetts team has improved somewhat on the Andersons' concept by additional design features.

Professor Heronemus describes the plant as looking like a giant pipe organ or automobile radiator. This "radiator," or network of heat exchange pipes, would be attached to a semisubmerged shiplike power plant made of reinforced concrete. The over-all shape of the plant would be akin to a "huge underwater kite," he says, about 165 feet long by 275 feet wide.[98]

The plants would be designed to produce more power by using the warm water of the Gulf Stream to circulate water across the boiler surface of the plant. The Gulf Stream, the Atlantic Ocean's great river of warm water that sweeps northward along the East Coast of the United States, has the astronomic force, or flow, of 74 billion pounds of water per second. By using the Gulf Stream's warm water and constant flow, some water for the plant's boiler would be pumped naturally by the Gulf Stream rather than by internal pumps, which would minimize the parasitic power of the system.

Additionally, the constant flow of the Gulf Stream into the water pipes would minimize fouling of the plant's metal parts by marine organisms. The study indicated that common marine organisms would be almost completely ineffectual in attaching themselves to the plant's components. This factor would virtually eliminate otherwise high maintenance costs necessitated by chemical or manual cleaning of the organisms from the plant's surfaces.

Other suggested improvements include incorporating underwater electrical cable with the plant's underwater pipe that brings cold water to the condenser. This system would combine the functions of water pumping, electrical transmission to the shore, and the seabed anchor necessary to secure the plant in the Gulf Stream.

The University of Massachusetts proposal noted that the extraction by sea-thermal power plants of 1 BTU of energy per pound of water passing through the Gulf Stream near Miami would yield the staggering sum of 235 trillion kilowatt-hours of electricity—more than 156 times the 1.5 trillion kilowatt-hours of power consumed in the entire United States in 1970!

According to project director William Heronemus, the sea thermal plants being advocated are in the 100,000-to-200,000-kilowatt power range, at a cost of less than $400 per kilowatt. This figure, though higher than the Andersons' estimate of $166 per kilowatt, may perhaps be more realistic, since it includes the cost of electrical transmission. In any case, the cost figure appears competitive with today's conventional power plants.

Tapping the kinetic, flowing energy of the Gulf Stream was also suggested by the Heronemus team, which proposed analyzing the prospects of placing "underwater windmills" on the ocean floor at some point near the southeastern coast of the United States, to obtain electrical power from the swift current of the Gulf Stream. Water velocities expected to be encountered by the undersea generators on an axis to the Gulf Stream would be on the order of about 5 to 7 feet per second.

One such undersea generator proposed in the report would have 16 slender blades, each 60 feet long. The machine could generate 5,000 kilowatts of electrical power in a water current velocity of 7 feet per second. A larger machine having 4 generators, each carrying 12 240-foot-diameter blades, could generate 24,000 kilowatts of power. The proposal suggests that 12 such machines could be emplaced across the Gulf Stream at one location, and additional installations added at 1-mile intervals for an over-all distance of 350 miles. This vast network could conceivably produce 100 million kilowatts of electrical power. The environmental effects of such an arrangement in the center of the Gulf Stream are unknown.

This conceptual power plant, called a "Florida Current Kinetic Energy Machine," is still on the drawing boards. Little is known about this process of extracting kinetic energy from water (with the exception of tidal power turbines in use at France's Rance River plant), and serious difficulties might be experienced in mooring such a plant to the ocean floor. In addition, the generator's blades, which would be constructed of Fiberglas or aluminum, would be quite costly. Nonetheless, if successfully developed at low energy and environmental costs, the plant could provide a significant source of electrical power.[99]

Another proposal for the development of sea thermal power has been advocated by Dr. Clarence Zener of Carnegie-Mellon University in Pittsburgh, Pennsylvania. Dr. Zener has designed a modular prototype sea thermal plant, which would be only one part of a very large plant. By utilizing the modular concept, he contends that costs can be reduced, eliminating the expensive one-of-a-kind construction of conventional power plants or of sea thermal plants of one specific large scale. In a Carnegie-Mellon proposal for funding to the National Science Foundation, Dr. Zener and his associates stated that the Andersons' calculations of sea thermal power plant costs were too conservative. "On a truly large scale," he contended, "the capital cost can be considerably less than that of a conventional power plant." The study indicated that a sea thermal power plant, if mass-produced in modular sections, would cost about $96 per kilowatt—as opposed to the Andersons' estimate of $166 and Heronemus' figure of $400 (including electrical transmission lines).

The Zener power plant would utilize ammonia as the heat transfer fluid instead of Freon or propane, as discussed in other proposals for sea thermal power. Dr. Zener's group has adopted the Andersons' proposed figure for estimated cost of electricity from the plant—less than 3 mills per kilowatt-hour of electricity.

However, electricity is not visualized as the power output in this concept. Instead, the Carnegie-Mellon research team proposed to the U. S. Department of Commerce in 1973 a long-range research program in modularized sea thermal power for the production of hydrogen fuel. Using the Andersons' figure of 2.85 mills per kilowatt-hour for electricity to feed the hydrogen/oxygen electrolyzers, they concluded that hydrogen fuel from a sea thermal plant would cost 42.5 cents per thousand cubic feet of hydrogen. This is equivalent to $1.28 per million BTUs of hydrogen fuel, which is comparable to the cost of liquefied natural gas imported to the East Coast of the United States from Algeria and the Middle East.[100]

Environmental Effects

The apparently appealing economics of sea thermal power plants should be weighed against a number of factors, including the effect of placing these bulky plants in biologically sensitive, productive tropical waters. If the massive emplacement of sea thermal plants were to produce 235 trillion kilowatt-hours of power from the Gulf Stream, as projected, the temperature of the Gulf Stream would be slightly lowered—about 1 degree, according to the University of Massachusetts proponents. This environmental effect would not only be felt in a local area, where various aquatic organisms would be affected by temperature changes, but such temperature changes might cause a shift in the actual direction and flow of the Gulf Stream.

Additionally, the emplacement of a large number of plants would involve mixing the normally differentiated warm- and cold-water thermoclines. As millions (and possibly billions) of gallons of cold ocean water were brought near the surface, unknown changes might occur in the species existing in the environs. The Andersons have estimated that "the average thermal energy storage of the ocean will increase slightly. A large slab of warm surface water is brought into the plant. This means that the surface layer over a large area is thinner, and solar heating effect will reach colder water. In other words, the absorptivity of the surface layer will increase."[101]

Along with the addition of colder water from the ocean's depths would also occur the addition of rich nutrients—organic material that could sustain other marine species near the surface. The cold, deep ocean water contains these nutrients, which are largely dead marine organisms. Natural upwellings of the nutrient-rich cold water to the surface water above occurs in a number of locations around the world.

Commercial fisheries near Japan, Peru, Guinea, and South Africa are based on these upwellings, which bring nutrients that feed tiny single-celled organisms called phytoplankton and certain forms of algae. This first link in the ocean's food chain in turn sustains other forms of life, upon which fish life feeds.

Since the algae found near the ocean's surface provide between 30 and 70 percent of the Earth's available oxygen, it is vital to know the biological consequences of altering the natural cycles by adding sea thermal power plants on a large scale. The addition of large amounts of cold water would change the salinity of the surface water, since the colder water is richer in salts such as nitrates and phosphates. The upwellings of cold water that supply many of the world's fisheries are localized phenomena, and little is known of the effects of man-made upwellings.

Would the sea thermal plants interfere with the natural biological cycles of the ocean?

Experiments are now under way to answer the question, and one research group is convinced that the environmental changes will be positive. This scientific team, from the Lamont-Doherty Geological Observatory of Columbia University, New York City, has studied the biological effects of ocean upwellings for several years and has built a test facility on the northern coast of St. Croix, U. S. Virgin Islands. The principal scientists in charge of the project, Robert D. Gerard and J. Lamar Woerzel, advocated a combined sea-thermal power system with a marine aquaculture "garden" in 1967 in *Science* magazine.

With funds from the National Science Foundation, their pilot project was initiated on St. Croix in 1970. The system does not generate sea thermal power, but a mile-long, 3½-inch-wide polyethylene plastic pipe delivers cold, nutrient-rich water from a depth of half a mile to 2 concrete pools at the research facility. The pools, which hold 16,000 gallons of water each, are used to raise a number of aquatic animals, including oysters, shrimp, lobsters, and fish.

Experiments planned by the group include the production of power and fresh water by the sea thermal process. The researchers believe that a prototype power plant of 7,000 kilowatts' electrical capacity could be built at a cost of $250 per kilowatt.[102]

Gerard told the Washington *Post* that the project "could have important economic and social benefits. Areas that are today unproductive and uninhabited could be supplied with abundant power, fresh water and high-protein food."[103]

In a January 1973 article in *Physics Today,* Dr. Zener commented on the environmental effects of his proposed sea thermal power system:

Let us now ask what are the qualitative effects of a lowering of the surface temperature of the ocean. The most direct is, of course, a lowering of the tropical atmospheric temperature. A more subtle effect is uncovered when we realize that the lower tropical surface temperature has been caused primarily by a transfer of heat from the surface to the deep layers, rather than by a removal of heat from the ocean. The lowered loss of heat by evaporation and by radiation, the heat input remaining constant, results in a net heat input from the sun. This net heat tropical input must be dissipated outside the tropics, presumably by increased convection currents.

Most people in the world would probably welcome a somewhat warmer ocean outside the tropics. Climatologists in particular will welcome an increased transfer of heat from the tropics to the temperature zone, for they are worried that the present interglacial period may be coming to an abrupt end, and that such an end may be accompanied by a marked drop in mean temperature over a period as short as 100 years.[104]

Unique Applications of the Sea-Thermal Concept

Several unique applications of the sea thermal power-generating technology have been proposed that would not utilize the temperature gradients in the ocean directly to heat and cool the evaporators and condensers of the power plant.

Scientists at the Center for Engineering Research at the University of Hawaii have investigated combining the sea thermal plant concept with geothermal heat energy. Under the aegis of Dr. Howard Harrenstien, director of the center,* several small-scale experiments were conducted in the early 1970s to ascertain the availability of geothermal heat reservoirs off the coast of the island of Hawaii, as well as the possibility of channeling nutrient-rich deep ocean water into ponds on the Hawaiian coast for aquacultural cultivation of food products from marine organisms.

The preliminary studies indicated that the best prospects for meeting Hawaii's future power needs with minimal pollution would be to use cold ocean water to cool the condensers of the sea thermal power plant, and hot geothermal steam to vaporize the heat-transfer fluid within the plant. The advantage of the combined system would be to improve the efficiency of the sea thermal plant by using higher temperatures (geothermal) than the natural gradients provided by the ocean. Instead of

* Dr. Harrenstien is now dean of the School of Engineering at the University of Miami at Coral Gables, Florida, where he is actively pursuing research into various applications of solar energy utilization and the development of hydrogen fuel applications.

the 80° F heat that the warm ocean layers can release to the plant's evaporator, the geothermal reservoirs could provide heat in the vicinity of 180° F—and possibly higher. Thus the plant would operate within a temperature range from 40° F (the cold ocean water) to 180° F, a heat differential of about 140° F. Since the conventional plant has only a 40° F heat differential to utilize in the process of converting power, the additional input of the added geothermal heat could greatly increase the engineering and economic potential of the plant. The conceptual plant would not need to be as large as an equally rated plant located in the tropical ocean, since the greater temperature differential would require less bulky heat-transfer surfaces on the turbine's evaporator and condenser. This translates into lower costs and a more efficient power plant.

The Hawaii scientists describe it as a "geo-marine system." In a proposal to the National Science Foundation requesting funding to further explore the concept and to build a prototype plant, they summarized the results of their early work:

> Studies to date show that thermal and chemical pollution control aspects of the geo-marine system also may yield valuable by-products. Thermal pollution control is inherent in the system. The water can be discharged from the condensers at virtual ambient ocean surface water temperature. The proposed research appears to offer great rewards for even limited success.[105]

The general idea is quite appealing and might be adapted to many coastal areas of the world, where geothermal power sources are available—and where ocean water can be supplied for plant cooling. Especially appealing is the concept of using this combined power generation method in California, where in many areas both geothermal reservoirs and deep Pacific Ocean water are available.

Another promising concept based on the technology developed by Georges Claude and others for sea thermal power production is that of using the temperature gradients found in shallow water ponds. Work conducted in Israel during the 1950s was designed to utilize the energy absorbed by water in highly saline (salt-containing) ponds. The Israeli investigations centered around a curious phenomenon noted by Hungarian scientists some years before. At a certain saline lake in Hungary, temperature increased with depth—exactly opposite to the temperature decrease with depth in the oceans. The Israelis conceived of a scheme whereby chemical salts could be added to shallow ponds, duplicating the process observed in the Hungarian lake. After the sun had heated the ponds, they calculated that a sea-thermal-type power plant could use the hot bottom layer of the water as a heat source for the vapor turbine.

Several small-scale experiments were conducted with small ponds in Israel, to which various salts were added. Dr. Harry Tabor, director of the project, called the efforts encouraging. A temperature differential of about 60° C was recorded between the 92° C (197.6° F) water on the bottom of the pond and the 28° to 32° C (82.4° to 89.6° F) water near the surface. As to the economics of power conversion from a larger installation, Dr. Tabor estimated the following costs:

> A 1-square-kilometer pond operating at 1.5 percent over-all conversion efficiency will, under local conditions, produce 30 million kilowatt-hours of electricity. At 1 cent per kilowatt-hour the annual value is $300,000. Of this, 52 percent is required to cover the capital charges and the operating costs of the station. This leaves $144,000 net value. A generous allowance for the cost of maintaining the pond and its machinery is $50,000 per year. The net value is reduced to $94,000. This is the return of the pond investment if the power cost is to be the same as that for a fuel-fired station. . . . Even if the pond and its associated equipment cost up to 4 times the cost of a "bare" pond, the cost of power produced will still be competitive with power produced from fuel.[106]

Dr. Tabor added that the pond could also be used to provide fresh water, through a combined power and desalination plant. A pond of this size could produce 125 million gallons of water per year, for an additional $80,000 to $100,000 investment. Other uses of the solar pond might include supplying low-grade heat for an associated industry or for nearby buildings.

Looking back over the Israeli solar experiments in many areas, including the ponds of the 1950s—most of which were abandoned as the country switched to fossil fuels—Dr. Tabor says that

> the time has come to reopen the solar pond project. This was terminated not because of proven technical nonfeasibility but on purely local economic considerations [cheap fossil fuel energy].
>
> Now that the economics have changed and the interest is likely to be wider, the expenditure of money needed to determine feasibility may be justified.[107]

Solar pond power and water desalination technology might be an ideal solution for power and water needs in the sunny coastal states in the United States, where an abundance of water may be available. Unlike other large-scale sea thermal projects, this technology would always remain confined to specially suited locales, and the low-grade heat produced would be used in the neighboring areas. It is doubtful that solar pond technology would provide surplus power for transmission over long distances.

Conclusion

The concept of sea thermal power has been experimentally verified, and vapor turbines utilizing external energy inputs similar to those that would be required for sea thermal plants—but operating with higher-grade heat sources, such as geothermal heat—have been built. As is the case with other solar technologies, much work remains to be done to test prototype power plants, fully delineate the environmental impact of projected plants, and precisely determine the costs—both in energy and money—of a sea thermal plant. How much energy would be required to manufacture the enormous surface areas of the sea thermal plants—areas that call for energy-intensive materials such as aluminum? How long would the plants last in the hostile and unpredictable ocean environment? What effect would local climatic changes have on the power production of the plants?

Finding answers to these questions is vital to the future of the sea thermal concept. What is required is a large-scale research effort to fully investigate the technology and effects of the projected plants. Less work has been done in the sea thermal power area than in almost any other solar energy application. Reversing this trend may pave the way toward a significant source of clean energy in the future.

The Chimerical Promise of "Solar Bioconversion"

Since the early 1950s, considerable scientific interest in the United States has been devoted to large-scale production of fuels from green plants—the natural converters of solar energy in the Earth's biological systems.

The object of the research is to circumvent the several-billion-year fermentation period by which nature has provided us with fossil fuels—the decomposed hydrocarbon-rich remains of plants and animals. Scientists hope that such a shortcut might produce high-quality fuel in days rather than in the billions of years now required.

Energy specialists today refer to this general concept as "bioconversion"—the production of "energy crops" by controlling and boosting natural photosynthesis. Such controlled production of fuels from plants represents a most difficult—and possibly insoluble—problem in converting solar energy for use as a power source.

As we have seen, plants convert sunlight through photosynthesis into chemical energy at very low efficiency. The average agricultural crop converts into chemical energy (as food) only 0.1 percent of the original solar energy falling on it. An increasing number of biological scientists and engineers are hopeful that through controlled plant breeding and the industrialization of agriculture in selected areas, this energy conversion ratio can be significantly improved.

The 1973 report of a joint panel of solar scientists from the National Aeronautics and Space Administration and the National Science Foundation suggested that a "moderate portion" of the agricultural lands currently under cultivation in the United States could be used to provide the entire electrical needs of the country. According to the panel,

> the large-scale photosynthetic production of plant material at solar energy conversion efficiencies, e.g., 3 to 5 percent greater than usually observed in ordinary agricultural operations, 0.1 percent, would supply materials that could serve directly as fuels for production of part of man's energy needs or that could be subsequently converted into other forms of higher quality fuels. Concepts to be considered here are directed toward production of large amounts of land-grown products, e.g., trees or grasses, and also large amounts of water-grown products, e.g., algae or water plants.[108]

In addition, they suggested that presently available organic wastes—such as animal and agricultural wastes—and solid wastes from industry, farms, and cities could be converted to fuels.†

Thus the two potential bioconversion approaches are (1) the production of fuels from agricultural crops on a large enough scale to supply the energy needs of a large society and (2) the production of fuel from organic solid wastes already generated in society, including animal, vegetable, and human wastes.

Large-Scale Fuel Production from Agricultural Systems

Research in the use of crops for fuel has centered on laboratory processes that maximize the yield of organic material produced by selected plants. The problem of low efficiency in agriculture (low in the sense of energy conversion) is best illustrated by examining the energy yield of corn, one of America's key agricultural crops. On the average, an acre of corn will produce about a ton of dry organic

† As noted in the first chapter, these wastes may be burned for production of electrical power or turned into fuels such as natural gas—methane—in currently existing processes. Although the present contribution of the actual energy produced from these wastes is small, this process may play an important role in the future economy of the United States.

material per year. If this is burned, about one thousandth of the original solar rays falling on the crop will be realized as usable energy.‡ Corn production, however, is confined to temperate regions, where only a third of the year is devoted to its production. In tropical regions, a greater amount of fuel could be obtained. There are also other plants that can convert solar energy to organic material more efficiently than corn.

A comparison of organic production of selected crops shows a higher rate of biological energy conversion than corn. Some optimum recorded yields of dry organic matter in metric tons per acre are shown here for various plants in favorable location[109]:

Location	Crop	Yield (dry organic matter in metric tons per acre)
Mississippi	water hyacinth	4.5–13.4
Minnesota	maize	9.7
Georgia	salt marsh plants	13
Nova Scotia	Seaweed	13
Congo	tree plantation	14.6
West Germany	reedswamp	18.6
California	nutrient-rich sewage pond algae	20–30
Hawaii	sugarcane	30.4
Java	sugarcane	35.2
Puerto Rico	naplergrass	43

In tropical regions such as Java and Puerto Rico, conversion efficiencies of sugarcane and naplergrass enable them to produce up to 43 times the energy yield as the familiar crop of American midwestern corn. However, it would be difficult to raise such crops on a large scale in the United States for power production, since they are adapted to specific tropical ecosystems.

Agricultural Grains as Fuels

Notwithstanding the low productivity of agricultural crops in the temperate zones, suggestions have been made that crops be converted

‡ At this rate of "fuel production," ordinary agriculture could not begin to meet national energy requirements. According to Dr. Peter Glaser of Arthur D. Little, Inc., solar energy striking the United States is 500 times more than the expected U.S. energy consumption in the year 2000. If corn were converted to fuel at a solar-ray-to-organic-matter conversion efficiency of 0.1 percent, twice the total land area of the United States would be required for fuel needs alone!

into fuels, such as alcohol, which can be distilled from grain. An oil industry committee (no less!) reported to the U. S. Department of the Interior that idle lands in the United States might be turned into energy-producing areas.

They directed their analysis to land other than the 360 million acres —16 percent of all U.S. land area—under agricultural cultivation. Other land considered for fuel production would be land not currently used for crops, such as the acreage tied up by the Agriculture Department's subsidy program that pays farmers to leave the land idle rather than grow certain crops.

According to the report, "land lost to noncrop use could be significant in the long-range picture if diversion continues at the present rate. This loss is expected to total about 20 million acres by 1985. Fortunately, most of this has been land not suitable for cultivated crops. A logical sequence of energy conversion is to use the idle land to produce cereal grain, which is largely carbohydrate, and convert the grain by fermentation to ethyl alcohol—a convenient combustible fuel."

The report says that the conversion of this land to grain crops and the conversion of the grain to ethyl alcohol for use as a fuel for power plants or vehicles is technologically feasible. It states that 100 million of the unused acres, with an average yield of 70 bushels of grain—approximately 3,900 pounds—per acre, would produce 18 billion gallons of alcohol, equivalent to more than 10 percent of the 86 billion gallons of motor fuel consumed in the United States in 1970. (The energy equivalent of ethyl alcohol is 12,000 BTUs per pound, slightly more than half of the energy equivalent of gasoline.)

The report says that in regard to agricultural residues and fibers—e.g., straws, shells, corncobs, and similar materials—this residue amounts to 195 million tons annually, and has an energy value equivalent of 3,000 trillion BTUs. The waste is equal to about 4.3 percent of the 1971 energy consumption of 70,000 trillion BTUs in the United States. The report states that collection costs and delivery to regional power plants would vary from around $10 per ton to about $15 per ton. Heating values of the residues are about 8,000 BTUs per pound or 60 percent of the heating value of bituminous coal.

According to the report, if crops were grown specifically for fuel and produced 20 tons of dry matter per acre, production cost per ton would be on the order of $10. The heat of combustion per ton would be about 15 million BTUs. Thus, a ton of dry matter at $10 would be equivalent to coal at about $17 per ton on a comparative BTU basis. However, U.S. land that could produce 20 tons of dry matter per acre is limited.[110]

As is the case with a number of other proposals to use agricultural land for energy purposes, accurate cost-accounting of the energy needed to maintain, harvest, distill, and transport the grain alcohol or other fuel is not found in the National Petroleum Council estimates. This limiting factor is discussed later.

"Energy Plantation"—Trees for Electricity

One of the most provocative engineering approaches to the problem of harnessing solar biological systems to power production is that of burning trees in a conventional steam-electric power plant. The author of this concept is Dr. George Szego, president of the InterTechnology Corporation in Warrenton, Virginia. Dr. Szego, a prominent energy specialist and a founder of the Intersociety Energy Conversion Engineering Conferences held annually in the United States, has labeled this solar farming scheme an "energy plantation."

His calculations indicate that forests in sunny areas convert up to 3 percent of the available solar radiation to energy in the form of wood. The wood—or trees—would be burned in a conventional power plant that would convert the energy content of wood to electricity at 40 percent efficiency. To fuel a 1-million-kilowatt electrical power plant, the scheme would require something between 112 and 630 square miles of woodland, depending on the amount of sunlight striking the trees and on the trees' rate of photosynthetic efficiency. On the other hand, over the approximate 30-year lifetime of a conventional coal-burning power plant, about 40 square miles of land would be denuded to provide 74 million tons of coal. Dr. Szego cites this as a further argument for a wood-burning plant, since the fuel—trees—can be grown over and over, as opposed to coal, which is extracted only once, leaving wasted unproductive land.

The trees in the scheme would be harvested every 10 years. Assuming the electrical plant would be located in the center of the forested "fuel" area, logging equipment would bring trees in for burning on a crop rotation cycle. Since the land would be continuously reseeded in fast-growing trees for burning 10 years later, the process would be self-renewing over the lifetime of the plant.

As for the economics of the plantation, Dr. Szego says that the costs of producing trees as fuel for electric power plants might range from a little more than $0.03 per million BTUs under the most favorable conditions—to which case land and harvesting would cost $250 per acre—to more than $0.45 per million BTUs of energy, in which case land and harvesting would cost more than $1,000 per acre. Dr. Szego contends that other cost advantages will accrue due to the reuse for fertilizer of spent ash from the power plant; reuse of particulates for the same purpose; and the elimination of funds spent for sulfur dioxide removal systems for the plant's smokestack, since sulfur is not present in the combustible portion of the trees. Otherwise, the plant would be equipped with conventional pollution-control systems.

The major costs in this system, which are not readily apparent in the economic calculations, are hidden energy costs. To cultivate, maintain, and harvest the trees would require substantial power, as would the manufacture of the machines to do the job. Additionally, the plant might have significant environmental effects, particularly since the controlled cultivation of a single species of tree would result in a monoculture, which eventually would reduce the biological capability of the tree ecological system to survive.

Notwithstanding the hidden energy costs and the biological problems, Dr. Szego is convinced that "the cost of wood fuel in the worst case [the most expensive land and harvesting costs] is competitive with other fossil fuels," and that the plant would be cheaper than other power plants in use today, including nuclear plants.[111]

Fuel from Algae

Over the past few decades, a persistent crop choice of engineers and scientists for solar-fuel production has been cultivation of algae, the familiar single-cell organisms found in profusion around the world. These fast-growing organisms, which can be cultivated in the ocean or in special ponds, can produce a far higher yield of organic material per acre than almost any of the more advanced crops considered for fuel production.

Experience in algae cultivation for such a purpose dates back to 1951, when Arthur D. Little, Inc. (ADL) initiated a controlled algae growth project under contract from the Carnegie Institution of Washington, D.C.

ADL scientists constructed a small algae-producing pilot plant on a rooftop. The plant consisted of 600 square feet of polyethylene plastic tubes in which 1,200 gallons of a special nutrient culture medium were circulated to feed the algae—*Chlorella pyrenoidosa.* The complex system contained circulating pumps, heat exchangers, and devices for feeding carbon dioxide-rich air to the algae culture. From July to October of 1951, the total algae production was 75 to 80 pounds of dried organic material. On the basis of the experience gained in operating the pilot plant, they designed a larger plant, which might be used commercially. This plant would cover 100 acres and would produce 12,500 pounds of dried algae per day. Its total cost would be $3.2 million—or $30,000 per acre, which A. W. Fisher, Jr., of ADL noted "is extremely high in comparison with any conventional agricultural practice." In addition, he said that "even though the assumed yield is several times more than the best agricultural crop produced at present, the investment per unit of product will still be relatively high.[112]

This high cost amounted to $0.25 per pound for the dried algae, compared to $0.10 per pound for other agricultural nutrients produced for food. As for energy production using algae, Fisher calculated that the 10,000 BTUs per pound (about half the energy per pound of gasoline) energy content of algae would mean that the projected 100-acre algae power plant would yield a daily fuel product equal to only 5 tons of coal. This algae fuel would power no more than a relatively tiny 600-kilowatt power plant—at exorbitant energy costs and tremendous land waste, since the 600 kilowatts would represent only enough electricity to furnish a few dozen homes with power.

Fisher said of the experiment's merit that the lesson of high costs had shown that algae plants in the future might be used for food production but not for power production. However, even for food production, the cost of laboratory-grown algae was, and is, more costly than for conventional foods. The essential reason for this is the high energy cost of growing algae—the costs of building and maintaining the special facilities, the costs of operating the circulating pumps, harvesting the dried algae, and the like. These costs are investigated later in relation to the bioconversion of organic material for fuel production.

Another approach to the utilization of crops for energy production involves the duplication of this elaborate chemical synthesis of organic materials through the natural process of photosynthesis—or more precisely, photochemistry. The late Eugene Rabinowitch commented on the shortcomings of research to duplicate the photochemical process:

> Photochemical storage of solar energy has been going on in nature on an immense scale for millions of years: all human sustenance and industry depend on it. . . . It is a most sobering example of the limitation of today's science and technology that we cannot remotely imitate this figure [of 1 percent photosynthetic energy conversion]. Not only do we not know how to combine carbon dioxide and water, by means of visible light, to form sugars or starch (thus achieving with one stroke a feat of power engineering and a marvel of chemical synthesis), but we do not know how to use visible light to bring about *any* chemical reaction, however trivial, in which a significant amount of absorbed light energy would be stored in reaction products in a form suitable for practical purposes—producing mechanical or electrical work.[113]

Notwithstanding this somber introduction to the subject, a number of research laboratories have pursued photochemistry with the idea of ultimately producing a fuel. The theoretical attempts are largely based on the breaking up of certain organic chemicals by light, thus triggering a chemical reaction that absorbs heat. Allowing the reaction to operate in reverse releases heat, which can be stored and used for power. This seemingly simple approach has never come close to fruition on a realistic scale, although some promising laboratory experiments have been conducted.

Most scientific attention in the photochemical energy-storage field has been devoted to the chemical process of oxidation-reduction in photosynthesis, by which electrons are transferred from one molecule to another. Dr. Rabinowitch detailed the process:

> The molecule which has lost its electron is referred to as being reduced. The strength (or more precisely, the "free energy") with which a molecular species clings to its electrons is called its "reduction potential." Without the help of light, electrons can be transferred, in chemical reactions, only from molecular species with the higher to those with the lower reduction potential. In photosynthesis, however, electrons are raised by the action of light from molecules with a reduction potential of -0.8 volt to molecules with a potential of $+0.4$ volt.[114]

An innovative research program, funded by a $150,000 grant from the National Science Foundation to Case Western Reserve University in Cleveland, is aimed at combining a series of natural biological processes to produce hydrogen. Through the chemical production of photosynthesis, organisms such as algae would develop the "raw material" for conversion to hydrogen. Then, in laboratory cultures, hydrogen gas would be generated by adding special enzymes to the algae.

Dr. L. O. Krampitz, director of the project, points out that the production of hydrogen is a product of the natural metabolism of certain bacteria. In this bacterial process, an enzyme called hydrogenase triggers the chemical reduction of hydrogen ions to form hydrogen gas, according to the formula $2H^{+}+2e \rightarrow H_2$.

In order to produce large quantities of hydrogen gas from photochemical reactions, Dr. Krampitz is studying the development of organisms that make the key enzyme hydrogenase, in order to produce hydrogen. In laboratory experiments, he has produced hydrogen by adding hydrogenase from common bacteria called *E. coli* (*Escherichia coli*—anaerobic bacteria found in the digestive tracts of animals, including man) to spinach cultures. The two chemical reactions involve (1) conversion of sunlight by the spinach cultures into complexes of proteins, the essential chemical building blocks of life; and (2) the liberation of hydrogen gas from the cultures by the addition of the enzyme hydrogenase.

An intensive search is under way to find a source of hydrogenase, which is produced now by anaerobic bacteria such as *E. coli*—bacteria that cannot survive in the presence of oxygen. The development of mutant bacteria strains that grow in the presence of oxygen and can produce hydrogenase is being studied. Dr. Krampitz hopes that such bacteria can be found so that their hydrogenase can be added in cultures of common aerobic blue-green algae, which grow only in the presence of oxygen. The algae would then provide sunlight-induced chemicals, which would be converted to hydrogen gas by the contribution of hydrogenase by the bacteria.

In his request to the National Science Foundation for funding, Dr. Krampitz noted that blue-green algae are similar to bacteria, and thus could be manipulated by special laboratory genetic techniques—as is now done with bacteria in research applications. One significant drawback to the success of his work is the closed nature of the laboratory system—confinement that might lead to unexpected biological problems that would subvert the whole process. The proposal suggested corrective measures for this:

> Since some of the blue-green algae are susceptible to virus infection it is interesting to speculate that one might perform some genetic engineering on them. For example, there are numerous examples of the genetic exchange within bacteria of chromosome material between different species and even inter-generic exchange with the result that the recipient organism has acquired activities of the donor system. Some of these exchanges have been brought about by manipulation with bacterial viruses. It might be possible to introduce into a blue-green algae with one of its viruses hydrogenase activity of the type desired for coupling with photosystem I and II for hydrogen evaluation.[115]

Dr. N. O. Kaplan of the University of California at La Jolla pointed out at a National Science Foundation-sponsored conference on the subject that a potential problem in any biological energy conversion system like this would be the creation or discovery of techniques to stabilize the enzymes (such as hydrogenase) used for fuel production. He said that key enzymes might be trapped in glass beads, which would be immersed in the cultures. Such a system might then become a self-sustaining "biological fuel cell."[116]

Dr. Alexander Hollaender of the University of Tennessee at Knoxville summarized the results of the NSF conference on biological energy conversion prospects, pointing out that current knowledge of the exact biological processes is limited, but that the prospect of large-scale fuel production is tempting.

Further, he noted that intensified research is required to achieve:

1. A better understanding of the photosynthetic and photochemical reactions themselves.
2. Development of new strains of algae, microorganisms, and bacteria.
3. Development of enzyme stabilization techniques.
4. Determination of the feasibility of coupling photosynthetic reactions with microbial (bacteria, etc.) reactions.
5. Investigation of the immobilization of biological reactants (such as the enzyme hydrogenase) for use as a "biological fuel cell."

Dr. Hollaender noted that "there are still problems in the area of fundamental research," and suggested that to make the concept a reality would require a multifaceted experimental approach on a national level, involving scientists specializing in genetics, biophysics, biochemistry, physiology, microbiology, and other disciplines. The complexity of the experimental problem, he said, might also be ameliorated by feeding all information being developed in laboratories around the country into a central data bank—so that all parties could keep up with current findings.[117]

The Problems of Large-Scale Bioconversion

Some of the approaches previously discussed for providing a "new" energy source by harnessing photosynthesis on a massive scale are based on the notion that solar energy is a *free* energy source, and that all that is needed is an efficient way to speed up the biological system in order to get the energy. This simplistic approach may be in conflict with the essential laws of biology and the laws of thermodynamics.

As was discussed in the first chapter, the laws of thermodynamics clearly show that you can't get something for nothing, and that energy is constantly degraded from a higher level to a lower level. Some of these schemes, while reminiscent of attempts to achieve perpetual motion, actually require more energy to cultivate and harvest the special-fuel crops than is obtained from the "free" solar energy. In addition, biological laws indicate that survival of an ecosystem—be it a farm pond or the entire Earth—is dependent on variety, a diversity of species, whereas most large-scale bioconversion approaches would introduce a plant (or organism) monoculture, i.e., a single *species* that may become especially vulnerable to destruction by other natural agents, such as viruses.

To date, no large-scale schemes for harnessing power from solar biological systems have come close to recognizing these principles. Particularly vulnerable are approaches that would harness solar bioconverters such as algae in climate-regulated tanks.

Dr. Howard T. Odum of the University of Florida has concisely summarized some of the misconceptions of controlled biological energy conversion in his book *Environment, Power and Society*.[118] First, Dr. Odum said, is the fallacy of attempting to increase the efficiency of the photosynthetic process by which biological organisms convert sunlight to chemical energy. This efficiency can be boosted, but he also reminded that the increased efficiency can't be kept up indefinitely. "As an appreciable yield develops [e.g., from algae cultures], energy laws require a decline," he says.

In the basis of the results of the Arthur D. Little algae culture pilot plant project, Dr. Odum calculated that the energy required to build and maintain the plant, combined with the energy necessary to control the growth of the cultures and harvest the dry organic material *actually exceeded* the yield of energy (in organic material) from the plant. In other words, it took more energy (in the form of fossil fuels) to run the facility than that which the algae produced! This second fallacy, he noted, was

> considering the algae free from its supporting ecological system which in nature requires mineral cycles, the diversion of much energy into various networks for stability, and the maintenance of extensive work for survival. . . . In the laboratory tests, all these elements were supplied by the fossil fuel culture through thousands of dollars spent annually on laboratory equipment and services to keep a small number of algae in net yields. Scaling this expense up naturally produced fantastic costs. Until all these dollar costs were considered, the advocates imagined that they had higher efficiencies of net yield. . . . If the scientists involved had been more broadly trained to see that algae are part of an ecological system, whether they are in laboratory glassware or in the sea, they would scarcely have had the nerve to proclaim their physiological researches as being of such importance to survival.[119]

Outside of the laboratory environment, similar constraints apply to the use of marginal farmland for solar "energy crops." First, the most productive farmland in the United States is already under cultivation. Part of the good farmland not under cultivation could more readily be used for cultivation of food than synthetic fuels. Available land to sustain solar fuel crops is not readily available, and what land is available might more profitably be used for other methods of solar energy conversion—methods using mechanical, not natural, power plants to produce fuels such as hydrogen at higher efficiencies than controlled photosynthesis.

Solar Bioconversion and Waste Disposal

The future does not appear so dismal for some other prospects of bioconversion applied to fuel production, however. Several promising technologies can convert organic wastes to fuels such as oil and methane (natural gas). These approaches differ from the use of virgin land for industrialized solar crop production. Instead, bioconversion techniques would be used to reduce America's daily mountain of garbage to combustible fuels, foods, or fertilizer.

The detritus of contemporary American society is staggering. Each year, more than 2 billion tons of organic solid wastes and 1.1 billion tons of inorganic (mineral) wastes are generated. The organic wastes —indirect products of photosynthesis—can be easily converted to fuels. Other than on a small, pilot-plant scale, however, organic wastes are not now recycled for energy, but relegated to the waste heaps, the land fills, and other hiding places where they can't readily be seen by consumers. Clearly, such days are over. Land is unavailable for land fills in many areas today, and will surely become more scarce for this purpose in the future.

A 1973 study made by the National League of Cities and the U. S. Conference of Mayors indicated that major U.S. cities are virtually strangled in solid wastes. Almost half—46.5 percent—of the nation's cities will run out of places to dump trash by 1978. The study noted that the solid wastes problem was particularly concentrated in the high-energy-density cities, where the amounts of solid wastes discarded have doubled in the past 20 years, as opposed to a nationwide doubling of solid wastes in the past 50 years.[120]

Abundant opportunities exist throughout the country for converting wastes to fuel and fertilizer. The processes involved are part of the solar-biological system that produced the wastes, since the waste material is the end-product of the photosynthetic light reaction. The maximum use of these solar energy by-products through resource recycling is not only vital to the preservation of environmental quality, but provides key lessons in understanding the interrelationship of the natural energy system (food) and the synthetic energy systems (fuel for machines).

Communities—whether villages or great cities—may find waste dumps becoming gold mines. According to Eric Zausner, formerly of the President's Council on Environmental Quality (CEQ), using wastes as fuel (either by burning them directly in a power plant or by converting to liquid or gaseous fuels via bioconversion processes) can supply up to 10 percent of any community's energy requirements.[121]

A detailed study of the possibility of converting some of the nation's cascade of solid waste into food and combustible fuels was made by the Washington-based research firm International Research and Technology Corporation (IR&T) for the Environmental Protection Agency (EPA). Focusing on urban waste, the study summarized the sources of organic solid wastes and concluded that the most economic technology both to help convert large quantities of wastes to energy and to reduce the waste burden was pyrolysis of mixed municipal wastes for fuel (discussed in the following pages), which would then be burned as a supplemental fuel in coal plants. This priority goal, the study concluded, should command the bulk of the agency's research funds in this area.[122]

The Council on Environmental Quality commissioned the Midwest Research Institute to report on the status of these various methods of using solid wastes for fuels. The report, entitled *Resource Recovery, the State of the Technology,* concluded that, of all technologies studied, the cheapest and quickest means for a community to utilize wastes for energy would be to burn them in conventional power plants and industrial furnaces. The next most promising way to immediately utilize wastes would be materials recovery, pyrolysis, composting, steam generation with resource recycling, incinerating wastes with recovery and recycling of materials, and electricity generation with no recovery.*

Several cities have begun adapting large-scale programs of energy and resource recycling for urban wastes. One of the most advanced systems is that employed in St. Louis, Missouri, where about one third of the city's thousand tons a day of garbage is shredded and mixed with pulverized coal for use as power plant fuel. Tests were conducted in 1972 and 1973 by the city and the electric utility (Union Electric Power Company) to determine the economics of the process. On the basis of the tests, the Environmental Protection Agency, a major source of funding support for the project, pronounced it a great success.[123] A number of other cities are using variations of this approach, including Nashville, Tennessee, where a power plant burns wastes in order to generate steam for heating and cooling surrounding buildings.

* The report did not cover the advanced technologies of bioconversion discussed here because they are still on the horizon in terms of practical utilization today. The technologies it recommended are not the most efficient or logical methods for recovering energy and materials, but they are the quickest in terms of commercially available technology.

Several technologies considered by these studies involve conversion of organic wastes to combustible fuels such as oil and gas. Three of them appear technically promising. Two of these processes, hydrogenation and pyrolysis, use heat reactions to extract gas and oil from waste materials, and the third, anaerobic fermentation, utilizes anaerobic bacteria to digest the contents of organic material to produce fuel.

Hydrogenation and Pyrolysis

Two of the most promising approaches, hydrogenation and pyrolysis, have been developed by the U. S. Interior Department's Bureau of Mines at its Pittsburgh, Pennsylvania, Energy Research Center.

Bureau of Mines scientists Dr. Herbert Appell and Dr. Norman Wender were working on technologies to convert coal to high-quality fuel oil when, in 1970, they accidentally discovered that one chemical process under investigation was equally applicable to oil production from organic wastes. The process, called hydrogenation, is remarkably simple. The organic material is put in a reaction vessel to which are added heat at 380° C and quantities of carbon monoxide. Within 20 minutes, almost half the waste material is converted to fuel oil. The oil, when produced from rich animal wastes, has a good heating value—15,000 BTUs per pound, compared to normal fuel oil, with an energy value per pound of 20,000 BTUs.

Since the oil has virtually no sulfur—only 0.33 percent—the problem of sulfur dioxide pollution is virtually nonexistent. The temperature of the process heat required to make the oil is low in comparison with conventional power plant boilers, and some of the synthetic oil can be used to power the whole operation.

The second Bureau of Mines method of converting organic waste to fuel is called pyrolysis, a word derived from the Greek roots *pyr,* meaning fire, and *lysis,* the act of loosening. Organic wastes high in cellulose content, such as cattle manure, are heated to 900° C in a closed vessel at atmospheric pressure for about 6 hours. This high-temperature process, also called destructive distillation, yields 3 separate fuels—gas, oil, and combustible solids—with heating values comparable to those achieved in the hydrogenation process. All 3 recycled fuels can be readily combusted in conventional engines.

By 1973, the Interior Department had awarded the Pittsburgh Energy Research Center more than $750,000 to refine the hydrogenation and pyrolysis processes so that a pilot plant could process a ton of manure daily. The center's longer-range plan calls for building a $1.75 million plant for converting wood and logging wastes to oil. According to Dr. G. Alex Mills, chief of the Bureau of Mines' Coal Division, the proper development of these initial schemes could put the technologies into wide commercial use of 1980.[124]

A number of American corporations have researched and built similar pyrolysis waste systems, including Monsanto Chemical Corporation, St. Louis, Missouri; Union Carbide Corporation, Tarrytown, New York; Battelle Pacific Northwest Laboratories, Richland, Washington; and Garrett Research & Development Company, La Verne, California. The Garrett Company, a division of Occidental Petroleum Corporation, has included a pyrolysis unit as part of a complete system designed for disposing of municipal wastes. The Garrett urban waste system, which cost the company $3 million to develop, will be used to recycle the 200 tons of waste per day produced by 2 small Southern California communities, Escondido and San Marcos. Each ton of this urban refuse will produce a small quantity of fuel gas, 1 barrel of oil (of one third less energy content than a comparable barrel of oil made from the Bureau of Mines process), and recycled metals and glass. The U. S. Environmental Protection Agency (EPA) is subsidizing $3 million of the $4 million plant costs in the California project.

Garrett officials estimate that the plant's operating costs of $8 per ton will be 20 percent lower than the costs of disposing of urban wastes in conventional systems. They contend that in a plant 10 times this size, capable of handling the wastes of a city of 500,000 people, each ton of waste could produce a profit of $1 in fuel and recycled materials.

Already, the San Diego Gas & Electric Company has contracted to purchase oil from the Escondido-San Marcos pilot plant, and the utility has spent $150,000 on new facilities necessary to handle the recycled fuel.

Another large pyrolysis plant has been designed for Baltimore, Maryland, by Monsanto Chemical Corporation. The city announced plans to purchase the $14 million facility with financial assistance from the Environmental Protection Agency. Instead of oil, the plant was designed to produce low-heat-value synthetic gas for sale to the Baltimore Gas & Electric Company.[125]

Bacterial Bioconversion of Waste

One of the best-known and most thoroughly tested technologies for converting organic wastes to fuel is the process of natural fermentation, in which anaerobic bacteria (bacteria that grow in the absence of oxygen) convert the natural substances to methane (natural gas) and carbon dioxide. This is precisely the process—though on a drastically reduced time scale—by which nature decomposes organic materials to produce our conventional fossil fuels. Commercial installations using this principle were demonstrated as early as 1871 in England, and a large plant was built in Bombay, India, in 1905 to produce gas and fertilizer from organic wastes.[126]

This process of anaerobic fermentation is both simple and inexpensive. In a typical conversion plant, bacteria break down the waste matter at relatively low temperatures—between 15° C and 50° C (i.e., between 59° F and 122° F)—and at atmospheric pressure. According to a leading proponent of the technology, Dr. Martin Wolf of the University of Pennsylvania's National Center for Energy Management and Power, the bacteria that convert the organic materials to methane are unknown. All that is known is that there are two kinds of bacteria involved. "The first type [of bacteria] breaks down the organic solid material and generates organic acids, hydrogen, and carbon dioxide. The second bacteria lives on the waste products of the first type and makes methane in the process." According to Dr. Wolf, it is not clear whether the second group of bacteria use primarily the organic acids, the hydrogen, and the carbon dioxide, or the three together.[127]

Notwithstanding the fact that scientists still know very little about the process, it is in widespread use in several countries.† During the Second World War, the fuel-starved German Government developed several technologies for bioconversion of wastes to methane for use in automobiles and other vehicles. For the past two decades, Indian scientist Ram Bux Singh has built and designed thousands of the plants—which he calls bio-gas plants—to provide fuel and fertilizer for the people of India. Under his direction, a large bio-gas plant in Delhi, India, has been built that produces fertilizer for free distribution to local farmers, while also providing methane gas to operate 4 400-horsepower engines for generating electric power.

He points out that in order to operate a methane-producing plant, it is necessary to understand the precise constituents of the organic material fed into it. Some organic materials make better fertilizer than they do gas. In an interview in *The Mother Earth News,* Singh explained:

> Seven cubic feet of methane gas can be generated from one pound of dry leaves but only one cubic foot of gas will come from one pound of cow dung. The cow dung, on the other hand, is just that much richer a fertilizer than the leaves. You can say, then, that the cow has digested

† Since very little research effort has ever been devoted to understanding this process—or, for that matter, any solar energy-related technology—it is little wonder that its basic principles remain virtually unknown.

> the leaves and partly turned them into plant food. When the cow manure is then composted in a bio-gas plant, the bacteria there merely further process—or refine—the former dry leaves into a still richer plant food. It is all very natural.[128]

Singh's work in India has largely been responsible for a sudden U.S. upsurge in interest in the methane generating system,‡ although a number of sewage treatment plants in the United States have long used the process. Most of them were built in the early years of the century and were not designed for fuel production—or waste recycling—but only as a method of sewage treatment, although some of them capture the methane that is produced and use it to power all or part of the plant's equipment. This practice has posed many problems, however, and the current trend is away from sewage treatment by anaerobic digestion.

Singh says that the bio-gas technology is not readily applicable to the sewage effluent of a large city because the waste matter is accompanied by low-grade organic material—such as paper, which is not easily digested into gas—as well as large quantities of water, which literally swamp the digesters, reducing and sometimes completely crippling their efficiency. The process could be used, however, Singh contends, if the wastes were treated beforehand with nitrogen. "The anaerobic bacteria that do all the work in a bio-gas plant consume carbon about 30 times faster than they use nitrogen," Singh explained. "They work most efficiently, then, when the waste fed to them has that balance. When the carbon is 30 parts and the nitrogen is one part, the material put into a bio-gas plant will digest very rapidly and will produce much gas and good fertilizer."[129]

A number of small-scale plants for farms and homes have been built in the United States using Singh's designs. Certain necessary adaptations to these plants include the use of insulation and the addition of heat to keep the bacteria cultures at optimum growing temperatures.

On a larger scale, the National Science Foundation granted Dr. Wolf's research team at the University of Pennsylvania $600,000 to investigate the potential of the methane process in converting urban wastes to fuel. Working in conjunction with a research team led by Dr. George Christopher at the United Aircraft Corporation research laboratories in East Hartford, Connecticut, the Pennsylvania group is operating several laboratory-size digesters in an attempt to ascertain

‡ Singh's books, *Some Experiments with Bio-Gas* and *Bio-Gas Plant*, contain much helpful information on the technology. They are available from *The Mother Earth News*, P. O. Box 70, Hendersonville, N.C. 28739.

the feasibility—both engineering and economic—of using bacteria to convert urban sewage and other organic wastes to fuel on a large scale.

A major problem they foresee is the disposal of sludge, the undigested waste material that may constitute as much as 40 percent of the initial wastes fed into the digesters. Possible solutions include converting the sludge into fuel or fertilizer, as in Ram Bux Singh's bio-gas process, or drying it for burning. Were the sludge merely burned, however, substantial environmental and economic headaches could ensue. Dr. Wolf is optimistic about the future of the process, and says that 10 to 20 percent of the present U.S. natural gas requirements could be met by anaerobic digestion of wastes.

A variation of the anaerobic bacteria bioconversion approach has been pursued by scientists at the Sanitary Engineering Research Laboratory of the University of California at Berkeley. The experiments involved producing animal food by cultivating algae in ponds containing nutrient-rich sewage wastes. Dr. Clarence Oswald is convinced that modifications of this work will permit the development of a two-stage system for methane fuel production that will also serve as a means of disposing of organic wastes. First the organic material—domestic sewage, feedlot wastes, cannery waste, etc.—is fed into simple earthen ponds containing algae. The ponds are then covered with either a floating or a submerged cover to provide an oxygen-free environment for the essential anaerobic fermentation by which methane is produced. The by-product of the process, organic sludge, is added to new supplies of sewage to perpetuate the reaction.[130]

Dr. Oswald has calculated the costs of the bioconversion system relative to the costs of conventional sewage treatment in the United States. "In conventional activated-sludge sewage treatment," he found, "1 kilowatt-hour is required to provide 3 pounds of oxygen for waste oxidation; whereas, through use of solar energy, expenditure of 1 kilowatt-hour of electrical energy will catalyze the production of more than 100 pounds of photosynthetic oxygen."[131]

On the basis of their research, Dr. Oswald and an associate, Dr. Clarence Golueke, suggested that with modifications and improvements of the sewage-bacteria-methane system, methane could be produced for production of electricity at suitable locations at extremely low costs. In electrical power terms, they calculated that each acre of the plant's network of ponds would represent a capacity of 10.8 kilowatts. In 1960 dollars, they computed that power costs would be in the range of 10 to 20 mills per kilowatt-hour. However, the monetary savings achieved by "free" sewage treatment could lower the power costs to the extremely low figure of 1 to 2 mills per kilowatt-hour.[132]

Availability of Wastes for Bioconversion Processes

The technologies of pyrolysis, hydrogenation, and anaerobic digestion (with or without the addition of a further algae fuel step) can be utilized to generate fuels from a tremendous variety of organic wastes, ranging from paper to domestic sewage. The basic problem is collecting the wastes with a minimum expenditure of additional energy or costs. The Bureau of Mines has published estimates of available, collectable organic wastes that might be used for fuel production. Following is a list of wastes generated in the United States in 1971, with estimates of the amount of fuel that they might have yielded had they been converted to fuel:[133]

Waste Source	Total Available Amount of Waste Generated in Millions of Tons	Amount that Could Have Been Collected for Fuel in Millions of Tons
Manure	200	26
Urban refuse	129	71
Logging, wood residue	55	5
Agricultural crop and food wastes	390	22.6
Industrial waste	44	5.2
Municipal sewage	12	1.5
Miscellaneous organic material	50	5
	880	136.3

From the total available organic wastes in the United States in 1971, about 1 billion barrels of oil and 8.8 trillion cubic feet of methane (natural gas) could theoretically have been derived. Yet, the total fuel that could have been converted from the *collectable wastes* is estimated at only 170 million barrels of oil and 1.36 trillion cubic feet of gas. The total national waste pile turns out to be too scattered and diffuse to be easily harvested. However, what is available for conversion would have supplied 3 percent of the oil demand and 6 percent of the natural gas demand in the United States in 1971.[134]

In terms of national energy consumption, the conversion of wastes to fuel does not appear particularly significant; but looked at in the context of international oil pressures, 170 million barrels of oil emerges as a more exciting figure, since it is equivalent to 12 percent of the 1971 foreign oil imports. That amount of oil could have replaced half the U.S. domestic supply of residual heating oil which has led to speculation that the "oil wells" could be "drilled" in waste heaps rather than on the outer continental shelf, and that environmental problems today would prove to be sought-after fuel sources tomorrow. With the addition of advanced cycles such as that developed by Dr. Clarence Oswald, algae "booster" cycles have been proposed to increase fuel yields from anaerobic bacterial digester technologies.

The best starting places for the bioconversion waste disposal systems are locations where the wastes are already highly concentrated, such as cities and cattle feedlots. Pollution from feedlots in the United States has reached catastrophic proportions. Modern commercial cattle production is far different from the idyllic days when the animals roamed the grazing lands. Today, cattle are concentrated in small lots, where they can be fattened and marketed much more quickly.*

The quantity of wastes from the cattle feedlots is equivalent to the pollutants generated by 75 million people. In the words of environmental writer Gary Soucie, "some of the feedlots supplying us with steaks and hamburgers have sewage problems the equivalent of 1 million people living on 320 acres, a population density 26 times that of Calcutta."[135] The wastes contain thousands of chemical compounds, some of which are quite dangerous. For example, nitrate from the wastes can be converted by soil bacteria to nitrite, much of which then finds its way into the water supply and to humans. This poison combines with blood hemoglobin to form methemoglobin, which lowers the oxygen-carrying capacity of the blood.[136]

Clearly, the nitrates from the feedlots could serve a higher purpose than monumental pollution and human death; and bioconversion techniques seem promising as one alternative to the immediate problem.

Interest in the conversion of chicken manure to methane prompted the Maine Department of Economic Development to embark on a research program to assist the state's chicken farmers in producing commercial gas and fertilizer. Aided by a $45,000 grant from the federal Environmental Protection Agency, the Maine environmental agency is studying the anaerobic digestion processes with the hope that, in addition to methane, natural fertilizer can be produced and perhaps shipped in unit trains to the midwestern states.

Disposal of the chicken manure is a serious problem in Maine, where chicken farms discharge 500,000 tons of manure yearly. The U. S. Department of Agriculture warned the city of Belfast, Maine, that its wa-

* This form of agribusiness owes its existence to the cheap subsidy of fossil-fuel energy. Only through cheap energy can the cattle be maintained in these confined spaces. Whereas the livestock formerly roamed the land, taking food from the green plant and leaving manure to complete the natural nutrient cycle, today cattlemen have replaced this natural food cycle with a synthetic fossil-fuel cycle.

ter supply was in "critical" danger from nitrate runoff from chicken wastes. To alleviate the problem, the Maine environmental department is building 2 pilot plant digester units that can produce 32,000 cubic feet of methane from an 8,000-pound-load of chicken manure.†[137]

With significant research breakthroughs, some future day may indeed see the possibility of fuel production from solar energy farming; but today, the best we can hope for is the reduction of fuel use through biological resource recycling with such techniques as composting, sewage farming, returning manure to the soil as natural fertilizer, and conversion of animal wastes to food products.

Composting

This technique involves digestion of waste by aerobic bacteria—bacteria that thrive in the presence of oxygen. While its use is limited in the United States, composting is in widespread use in Europe and a number of Asian countries for converting municipal refuse and other organic wastes to fertilizer.

A recent report of the International Research and Technology (IR&T) Corporation concluded that compost derived from municipal wastes, sewage sludge, or manure materials not only offers exciting potential as an organic nonfarm fertilizer, but also that composting is a more efficient waste-disposal method than production of methane by anaerobic digestion. About half of all municipal refuse could be converted to compost. This means that, in addition to producing a commercially profitable and environmentally compatible fertilizer, composting of city waste would double the life of a sanitary land fill and halve the related costs of land acquisition, according to the IR&T report.[138]

Compost can be used as material for land fills or as a new soil replacement for reclaiming depleted strip mines. It can also be used to supplement up to 10 percent of the diet of cattle.

Sewage Farming

Sewage farming, or sewage irrigation, which is practiced in many countries, consists essentially of channeling sewage water to irrigate farm crops. Some of the most extensive and successful operations were

† The production of fuel from chicken wastes is nothing new to Harold Bate, a British farmer who has operated an automobile since the mid-1950s on methane from his farm's chicken manure. Bate's homemade digester design is available by mail order for less than $3. For $33, he'll send a kit with instructions and a special gas regulator, which can be hooked up to a digestion tank. Write to: Mr. Harold Bate, Penny Rowden, Blackawton, Totnes, Devon, England TQ 9 7 DN.

initiated in Berlin in about 1850. By 1931, the Berlin sewage farms were supplying one fourth of the produce and pasturage for the dairy herd that provided one sixth of the milk consumed by the city's 4.8 million inhabitants, according to Jonathan Allen, writing in the journal *Environment.*[139]

In the typical European sewage farm, sewage water is distributed through ditches alongside the crops. After percolating down through the living filter of the soil, the resulting purified water is collected by a network of drainage pipes, and tiles and returned to the waterway for reuse.

In Israel, noted Allen, Tel Aviv sewage used in this manner has helped turn the Negev Desert into rich farms and orchards.

In the United States, sewage effluent was used only on a small scale in the late nineteenth and early twentieth centuries. More extensive use in this country has been avoided up to this time because of low population density, great abundance of land, an abundance of cheap chemical fertilizers, and the "squeamishness" of the American public "about eating food irrigated with material derived from human waste," according to Allen.

However, it is now under investigation at Pennsylvania State University, and in 1972, the Army Corps of Engineers undertook a study of a sewage-treatment system using sewage-farm networks for the Chicago-northwestern Indiana region, home of 13,500 industries now and an estimated 9 million people by 1990. According to Allen, the Chicago study suggests that new sewage-farm systems may be more economic than conventional sewage-treatment methods.

Direct Use of Manure as Fertilizer

This was accepted common-sense farming practice until sophisticated, energy-subsidized and energy-intensive farming methods such as feedlots and chemical fertilizers began making rapid advances. Looking into the future beyond production of fuel from wastes, it now appears likely that many of these traditional farming practices will be reestablished. Returning the cattle manure to the soil as natural fertilizer, for instance, would not only be desirable from an environmental standpoint, but it may indeed prove inevitable with advancing fossil fuel shortages. Nitrate pollution from the use of chemical fertilizers is a serious problem both to health and to the environment. Eliminating the centralized feedlot agribusiness practice would make possible a return to harmony in the environment as "old-fashioned" methods replaced the feedlots *and* the synthetic fertilizers.

Conversion of Animal Waste to Food Products

Wastes can be used for producing food as well as synthetic fuel. Scientists at the General Electric laboratories in Valley Forge, Pennsylvania, and Schenectady, New York, are working on processes that utilize thermophilic (heat-seeking) aerobic (oxygen-using) bacteria to convert cattle manure into protein-rich food. The research was motivated by the fact that meat consumption for human food is enormously wasteful of the natural food chain. A steer must consume more than 100 pounds of plant protein to produce less than 5 pounds of beef for human consumption. However, beef is now the major source of amino acids, which are vital to human survival. Since cattle can synthesize all 20 amino acids from the vegetation they consume, while man can produce only 8, the human diet must be supplemented with beef or a high-protein beef substitute. Single-cell proteins produced from the cattle manure might provide an excellent "new" source of amino acid-rich protein food.[140]

Use of organic wastes to produce single-cell protein is being pursued by a number of commercial firms and universities. The Department of Chemical Engineering at Louisiana State University has developed a pilot-plant process that converts bacterial protein from bagasse—the cellulose-rich by-product of sugarcane processing.‡ The International Research and Technology Corporation report to the Environmental Protection Agency on solid wastes commented that "the most valuable application of the pilot plant process as presently designed would be the production of animal feed in underdeveloped countries that have a surplus of cellulosic agricultural by-products, but only small hydrocarbon resources that could be directed to SCP (single-cell protein) production."*

The IR&T study indicated that the cultivation of protein food from municipal wastes was not a promising technology, but that protein production from animal manure was a logical one. "In particular," the report noted, "the feedlot owner/manure processor can control the characteristics of the waste stream, and all marketing problems are eliminated as the feed is generated right next to the animals that will eat it."[141]

Rather than pursuing this attempt to add still another step to our food chain (animal manure to animal food to beef for humans), perhaps a more rational step would be to question the logic of the beef food chain itself. The cattle industry is highly industrialized and is

‡ Bagasse is now being widely used as a source of recycled paper.

* Fossil fuels can also be used to produce single-cell protein through similar bacterial action.

heavily subsidized by fossil fuels. Reducing the consumption of beef and encouraging more vegetarian diets would accomplish more for food economy than constantly increasing the size of the cattle industry.

In terms of efficiency, the heavily meat-oriented American diet requires a greater subsidy of resources and energy than would a more vegetarian diet, because the protein from animals comes from a higher niche in the food chain. More synthetic fertilizer, more food, more fossil fuels for more mechanization, transport, etc., and more energy-intensive pesticides are needed to skim off the top layer of protein (in cattle) from the natural food chain than would be the case using protein-rich plants directly for human food consumption. Even a totally vegetarian diet can provide all the necessary nutritional needs with the single exception of vitamin B-12.[142] Agribusiness, with its focus on beef production, has developed into an enormous technological monoculture that seriously threatens, in some areas, the ability of the natural environment to support it.

Conclusion

The panel of solar scientists established by the National Aeronautics and Space Administration and the National Science Foundation estimated that by the year 2020, solar bioconversion techniques applied to industrialized agriculture and organic waste sources could produce 30 percent of the nation's gaseous fuel and 10 percent of our liquid fuel requirements—*economically!*[143]

This can be achieved only by maintaining the current level of inefficient waste production in the inefficient energy systems (both biological and synthetic). Recommendations such as this are based on strictly linear thinking, which does not take into account the nature of the biological system and its pervasive synthetic fuel subsidy.

Herein lies the lesson of promises to produce energy using bioconversion techniques: There can be no *free energy* from the natural biological system, because it is prohibited by the immutable laws of thermodynamics. Promises to "supplement" the nation's gas supply by creating new fuel sources from agricultural grains, algae, or animals that feed on plants, are based on faulty logic. Such promises ignore the nature of the biological energy system on which the synthetic energy system is based. Thermodynamic laws, which clearly demarcate the line between something for nothing and something at a price, cannot be wished away. To produce synthetic energy, for example, in the form of methane from agricultural bioconversion processes implies that the agricultural system itself will be allowed to remain centralized and biologically inefficient through extensive support by fossil fuels. The same applies to the conversion of chemical energy—food for human consumption—from the wastes of animals that are confined in the pens of the cattle producers. The so-called "free energy source" is available for conversion to gas or food because the biological energy system is no longer so purely biological. It has been transmuted into another appendage of the centralized high-energy culture.

The attempt to get more synthetic fuel from high-energy agriculture might be successful, but only at the expense of the natural biological system itself. We must rob Peter to pay Paul. Instead of making natural gas from the wastes, the wastes themselves might be used more wisely—for natural fertilizer. The agricultural energy system is part of the same world as the synthetic energy system. Tug on one and the other responds. In the case of bioconversion technologies, a great lesson can be learned: the lesson of the interconnected energy systems—nature's and our own. Having stripped the land for coal and drilled into the earth for oil and gas, man now returns with hands outstretched for the last glittering resource: fuels from agriculture. The dream collapses when it becomes apparent that the system can't be converted overnight into a synthetic new oil well or coal mine. What nature required billions of years to produce—the fossil fuels—can't be farmed in days from the green plants or the wastes of animals (including man) that eat the green plants.

Attempts to duplicate the natural biological cycle to provide cheap energy to spin more turbines and operate more vehicles have revealed the price of the energy-guzzling machines. There is no such thing as a free lunch. A complex and remarkable system is involved, which interconnects all sources of energy. The source of biological fuels and extracted fuels is one and the same!

Instead of trying to achieve the impossible—i.e., force the natural system into the technological mold—the technological mold must be fitted back into the biological system. The contest is not whether we can create synthetic natural gas, but whether we can decentralize agriculture and recycle resources. That is the promise of solar bioconversion—the ability to recycle resources, which includes making better use of energy resources. The additional energy wrested from the wastes of a city might be used for fuel conversion to run some of the city's systems, but it can never be more than a supplement to the energy demands of the city. The energy needs of the feedlots might be supplemented by utilizing the organic wastes of the confined animals for fuel, but an infinitely better solution is to decentralize the feedlot-agribusiness system and begin the return to a more natural biological cycle.

Conclusion: Large-Scale Solar Power Projects

The prospects for developing solar power on a large scale are indeed intriguing, but the promises and aspirations of a few scientists may be fraught with error—unintentional, but potentially disastrous in terms of social planning. The primary flaw in most proposals for harnessing solar power is that the proponents have not accurately accounted for the amount of energy required to build and maintain solar power plants. In the preceding analysis of solar bioconversion projects, Dr. Howard Odum's comparison of the energy needed to build and maintain complex laboratories for synthetic fuel production, compared to the fuel delivered from such schemes, is a clear analysis of possible technical miscalculations.

What is needed from the proponents of other solar technologies is a clear accounting of the energy required to build and operate plants using the sea-thermal, photovoltaic, or large-scale solar heat plants. Research should be quickly initiated to evaluate and assess the potential to society from each solar process, rather than relying on faith in corporate wisdom to select a rational alternative power process for society.

The levels of government funding for solar energy have never been great, and the few million dollars that the government now allots for solar research is clearly inadequate to the task ahead. If our society is to reap any benefit from solar energy in the power production area, an industry-government coalition should begin work now to avert the real possibility of social collapse when the fossil fuel levels decline and the only energy source available is nuclear fission.

7

Decentralized Uses of Solar Energy

A great drawback to the large-scale attempts to harness solar energy is the necessity to duplicate today's electrical transmission system, which burns fuels at one point and then distributes electricity to another. The sun's energy is already distributed—free—to all areas. The economic drawbacks of large-scale solar energy technologies are largely mitigated by incorporating solar collection and distribution on-site.

The following examination of decentralized uses of solar technology illustrates many of the potential uses of solar energy on a household and community level.

In the United States, apart from transportation, the largest single use of all energy forms is for space-heating buildings. About 20 percent of our over-all maintenance energy supply is devoted to this one need.[1] Much of the energy used for this purpose today is inefficiently supplied by electrical generating plants, which burn fossil fuels at an average efficiency probably no greater than 25 percent; this means that about three fourths of the precious stock of fossil fuel used to generate electricity (a very high-grade form of energy) is wasted as "thermal pollution," after which an additional 10 percent of the converted electricity may be lost through the inefficiencies of transmitting it to where it is needed. The electricity at the point of use is then reconverted to heat at temperatures below 100° F (very low-grade heat). Thus, at the generating plant, coal or some other fuel was burned at temperatures higher than 1,000° F to generate electricity that could be applied a few miles away to heat buildings to temperatures considerably less than 100° F!*

On the other hand, the average intensity of solar energy reaching the ground in the United States is about 17 thermal watts per square foot of area, which means a 24-hour average of 410 thermal watts per square foot. This is more than twice the energy needed to heat *and* air-condition an average house in the continental United States.

Put in terms of another energy equivalent, British Thermal Units (BTUs), a typical U.S. home of 4 family members with a roof space of 1,600 square feet today uses about 130 million BTUs of energy for space heating. Yet that same residence will receive during the year solar power 6 times greater than the energy needed for heating the house.

The use of solar energy to heat buildings is a more logical means of supplying low-grade energy than present inefficient methods, such as electric heating. Solar energy can be concentrated with low-cost flat-plate collectors covered with clear plastic or glass, to supply the heating needs of buildings. By supplying the energy requirements in this fashion, not only is the need for the central station power plant eliminated, but the energy source itself—solar energy—is almost precisely matched to the requirements of the structure. It seems a capricious waste to use these vital fossil fuel resources by burning them at high temperatures to supply homes with low-grade heat. The sun's energy may be diffuse, but on the other hand, it is ideal for application to heating needs as well as to other low-temperature energy needs in the home.

Solar collectors can be designed to supply heat at temperatures ranging from 100° F to 300° F on a year-'round basis. In addition to space heating, some of the other household applications for heat in this range include water heating, cooking, air conditioning, refrigeration, food freezing, and clothes drying. In 1968, energy supplied by conventional sources for these energy needs added up to the following amounts:[2]

* The comparison of the relative efficiencies of electric heating and heating with solar energy is more than an academic one. Extensive economic studies (reviewed in detail in subsequent pages) have shown that in many cases solar heating systems are less expensive to operate than electric heating systems.

Energy Use in U. S. Residences (in percent)	
Space heating	57.5
Water heating	14.9
Cooking	5.5
Refrigeration	6.0
Air conditioning	3.7
Clothes drying	1.7
Food freezing	1.9
Total	92.1 percent†

Thus, solar heat collectors could theoretically provide the energy needed for more than 90 percent of the energy needs of a typical American home. The use of solar space and water heating could supply over two thirds of the energy needs. Areas where solar collector systems might be used do not necessarily have to be sun-rich, as is the case with large-scale solar power plants. The following map of the United States indicates areas in this country where solar energy might be used for the space heating of homes and buildings:

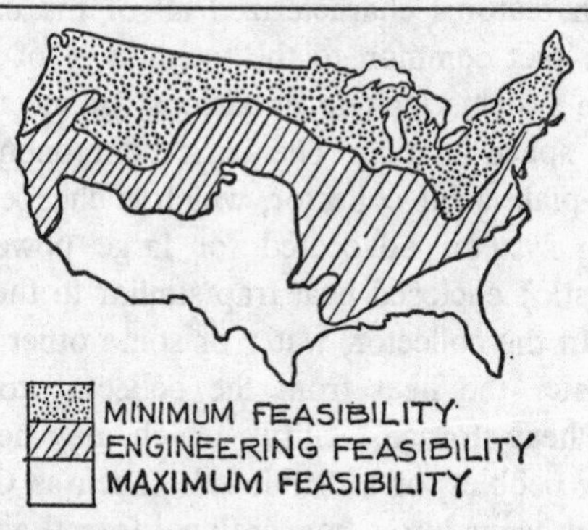

Feasibility of solar heating

This chart was originally published in the journal *Heating and Ventilating* in 1950, when interest in solar-heated houses was fairly widespread in the United States.[3] The nation is divided into three areas: "maximum feasibility," where the entire heating requirements of a house could be met by solar energy; "engineering feasibility," where solar energy in combination with a conventional auxiliary (electric,

† Other energy uses, such as lighting, appliances, television, etc., accounted for the electricity-related requirements that could not be met with low-level heat.

gas, oil) heater could supply most of the needs of the home; and "minimum feasibility," where solar energy would be useful primarily as a supplementary heat source in the spring and fall months.

This chart is a useful aid to the understanding of solar energy, but the 1950 study does not take into account a number of factors, including the use of solar-powered air conditioners for use in the sunny states; use of solar energy for cooking, clothes drying, etc.; and the development of advanced methods of storing energy for use at night and during periods of stormy weather, when available sunlight cannot add to the energy "store" of the system.

Basic Technology and History

Since the early 1940s, about twenty-five houses have been built to meet heating needs with solar energy. Some solar systems installed on the houses were nothing more than enlarged solar hot water heaters adapted to space heating. Consequently, they were unable to supply more than a fraction of the heat intended for the house. Other more elaborate solar heating systems supplied more than 90 percent of the heating needs of the experimental houses.

Certain common features characterized all of the early solar-heated houses, and are in fact common to the technology of all solar houses, whether or not the systems include provisions for meeting other energy requirements than space heating. The basic components of the system are the simple flat-plate solar collector, which is cheaper than expensive solar-concentrating systems advocated for large power plants; and a glass (or clear plastic) enclosed heat trap similar to the solar hot water heating collector. In the collector, water or some other fluid (or gas) is circulated to transfer the heat from the collector to a heat storage facility. From the heat storage facility—which may be a large tank of water, a bin full of pebbles, or a fusible salt (such as Glauber's salt)—air is forced through vents into a conventional forced-air heating system. In other systems, the solar-heated (and stored) water is ducted through pipes in ceiling or conventional wall radiators for heating the building. Most systems also utilize thermostats for temperature control and a small auxiliary heating unit, fueled with gas or oil (or perhaps an electric heater) as a backup system when the solar heater is not operating. Following is an illustration of a typical solar home-heating system.‡

‡ This schematic illustration is the core of the solar house—or building—power system. With modifications for heat utilization in other energy areas, such as air conditioning, cooking, drying, etc., the basic system can meet other needs.

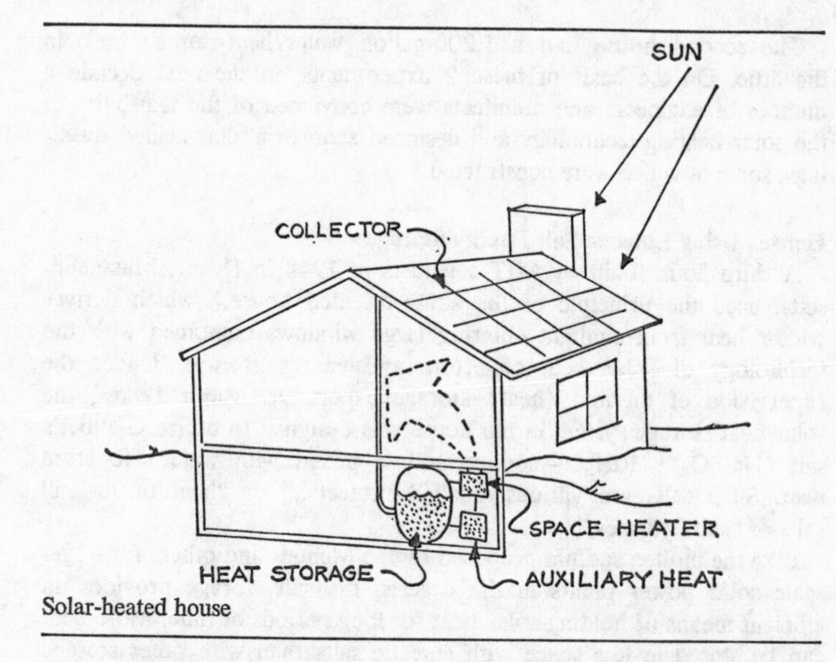

Solar-heated house

In evaluating the technological and economic feasibility of solar-powered dwellings, it is useful to examine some of the houses built in the United States over the past quarter century. Experience gained in the construction of these houses is instrumental in pinpointing the problems in the construction of solar-powered houses today.

Other than experiments conducted by professor F. W. Hutchinson at Purdue University in the early 1930s on houses that were designed with large south-facing windows to supplement winter heating requirements, no solar-heated houses containing energy storage units were built in the United States until the Massachusetts Institute of Technology (MIT) constructed a test structure in 1939–40.* Under the direction of Dr. Hoyt Hottel, a leading solar scientist at MIT, the 2-room structure was fitted out with a modified solar hot-water-type collector and a huge 17,000-gallon insulated water storage tank in the basement. After the Second World War, Dr. Hottel's group refined the design of the first structure and built a second house, which was occupied by students who kept measurements on the heating capability of the unit.

* The individual responsible for this program was Boston businessman Godfrey L. Cabot, who granted MIT more than $600,000 in 1938 to develop solar energy. Cabot lived more than 100 years, and retained his keen interest in solar energy until his death in the early 1960s. The Cabot solar program is still alive—though little more than nominally—at MIT.

The second house had a 1,200-gallon water/heat-storage tank in the attic. On the basis of these 2 experiments, in the next decade a number of engineers and architects were convinced of the feasibility of the solar heating technology and designed scores of solar-heated dwellings, some of which were constructed.[4]

Houses Using Eutectic Salt Energy Storage

A third house built by MIT scientists in 1948 in Dover, Massachusetts, used the principle of the south-oriented house,† which derives winter heat from sunlight entering large windows combined with the technology of solar heat collectors and energy storage. Under the supervision of thermal (heat) storage expert Dr. Maria Telkes, the solar heat storage system in the house was designed to utilize Glauber's salt ($Na_2SO_4 \cdot 10H_2O$—sodium sulfate mixed with water) to store heat. Such salts are variously called "eutectic," or "heat of fusion" salts or "salt hydrates."

Like the molten sodium proposed by the Meinels and others for large-scale solar power plants in the deserts, the salt storage provides an efficient means of holding solar heat for long periods of time. More heat can be stored in less space with eutectic salts than with other storage media, such as water or rocks.

Glauber's salt melts when it is exposed to heat. When the external heat (solar) is no longer available, the salt hardens and releases heat. In the Dover house, air from the vertically mounted solar heat collectors was circulated by electric fans to 3 bins containing 5-gallon cans of the Glauber's salt. Dr. Telkes designed the heat storage system unlike that for most other solar-heated houses, which have storage tanks (of rock, water, or eutectic salts) in the basement—and, at times, in the attic.

She explained this unique storage method:

> By placing the heat storage bins between the rooms the heat leakage is to a great extent directed into the rooms and the walls of the bins are changed into radiant heating panels, supplying a background heating effect . . . sufficient to keep the rooms comfortable, when the outdoor temperature is milder. When more heat is required, thermostatically operated fans circulate the air in the bins through the rooms to supply additional heat; the colder air of the rooms returns to the bins through louvres.[5]

Approximately 21 tons of the Glauber's salt was used to store heat gained from the solar collectors, which covered 720 square feet—the

† See Afterword.

south-facing second-story wall of the house. A February 1949 test of the heating capability of the house showed that the solar collectors, converting sunlight to heat with a 41 percent efficiency, captured 8.6 million BTUs of heat energy, and the south-facing first-story windows of the house captured nearly 3 million BTUs. This was sufficient energy to keep the house comfortably heated.

The total cost of materials used in the solar heating system of the house amounted to $1,865. Of this, the solar collector cost $540, with backing insulation adding $400. The heat storage drums, Glauber's salt, and insulation cost $635. The other material—ducts, fans, and thermostats—cost $290. (These costs are for materials only and do not include labor.)

Tests of the Dover house were conducted for several years, and significant problems developed with the Glauber's salt. After the 1949 winter, succeeding seasons brought colder, cloudier weather, which required the use of the gas auxiliary heater. The Glauber's salt also underwent "phase separation," meaning that, during its periods of heat-cold cycling, it lost its capability to release heat effectively due to chemical changes.

Another solar-heated house was constructed under Dr. Telkes' supervision. It utilized similar techniques: a vertically mounted collector and eutectic salt energy storage. The house was built for the Curtiss-Wright Corporation in Princeton, New Jersey, and was used as a solar energy laboratory. The architect of the building, Aladar Olgyay, a brother Victor Olgyay) of *Design with Climate,* said that tests on the leading proponent of solar-oriented structures and co-author (with his building conducted in the late 1950s indicated satisfactory heating characteristics during 2 winter seasons.

The house had 1,200 square feet of floor area and an equivalent roof space; 600 square feet of solar collectors on the south-facing wall; and 275 cubic feet of fusion salt storage area. This amount of salt hydrates, or fusion salts, could store 2.5 million BTUs of heat—enough to warm the house for almost 2 weeks without sunshine!

On the basis of experience gained in testing the Princeton solar house, Olgyay concluded that a "design criterion" could be established for solar-heated houses. "A solar-heated house will break even, economically," he said, "if the installation of the system costs 2.3 times as much as . . . conventional heating equipment. If it costs less, it will save money."[6]

In 1953, a solar-heated house utilizing eutectic salt energy storage was built by Lawrence Gardenhire at State College, New Mexico. The

457-square-foot solar collector was built at a 45° tilt in 3 sections facing south and southwest. Air was circulated to a bin of 5-gallon cans containing the salt hydrate. Similar problems appeared with the salts, as in the early MIT experiment, however, and after a few cycles of melting and freezing, the storage system delivered only a fraction of the heat that it was supposed to store from the collector. This necessitated once again the use of an auxiliary gas furnace. Tests of the system showed that it supplied about half of the heating needs of the dwelling, limited mainly by the failure of the salt hydrate storage system.[7]

Dr. Telkes has remained unconvinced that the early experiments using eutectic salt heat storage systems proved that salt storage was undependable. She has more recently (in the 1970s) designed a number of storage systems for houses at the University of Pennsylvania and the University of Delaware that use methods of controlling the heat losses in the salts when they are exposed to the alternate heating-cooling cycles in a storage bin.

During the 1960s, she identified the basic problems with the early eutectic salt storage schemes. One of the problems with the salt hydrates was "incongruent melting," which occurred when, during the melting (heat-absorbing) process, some of the salt did not fully melt and became solid residue in the bottom of the container. One solution would be to mechanically stir the chemical solution to assure complete melting; but that, Dr. Telkes noted, "is highly impractical, especially in sealed containers. . . ." The problem should be corrected by adding other chemicals to the salt hydrate solution before it is sealed in the heat storage containers. "For this purpose," she said, "the salt hydrate is mixed with thickeners to change it into a gel when melted, preventing the settling of solid particles. Or the incongruent melt can be mixed with additives to prevent settling by keeping the solid particles in suspension," she added.

Another problem is "supercooling" of the salt hydrates, which occurs when the alternate heating and cooling cycles occur as solar heat is supplied to the sealed containers and heat is drawn from the containers. When this occurs, the solutions also do not deliver sufficient energy from the storage system. To correct this, Dr. Telkes has used chemicals—added to the salt hydrate solution before it was sealed in containers—that caused crystal formation, decreasing the supercooling effect. Such chemicals, called nucleating agents, increase the performance of the storage system. During the 1960s, Dr. Telkes found that the use of thickeners and nucleating agents dramatically improved the performance of the eutectic salt storage systems.[8]

Further improvement of the eutectic salt energy storage systems would lead to solar houses that could store energy at low cost, and the storage bins would require less space than use of rocks or water.

Solar Heat Pumps

Several solar-heated houses were built in the 1960s which utilize electric heat pumps, machines that utilize a small amount of external energy—fossil fuel or electricity—to "pump" heat from the outside environment (air, water, earth) to a higher temperature for use as heat in the house or building. The principle of the heat pump is similar to that of a refrigerator or air conditioner. A fluid having a low boiling point is compressed by a conventional compressor and circulated between an evaporator and a condenser. Heat energy is absorbed by the fluid being circulated through pipes within a closed space—inside a refrigerator or air conditioner, for instance—which causes the fluid to evaporate, and flow as a gas to the condenser outside the refrigerator or air conditioner, where the condenser coils dissipate the heat and cause the gas to condense back into the liquid phase, giving off heat—whereupon the cycle is repeated.

A heat pump, frequently called a "reverse cycle" air conditioner, has a reversible valve that can channel the heat transfer—or refrigerant—fluid in two directions. In the summer, the machine operates by cooling the interior space as a conventional air conditioner (with an input of external energy, which can range from electricity to solar heat). In the winter, the heat pump operates in reverse; it is used to warm the inside space by reversing the flow of the heat transfer fluid so that low-level heat from the atmosphere (or ground, or local water supply) is raised to temperatures needed for heating the house.

By the addition of a small amount of electricity (to power the compressor), the heat pump in effect uses a "free" energy source—heat from the earth, water, or air—for heating the house or building.

During the 1950s, a considerable amount of research was conducted aimed at applying the heat pump concept to solar house heating. Several houses and commercial buildings were powered by heat pumps using concentrated heat from solar collectors rather than atmospheric heat as a supplementary heat source for winter heating.

Probably the leading enthusiast of solar-electric heat pumps was the head of one of the nation's largest electric power companies—Philip Sporn, president of American Electric Power Company. Working with E. R. Ambrose, the chief of the company's air conditioning division, he conducted a number of tests on the solar power-augmented electric heat pumps. In 1950, solar collectors were mounted on a building in New

Haven, West Virginia; the collectors were designed to tilt at various angles relative to the sun's position. Four insulated 4-foot-by-7-foot solar heat collectors (112 square feet) made of steel plates and covered with 2 glass plates were used to trap the sun's heat in the experiments.

A refrigerant (or heat-transfer) fluid was circulated from the solar collectors to a water storage tank, where the solar heat was stored for use when it was needed as a heat source for the electrically operated heat pump. The collectors were highly efficient in absorbing and trapping solar heat. During the 5 years of testing (1950–55), efficiencies claimed for the unit ranged from 50 to 100 percent in converting solar radiation to heat. The solar collector was also used as a cooling device. By removing the glass plates from the front of the collector, cool night air (sky radiation) was utilized as "heat sink" during the summer months. Releasing heat to the night air cooled the water storage tank, permitting temperatures to be lowered within the confined space—in this instance, the living space of a house.

The solar heat pump tested by Sporn and Ambrose was a highly efficient means of trapping heat and using it for heating and cooling. The electric heat pump, using the solar collector, showed a coefficient of performance (COP) of 6 to 8. The COP of a heat pump is the ratio between the heat delivered by the unit (as heat into a home, for instance), and the power or work (i.e., an electrical compressor to pump refrigerant fluid) required to operate it. An average heat pump, not using collected solar energy as a heat source, may have a COP of 2. They claimed that the solar-electric heat pump was 3 to 4 times as effective in delivering heat.

Sporn and Ambrose noted that the performance of the solar collector would have been even better had it been in a more southerly location than West Virginia; but even so, "there were only 10 days from January to May 1955 when the heat pump did not operate because of insufficient solar and sky radiation."

Their enthusiasm for the solar-electric heat pump led them to speculate that a bright future was in store for heat pump technology both for heating and cooling structures *and* for hot water heating. Of the latter, they said that the heat pump water heater would be well-suited as "a self-contained quantity production item, and with proper development could have wide public acceptance in many areas. Both types of systems have the common advantage of being able to employ a relatively free long-life, hermetically sealed refrigerating system."[9]

Nothing much came of the experiments conducted by Sporn and Ambrose, except that the experiments demonstrated the technical soundness of the combined solar-electric heat pump system. In the

1960s, Ambrose noted that only a few actual systems were built, due to the high initial cost of the solar collectors combined with the heat pump. He pointed out that the use of the heat pump could provide significant improvement in conventional solar heating (or collecting) systems because the heat pump could utilize lower temperatures than most conventional solar heating systems could. This feature means that much of the heat normally wasted by a conventional collector could be effectively used. For example, a conventional solar heating system can't use some of the solar heat derived from the collector during early-morning and late-afternoon hours (or during cloudy periods)—but a heat pump combined with a solar collector can effectively use the low-temperature heat.[10]

An electric heat pump system was designed at the University of Kentucky during the 1950s that utilized two sources of energy: solar collectors and the heat storage capability of the ground itself. Both of the sources would supplement the operations of electrically driven heat pumps. A resulting study—conducted under the supervision of Dr. E. B. Penrod, head of the Department of Mechanical Engineering at the University of Kentucky—showed that a solar collector was an excellent heat supplement to a conventional electric heat pump.

During most bright, sunny days in winter, the ground itself would serve as a good heat source (or natural solar heat collector) for the heat pump, since as the sun warms the earth, the ground retains much of the heat. But during periods of cloudy weather, when temperatures plummeted, a conventional solar collector would serve as an "added" source of heat for the heat pump.

Additionally, the study showed that the collector could be used in a modified form as the condenser side of the heat pump, for air-conditioning purposes. Conventional heat pumps using a ground coil suffered from a common problem: "During summer months . . . the earth is a relatively poor heat sink due to the fact that moisture migrates away from the ground coil and the soil may not cling so tightly to the pipe," the study noted. This problem meant that less heat energy could be dissipated from the heat pump through the ground coil. Use of the solar collector could solve this.

The study concluded that a 400-square-foot solar collector made of blackened aluminum with 2 glass cover plates could be used in conjunction with a heat pump (circulating air as a heat-transfer medium) to provide cheaper winter heating than natural gas (assuming that costs of electricity for the heat pump were 1.5 cents per kilowatt-hour).[11]

Other detailed designs of solar heat pump systems were made by engineers Richard C. Jordan‡ and J. L. Threlkeld at the University of Minnesota at Minneapolis.

In conjunction with Dr. Harlan McClure of the University of Minnesota's School of Architecture, professors Jordan and Threlkeld calculated the engineering and economic aspects of solar heat pump systems for several house designs. They noted the problems of design, particularly since the houses would be one-of-a-kind, carefully planned to integrate with specially oriented solar collectors:

> The design of a solar house offers formidable problems to the architect. A large lot is necessary to accommodate a house of such a long, narrow nature; sufficient space in front of the collector is required to prevent shading from neighboring buildings and to prevent reflection from the collector from disturbing inhabitants of nearby buildings. In northerly regions the collector area may be so large that it dominates the exterior of the house and also complicates the admission of sunlight to the interior. . . . Regardless of the geographical location, any design must be a compromise between a well-engineered solar utilization system and a house whose livable qualities can compete with more conventional types of architecture.[12]

In 1954, they presented calculations of a prototype solar house to the annual meeting of the American Society of Heating and Ventilating Engineers. Characteristics of specially designed solar collectors in combination with heat pumps were given for 4 U.S. locations: Madison, Wisconsin; Lincoln, Nebraska; Nashville, Tennessee; and New Orleans, Louisiana. The conceptual 2-bedroom house would utilize fusion salt storage, a south-facing vertical (wall-mounted) solar collector, and an electric heat pump in each location. The heat-storage capacity of the fusion salt bin (in the basement) averaged about 8½ days for all locations. The necessary area of the solar collector surface ranged from a low of 226 square feet needed in the New Orleans location, to 715 square feet in Lincoln, Nebraska, to more than 1,000 square feet of collector area required to supply solar heat in Nashville, Tennessee, and Madison, Wisconsin. The solar collector was designed to utilize 2 glass panes to trap heat in the northern locations in the United States and one glass pane in the southern locations.

They concluded that the solar heat pump systems would result in significantly lower operating cost than conventional heating systems using fossil fuels in 2 of the locations: Madison and Nashville. They of course recognized the problem of high first costs and the custom-

‡ Dr. Jordan, one of the pioneers of solar energy utilization in the United States, is in charge of the large-scale solar power plant project at the University of Minnesota, in conjunction with Honeywell, Inc. See Chapter 6.

built nature of the system. "However, with standardization of collectors and heat-storage units, it would be possible to reduce these costs," they said.[13]

Based in large part on the early theoretical studies and tests of heat pumps—under laboratory conditions—a number of solar-heated structures were constructed in the 1950s using the newly developed electric heat pumps.

One of the more noteworthy experiments conducted during the period, involving the use of the heat pump with solar collection, was that of the Bridgers and Paxton engineering firm in Albuquerque, New Mexico. Partners Frank Bridgers and D. C. Paxton constructed an office building with a solar heat pump system, completed in 1956 and tested during the 1956–57 winter season. The building had about 3 times the floor area of an average house—4,300 square feet—and was equipped with an 830-square-foot solar collector connected to a 6,000-gallon hot-water storage tank in the basement. The collector was tilted at a 60° angle from the horizontal to maximize the winter collection of sunlight. Heat was supplied to the building directly from the solar hot-water system by means of water pipes in ceiling and floor panels. When the solar collector did not supply hot water for the heating—i.e., when it dropped below 90° F—the heat pump took over, drawing energy from the water storage tank and pumping it to a higher (110° F) level for use in heating the building.

The cost of heating the building with the combined collector-heat pump was $78.46 (in electricity at $0.02 per kilowatt-hour). The solar collector supplied about 63 percent of the heat directly for heating the building, and the heat pump was used the remainder of the time. Even when the heat pump was operating, however, it utilized the stored solar heat in the water tank, so its operation required only 8.2 percent of outside electrical energy. Therefore, solar energy supplied more than 90 percent of the building's heating requirements. Bridgers and Paxton estimated that an inexpensive natural gas furnace would have cost them almost $170 in gas fuel charges for the winter season (which they noted experienced less sunshine than normal). The expense of the solar heat pump system amounted to half the costs of a gas heater.[14]

Another solar heat pump installation was used to power the solar energy laboratory at the University of Arizona in 1959. Built under the supervision of Raymond Bliss of the Institute of Atmospheric Physics, the solar collector on the roof of the building covered more than 1,600 square feet—slightly larger than the floor area. In an unusual move, the solar-absorbing surface was painted dark green instead of black—for aesthetic reasons.* The heat pump system was similar to that used by Bridgers and Paxton in Albuquerque in that the water storage tank was used as a heat source for the electric heat pump when it couldn't supply warm water for heating the building. Heating of the building was similar in both cases; the Arizona building used a complex arrangement of water pipe circuits in the entire ceiling of the building to supply heat.

Unlike the Albuquerque building, a unique summer cooling method was utilized in the Tucson solar laboratory. The collectors were designed without glass covers, so that water could be pumped through the storage tank at night, to be cooled by natural evaporation in the dry Arizona air.† The water was returned to the storage tank and used the next day to cool the building. If the water was too hot to use directly for this purpose, the heat pump was used in conjunction with the stored water to achieve the cooling effect.

The system's cost was estimated at about $215 per year for electricity to power the heat pump—or about one half the energy costs of a conventional gas furnace and an electric air conditioner.[15]

Heat pumps were also utilized in conjunction with solar heating systems in Japan in the 1950s. Solar energy pioneer Masanosuke Yanagimachi built 4 solar-heated homes in the Tokyo area. In 1958, he completed a 2-story residence for his own family, which utilized a 2,460-square-foot solar collector covering the entire roof, in conjunction with 3 electric heat pumps. The central heat pump was used for house heating, and the others were used for heating water and a Japanese bath.

As was the case with the solar heat pump experiments in the Southwest, the solar collector was not covered with glass cover plates —so that it could be used at night for evaporative cooling. Water was circulated from the solar collector into a storage tank in the basement. From there it was pumped into ceiling pipes for heating.

Yanagimachi experienced some problems from leaking Freon gas (the refrigerant agent) in the heat pump, and problems with corrosion in the aluminum solar absorber plates. Otherwise, he reported that 2½ years of family living with the system had indicated great success.

"At the present stage," he said, "it is premature to appeal to the public with this solar system. However, it will become essential for

* One wonders if the dark green surface isn't as good as the solar-absorbing surface of the dark green leaves of trees, nature's well-adapted and proven solar collectors.

† This system of evaporative cooling was a sophisticated variation of "desert coolers" commonly used in dry climates for evaporative cooling.

almost all residences to have air-conditioning installations in the very near future. Then . . . it will become economically feasible to have such a solar system in every residence."[16]

Other U. S. Solar Houses

Returning to the United States, we find the development of yet another solar house at the Massachusetts Institute of Technology. Based on the three solar houses constructed by MIT in the 1940s, the fourth structure was built in 1958 at Lexington, Massachusetts. The 2-story house was designed with a large, sloping south-facing roof, which housed a large solar collector comprised of 60 solar hot-water heating panels. The total solar collector area was 640 square feet, and the integrated collector/roof was inclined at a 60° angle (from the horizontal) for trapping the winter sun.

The heating system was fully automated with thermostatic controls, and utilized forced-air heating, with the heat supplied by a 1,500-gallon hot-water storage tank in the basement of the house. An auxiliary oil heater was used to supply heat when the solar heat storage was not adequate for the job. The measured performance of the system indicated that the solar heating part supplied about half the heating needs of the house, a not-too-dramatic figure. The solar collector was covered with 2 glass plates, necessary for trapping heat in the cold New England climate; but even so, its over-all efficiency was just over 40 percent in converting available sunlight to heat.

MIT's C. D. Engebretson commented on the institute's decision not to utilize a heat pump system in the house, which might have boosted efficiency of the solar heating system. He said that since the house was not designed to utilize air conditioning—not required in the New England climate region anyway—the economics of the heat pump would not justify its use. Also, the temperatures, even in winter, in the solar heat storage tank would range so low that the heat pump would not be able to function effectively much of the time. "These temperatures [in the MIT solar house] will vary with individual systems," he said, "but they show quite clearly that there is a point below which the cost of reclaiming energy exceeds the value of the energy itself."[17]

This was the last solar-heated dwelling built by MIT. Even though funds still remain at the institute from Godfrey Cabot's generous grant for pursuing research in the field, MIT has done very little in the 1960s and '70s to further refine the technology of solar houses built in the previous decades.

According to a report on energy research needs prepared jointly by

Resources for the Future, Inc., and MIT, the prospects for combined solar energy systems with conventional fuel auxiliary systems may be more favorable in the future. Prepared under the direction of Dr. Hoyt Hottel, in charge of MIT's solar energy research, the report suggested to the National Science Foundation that:

> Little additional developmental effort is needed [for solar space heating of houses]. Any improvements in ruggedness, reliability, efficiency and cheapness of roof collectors will . . . hasten the day of acceptance of solar houses. . . . However . . . a constraint is put on the appearance and orientation of the house. Furthermore, solar heat is more expensive the smaller the installation. . . . For these reasons, and particularly because many householders still install heating on the lowest cost basis, solar heating cannot be expected to have a large influence on the U.S. energy scene for many years.[18]

All the previously described American solar-heated houses and buildings built in the 1950s have long since been either razed or modified to utilize conventional electric, oil, or gas heating systems. Only a few houses survived this initial era of solar experimentation. They were built—and designed—by two pioneers of the period, engineer George Löf of Denver, Colorado, and Harry Thomason, a patent attorney in Washington, D.C.

The Löf House

Dr. George O. G. Löf of Denver, Colorado, is one of the original U.S. proponents of the concept of solar energy utilization in residences as well as for large-scale applications.‡ He has designed and built 2 houses in Colorado utilizing solar heat collection, with energy storage in bins of small rocks. The first house was built in Boulder, Colorado, in 1947. The small bungalow, with 1,000 square feet of floor area, was partially heated by solar energy received by a 463-square-foot roof collector. The solar heat, stored in 8 tons of rock, saved only about 20 to 25 percent of the gas fuel used as the standard heat source.[19]

Dr. Löf's second solar-heated house was completed in 1958 and was financed by the American-Saint Gobain Corporation (which was also a partner with the University of Colorado in the first Löf solar house).

The attractive 9-room, single-story home serves both as a solar laboratory and as Dr. Löf's residence. The home has 2,050 square

‡ Dr. Löf is in charge of the Colorado State University/Westinghouse Electric Corporation study of solar thermal power plants; see Chapter 6.

feet of living space and an additional 1,100 square feet in the basement. A forced-air heating system is served by 2 300-square-foot solar collectors mounted at a 45° angle (from the horizontal) on the roof. The 2 collector rows contain 40 individual glass-covered panels, arranged in a special series so that air is circulated by a rock storage bin in the house to "cold" solar collector panels and then, as it is preheated by solar radiation striking this panel, the air is circulated to the second series of "hot" panels, which contain 2 glass cover plates, as opposed to the cold panels, which are covered with single glass plates.

The unusual heat-storage bins, which contain 23,460 pounds of 1½-inch-diameter granite pebbles, are 2 vertical fiberboard cylinders 3 feet in diameter extending from the basement to the roof.*

The heating system is carefully instrumented and is thermostatically controlled so that heat can be supplied either by direct use of hot air from the collectors; by air circulated through the heated pebbles; or from an auxiliary natural gas heater. Solar energy is used almost solely for winter heating, and the only use of solar heat in the summer is for preheating water furnished to a natural gas water heater.

Early tests of the system showed that the "net" efficiency of the system—including heat received by the collectors as well as heat stored and delivered by the forced-air system—was about 35 percent. The actual savings in fuel for the 1959–60 heating season amounted to $92.80; had the system used an auxiliary oil furnace, the savings (in oil costs) would have been even higher—about $130.

The system has been in successful operation for more than 15 years, and Dr. Löf still considers it an experimental unit. Some problems noted in the operation of the system have ranged from faulty fittings in the solar collector—resulting in loss of solar heated air—to roof leakage due to the weight of the collector panels.

Of the architectural design of the house, Dr. Löf's initial report on the system observed:

> In contrast with other solar-heated dwellings currently or previously tested, the plan and architectural design of the Colorado House were independent of the solar heating feature. In other words, if conventional heating had been installed, the same house would have been built.[21]

* Dr. Löf comments that, instead of the tubes—which extend through his living room—being considered an oddity in the house, he finds them aesthetically attractive. When confronted with the prospect of hiding them from sight, he decided: "Hide 'em, hell; I painted them red."[20]

The Thomason Solar Houses

Three solar-heated houses were built in Washington, D.C., in the late 1950s and early 1960s by Harry H. Thomason, a patent attorney with the Army Signal Corps and former refrigeration engineer. Thomason's first house utilized the entire south-facing half of the roof as well as the south wall for solar absorber space. The floor area of the house is 1,500 square feet, and the collector surface is 840 square feet.

The solar-generated heat is stored in a combination water-rock storage bin. Solar-heated hot water is pumped through the collectors in copper pipes to a 42-gallon water preheater drum, which is contained inside a larger 275-gallon drum for additional insulation. The hot water is then pumped from the preheater tanks and is stored in a 1,600-gallon tank, which is further insulated by 50 tons of stone. Heat is transferred from the large water tank at night by convection and conduction to the stone, which provides additional heat-storage capacity.

Thomason reports that the cost of the solar collector averaged about $1 per square foot, and that the entire cost of the solar heating system was only $2,500 at the time (the late 1950s). The house is served by an auxiliary oil furnace, and the whole heating system is thermostatically controlled. During the 1959–60 winter, the solar heat supplied 95 percent of the heating requirements. The supplementary heating bill for oil amounted to only $4.65. The next winter was the coldest the Washington area had experienced in 43 years, yet the supplementary fuel bill was only $6.30. Thomason calculated that his collector converted between 50 and 60 percent of the available sunlight into heat.

During the summer, the collector is used as a cooler. By circulating cool water from the basement storage tank through the collector at night, a limited form of evaporative cooling is achieved in the house.

Since the Washington summer is often quite humid, evaporative cooling is not as useful as in a dry climate, where such coolers are very effective. Nonetheless, Thomason reports that the night water circulation in the collectors kept the house 8 to 15 degrees cooler than outside temperatures during the summer. Solar energy from the collectors was also used to heat a small backyard swimming pool.† Thomason has built two other houses using essentially the same

† Plans of Thomason's heating system are available from the Edmund Scientific Company, Barrington, New Jersey.

system. However, limited economic or engineering data are available on them.[22]

The solar houses of the 1950s demonstrated a number of important principles about the use of solar heat. The projects showed that the sun could provide the energy required for space heating, but in almost every instance the experimental house failed to survive because solar-powered equipment could not compete economically with cheap fossil fuels.

After all, it is easy for a home-owner to select a conventionally fueled central heating system costing only a few hundred dollars when he or she is confident that fuel costs will remain relatively cheap. On the other hand, a solar heating system is an unknown quantity—and in the case of the solar houses of the 1950s, an unknown quantity that cost several thousand dollars and required considerable maintenance: The solar collectors had to be cleaned periodically; water, fusion salt, or rock storage had to be monitored; the solar-electric heat pumps had to be attended, and—as was the case with all-electric heat pumps of the period—their performance record was spotty and unreliable. Aesthetically, many of the solar houses of the period presented an unappealing glass-covered black expanse to passersby; and, for neighbors, jarring reflections created some disturbance.

Financially, the systems would have proved burdensome to the owners, since each installation was a different, hand-crafted installation that required custom care. The heating system and house design that were appropriate in Albuquerque wouldn't have worked as well in Kansas City or New York—if for no other reason than because of the different latitudes, which means different degrees of solar energy available for conversion to heat.

Even assuming that one of the solar heating systems could have been purchased on the open market, different solar collector designs would have been necessary for the specific climatic requirements of each area. The only commercially available solar equipment available in the 1950s and '60s were handcrafted solar water heaters sold by the few remaining solar water heater companies in Florida. Because of the cottage industry nature of their construction, costs were uncompetitively high—several hundred dollars for a solar water heater collector, which in turn was not easily adaptable to a complete house-heating system.

Even more important, the solar experiments of this era did not result in a rush to solar houses for another very significant reason: the technological problems revealed in the early designs.

Dr. Maria Telkes says that she did not solve many of the problems associated with the use of fusion salt storage chemicals until the mid-1960s. "The basic thing I learned [from the construction of two solar houses] was how to make a better collector," she explained in an interview. "I learned that better collectors can be made with efficiencies from 40 to 55 to 60 percent. The other thing was how to make them."

She pointed out that many of the houses were not properly situated on the building lot with regard to shading of trees, solar orientation, wind orientation, and other local climatic conditions. One widely touted house of the 1950s, she reports, "couldn't heat itself in the winter and was too hot in the summer. The architect put in too much [window] glass and the glass transmitted solar energy. He used just one single pane—not even a thermal pane [two glass sections with an insulating layer of air between them]. During the summer it transmitted too much [sunlight] and so it just never worked."[23]

All of these early design problems have led to significant improvements in the construction of all components of solar heating systems. Rather than proving the inadequacy of solar heating systems, the early experiments have revealed the flaws of specific systems—flaws that have been corrected in more recent designs.

A recent technical memorandum prepared by the Arthur D. Little consulting firm outlined the difficulties with the early experimental houses:

> . . . . Most of the solar heated houses were designed for heating only [as opposed to heating and air conditioning] and due to the relatively poorer utilization of the collection equipment (winter service only) were not as economically competitive as if they had incorporated cooling as well. Lastly these houses were frequently quite experimental in nature and in no instance were designed to be readily adaptable to industrial construction. Therefore, the prior experience which developed the technical feasibility of solar climate control did not adequately demonstrate its practical and economic viability.[24]

Economics

Today, many engineers and scientists working in the field of solar power for decentralized applications maintain that not only are new solar house designs sound from the standpoint of engineering, but from the standpoint of economics as well.

Two comprehensive studies comparing costs of solar-heated houses with costs of houses heated with conventional fuels have been conducted by Dr. George Löf in collaboration with economist Richard

Tybout, professor of economics at Ohio State University. A preliminary study published in 1970 on house heating economics was expanded, and in 1973 the authors published the results. The sophisticated survey, based on the performance of solar heating systems in a number of U.S. locations, showed that solar heat is economically practical now in most U.S. locations.[25]

The survey utilized more than 400,000 hourly recorded observations of solar radiation, temperature, wind regime, cloud cover, and humidity in each of 8 U.S. cities. The data were programmed into a computer and matched against performance capabilities of solar collectors of known design and their related heat-storage systems, along with the energy demand, insulation requirements, house size, and other important factors involved in designing solar-heated dwellings for each of the specific geographic areas.

The study made two major contributions: (1) optimization of the economic design of solar heating systems; i.e., the amount of space required for the collector and for storage, and the collector orientation that would be most economic in each location; and (2) establishment of realistic costs of solar heating vs. recorded costs of conventional fuels used in house heating today.

The study compared the costs of solar heating systems supplemented with electricity and with conventional oil and gas-fired furnaces, to the costs of nonsolar-supplemented electric, oil, and gas-heating systems. Costs of the solar heaters were based on amortizing the high initial costs at 6 percent interest over a 20 year period.‡

The initial costs of the solar heating systems were estimated for 2 different sizes of home: (1) a large house in the "upper-middle-income" bracket, with a heat demand of 25,000 BTUs per degree-day of heat; and (2) a house in the "middle-income" range, with a heating demand of 15,000 BTUs per degree-day. Controls for the solar heating system were estimated to cost a total of $375, exclusive of a conventional house heater (used as an auxiliary with the solar heating system) and associated heat distribution pipes or ducts.

These costs are broken down as follows:

- Water storage system—one-half cent per pound of water stored multiplied by the number of square feet of solar collector space used.
- Controls (thermostat and associated equipment)—$150 for all systems, regardless of size.

‡ Even using a higher amortization rate than 6 percent, costs of solar heat are appreciably cheaper than those for electric heating today.

- Pipes and fittings—$100 plus an additional 10 cents times the number of square feet of solar collector area.
- Motors and pumps (for circulating water or forced air)—$50 plus 20 cents times the number of square feet of solar collector area.
- Heat exchangers (for transferring solar-heated water in the house) —$75 plus 15 cents times the total square feet of collector area.

The cost of the solar collector were estimated in two ranges: (1) the estimated costs of solar collectors manufactured today in factor lots suitable for installation—concluded to be $4 per square foot of area; and (2) the cost of a solar collector manufactured on a truly mass-production basis to fulfill the demands of a future upsurge in orders for solar-heated homes—estimated to be about $2 per square foot of solar collector.

The 8 U.S. locations considered were selected as representative of the various climate zones: Miami, Florida—tropical savannah climate; Albuquerque, New Mexico—tropical and subtropical steppe; Phoenix, Arizona—tropical and subtropical dessert; Santa Maria, California—Mediterranean or dry summer subtropical; Charleston, South Carolina—humid subtropical; Seattle, Washington—marine West Coast; Omaha, Nebraska—humid continental, warm summer; Boston, Massachusetts—humid continental, cool summer.

Results of the study indicated that in 6 of the 8 cities surveyed a good solar house system would provide cheaper heat than *electrical* heat at prevailing power rates. In only 2 cities, Miami and Seattle, would economics favor electricity over solar energy for home heating. In Miami, it turns out that home heating is not required enough of the year to warrant the high capital costs of solar space heating systems; and in Seattle, very cheap electricity available from the government-subsidized hydroelectric dams of the Pacific Northwest make solar heating uneconomical in comparison.

On the other hand, the solar heating systems could not compete effectively with conventional home furnaces fueled with oil and gas at prevailing fuel costs, so that solar heating systems were found to be cheaper in only a few of the locations studied.

However, because the costs of oil and gas have risen appreciably since the study was made, even this economic indicator has changed, and solar home-heating systems might be economically competitive in more areas today.

The study was made to ascertain the economic feasibility of solar energy—not necessarily what the most effective solar heating system would be, but what the least expensive and most economically competitive would be. This aspect of the study affected the design objectives of the houses. For example, the heat-storage requirements covered only 1 to 2 days of winter demand, and the entire solar collector heating system was designed to supply about half of the house heating needs in most locations. In only a single very favorable location—Santa Maria—would the small solar heating system supply three fourths of the winter heat requirement, using a relatively small collector of 261 square feet. In Charleston, a 208-square-foot collector would supply 55 percent of the heating need; and in Omaha, a 521-square-foot collector would supply 47 percent of the heating needs.

The authors found that only a single glass plate would be needed to cover the solar absorber plate in the Phoenix and Miami locations, but 2 plates would be required in the other cities.

They said that "it is probable that solar heating costs will decrease somewhat as improvements are made," and that "competitive solar heat will become increasingly possible as these trends [the spiraling costs of conventional fossil fuels] continue."

They further concluded that:

> Conditions conducive to economical solar heating are moderate-to-severe heating requirements, abundant sunshine, and reasonably uniform heat demand during the period when heat is needed. The higher the cost of conventional energy for heating, the more competitive a solar-conventional combination becomes.[26]

The value of this economic survey is that it emphasizes the principle of life-cycle costing—i.e., the cost and maintenance of materials that go into the house (including the heating system) should be amortized over the lifetime of the house. Today, most houses and buildings are constructed without great concern for the durability of their components. The only criterion is low first cost. As has already been noted, this viewpoint was made possible largely by the abundant, cheap fossil fuel energy resources that have shaped the energy monoculture of the current U.S. society. As long as such cheap fuels—and, concomitantly, cheap machines and building materials made possible by cheap energy—were available, why should builders and owners be concerned about fuel costs over the lifetime of a house? Why should they care about the possibility of replacing a cheap electric water heater every 5 or 6 years? As long as long-range availability of cheap fuels was assured, there has been no need for concern. Today, however, the picture has changed. The rosy days of cheap energy are over.

Thomas A. Robertson of the Energy Center at the University of

Florida discovered just how quickly the attitudes toward life-cycle costing have changed in regard to solar energy. In the summer of 1973, Robertson conducted a survey of houses with solar water heaters in Miami, Florida. As has been noted, while thousands of solar water heaters were built in Florida in the earlier years of the century, the industry today has dwindled to a few companies building only a few heaters per year.

Robertson talked with a number of residents of homes with solar water heaters in Miami and discovered an interesting fact: People who have purchased houses in which solar water heaters had been installed decades before, tended to have them repaired and remained quite happy to use them. On the other hand, landlords who *rented* houses containing old solar water heaters tended to install electric water heaters (bypassing the solar unit), letting the occupant pay the electric bill and relieving themselves of the responsibility for maintenance and other costs.

The survey clearly indicated that individuals interested in the lifetime costs of water heating—the resident owners—repaired and used the solar units; but landlords who couldn't care less about life-cycle costs, since the tenant would have to pay the electricity bills, preferred the more convenient electric heaters.

The nationwide application of solar energy for the purpose of water heating alone could have a tremendous impact on national energy demands. At present, the energy requirement for water heating represents about 3 percent of the nation's energy budget. In some areas, it is up to 6 percent. However, if all the water heating in the country were accomplished by electricity, this use alone would consume about a third of the nation's entire electrical generating capacity, according to Dr. Jerome Weingart, formerly of the California Institute of Technology's (Caltech's) Environmental Quality Laboratory in Pasadena, California.

Dr. Weingart studied the problems of solar water heating—vis-à-vis conventional water heating—in collaboration with a group of scientists from Caltech and the nearby federally sponsored Jet Propulsion Laboratory.

Dr. Weingart predicts that the use of solar water heaters supplemented by natural gas furnaces or by electricity will become economically competitive with conventional fossil-fueled or electric water heating systems in this decade, even though the solar water heaters will cost a minimum of $100 to $150 per unit. He notes that of all solar energy technologies, solar water heaters have demonstrated economic feasibility in the past, and offer the prospects of immediate commercial success today—assuming they can be properly developed and mass-produced. One likely avenue toward accelerated development is the installation of modular solar water heater systems in mobile homes, which now approach 50 percent of national residential construction.[27]

The Caltech scientists with whom Dr. Weingart has been associated are involved in a current project funded by a natural gas company in Los Angeles, a project designed to launch the rediscovered solar water heater technology with a supplemental natural gas auxiliary heater. According to another scientist in the group, Dr. Richard Caputo of the Jet Propulsion Laboratory, a basic drawback to the reintroduction of this technology is the current high cost of solar collectors.

The solar water heaters which they designed for a 30-unit apartment building added up to $15,000—which works out to $500 per unit, many times higher than any economic estimates the group had made. He reported in 1973 that an elaborate investigation was being made of the possibilities of mass production in bringing the initial costs down. But, he added, "we really don't understand where all the costs are."[28]

Another economic problem is the rising cost of borrowing money. In 1973 alone, a series of increases in the government's prime lending rate raised the costs of loans to unprecedented levels. This rise in the lending rate could offset much of the projected gain in savings by the use of solar energy equipment for household heating and other applications.

Solar Air Conditioning

A number of other factors may further increase the economic benefits to be derived from the use of solar energy in residences and buildings. The most important of these factors is the use of solar air conditioning, which will permit year-'round use of the solar systems.

The only use of solar energy to cool the solar houses of the 1950s and 1960s was by allowing water to circulate in reverse—from the storage tank back into the collector at night, giving off heat to the cool night air. The cooled water could then be used during the day to reduce temperatures inside the house. However, this works well only in a dry climate. In a humid climate, this type of cooling is not particularly effective, because it cannot control humidity.

A number of tests have been conducted using more sophisticated solar-powered air conditioners. Unlike the electric heat pumps in the earlier solar experiments, which utilized solar energy as a supplementary heat source for heating in the wintertime but used electricity to cool the buildings in the summer by conventional compressors, a completely different kind of heat-pump air-conditioning system has been tested, which uses solar energy as the winter heat source and as the summer heat source to operate the air conditioner in the house.

This device is not an electrical heat pump, but a heat-actuated heat pump. It uses heat supplied by the sun (or by conventional fuels as well) to vaporize the refrigerant fluid in the air conditioner—as opposed to the electrical heat pump, which compresses the refrigerant fluid mechanically with the compressor.

The most significant series of tests with this type of solar-powered equipment has been conducted by Dr. Erich Farber at the University of Florida over the past two decades.

As Dr. Farber noted, there is a distinct advantage to this equipment: ". . . Refrigeration, cooling and air conditioning are needed most when the sun shines hottest—concurrently with the time the greatest amount of energy is available. This makes solar energy an ideal source of energy for refrigeration [or air-conditioning] equipment."[29]

Dr. Farber and his associates built a number of solar heat-actuated refrigeration and air-conditioning machines during the '40s and '60s. A solar-powered air conditioner large enough to cool a house was built in 1964. The unit had a 3½-ton capacity, and used 400 square feet of solar collector surface. Since the air conditioner was designed to operate on low-temperature solar energy, the solar collector configuration was simple: 8 panels of the basic solar hot water heater absorber design—blackened metal and copper tubes covered with a single glass plate.

The machine utilizes ammonia and water as the refrigerant fluids. The cycle of operation is as follows: First, water is circulated through the solar collectors, gaining heat, which turns the water to steam, which then passes through coils inside the air conditioner's generator, filled with ammonia and water. The ammonia in the generator vaporizes with the addition of the heat from the solar-heated steam. The ammonia-water mixture then enters as a fine mist into an absorber tank, where it encounters and absorbs more ammonia vapor. In a series of additional stages, the ammonia vapor is separated from the liquid water and is carried to the plant's condenser, where it is condensed into concentrated liquid ammonia. From there it is circulated to the evaporator, where it is again vaporized. The evaporative process cools the gaseous ammonia to a chilly vapor. The chilled vapor then cools water circulating in coils through the evaporator. The chilled water is in turn used to cool the air, which is circulated by the air conditioner.

The advantage of this system is that it can be used at the low temperatures easily achieved by a simple flat-plate solar collector. It was operated experimentally in a range between 160° F and 190° F. Dr. Farber estimated that a house using such a system would require about 1,000 square feet of collector area. "Furthermore," he said, "if the solar collectors are placed on the roof, the cooling load on the house is decreased tremendously, thus reducing the cooling requirement."[30]

Other tests were conducted at the University of Wisconsin in the early 1960s on solar-powered absorption air conditioners. The Wisconsin studies used commercially available heat-actuated air conditioners built by Arkla Industries Inc. for use with natural gas as the heat source.* Instead of ammonia and water, the Arkla absorption cooling unit operated on a solution of lithium bromide and water. Instead of a natural gas flame, a small solar collector (107 square feet) was used to supply heat for the 3-ton unit. However, because the solar heat supply resulted in a lower temperature (175° F) than the unit was designed for, the rated cooling capacity was trimmed by a third, to 2 tons.

Dr. George Löf, one of the directors of the project, commented that energy-storage methods were investigated to allow for long-term operation of the air conditioner. For office use during daytime hours, he said, "storage for the rather heavy cooling load in the late-afternoon hours is desirable. Energy will be stored either in the form of hot water for use in the refrigerant generator when the solar energy has increased in the afternoon, or in cold water for use in the air-cooling unit as needed." Dr. Löf added that success in the Wisconsin and Florida solar air-conditioning experiments might lead to the use of an advanced system that would require the use of a 300-to-500-foot solar collector for supplying the cooling needs of an average U.S. home.[31]

Another recent approach to the problem of cooling with solar energy has been pursued by Professor Robert K. Swartman at the University of Western Ontario, London, Ontario, Canada. Professor Swartman and associates have developed a similar type of solar refrigeration machine, which uses ammonia in combination with sodium

* In 1973 discussions with the author, Dr. Philip Anderson, vice president of Arkla Industries Inc., reported that the small units are no longer manufactured by the company.

thiocyanate. Tests in 1971 showed high operating performance, and Professor Swartman plans to modify the unit for use as an icemaker in underdeveloped countries that boast a great deal of sunshine but few other resources.†

Professor Swartman claims that the ammonia-thiocyanate machine is both simpler to construct and operate, and more efficient, than other types of solar-absorption cooling machines. The working fluids in the other machines "have been relatively inefficient in performance or rendered the machines complicated and expensive to construct," he said. His 1971 tests showed that the unit's actual performance closely matched theoretical calculations. Unfortunately, he was unable to raise the necessary funds by 1973 to ship the machine to Africa for its intended use as an icemaker.[33]

In areas where solar air conditioning would not be necessary, but air dehumidification would, solar-powered dehumidifiers might be used. Tests conducted in the 1950s by the American-Saint Gobain Corporation showed that solar-absorption cooling units could be modified to serve as dehumidifiers. The tests employed a small 110-square-foot solar collector, which supplied heat to the dehumidifier in which the working fluid was triethylene glycol. At low temperatures of 140° F to 175° F, the solar heat could effect the removal of 6 to 14 pounds of water per hour from experimental fluids. This indicated that a solar dehumidifier using a few hundred square feet of collectors would perform favorably for household use in competition with fuel-fired units.

In humid areas, such a device might enable a solar collector designed for nighttime evaporative cooling to also function as a daytime cooler. Since the humidity could be controlled, the effectiveness of the nighttime cooling could be extended beyond the dry-climate areas to which its effectiveness is limited today.[34]

Solar Houses and Buildings of the 1970s—Everett Barber's Designs

The belief that the use of solar energy for heating and cooling would be particularly economic in large installations led a young National Bureau of Standards (NBS) engineer in Washington, D.C., to design a solar power system for a government-owned building in

† The original unit was developed by Professor Swartman in conjunction with Ron Alward, a young scientist now at the Brace Research Institute in Montreal. Alward served in the 1960s as a volunteer worker in Africa, where he was struck by the plight of people with no way to refrigerate—hence to preserve—food. He returned to Canada committed to the development of solar energy for use in underdeveloped countries, and teamed up with Professor Swartman at the University of Western Ontario.[32]

Washington, D.C., in the early 1970s. Large-scale solar energy applications would be economically attractive, he reasoned, because of the ready commercial availability of large heating and air-conditioning system components that could be adapted to a solar heat source.

As an avocation, NBS engineer Everett M. Barber, Jr.,‡ had built and experimented with solar flat-plate collectors for several years. One of his small test collectors employed a plastic acrylic cover plate, a blackened copper plate underneath to absorb heat, and an advanced "honeycomb" mesh of shiny aluminized Mylar strips arranged vertically between the cover plate and the absorber. Such a honeycomb design significantly reduces the convection of heat from the collector, improving its energy-conversion efficiencies.* With this collector, Barber produced temperatures higher than 300° F—heat sufficient to operate an array of machines ranging from heaters and cookers to more sophisticated air-conditioning equipment.

By 1971, Barber was convinced that combining his knowledge of mechanical systems for heating and cooling buildings with his experimental work in solar energy could lead to a large-scale demonstration project. Reading a local Washington, D.C., newspaper one morning, he happened across a notice that an architectural meeting was being held at a local amusement park owned by the National Park Service. He attended the meeting, met several Park Service officials, and discussed the prospects of converting a Park Service office building at the park to solar-powered heating and cooling. The resulting interest on the part of the Park Service convinced Barber that he should investigate the precise energy needs of the building in order to determine what would have to be done to meet them with solar power.

He first gathered available U. S. Weather Bureau data to determine the average seasonal amount of solar radiation in Washington. A sophisticated computer program enabled him to ascertain the amount of solar heat energy that would have to be collected to heat and cool the 2-story building located at the Glen Echo Amusement Park. After numerous telephone calls and letters to manufacturers around the country, he compiled a list of available heating and cooling equipment that could be used in conjunction with a solar collector of his honeycomb design.

Barber found that solar energy could economically provide almost 80

‡ Barber has established a company to manufacture and distribute solar energy collectors. The address is Sunworks, Inc., 669 Boston Post Road, Guilford, Connecticut 06437.

* The original proponent of the honeycomb design was K. G. T. Hollands, in an article, "Honeycomb Device in Flat-Plate Solar Collectors," *Solar Energy*, Vol. 9, No. 3 (1965).

percent of the winter heating requirements of the building with a 3,000-square-foot roof collector, on the basis of a conservative assumption that the solar collector would be 50 percent efficient in trapping heat.

He found that the total cost of installing solar heating and cooling equipment would be $46,000. This would include a new rooftop cooling tower, 2 water heat storage tanks, an absorption air-conditioning system, improved duct work for circulating heat for the building, and the solar collectors. At $4 per square foot, the 3,000 square feet of collectors would represent $12,000 of the total.

According to Barber's calculations, the building—as it stands today, with fuel oil heating and electric air-conditioning—will cost a minimum of $30,000 to heat and cool over a 10-year period, assuming no rise in fuel or electricity costs, zero cost of maintenance, and no inflation. Furthermore, many of the modifications required for adapting to solar heat and air conditioning would have to be made in any case. If the building were to be electrically heated and electrically air conditioned, the yearly cost of operation—subject to the above constraints—would cost between $8,000 and $9,000 *per year* to operate—upward of $80,000 in a decade.

A look at the solar-powered building modifications shows that the accumulated energy costs with 3,000 feet of solar collectors and improved insulation would drop the operating costs of the building over 10 years to less than a tenth of the conventional system cost of $30,000. Even in 18 years of use, total operating costs would not rise above $5,000, Barber calculated. He estimated that payoff time for the solar installation would be between 6 and 11 years.

The key to success in the year-'round solar system in the building, he found, was the fact that the existing heat-actuated air-conditioning equipment was available from Arkla Industries, Inc. Barber's plan for the Glen Echo building called for the use of 23 tons of air-conditioning capacity, and the company reported to him that they would modify a fuel-fired absorption chiller (as they are called) for use with the solar heat source. The proposed combination of winter heating and summer cooling would work out well in this situation, because the 3,000 square feet of collector area was sized properly for the winter heating job—leaving more than enough area for summer cooling when more sunlight is available, lessening the need for a large collector area.

Barber points out how significant the availability of this equipment is for the designer of a large-scale solar-powered building. The 23-ton unit he planned for the Glen Echo building "is the smallest liquid absorption chiller that's available now, without fairly significant modifications. In other words, you cannot cool a house right now with an absorption chiller that's an off-the-shelf item. . . . You *can* cool a building that requires 23 tons or more. Once you get above 100 tons, then there are 5 or 6 manufacturers of absorption chillers from 100 tons up to 2,000 tons, so there's no problem [in adapting large buildings to solar collector heating and cooling systems]."

The attempt has an unhappy ending, no matter how promising the potential energy and economic savings to the government depicted by Everett Barber may seem. After a series of meetings in 1972, the National Park Service rejected the project, citing an unwillingness to experiment with a concept so new and untried.[35]

Shortly after his failure to convince the National Park Service of the exciting potential of using solar energy, Barber resigned his position with the National Bureau of Standards, where he had also unsuccessfully attempted to initiate a solar energy development program, to teach a full-time course in solar energy and energy conservation technologies at the Yale University School of Architecture.†

In cooperation with Fred S. Dubin, the energy conservation specialist cited earlier for his work in developing decentralized Total Energy systems, Barber has designed a solar heating system for use in a building in Manchester, New Hampshire, being constructed by the federal government's purchasing organization, the General Services Administration (GSA), to demonstrate optimized energy conservation systems.

Barber has devoted considerable effort to analyzing ways to incorporate solar collection systems in new houses—including holding discussions with officers of banks and mortgage companies to identify the problems in financing the systems. He says that over a 25-year house lifetime, if the costs of a good solar heating system—about $3,400—were amortized at an 8 percent interest rate, about $13 would be added to the monthly payments (there would also be a little less than $200 extra added to the down payment under Federal Housing Authority [FHA] standards, which require a 10 percent down payment).

On this amortization schedule, a solar system would represent a savings of $18.67 per month over an electric heating system, and just under $10 per month over fuel oil heating.

Largely spurred by the efforts of Everett Barber, Dubin has recently designed several potential solar heating systems for houses and buildings, in conjunction with other energy-conservation equipment. Dubin's work on energy conservation techniques for federal goverment buildings (described earlier) includes several novel designs for integrating solar collectors with other mechanical equipment.

† The National Park Service has since decided to use Barber's collectors on the Glen Echo building.

The Wilson House

Two avid proponents of solar houses are architects P. Richard Rittelman and Alva Hill of the Pittsburgh, Pennsylvania, architectural firm of Burt, Hill and Associates. The two architects became interested in the use of solar energy for heating and cooling purposes in 1971, when they proposed the design of a solar-powered school building in Pennsylvania, a building that was to be funded by a consortium of glass and metal companies as an attempt at energy conservation in architecture. Although their proposed school was designed, it was not built.

This did not deter them, and the two continued to pursue the solar area. Rittelman, in fact, served on the 1972 panel of solar scientists commissioned by the White House to evaluate the potential of solar energy. During that year, he also designed a solar-heated home for a former employee of the National Park Service, Mrs. A. N. "Nicky" Wilson, a long-time believer in the use of solar energy as a solution to energy problems, who wanted to build the solar house on her farm at Martinsburg, West Virginia.‡

Rittelman's design incorporates a number of unique features, including a greenhouse for house plants; a 45°-angled solar collector, which will supply more than three fourths of the house needs; the possible use of a small electric wind generator for electricity; and a Swedish aerobic waste digestion system, which converts human and kitchen wastes to organic compost for gardening. Unlike the anaerobic digestion systems described earlier, the Swedish composter—manufactured by Clivus Company—works by harnessing the digestive powers of oxygen-seeking bacteria. Other house innovations include the use of an experimental array of silicon solar photovoltaic cells (supplied by Dr. Joseph Lindmayer's laboratory at the Clarksburg, Maryland, COMSAT Corporation) to test the feasibility of using a small quantity of electricity for household needs—such as operating the electrical pumps that circulate hot water in the solar-energy collector-storage system.

Mrs. Wilson has attempted to get research funds from the National Science Foundation (NSF) and other government agencies to help finance the experimental house. As this book was completed, she had not secured financing for the house.

Wolf von Eckardt, Washington *Post* architectural critic and author, wrote that the house design with its modern architecture "integrates a

‡ Mrs. Wilson derived much of her interest in solar energy from a National Park Service colleague, John Hoke, a solar energy advocate and author of an excellent children's introductory book entitled *Solar Energy*, published by Franklin Watts, Inc., New York (1968).

modest house—a Frank Lloyd Wrightean 'cathedral' living room, a study, three bedrooms, and a garage which is connected to the house by a greenhouse-breezeway—with a small solar energy plant. . . ." Going farther, he predicted that houses of such modern and functional architecture as the Wilson home "may turn out to be as important in the history of architecture and, in a sense, of modern civilization, as Sir Joseph Paxton's Crystal Palace at the London World's Fair of 1853 and William Le Baron Jenney's Home Insurance Building at Chicago of 1883."[36]

Somewhat more modestly, the house's architect, Richard Rittelman, points to another feature, which may prove even more likely to bring about the increased acceptance of solar houses. His design for the solar heating system provides a completely independent control and storage system—in the basement of the Wilson house—which can be manufactured for any solar house, and installed in a basement energy storage room. Unlike the customized installation of other solar power systems, his design can be built into the house and hooked to the collector by joining together a few pipes. This would make it possible to design special collector systems for houses across the country to meet the specific solar energy needs of a given region; and independently of the collector and architectural design of the house, the mass-produced central storage system could be "plugged into" the finished house, reducing costs.

"Industrialization" of Solar Energy

Testing of solar heating and cooling equipment in university laboratories is one thing; developing integrated equipment of rugged, workable design to serve a potential new consumer market is another. "Industrializing" solar energy equipment has long been the dream of scientists in the field—yet few of the solar pioneers have had the business experience and foresight to span the critical gap between research and development, and securing consumers' acceptance for solar equipment in mass markets.

Dr. Peter Glaser, of Arthur D. Little, Inc., Cambridge, Massachusetts,* has long been convinced of the value of earlier experimentation aimed at greater utilization of solar energy on the individual household level. Of the previous solar house experiments, "most of the installations have been technically successful, and extensive performance data have been obtained on various modifications of the air and water heating systems," he observed.

* See Chapter 6 on the Glaser solar satellite concept.

Going farther, however, Dr. Glaser is interested in the all-important economic aspects of solar houses:

> In most United States locations, residential heating with solar energy would be somewhat more costly than present conventional means because of the relatively high cost of equipment still under development. If produced on a large scale, however, solar house-heating systems could be competitive, particularly when all the hidden environmental costs are accounted for.[37]

Dr. Glaser's conviction that all that would be needed to bring solar house equipment from the laboratory to the market is mass production has led him in the past few years to pursue a number of major American industries and trade associations in an attempt to convince them to underwrite a development program for the decentralized applications of proven solar technology.

His first attempt to interest a trade association in the concept was initiated in 1971, when he contacted the American Gas Association (AGA) in an effort to launch a joint Arthur D. Little, Inc./AGA solar development effort. Under a proposed Arthur D. Little contract, the gas industry group would spend $197,000 to build and test an upper-middle-income-level home using solar heating and absorption cooling equipment in conjunction with an auxiliary natural gas furnace. Construction would require 11 months, and for 2 additional years, Arthur D. Little personnel would conduct periodic checks of the house—which would be occupied by tenants selected by the American Gas Association—to ascertain the performance of the solar equipment and to evaluate the economics of the system.

To buttress their case, Dr. Glaser's team at ADL prepared a technical memorandum on the engineering aspects of solar house technology—singling out the Washington, D.C., area (where the American Gas Association is located), and comparing known solar equipment performance with expected performance of a house in the Washington area.[38] Discussions with the AGA staff proceeded smoothly, but the project bogged down in the spring of 1972, when the AGA's board of directors rejected the proposal, citing uncertain economics of the solar/gas-powered house as a major factor.

This setback did not deter Dr. Glaser or other team members at Arthur D. Little. For more than a year, they forged ahead with plans to expand the project and interest a consortium of industries in the development scheme. By the spring of 1973, Arthur D. Little announced publicly (including a display ad in *The Wall Street Journal*) a conference in Fort Lauderdale, Florida, that would bring together interested industries in what they described as "A Program to Develop Solar Climate Control Systems." The 2-day March conference session was packed with representatives of numerous corporations, ranging from the glass and metal industries to International Telephone & Telegraph and Sears, Roebuck. In addition, several representatives of prominent environmental groups attended. Dr. George Löf, a special solar consultant to the project, discussed the prospects for solar house economics, on the basis of his pioneering work with Dr. Richard Tybout.

Arthur D. Little unveiled its plans to "industrialize" solar house technology, and called for a 2-phase program—each phase to last 9 months. The first phase would identify the technical, economic, and marketing prospects for solar equipment for 4 geographic regions of the United States. Later, in Phase II, the accumulated data would be used to formulate specific production and marketing plans for the use of sponsoring industries.

One of the ADL spokesmen at the meeting, D. Elliot Wilbur, a housing and construction industry consultant, made it clear that "the essence of what we're saying at this conference is that we cannot see solar energy strictly from the point of view of the technician. . . . The real key is, we're not trying to build a solar house. We're trying to build [a solar] industry."

Wilbur went on to point out 5 major reasons for the probable success of a new industry effort in the area at this time:

1. Air conditioning would be incorporated with heating in the house design. He pointed out that 40 percent of new houses and 80 percent of new apartments are air conditioned when they are built.

2. In the past, he noted, "solar houses have been built to prove a point from the technological view, so they had large collectors, large storage capacities, to get everything from the sun." The ADL/industry project would be oriented, on the other hand, toward using supplemental heat from conventional oil or gas furnaces, or electric auxiliary power, to bring initial capital costs down.

3. The way construction is financed, he said, can provide a boon to solar space-conditioning equipment, since the equipment can be amortized as a part of the initial construction cost. "We have to play the game of housing construction the way it's played"; and the game, he said, is "a financial mortgage-based business."

4. The fourth reason for success in the project would be that the solar house equipment would be designed "for the construction industry, not space engineers. . . ." The solar equipment would be designed to fit into various standardized types of housing plans. The solar houses

would not be individualized architectural projects, but would consist of interchangeable systems that could be used in conventional mass-produced houses. "In other words," he added, "if we want a Cape Cod, if we want a Levitt house, if we want a Techbilt, the [solar] system must be flexible and not impose its own style on the house."

5. The final reason for prospects of success today, Wilbur said, is the nature of the developing centralized building industry. He noted that today, 3 percent of the houses being built in the United States are by builders who build 1,000 or more houses a year. But by 1980, 25 percent of houses are predicted to be built by home builders of 1,000 or more. He suggested that this increasing consolidation in the industry brings "sophistication in decision making, and with this consolidation comes different kinds of brains working on the problems of housing." Along with the consolidation of the industry, he added, has come the rise of apartment buildings over houses, which affords an excellent opportunity for the use of solar equipment, since potential objections over the aesthetics of large glass solar collectors would be minimized. "People will rent something they wouldn't buy," he concluded.[39]

The enthusiasm of the industry participants was sufficiently aroused at the conference to warrant initiation of the first phase of the project. Arthur D. Little announced in May 1973 that the project—which was to cost between $300,000 and $500,000—was off to a start. Participating industries included electric utilities, oil companies, and chemical companies—as well as the firms that stand to gain the most from success: the air-conditioning, glass, and metal industries, and a handful of small companies specializing in solar energy development.

And the elusive goal? "Arthur D. Little anticipates that new markets for solar climate control systems will approach $1 billion worth of equipment over the next 10 years," the company predicted in announcing initiation of the project.[40]

Solar House Development at the University of Delaware

The most ambitious university program in the United States aimed at developing the full utilization of solar technology is being conducted at the University of Delaware under the direction of Dr. Karl W. Böer, a specialist in the conversion of electricity by the photovoltaic solar cell process.

Over the past several years, Dr. Böer has expanded the development of a small group of energy researchers into the Institute of Energy Conversion. Today, the institute boasts a budget of $500,000 and a staff of over 20 full-time scientists and engineers devoted to the development of numerous technologies that they hope will lead to the commercialization of solar house technology by 1980.

Dr. Böer, a long-time expert in the conversion of solar electricity with cadmium sulfide solar cells, has attracted several top solar cell scientists to his laboratory—including Dr. Maria Telkes. His program is based on the coupling of the two technologies—generating electricity from sunlight, and heating and cooling houses with the sun's heat. To accomplish the goal, the institute's scientists are trying to perfect a method of mass-producing cadmium sulfide solar cells so they can be produced cheaply enough for purchase by prospective home-owners. By 1973, a prototype assembly line was built at the institute to produce cadmium sulfide cell material.

Dr. Böer has thus far limited his investigations of solar-generated electricity to cadmium sulfide solar cells because he believes cadmium sulfide can be produced more cheaply than the more sophisticated silicon solar cells. "The cheapest silicon we can make now is $20 a watt, and even if we take the Tyco process† we are estimating very hopefully we can get something which is in the neighborhood of $4 a watt. That's 20 times more expensive than our process." He adds that he might switch to silicon production if better [and cheaper] mass-production technology were developed.

On the other hand, even though the cadmium sulfide material is cheaper to produce for solar cells—at least at the present time—it suffers from 2 intrinsic problems: First, it is less efficient in converting sunlight to electricity; even the best cadmium sulfide solar cells only convert sun energy to electricity at 7 percent efficiency—compared to an average silicon cell conversion efficiency of 11 percent, and a COMSAT violet cell efficiency of more than 14 percent.‡ Second, cadmium sulfide solar cells have never shown great stability or durability in use; that is, they tend to lose power—resulting in conversion efficiencies that drop to 3 or 4 percent.

A combination of these factors means that a good deal of surface area must be covered with the cells to produce significant quantities of electricity. Dr. Böer's plan to develop solar houses seeks to deal with this by incorporating into the house roof both the solar electric cells and a conventional solar heat collector. The solar heat collector is being designed by Dr. Telkes, in cooperation with another internationally known solar engineer, Dr. K. M. Selcuk, who has built a number of machines that use solar heat for mechanical power production.

The combined solar electric cell/flat-plate heat collector is being de-

† See Chapter 6.
‡ See Chapter 6 on the "violet cell."

veloped at the institute in a module form, adaptable for mass production. Each module is 4 feet by 8 feet (32 square feet), and contains a cadmium sulfide backing strip enclosed in a conventional solar collector plate covered with a clear, tough plastic called Abcite so that heat can be pumped out of the collector. The direct-current (DC) electricity will be channeled to storage batteries in the house basement, where it can be stored and later converted to alternating-current (AC) electricity for operation of appliances. Solar heat will also be stored in the basement, using fusion salt storage developed by Dr. Telkes.

The technologies developed by the Delaware group are still in their infancy, although an experimental house was constructed in 1973 to test various components of the system. The solar house technology under development is aimed at eventual use by the middle-to-upper-income market. "Let's take a single-family house which may cost $70,000," Dr. Böer said in a recent interview. "The condition we have established is that the cost of the house should not be larger than about 10 percent more than the cost of another house in that price range."[41] In addition, he said, "the cost of energy should be compatible with the cost of energy currently." He indicated that the costs of solar electricity and solar heat should not exceed the current costs of $0.02 to $0.03 per kilowatt-hour of electricity from conventional sources. The other point he stressed is that the solar energy systems must be designed for reliability. "You flip a switch and on comes light—no questions asked—whether it's good or bad weather."

Whether or not the design goals can be met will not be known for some time. One of the factors that may lead to the successful development of his approach is that the program is designed for a high-cost market. A house costing $70,000 can absorb more of the initial costs of a sophisticated solar power system than could a house costing $20,000.

In preliminary discussions with electric power companies in Delaware and adjoining states, Dr. Böer discussed the possibility of using solar electric cells on houses as a "load leveling" strategy. He found that demands of the suburban home for electricity are greatest when solar energy is most abundant. By using heat and electrical energy storage in houses, solar energy could be used to reduce the impact of electrical power shortages felt by the utilities. The solar houses would become a sort of storage battery for the utility grid: when demands were great, they would operate on their own power; when this home-grown power was not sufficient to operate the house (at night, for example), the extra electricity needed would be purchased from the utility during a period of low over-all drain on the electrical grid.

Dr. Böer calculated that 400,000 solar houses of his laboratory's design built in New Jersey could make a significant difference in electrical supply in that critical high-energy-use area. The houses, producing their own power during the afternoon electrical peak use period, would save the utility more than 1 million kilowatts of electrical power that otherwise would be needed to meet consumer demands for current, he said.

Dr. Böer is confident that his approaches will lead to commercial success. Toward that end, he has established a company—Solar Energy Systems, Inc.—to market the expected commercial solar products.[42]

Government Interest in Solar Power

An increased awareness of the role solar energy can play in housing technologies has been slow in coming to the agencies of the federal government. Serious interest in solar energy has been shown by only 2 agencies, the National Science Foundation and the National Aeronautics and Space Administration.

The recent panel of solar scientists commissioned originally by the President's Office of Science and Technology (OST) was comprised of a number of NASA and NSF scientists, and the report of the panel was published under these agencies' auspices. The panel suggested a development program—to be undertaken by the government—for proving the technical feasibility of heating and cooling buildings. They suggested that the government spend $74 million to develop solar energy systems for buildings (collectors, energy-storage units, coolers, engineering control systems, and architectural systems). In addition, they suggested federal expenditures of $24 million to demonstrate the building concepts. The whole program would take 10 years, they said, adding that the probable impact of the effort would result in 10 percent of all new buildings using solar components in 1985, and 85 percent of all U.S. buildings using solar energy by the year 2020. By 2020, the accumulated fuel savings from use of solar energy instead of conventional fuels would amount to the sum of $16,300,000,000.*

Notwithstanding the agency proposals and the accumulated, echoing rhetoric of the absolute necessity and the national need to develop solar-powered buildings, the only funding by the federal government for this purpose was about $1 million allocated by the National Science Foundation. The primary goal of the NSF program was to perform more experiments on integrated heating-cooling systems. Dr. Alfred Eggers, Jr., NSF's assistant director for research applications, told the

* On the basis of fuel costs in the range of $1.50 per million BTUs of fossil fuels delivered to residential customers. This is about 20 percent higher than today's cost to the consumer of imported natural gas.[43]

House Science and Astronautics Committee that "proof of concept" experiments on solar air conditioners would be initiated in fiscal year 1974.[44]

Even though the National Science Foundation has expressed considerable interest in experiments to develop decentralized solar energy technologies, very little funding (compared to other energy technologies) has been made available for construction of structures. The maximum amount devoted to solar heating and cooling experiments is less than $3 million, hardly enough to be considered substantial. What is needed from the federal government is a commitment to actually construct solar-heated and -cooled buildings for its own use, not to engage in lengthy research projects to ascertain the "feasibility" of a technology that is already well-established!

Low-Cost Solar Technology—the Work of Harold Hay

The use of solar energy for household heating and cooling need not involve an elaborate capital investment. Much of the government, university, and industry effort has been devoted to developing solar power systems for middle-class dwellings. One man who has set out to develop less expensive but well-engineered systems is chemist and building materials adviser Harold Hay of Los Angeles, California.

Hay's interest in solar energy dates back to the early 1950s, when he built a small house for the Indian Government—a house designed to use new building materials such as hardboard to provide a well-insulated and efficient structure at low cost. While involved with the construction of the house, Hay became interested in a unique concept whereby he could utilize solar heat during the day and evaporate cooling at night by moving the walls or roof of the house physically—a concept he calls, appropriately, movable insulation.

As has been noted, many of the early solar house experiments used the solar collectors to provide evaporative cooling—by running water through them at nighttime and storing it in a water tank for cooling the house during the day. Harold Hay considered this far too expensive a means of getting the job done. His idea was to use water directly on the roof—eliminating the expense of an elaborate solar collector, a water-storage tank, or the associated pumps and controls needed to operate them.

He refined his concept for more than a decade, and got a chance to demonstrate it in the mid-1960s. He teamed up with another solar energy specialist, John Yellott, of Phoenix, Arizona. In 1966 and 1967, they conducted experiments using water basins as the actual roofing materials for an experimental house/solar energy laboratory. The small structure they built—which was only 10 feet wide and 12 feet long—had 7-inch-deep plastic-lined roof ponds installed between the roof beams. The water ponds (lined with 10-mil black polyethylene plastic sheets) were designed to serve 3 goals: insulating the roof (with an additional backing of foam plastic urethane underneath the ponds); heating the house by solar warming of the ponds in winter; and cooling the house by exposure of the water to night-sky evaporation in summer. The secret of the scheme was the use of movable insulation: The actual roof cover—above the ponds—was composed of 3 stackable panels of urethane insulation, which could be moved over the water ponds by means of a pulley in the house.

The seasonal operation of the panels was as follows: During the summer, the roof panels would be removed to expose the ponds at night to be cooled by the evaporation of water in the night sky. In the morning, the ponds would be covered again; and since the roof panels were well insulated, the cooling effect of the water would serve to keep the house cool during the day.

In winter, the cycle was reversed. During the day the ponds were exposed to the sun, providing heat by solar radiation. At night, the ponds would be covered, and the heat stored during the day would radiate downward from the ceilings, warming the house. During the fall or spring seasons, the roof panels could be varied to cover all or part of the water ponds to provide only the amount of heating or cooling that might be needed.

The house/laboratory was carefully fitted out with monitoring instruments, and the inventors reported that the structure remained between 70° F and 80° F over 90 percent of the year. The remaining days caused the house temperature to drop at times to 68° F in the winter and to rise to 82° F in the summer, but at no time was the inside comfort level sacrificed more than these few degrees.

Drawing on his experience with this house, Hay spent the next several years refining the concept, and in 1973 he initiated construction of a larger test house near San Luis Obispo, in cooperation with California State Polytechnic College. The house, funded by a $35,000 grant from the federal Housing and Urban Development (HUD) agency, is designed for use in the dry California climate.

Hay modified the roof design of the California house so that the water was contained in long plastic bags—resembling large water beds—between the roof beams. The insulating roof is designed with automatic controls so that thermostats can automatically regulate the position of the insulating panels.

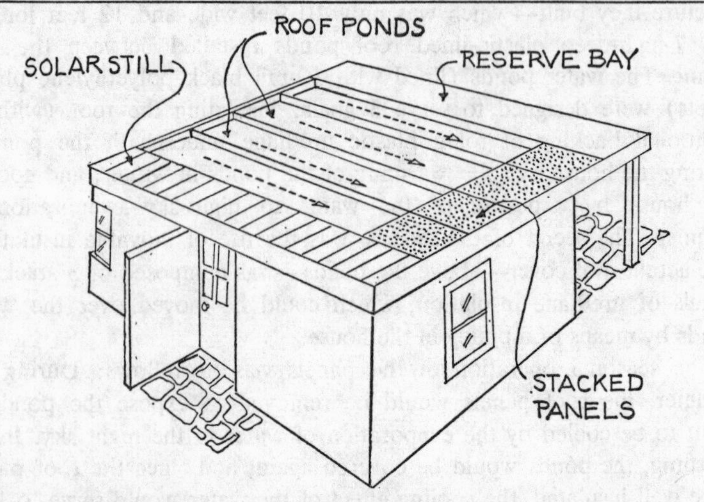

Harold Hay's naturally air-conditioned solar-heated house

As to the question of energy storage, Hay says his earlier experience indicates that energy storage of winter heat in the roof could last for a week. The house is expected to suffer no more than a few days of sunless weather.

Hay describes the climate and the problems expected in the California location:

> It is mountainous country, which is one of the interesting things. It's in a pocket where it may run up to 117° F, which is the highest temperature in the daytime; but the nights are cool. They get down to 60° or even 50° [F] . . . and while the extremes are such that one would like to run an air conditioner in the daytime and a cooler at night, and an air conditioner the next day and a heater that night, this house basically will go through that even without the use of movable insulation.[45]

The real test, though, will be in winter, when temperatures drop to as low as 10° F. "So it's going to get a much more difficult solar energy test," he says. "I want that. Even if it fails . . . you learn at that."

Hay has established a company called Skytherm to develop components for his intriguing "natural air-conditioning system." If the new experiments successfully demonstrate the concept, this method of solar heat and night cooling might be quite effective in the dry-climate regions of the United States and other parts of the world. Even better, such a system will be much cheaper to construct than more sophisticated glass and metal collectors, with their related assortment of tanks, pumps, and mechanical equipment.

As to the problem of having a large volume of water on the roof, Hay says that insurance companies give it premium treatment—in case of fire! In addition, he says that the necessary support for the added weight of the water can be designed into the original architectural plan without problem.

Steve Baer's New Mexico Zome

A startlingly futuristic home utilizing an approach similar to that of Harold Hay's Skytherm concept has been built near Albuquerque, New Mexico, by Steve Baer, youthful founder of the Zomeworks Corporation, which he formed for the purpose of developing and marketing solar houses.†

Baer has been involved with solar energy as a builder-inventor for several years, and received considerable recognition for his contributions to the *Whole Earth Catalog* on the technology of solar houses and geodesic domes. In 1972, he completed construction of a house that looks like a cross between a geodesic dome and an enormous cut gem rising out of the New Mexico hill country.

The house is constructed as a modular system, with the outside south walls of the modules consisting of movable insulated panels made of stryrofoam and covered with aluminum skin. Baer calls the modules "zones"; and the name of the entire structure, or "zoned home," he has shortened to simply "zome."

Behind the movable panels that comprise the exterior south walls of each zone are inner walls of double glass; and behind the glass walls are stacked rows of 55-gallon drums filled with water. The drums are painted black on the ends facing the windows and white on the ends extending into the interior of the structure.

Heating the "zome" is based on Harold Hay's Skytherm concept. The drums are exposed to the sun during winter days by opening the insulated zone panels to let the sun shine through the double-glass walls to heat the water-filled drums. At night the panels are closed, and the heated drums furnish heat to the interior by radiation.

Another innovation in Baer's structure is the use of a louvered skylight, which he calls a skylift, a solar-heat-actuated glazed opaque skylight that opens to let the sun in but closes automatically when there is no sun.[46]

‡ Steve Baer, Zomeworks Corporation, P. O. Box 712, Albuquerque, New Mexico 87103.

Day Chahroudi's Solar House

Another variation of home design using natural heating and cooling—and an even cheaper one—is that conceived by Day Chahroudi, a young solar energy enthusiast who has worked with Steve Baer. Chahroudi has designed a small (950-square-foot) structure that he calls a "Biosphere" that harnesses solar energy and stores it in the ground. Perhaps its most attractive feature is that it can be constructed for about $2,000.

The Biosphere, he says, "is an integration of a house, a greenhouse, and a solar heater. It provides heat and fruit and vegetables all year-'round and consumes almost no water. This makes lots of cheap land habitable." His basic design calls for the construction of a movable south-facing wall of solar panels made of inexpensive clear vinyl plastic sheets secured by wood. Heat storage for the panel-acquired solar radiation is the earth beneath the house, which Chahroudi refers to as a "wet dirt heat battery." Solar-heated air from the wall collector is blown into the open passages of cinder blocks that tunnel beneath the earthen floor.

This remarkable low-cost structure was first pictured in *Domebook II*—a publication packed with useful information on the construction of geodesic domes and a first cousin to the *Whole Earth Catalog*. According to Chahroudi, 2 people can make the structure with easily available materials at a total cost of $2,000 and an investment in time of a month.[47]

Applied Low-Cost Solar Energy

Experiments with low-cost solar houses have been performed across the country by young people interested in cheap household power for homesteads and communes. Much of this work is reported in *The Mother Earth News,*‡ and in a publication specially devoted to natural energy sources—a bimonthly magazine called *Alternative Sources of Energy**, published jointly in Wisconsin and New York. The editors of

‡ P. O. Box 70, Hendersonville, North Carolina 28739.

* *Alternative Sources of Energy* is available for $5 per year (bimonthly) from Route No. 1, Box 36B, Minong, Wisconsin 54859. Other groups that are actively pursuing low-cost applications of solar energy technology include the Brace Research Institute of McGill University in Montreal, Canada (some of the activities of the Brace Research Institute group are covered in the following chapter, on wind power); the New Alchemy Institute in Woods Hole, Massachusetts (P. O. Box 432, Woods Hole, Massachusetts 02543); and Environmental Action of Colorado, a group that publishes a directory of solar heating and cooling specialists in the United States (Environmental Action of Colorado, 1100 14th Street, Denver,

these publications are involved in converting low-cost household systems using solar power, wind power, and other natural sources of energy. Eugene Eccli, an editor of *Alternative Sources of Energy,* coordinated a student program in environmental studies at the State University of New York at New Paltz, in which solar heaters and electric wind generators were constructed by students.

Conclusion—Decentralized Uses of Solar Energy

The application of solar energy technologies on a low-cost basis offers the most direct and logical means of harnessing sun power. In the coming years of energy and material shortages, there may very well be no major effort in large-scale uses of solar energy, because there will be no sophisticated technology base in Western societies to engage in the effort, due to social disruption caused by energy and material problems.

Fortunately, however, there will be ample opportunities to apply the "low technology" solar energy principles, and the communard dropouts of today's high-energy society may well prove to be the messiahs of the future, as their development of low-cost solar energy technologies and decentralized lifestyles will be the best survival options for a majority of citizens of today's rich nations. The wealth of today's technology is based on fossil fuels and cheap resources; and as these decline, the wealth of the future may be based on the technologies looked down upon by today's standards.

Colorado 80202). Information about the activities of solar scientists in the world at large may be obtained from the International Solar Energy Society, P. O. Box 52, Parkville, Victoria, Australia 3052. A good source of information on solar energy activities in the United States, including low-cost solar energy technologies, is the monthly *Solar Energy Digest,* P. O. Box 17776, San Diego, California 92117, for $27.50 per year.

8

Energy from the Winds

The role of wind energy has historically been a major factor in the development of human civilization, with wind powering the early sailing ship as well as the first major source of mechanical power, the windmill. This chapter is devoted to an examination of efforts to use modern technologies to harness the winds in ingenious ways as a major energy source in coming years. As is the case with solar energy technologies, harnessing wind power depends on society's willingness to use other energy sources, such as the dwindling fossil fuels, as well as metals and other material resources, to construct the wind equipment that would make possible an era of renewable power.

What Is the Wind?

Wind is defined as the movement of air—usually horizontal—as air currents sweep across the Earth's surface. The winds are a by-product of the tremendous quantity of solar energy intercepted by our planet, energy estimated to be on the order of 170 trillion kilowatts striking the upper atmosphere of the Earth.[1] Wind is produced by the Earth's built-in "heat engine" as air is alternately cooled and heated by the direct impact of solar energy. The atmospheric redistribution of this energy as it is absorbed and later released by the land and ocean surfaces of the Earth also affects the creation of the winds.

The principal supply of "fuel" for our planetary heat engine is in the form of water vapor. Water vapor is actually fuel storage, because the energy is latent. When water is alternately evaporated and condensed, energy is transferred to the atmosphere. The most evident form of this energy is the wind.

In meteorological terms, this global atmospheric transfer or circulation of energy takes place in "cells," which release the potential energy (supplied by the sun) as kinetic energy—the winds. Three such cells circulate air in each hemisphere; the power of this circulation may be gauged by the estimated circulation of 200 million metric tons of air per second in the tropical "Hadley" cell.*[2]

Although we occasionally experience the force of this tremendous energy, as in cyclones and hurricanes, a number of moderating influences temper the force of the winds and enable us to tap a significant amount of energy—at less than hurricane force.

The principal causative factor for the winds we experience at the Earth's surface is a change in air pressure, which alters the speed (or velocity) and direction of the wind. (All the Earth's cycles are interwoven, making it difficult to separate individual causative factors. For example, the air pressure changes indicated here also cause changes in temperature and precipitation.[3])

Winds caused by these pressure changes are moderated by the various features of the Earth's surface—trees, hills, mountains, deserts, bodies of water, and buildings—to name only a few. Wind flow is variable, and only general observations can be made of its *certainty* at most locations. Although winds can be predicted with great accuracy to blow in certain oceanic and mountainous areas, for most regions only estimates made over a period of time—a month, for example—are reliable.

Not only do winds vary seasonally, monthly, and daily, but differences in wind speed can be detected by measurements made only a few seconds apart. The same is true for distance. For this reason, it is important to measure the wind's behavior, pattern, and speed as accurately as possible, particularly with the object of producing power for human purposes.

For power purposes, the winds have historically been used for sailing ships and for pumping water by means of a windmill. Windmills were

* The cellular model was originally proposed by English meteorologist William Hadley in 1735, who thought that 1 cell was necessary in each hemisphere. Nineteenth-century American meteorologist William Ferrel proved that 3 cells were necessary.

built for centuries wherever it appeared that winds blew with some regularity during substantial portions of the year. It was noted by early millwrights that hills, ridges, and high places were favorable locations. Due to interest in the measurement of the winds for meteorological reasons, surprisingly accurate instruments for recording the speed and direction of the winds were developed and have been in existence for more than 300 years.

In addition, experienced seamen and other observers of the winds can estimate the speed of the wind within a tolerance of ± 10 percent. In 1805, Sir Francis Beaufort, hydrographer to the British Navy, compiled the most famous scale of wind strength, based on measurements as well as on records of personal observations. The Beaufort Scale has been adopted internationally, and is now in wide use.

By noting the reference speeds on the Beaufort Scale, we find that winds needed for producing power for purposes ranging from pumping water to generating electricity fall between the Beaufort Nos. 3 and 6. A gentle breeze, rated 3 on the scale, represents winds of from 8 to 12 miles per hour. Ratings 4 through 6 are for moderate, fresh, and strong breezes, with speeds of from 13 to 31 miles per hour. Higher numbers designate gale, storm, and hurricane forces, which carry plenty of potential wind power, but are too destructive to be harnessed for our use.[4]

Of all the winds that blow in the world, which ones are suitable for the production of power? The answer is not easy to arrive at, since the job of translating the scanty data that are available—despite the ready availability of sophisticated wind instruments—must take into account the varying nature of wind, the height at which one would erect a wind generator, and other very uncertain factors.

Available wind data are distributed in the United States by the U. S. Weather Bureau, and winds are classified into several groupings, including one called energy winds. These winds have speeds ranging from 10 to 25 miles per hour, and blow, on average in the United States, about 2 days out of the week. The more common winds, having speeds of between 5 and 15 miles per hour, blow the other 5 days, and are called the prevalent winds.[5]

The World Meteorological Organization at Geneva, Switzerland, surveyed the world's existing data on the nature of winds for the United Nations' meeting on new energy sources held in Rome in 1961. The organization calculated that 20 billion kilowatts of power—expressed in electrical capacity—are available from the winds at various locations on the earth's surface![6]

Use of the Winds for Power

The following sections examine the conversion of the wind's kinetic energy to useful power for our uses. These three sections begin with the oldest and most widely used application of wind power: wind-driven ships. The second section describes the use of wind for irrigation, water pumping, grain grinding, and miscellaneous uses of mechanical power. The third section describes the use of the wind for conversion to electrical power.

Wind for Ships

Although large oceangoing sailing vessels have not been built since the beginning of this century, there are signs of interest in a return to wind power as the primary engine for seagoing vessels. Let us examine some of the reasons for the decline and sudden rebirth of interest.

From ancient times up until the nineteenth century, the manufacture and use of sailing ships determined the economic and political power of nations. The first known use of sailing ships was by the Egyptians in 2800 B.C.

Later governments utilized sea power to great advantage, particularly after the development of accurate instruments for navigation, such as the astrolabe and the compass, which enabled men to venture across the oceans. With the development of the sextant and the chronometer in the eighteenth century, naval vessels as well as trading ships came into their own. By the nineteenth century, the large (some exceeded 2,000 tons) and efficient clipper ships were developed. They were the fastest ships that had ever been built, and could travel more than 400 miles a day in good winds. Their design derived from Yankee schooners built in Baltimore as privateers in the War of 1812.[7]

The coal-powered steamships, which had been developed in Britain and America in the late eighteenth century, spread rapidly in the following century, with the increase in international trade spurred by the Industrial Revolution. Growth in international commerce quadrupled the tonnage of the world's merchant ships between 1800 and 1860, and by the turn of this century, the total ship tonnage of the world's merchant marines was more than 20 million tons. Two thirds of that was represented by steam ships.

The great sailing ships lingered on a few more years. Among the last to go were a few large wooden schooners built in the United States and Canada. They continued to carry cargo up and down the North American East Coast until the 1930s.[8]

One wonders why the sailing vessels did not survive in a modified form. Why, when the winds are free and fuels costly, could not a ship economically transport goods across the oceans using both sail and auxiliary motor power? James Carlill addressed this subject in an analytical article in the *Edinburgh Review* in October 1918. He reported that:

> Many experiments, now forgotten, were tried with this object. The Navy was very unwilling to part with the power which had so long been its prime mover. Even merchant ship owners were reluctant to part with the clippers which had built up their commerce. But in the course of experiments to combine sail and steam it became apparent that the space demanded by engine and boilers and bunker coal, together with the quarters of an engineering staff in addition to the large number of hands required by a full-rigged ship, were too great a tax on the cubic capacity of the hull and left little space available for cargo.[9]

The specific reasons for the disappearance of the merchant sailers had as much to do with the ship's own technical drawbacks as with the impinging development of sophisticated motor propulsion. Basil Greenhill, the director of Great Britain's Maritime Museum, who has carefully examined the decline of the sailing ships, divides the merchant sailers into two categories: British and European ships made of steel; and wooden ships—predominantly small craft from Europe and America, but also large American schooners.

The British and European steel vessels were square-rigged—i.e., the sails were mounted on spars attached to the masts so they could be trimmed, or adjusted, in the wind. The ships were expensive to build and maintain, and were not efficient users of wind power. In addition, "they made the minimum use of available labor-saving devices and even though a series of special winches and other equipment had been devised by an enterprising British master mariner, hardly any British ship owner used them," Greenhill noted.[10]

On the other hand, the wooden ships of the era were rigged, or outfitted, more efficiently, with expanded versions of the sails that were used on certain old-fashioned and very efficient yachts known as "old gaffers." They were cheaper to maintain, and required smaller crews than the steel ships. American schooners in the early twentieth century could carry about 250 tons more cargo per crew member than any other ships before them—either power or sail.[11] Greater knowledge of the winds enabled the ships to set more reliable timetables than their predecessors. Such factors contributed to impressive improvements in efficiency all around—yet these ships also declined, probably due to (1) their wooden construction, which was out of step with the times technologically; and (2) the exorbitant insurance rates against the ever-present threat of water damage from leaking wood.

Greenhill argues that by adapting the lessons learned in the course of the development of the wooden schooners, along with application of current aerodynamic knowledge to better sail design, a bold new era is now in store for sail:

> Today it would be possible to launch a design and experimental program leading towards the production of a vessel which would have a far higher performance, in terms of speed for a given state of wind and sea and ability to make progress in contrary winds, than the vessels of the past.[12]

According to his formula, such ships, using steel and metal alloys, would carry vastly improved rigging and would employ low-power electric motors to trim semirigid sails. Surprisingly, he does not advocate the use of any auxiliary power on the vessel—rather, it would be dependent on the use of tugboats when approaching and departing harbors.

Dyna-Ships

A variation of this approach is today turning into reality in West Germany at the School for Naval Architecture of the University of Hamburg. On the drawing boards are plans for a fleet of "Dyna-Ships," modernized four-masted clipper ships that utilize the basic lessons summarized by Basil Greenhill.

Credit for the idea for development of the Dyna-Ship belongs to a seventy-two-year-old naval engineer and former Shell Oil Company designer, Wilhelm Prölss. Prölss spent more than two decades surveying reasons for the decline of the era of sail, concluding that one of the major reasons was that the ships' crews were unable to effectively steer the ships into areas of higher winds—thus enabling shorter, more efficient cruises. This problem can now be ameliorated by the use of better navigation instruments and global weather forecasting. He reasoned that with the addition of modern aerodynamic skills, more efficient sail structures could be developed to trap a greater percentage of the wind's energy.

Finally, on New Year's Day 1956, he began work on his model of the modernized clipper—the Dyna-Ship. Within a year and a half, he completed his designs and built a wooden model. Next, he convinced the School for Naval Architecture of the University of Hamburg to fund further research, including elaborate wind-tunnel tests of the Dyna-sails. Writer Jim McCawley effectively described the design:

> The ship that Prölss had designed carried masts that soared 200 feet. There were no lines, no shrouds, little work for a crew. The yards in the square-rigged system were made of curved stainless steel fitted with tracks on which the sails would roll out from the center of the mast. Prölss had, in effect, designed a continuous airfoil from the top of the mast to the bottom, the angle of which was set by turning the mast hydraulically via remote control on the bridge.[13]

Tests conducted at the university showed that winds in the North Atlantic would be sufficient 72 percent of the time to propel the ship at speeds up to 20 knots. For a full-scale, fully automated, 17,000-ton Dyna-Ship equipped with auxiliary engines, fuel consumption would be only 5 percent of that used by an ordinary freighter. (The auxiliary engines would be used for maneuvering in harbors and for stretches of ocean with little or no wind, such as the notorious "horse latitudes" near the Equator.)

In addition, the university reported that due to the Dyna-Ship's large cargo hold, and its low operating and construction costs, it would yield a 30 percent higher return on investment than a comparable conventional ship. Shipowners, however, remained skeptical.

Finally, Rudolf Zirn, a wealthy Bavarian lawyer and environmental activist, commissioned the construction of the first Dyna-Ship for $5 million. It was scheduled for completion in 1973, and will be used for cruises to some of the remaining wild areas of the globe. Another commitment for construction was pledged by Lübeck shipbuilder Friederich Beutelrock, who inherited his father's shipping firm. Beutelrock will test the capabilities of the Dyna-Ship on international freight runs.

Mechanical Power from the Wind

The use of the wind as a direct source of power for mechanical purposes was not known in the ancient world, even though Hero of Alexandria referred to its potential in his writings. The Roman engineer and architect Vitruvius advocated an advanced design of the common Egyptian water mill to be used for grinding corn, but centuries passed before similar principles were used to trap the wind's power by machines.†

The first uses of the wind for mechanical power appear to have been developed in Persia, where, in the province of Segistan, water was

† In the first century B.C., Vitruvius established firm principles on the siting of houses and other structures to take advantage of the natural cooling properties of the wind in his work *The Ten Books on Architecture.*[14]

pumped for irrigation by windmills. Between the seventh and tenth centuries, windmills were firmly established in Persia.

The principle of a mill's operation is simple. The pressure of the wind on the blades, or "wheel," turns a rotating shaft, which operates a pump. This principle has not varied for centuries, even though mills have become more efficient.

The first account of windmills in the Western world was in the twelfth century, when, in 1105, a French permit was issued for construction of windmills. In 1180, a Norman deed reports the existence of a windmill in Britain.[15, 16]

By the thirteenth century, windmills were common in northern Europe. Borrowing from the Persian mill designs, which incorporated sails to trap the wind's force, European mills were built with efficient canvas sails. Within two centuries, elaborate structures were developed to transfer the power produced by the spinning sail directly to a shaft, which turned machinery underneath the mill. Beforehand, the entire mill had to be turned to face into the wind.

Dutch windmills of the fourteenth century were called tower mills, having a fixed body, the tower, and a top, which could be rotated into the wind. The primary problem with the European windmills was turning the windmill when the wind's direction changed. The mill, often a massive, bulky structure, did not always point into the wind; so, each time the wind changed, crews of men had to physically turn the mill into the wind again with the help of crude gears and winches.

To alleviate this problem, the Englishman Edmund Lee invented the "fantail," which was patented in 1745. His fantail was a small wind wheel with radial vanes arranged perpendicular to the main sails of the mill. At a typical 3,000-to-1 gear mechanism, it could orient the sails of the windmill into the wind automatically.[17] A simpler version of the fantail is a single vane, which can be seen today on existing windmills scattered throughout the western and central United States.

Sails for the windmills typically consisted of a wooden frame stretched with canvas. The canvas was either stretched out or rolled up by the miller to compensate for differing wind strengths. In the eighteenth century in England, several design improvements enabled the sails to be furled, or reefed, in variable winds automatically. The most important were the "spring sail" and the "reefing sail."

The primary application of windmills during this period was for corn grinding, water pumping, and irrigation. Although not particularly powerful by modern plant standards, English windmills having 30-foot-diameter sails could produce 44 horsepower in a 20-mile-per-hour wind with an over-all efficiency in converting wind energy to pumping power of 28 percent.

The average horsepower of small European mills ranged from 5 to 10 horsepower. In fact, the windmill is often credited with sparking the early Industrial Revolution. According to the *Oxford Short History of Technology,* "as water wheels and windmills were the only important prime movers of early times, it is fair to say that the industrial revolution was launched with power units generating no more than 10 horsepower."[18]

Wind was used on a large scale in nineteenth-century America, as settlers spread out over the continent. Statistics from an energy study conducted in the 1950s by Resources for the Future, Inc. (RFF), of Washington, D.C., a Ford Foundation-supported research group, indicate the importance of wind power during that period. RFF's economists concluded that in 1850, the use of windmills in America represented about 1.4 billion horsepower-hours of work. They translated this as the equivalent amount of coal that would have had to have been burned to get the same amount of energy—in this case, 11,830,000 tons of coal! By 1870, the amount of power produced by the windmills had been cut in half—to the equivalent of 5.9 million tons of coal. It was during this crucial 20-year period that the steam engine had come to stay.[19]

The two most successful historical examples of the water-pumping windmill are the famed Dutch mills used to reclaim much of Holland from the sea; and the widespread use of windmills in the United States in the nineteenth and twentieth centuries. During the period of the windmill's decline in Europe due to the availability of cheap coal to fuel the new steam engines, America faced a new problem. The settlers of the Great Plains and the West had no steam engines, and consequently had to utilize whatever additional power was available to ease their toil. The predominant machines used to extract power were windmills and water wheels.

During the latter part of the nineteenth century, more than 30,000 windmills operating in Denmark, Germany, the Netherlands, and England produced the equivalent (in mechanical power) of 1 billion kilowatt-hours of electricity.

At first, the American mills were built of whatever materials were handy. In the nineteenth century, a great variety of mills could be seen on the Plains, ranging from devices resembling the paddle wheel of a riverboat flapping in the breeze to the prototypes of the familiar multi-vaned wind-powered water-pumper with the projecting tail vanes.

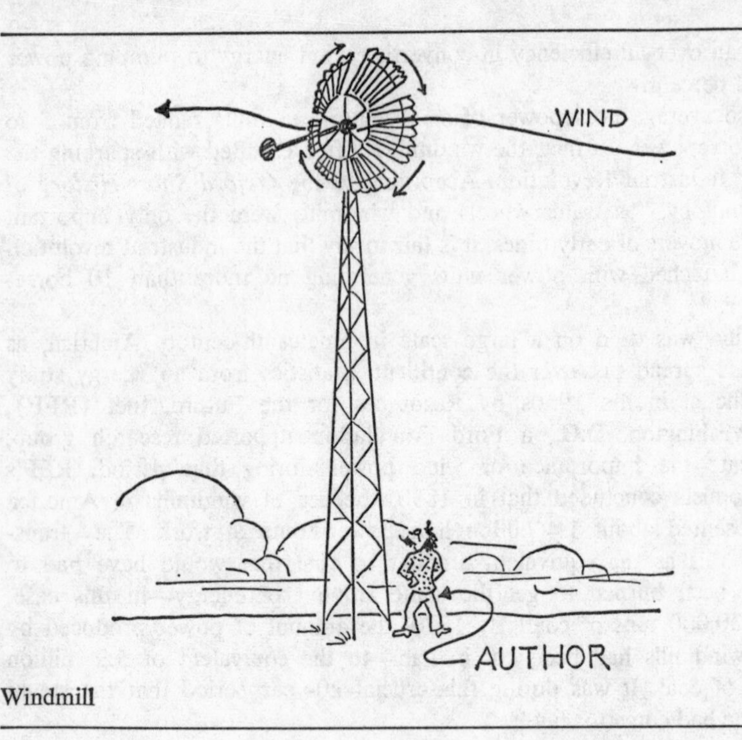

Windmill

By the end of the nineteenth century, a mature industry had developed in the midwestern United States to equip homesteaders and ranchers with mills. Between 1880 and 1900, the combined capital investment of the American windmill industry increased from less than $700,000 to $4.3 million.[20] At the same time, fierce competition spurred the development of lasting and efficient machines.

One large company, the U. S. Wind Engine and Pump Company of Batavia, Illinois, conducted intensive experiments in 1882 and 1883 on various kinds of windmills to determine the best possible machine for use. Part of their investigation involved building windmills to the specifications offered by Englishman John Smeaton, in his work of the previous century. Smeaton observed in 1759 that fewer sails were needed to extract the equivalent amount of power.

The final product of the experiments was a 6-bladed water pumper that they claimed had an efficiency of more than 44 percent in converting the wind's energy intercepted by the sails into work for pumping water. This mill, built with 12-foot-diameter blades, could extract 1.6 horsepower from a 20-mile-per-hour wind. But it could also operate at much lower wind speeds, and would produce .025 horsepower in a 5-mile-per-hour breeze.

A great advantage of the American windmills was their durable construction. The blades and most of the other parts were made of metal, which had a longer life than canvas sails or wooden parts. The over-all efficiency of power conversion is not high with the water pumper, but it can operate at low speeds, since they have a higher starting torque and a steady speed at low velocities. More efficient mills have been devised, and wind-powered electric generators have been developed, which will be described later; but they have had to be designed to operate at higher starting speeds, thus losing the fraction of power available from the wind of low velocity.

American water-pumping mills were exported around the world in the late nineteenth and early twentieth centuries. In 1903, the State Department commissioned a special report from its consulates in foreign lands to determine possible markets for the American windmill industry. The results showed that American windmills were already in use in many lands—with the exception of Europe, where a native industry existed. One enthusiastic American consul in the Mexican state of Yucatan, Edward Thompson, described how he had ordered American mills to demonstrate their superiority over a few European wooden mills in use there already. After ordering and emplacing his new American mills in Mérida, the Yucatan capital, he reported that

> I turned an enthusiastic preacher on windmills and illustrated my arguments with my own material examples. Having turned preacher, I naturally frequented high places, and very probably became something of a nuisance. . . . I also carried my friends and would-be proselytes far above the city's housetops . . . and they saw in the distance the two bright windmills of metal working merrily in a mild zephyr, while the ponderous wheels of the others were either sulking silently or turning croakingly. They came down convinced.

Within a few years, more than 1,000 Americans windmills had been sold in that area—more than 95 percent of all windmills there, Thompson estimated‡[21]

The American windmill industry continued to grow in the early twentieth century, until other sources of power invaded the prairies. In the 1920s, companies began to develop wind-powered electric generators. By the 1930s, the death knell was sounded for wind machines of both the water-pumping and the electric-generating variety. The

‡ The windmills can still be seen working merrily in Mérida, and are pictured in Aden and Marjorie Meinel's book *Power for the People.*

Rural Electrification Administration (REA) was established in the Roosevelt administration to provide federally subsidized power to America's farmers in regions remote from privately financed power plants. Through federal laws guaranteeing 2 percent loans and by eliminating income taxes to rural power cooperatives (co-ops), electricity was brought to the West.* Although the legislation was enacted in the '30s, the demise of the windmills and wind generators was postponed for another two decades—the time required to string electric wires throughout the central and western states.

Under a different philosophy, the federal government might have developed special loans and tax incentives to enable farmers to purchase wind-powered machinery—but history was written differently.

Only 2 U.S. manufacturers still make water-pumping windmills: Heller-Aller Company of Napoleon, Ohio, and Dempster Industries in Beatrice, Nebraska. Total yearly production is less than 2,000 from both firms, and Dempster's assembly line is only run 2 months out of the year. Robert Murray, Dempster's sales manager, notes that "we're not killing the windmill. It's just dying."

The current windmills are available in several sizes, ranging from 6-to-14-foot-blade diameters. Dempster manufactures a 33-foot-high windmill with an 8-foot-diameter wheel. It costs $839 plus shipping and installation. Electric water pumps, on the other hand, are priced from $150 to $350, plus another $100 for accessories. This, of course, does not include lifetime electric power costs.[22]

Responding to a New York *Times* article on the disappearing mills, Professor Stephen Unger of Columbia University in a letter to the *Times* published January 3, 1973, challenged the economics of electric pumps pitted against windmills. If a windmill costs $1,200 installed, he said, and lasts 40 years compared to a $350 electric pump that lasts 5 years, then the windmill's operating life costs are lower. Computing interest at 8 percent and a 20-year amortization period, the windmill would have a yearly operating cost of $120. The electric pump, with an annual $96 energy bill, would cost $182 per year to operate.

> Thus, the "efficient" electric pump costs about 50 percent more on an annual basis than the old-fashioned windmill. . . . when we consider such factors as maintenance costs, inflation (which will make the successors to the initial electric pump increasingly expensive), and the likelihood of rising electric power costs, the case for the windmill becomes even stronger.[23]

* President Nixon eliminated this loan guarantee in 1973 by a controversial executive order, but it was restored by Congress.

Electricity from the Wind

Following the development of efficient electrical-generating equipment in the mid-nineteenth century, a number of European investigators as well as a few Americans experimented with wind as a driving force to operate the electrical equipment.

In a conventional power plant, generators that produce electrical current are turned by turbines. Steam produced by burning fuel—coal, for example—turns the turbine blades. Generating electricity by wind power is accomplished by connecting a rotating shaft directly to an electrical generator. The rotating shaft, in turn, is connected to the hub of the wind rotor, which has 2 or more blades.

For purposes of comparison, windmills are herein defined as wind-powered machines that deliver mechanical power used for pumping water, etc.; and wind generators and wind turbines are defined as wind-powered machines that deliver electricity. By their very nature, the wind generators differ from windmills. Windmills used for water pumping are usually designed with a large number of blades. This design permits them to operate efficiently in winds of low velocity. They are usually placed on low towers.

Wind generators, on the other hand, have fewer blades and are designed to operate in winds of greater velocity—a usual minimum of 15 miles per hour. They are also usually placed atop higher towers. Even small generators of a few thousand watts' output are customarily built on 40-to-60-foot towers.

To extract the maximum amount of power available in the wind, a wind generator's design must be carefully tailored to be efficient. Beginning with John Smeaton, who observed in the eighteenth century that after a certain point additional blade area did not add to the power output of a windmill, a number of principles were discovered, and applied to the operation of windmills. With the more recent concentration of scientists and engineers on the exciting potential of electricity production from the winds, additional principles have been discovered.

Law of the Cube

The most important principle in the relationship of any wind machine to the power of the wind is the "law of the cube".[24] This mathematical formula indicates that, at any given moment, the power content of the wind is proportional to the cube of the wind velocity. In other words, when wind speed doubles, power output from a wind machine will increase 8 times. This principle applies to all wind machines—both fan-type windmills and electric generators. If a wind generator produces 1 kilowatt of electricity in a 15-mile-per-hour wind, for example, it will produce 8 kilowatts in a 30-mile-per-hour wind. Of course, the reverse is true, also, and a drop in wind velocity can wreak havoc on the power production of a wind generator.

This basic law is the central limiting factor in the siting of wind generators. Obviously, the increase of a few miles per hour in the available wind speed can increase the machine's output of power by a much greater amount. The World Meteorological Organization's estimate of available global wind power—20 billion kilowatts—referred to earlier must be understood in light of the cube law.

At a given location, for example, wind velocity is much greater 100 or 200 feet above the ground than it is at 60 feet above the ground. Consequently, a wind generator built on a 200-foot-tower will extract much more energy than the same-size generator built on a 60-foot tower—and the lowly windmill pumping water at 10 feet above ground level won't even be in the race!

In addition, there are mechanical and size limitations on the machines themselves. The power of a wind machine is proportional to the square of the diameter of the blades. In other words, if the diameter of the blade is doubled, the power output is quadrupled.[25] A wind power plant also can't get all the power the wind has to offer, as is seen by Betz' Law.

Betz' Law

The theoretical maximum amount of the wind's energy that can be extracted by a machine was calculated to be 59.3 percent by the German engineer A. Betz in 1927.† In practice, the limit drops below that. According to British wind power expert E. W. Golding, "the aerodynamic imperfections in any practical machine and mechanical and electrical losses" limit the efficiency factor to 40 percent.[26]

Most wind generators are "rated" at a certain wind speed, which is the lowest wind speed that will generate the maximum amount of electric power. At higher wind speeds, wind plants have controls that limit the power output to the rated power level. In various plants, this may range from 15 to 30 miles per hour.

In light of the limitations that Professor Betz outlined, wind generator designers have had to be quite resourceful in designing the blades as efficiently as possible to trap the maximum potential of the wind's

† Betz' calculations were based on applying momentum theories developed for ships' propellers to aerodynamically designed wind generators.

energy. With the development of the airplane in the twentieth century, modern aerodynamic theory developed for application to aircraft propellers was applied to wind generators, along with scientific knowledge relating to ships' propellers. The most important rule that has been followed in this regard is called the tip-speed ratio.

Tip-Speed Ratio

The trim aerodynamic blades of a wind generator extract more power from the wind at their tips than they do near the hub of the blade. This is expressed mathematically as the tip-speed ratio, which expresses the difference between the rotational speed of the tip of the blade and the actual velocity of the wind.

Windmills characteristically have very low tip-speed ratios, ranging from 1 to 3, but high-speed wind generators have ratios ranging from 6 to 8 and higher.‡[27]

Most modern wind generators employ advanced aerodynamic propellers, usually 2 to 3 to the blade structure. Light weight is an important consideration, since the greater the weight of the blades, the greater the centrifugal force that must be overcome as the blade turns in the wind. The blades have been constructed of various materials, including wood, but particularly Fiberglas.

Power from the blades is transmitted to a shaft, which turns an electric generator. Most wind generators, large and small, utilize a transmission with gears to increase the speed of the electrical generator in the plant relative to the slower speed of the blades turning in the wind.

Commercially available wind plants typically have a 1- to 5-kilowatt size range, and supply enough power to fulfill the needs of a farmhouse or specialized power needs in remote locations, such as for a harbor beacon.* Such generators supply direct-current (DC) electrical power, which is not in wide use. DC power is suitable for remote locations and for special battery-charging systems, but alternating-current (AC), electricity is essential for the operation of modern electrical appliances, machines, and equipment. It is not possible to connect a wind power plant directly to existing alternators that convert DC current to AC,

‡ It should be pointed out that the old-fashioned water pumpers, while not having high tip-speed ratios—thus higher efficiency—*do* have compensating features. They maintain a constant speed at low wind velocity and have a higher starting torque than the more efficient wind generators, which better suits them for their job.

* During World War II, the U.S. military experimented with wind plants for operating radios that were dropped with paratroopers behind enemy lines.

since the wind speed is subject to fluctuation. An alternator generates 60 cycles of electric current per second, meaning that the current reverses direction 60 times per second. A wind machine cannot produce a steady 60-cycle electric rhythm, so other devices, called "mechanical rotary inverters," must be used to convert the wind-generated power to AC current, if AC current is desired. Unfortunately, the average efficiency of such devices is only 60 percent, meaning that 40 percent of the original DC power is lost in the process of conversion. More expensive inverters are available that utilize advanced electronic technology and produce conversion efficiencies of closer to 80 percent, cutting the power loss significantly. Large wind generators have been built with even more sophisticated AC power systems.[28]

Specialized electrical equipment that can be used with wind generators to produce AC power has been experimentally developed by researchers at the University of Oklahoma.† The only commercially available AC generator adaptable for a wind generator is manufactured by Precise Power Corporation of Bradenton, Florida. John Roesel, Jr., inventor of the device and president of the company, is enthusiastic about the possibility of using it in connection with wind equipment. Present uses for the AC generator are primarily for small-power plant controls.

A conventional AC generator employs a large permanent magnet that rotates, producing a magnetic field and a variable voltage. Roesel's generator uses a more sophisticated electronic system to control a magnetic rotor. The electrical output of the system is controlled, not by the speed of the device, as in conventional generators, but by an electronically operated oscillator.

In an April 1973 interview, Roesel predicted that adaptation of his AC generator would make wind power feasible on both a small and large scale. A 5-kilowatt wind power system incorporating complete AC power and a special storage device called a flywheel (described in the following section) would cost approximately $5,000, of which the AC generator itself would represent about $400 to $600.[29]

Storage of Wind Power

As is the case with solar energy, wind power is not always available when needed, which necessitates storing the surplus power for later use. This is especially true in the case of small wind-energy systems. In a home-size wind power plant producing DC power, storage is accom-

† Further details are given later in this chapter.

plished by charging batteries for later use. Several other methods of storing this power are also reviewed in the following pages.‡

Batteries

Lead-acid batteries (originally called accumulators) that store wind-generated electricity chemically are the most dependable form of energy storage for a small system, and they can go through thousands of cycles (charging and discharging) without significant deterioration. However, the batteries are expensive. A full set of batteries for household energy storage may cost $500 to $1,500 in a complete wind power plant system having only a few kilowatts' capacity. Depending on their storage capabilities, the batteries may cost as much as the wind generator itself.

In the windy Canadian province of Saskatchewan in the late 1930s, tests on a small American wind generator (Wincharger) showed that the batteries stored 86.5 percent of the power actually intercepted by the blades of the machine. In 12 months, the machine captured 1,105 kilowatt-hours of electricity from winds averaging 10.2 miles per hour. The batteries later delivered 956 kilowatt-hours.[30]

E. W. Golding pointed out in his book *The Generation of Electricity by Wind Power* that in areas with a lower annual wind-speed average, battery efficiency would be far less. He pointed out that in areas with annual wind speeds of 6 to 7 miles per hour, the battery would begin charging at about 4 miles per hour, and the wind plant would reach full power at wind speeds of 15 to 20 miles per hour. A wind plant in a better location would reach full output between wind speeds of 20 and 25 miles per hour. This higher performance would increase not only the battery's efficiency but also its operating life.

In addition, Golding said, "long periods with no charging may cause loss of charge in the battery . . . so that its average efficiency may fall to around 50 percent. This doubles the cost per kilowatt based on the annual energy produced by the generator.[31]

Chemical energy storage by various kinds of batteries is not economic for large wind stations yet, although research is progressing to bring costs down. For a large installation, $8 worth of lead would be needed to store a single kilowatt of wind-generated electricity. Newer battery systems, called molten salt batteries, would cost less. These batteries are under development by several U.S. companies as well as by the Atomic Energy Commission. They operate at much higher temperatures

‡ Storage concepts and methods are explored in greater detail in the chapter on energy storage technologies.

(500° F) than conventional lead-acid batteries, which operate at room temperature, but a kilowatt of electricity could be stored for only $0.55 worth of sulfur and sodium—the active agents.[32]

Pumped Water and Air Storage

In a large wind-power system feeding electricity into the power lines of an electric utility, energy can be stored mechanically (or hydraulically) by pumping water or air into a reservoir, then releasing it at a later time to drive an engine operating an electric generator.

The most common method in the past has been water storage, or "pumped storage," which involves pumping water uphill into a reservoir during periods of low demand, and later releasing it to drive a turbine—like those found in hydroelectric plants—when additional electricity is needed.*

A more likely form of storage would be the use of compressed air forced into an underground reservoir, and later released to drive a specially designed engine, or to supplement the operation of a gas turbine. Air-storage plants have been built by Stal Laval Company of Sweden.[33]

Flywheels

Wind power might also be stored by the modern application of the well-known flywheel, which stores kinetic energy and is similar, in principle, to the familiar spinning top or toy gyroscope.

Like a gyroscope, the flywheel can be spun by the input of energy (in this case, wind-generated electrical power), and energy storage is in kinetic energy—motion. Modern flywheels range in shape from oblong bars to heavy circular solid wheels of steel and other dense materials (including pressed wood and reinforced Fiberglas). Flywheels as part of a motor vehicle energy storage system have been used in Switzerland to power a bus,[34] and experiments are under way by Lockheed Corporation in the United States for flywheels to supplement energy in the automobile.[35]

Florida engineer John Roesel, Jr., inventor of the constant-output AC generator referred to earlier, has built a number of flywheels for use in electric utility power equipment,[36] and plans to incorporate flywheels

* Pumped storage is costly and, in many cases, environmentally disastrous, as is seen in the chapter on storage. Pumped storage facilities are planned by the Federal Power Commission for numerous mountain valleys in the Appalachian states to help "level" the daily electric loads of eastern utilities. The facilities have the disadvantage of threatening to turn the valleys into, not artificial lakes, but mudholes.

made of steel or wood† into small wind-power plants as energy storage devices. In his projected system, the blades of a wind generator turning in the wind at 120 revolutions per minute (rpm), for example, would be stepped up through a motor generator to 24,000 rpm to spin a flywheel for energy storage.

A properly designed flywheel, made perhaps of pressed bamboo and enclosed in a vacuum chamber to minimize friction (the only energy loss in the system), is not only much more efficient than battery storage but is potentially cheaper. Compared to lead-acid batteries, which store only 8 to 10 watt-hours of electricity per pound of material, the bamboo flywheel Roesel envisions would store more than 20 watt-hours per pound.[37] Such a wind power system on a home scale would be capable of storing enough energy to provide power for more than a week of windless days.

Production of Fuel as Storage

Equally exciting as the flywheel concept of wind energy storage is the production of a synthetic fuel: hydrogen. Hydrogen, the Earth's most abundant element, has many of the fuel properties of natural gas (CH_4—carbon and hydrogen), and it can be produced by the electrolysis of ordinary water. A wind generator would furnish the electrolyzer with electricity to break the water molecules down into their constituent parts: hydrogen and oxygen. The hydrogen would then be compressed and stored for later use as high-grade fuel.

This "modernistic" approach to the energy storage problem was first applied in connection with wind power before the end of the nineteenth century. In the 1890s, Danish professor Poul La Cour, whose pioneering wind power experiments are described later, produced hydrogen and oxygen fuel by using wind electrical power to electrolyze an aqueous solution of sodium hydroxide. The local high school at Askov, Denmark, where he taught, was illuminated by "Drummond light," which was an oxygen-hydrogen flame directed on a zirconium element, causing it to emit a brilliant light.[38]

Machines for electrolyzing water and producing hydrogen are widely available commercially. Hydrogen has already been used as a supplement to conventional natural gas as the coal gas that fueled many U.S. homes of the 1930s. There are no significant technical bars to the widespread use of this fuel, which in fact represents a logical economic solution to the problem of wind energy storage.

† Research scientist David Rabenhorst of The Johns Hopkins University, Baltimore, Maryland, has suggested the use of wood and bamboo for flywheel material.

No Storage

In addition to these proposals and methods for various forms of wind energy storage, a number of researchers have suggested the direct coupling of wind plants into a regional utilities grid to avoid the necessity of storage.

Ernst Cohn, Manager of Solar and Chemical Power Systems for the National Aeronautics and Space Administration, has summarized the arguments for this approach:

> Pick the areas in the U.S. where they have the right kind of wind velocity, and the right kind of wind durability, and put your windmills up and feed their power into a grid. You use them as base power, with no storage. . . . On a day when you have no wind—which will probably not happen for all windmills anyway . . . you use conventional fossil fuel power, and peak power plants.

Such a system, Cohn predicted, "would prove competitive today."[39]

Evolution of Electric Wind Generators

Poul La Cour started experimenting with wind power in 1891 and, until his death in 1908, he supervised tests on wind generators for power production. His tests on windmills at Askov, Denmark, indicated a number of important principles of aerodynamics that have been applied to countless successive generations of wind power plants. Because of his scientific experiments, local folklore depicted him as one who was able to "turn rain and wind into light and power."[40]

His experiments showed that propeller-shaped "wings" were the best design, but these were not always used in the latter (wind-generator) tests. One of La Cour's modern successors at the Askov Experimental Mill, J. T. Arnfred, noted at the United Nations' 1961 conference "New Sources of Energy" that:

> It is more important to achieve a simple, robust and dependable mill which gives the greatest possible kilowatt-hours a year in proportion to initial and running costs. In addition, none of the windmill builders of those days would be able to construct a well-shaped propeller-type wing, just as regulation of the large expanse of wing surface in relation to the wind velocity was difficult in those days.[41]

According to Arnfred, La Cour searched in vain for an internal-combustion engine that could be run on the oxyhydrogen produced by electrolysis, but since none could be found he used the wind-generated electricity to charge lead batteries and power ordinary incandescent lights. By the turn of the century, La Cour's experimental windmill supplied not only the high school, but most of the village with electricity.‡

La Cour's basic designs for wind generators were used for the next half century for various applications in Denmark. Essentially, his optimum design calls for a 4-wing, multibladed mill, in which the wing surface consists of transverse blades. The blades are kept in position by means of a counterweight, which is balanced to allow the vanes to open when the wind velocity reaches the greatest intensity that can be utilized. This mechanism prevents the blades from being destroyed in high winds and storms.

Following the successful development of Professor La Cour's wind electric plants in Askov, an organization called the Danish Wind Electricity Company was established in 1903. Within 4 years, they had more than 27 wind power stations. By 1916, more than 1,300 wind generators had been built, but shortly thereafter construction declined. During the First World War, few wind plants were built in Denmark due to fuel shortages.

The Askov wind plant* continued to supply the village with electricity until 1960, when hydroelectric power was introduced. Danish experimentation with wind generators also continued unabated until the 1960s. One large plant, the 3-bladed Gedser generator built in 1942, had a blade area of 450 square meters and was rated as a 200-kilowatt plant. It produced more than 410,000 kilowatt-hours of electricity per year. Initially, it fed DC power into the electrical grid, but in 1955 it was rebuilt as an AC machine. Danish engineer J. Juul, addressing the UN's 1961 Rome conference on new energy sources, said that the Gedser plant was an economic success in Denmark, a country historically lacking in fuels. He argued that Denmark should supply at least 20 percent of its electricity by wind generators, and listed a number of advantages.

- Wind is a "home source of energy" independent of foreign supply.
- Experiments showed the way toward integrating plants of the Gedser type into regional grids.
- Plants could be built more efficiently and cheaper by mounting wind turbines on each tower.

‡ Professor La Cour's electric wind plants were exported to Sweden in the early years of the twentieth century for use in supplying electric light and power to villages and large country estates. Three sizes of wind plants were sold, all designed for winds of 15 mph. Size and prices in Gothenburg, Sweden, in 1904 were as follows: 2-kw wind plant, $322; 4-kw wind plant, $616; and 6-kw wind plant, $885.[42]

* La Cour's original burned in 1929 and was replaced.

- Plants could be constructed from locally available materials (80 percent) and using local labor.[43]

Soviet Experience

As had the Danes, the Soviets became interested in wind power for electricity production, and constructed the world's first large-scale wind electric plant in 1931. Overlooking the Black Sea at Yalta, the 100-kilowatt generator had blades 100 feet in diameter. The area had a recorded annual wind regime of 15 miles per hour, and the generator developed maximum electrical output at 24.6 miles per hour, producing about 279,000 kilowatt-hours per year. It was used as a supplementary power source, and was connected to a fossil fuel plant at Sevastopol, 20 miles away.

Experience with this plant was not altogether satisfactory, due to problems in control and materials. American engineer Palmer Putnam commented that "the principal weakness of the Russian design—low efficiency, crude regulation and yaw control, high weight per kilowatt, and induction generation—had been imposed upon the designers by the state of the industry in Russia, where heavy forgings, large gears, and precision instruments were unavailable."[44]

Following the development of the Yalta generator, Soviet experience with wind plants between 1935 and 1955 was confined to plants generating about 30 kilowatts each.

Argentine engineer Narcisco Levy, on leave from the University of Buenos Aires, recently spent two years (1966–68) in Moscow at the All-Union Research Institute for Farm Electrification. According to his report on current wind research in the Soviet Union the trend is toward development of wind plants with a maximum power output of 15 kilowatts. According to Levy, the 30-kilowatt plants were built to feed electricity into regional grids, but were never put into full-scale production. The plants weighed about 13 tons each, and worked well, but were not particularly economical.

Soviet wind generators range in size from the "Sokol" (Falcon), a 3-blade electric wind plant with a rotor diameter of 39.4 feet and power output of 15.6 kilowatts, to the "Berkut" (Golden Eagle), a small 2-blade plant with a 13-foot rotor diameter, delivering 1.6 kilowatts. All the Soviet wind plants are designed for optimum wind speeds of 15.7 miles per hour, and the power is used to operate various types of pumps for irrigation and supplying well water.[45]

The Soviet plants also include classic 18-blade windmills rated at less than 1 horsepower (0.75 kilowatt). These windmills pump 2.6

cubic meters of water per hour from 8-foot depths, and are in industrial production.

According to Levy's report, the following criteria have been established for production of wind plants:

- Maximum size is set at 15-kilowatt electrical production with 42-foot-diameter blades.
- Electrical generators are mounted at the tops of wind towers, with power transmitted from the shaft to the load by pneumatic and hydraulic means.
- Current speed control, consisting of centrifugal and spring governors to change blade angle, is being changed to a special aerofoil profile to modify the airflow pattern to prevent blade stress at high wind speeds.
- Improved aluminum alloy and Fiberglas blades are under development.
- An automatic centrifugal clutch is being developed to allow power production at wind speeds as slow as 7 miles per hour. Direct-drive wind generators must overcome constant resisting torque, and cannot start at wind speeds of less than 15 miles per hour.

Levy reported that about 10,000 wind machines were in operation in the Soviet Union, primarily for pumping water and for other agricultural purposes. The Soviet engineers concluded that wind plants are more economical than other power sources for agricultural applications in areas where the annual average wind speed is more than 9.2 miles per hour.

Aside from the Soviet Union and Denmark, scientists and engineers in other European nations made significant strides in the development of electric wind generation on a medium to large scale. The 4 countries that have developed the greatest expertise in wind power are Germany, France, the United Kingdom, and Hungary.

Germany

Interest in Germany was spurred by the publication in 1932 of Berlin Professor H. Honnef's book *Windkraftwerke*—in which Professor Honnef proposed the construction of enormous towers 1,000 feet high supporting 5 wind turbines and having a rated capacity of 50,000 kilowatts. Professor Honnef's designs were not carried out in their full dimensions, but for 2 decades German engineers constructed smaller plants.

Professor Ulrich Hütter, an expert in aerodynamics and chairman of the Aviation Department of the Stuttgart Institute of Technology in

West Germany, designed several wind generators after the Second World War. In 1955, a 100-kilowatt experimental plant of his design was erected on an 80-foot tower about 80 kilometers from Stuttgart by the West German Wind Power Research Association. The 2-blade generator used a constant-speed AC governor to supply conventional alternating current, and the slender, efficient airfoil blades were made of glass-reinforced plastic. This machine—and some smaller versions—were operated for several years.[46]

Small—8-kilowatt—versions of Professor Hütter's design, called Allgaier wind turbines, have been commercially distributed internationally.[47]

France

During the 1950s, a 150-kilowatt wind generator similar in design to the Hütter plant in West Germany was built by the Neyrpic Association in France. Additionally, a number of smaller generators were built by the same group—all using reinforced plastic airfoil blades.

United Kingdom

A 100-kilowatt AC wind generator—called the "John Brown" generator—was constructed in Costa Hill, Orkney, Scotland, in the early 1950s. It had a 3-blade, 50-foot-diameter rotor and was rated for very high wind speeds of 35 miles per hour. During the 1950s, the English Electrical Research Association supported a wind power committee headed by a leading international wind power expert, E. W. Golding (author of *The Generation of Electricity by Wind Power*). Golding's work in the 1950s and early 1960s was primarily confined to a theoretical analysis of wind generators—mostly of small size—and there is no evidence of significant British interest in large-scale wind power.

Golding did, however, conduct a number of experiments on the potential of wind power in the British Isles, and found that in many areas of Great Britain and Ireland "there exist clusters of hills on each of which one or more wind turbines could be installed to provide, by local interconnection, a significant total capacity."[48]

Hungary

Engineer Aladar Kiss of the Budapest Scientific Society for Energy Economy reported in 1961 that a 100-kilowatt wind plant was under construction in Hungary. The plant was designed to produce 320,000 kilowatt-hours of electricity per year. It had 4 blades and was not designed to feather in high winds. Rather, it would turn perpendicular to the wind at high speeds.[49]

Non-European Countries

The United Nations-sponsored 1961 conference in Rome included discussion of large and small wind generators by several hundred attendees. In addition to the work in the Soviet Union, Europe, and the United States previously mentioned, the UN attendees discussed significant work on wind generation in Japan, South America, India, North Africa, and other areas.

Not much came of the conference, which was designed in large part to provide expertise in natural energy sources to countries rich in sun and wind but poor in economic resources. The essential problem remained then as now: Even though the poorer nations need power, they also lack money. Wind generation is still more expensive, initially, than conventional sources of power in most areas, which turn the tables in favor of cheap oil-burning equipment instead of more initially expensive wind turbines.†

The potential for both small- and large-scale production of power from the winds lies in the more developed nations that possess both capital and sophisticated technology. With the introduction of mass production, elaborate, automatic wind generators may become cheap enough for wide-scale use in the poorer countries, but not on a prototype basis—which was all that existed in 1961.

However, the economic picture is substantially different for the use of low-cost equipment, and some institutes and governments in a few countries have developed this sort of equipment.

Brace Research Institute

One of the leading international centers of research into natural energy sources—including wind power—is Brace Research Institute (BRI) at McGill University, Montreal, Canada. The institute was established in 1961 to investigate means of irrigating arid land for agricultural purposes. It was endowed by Canadian major James Brace, who died in 1956 leaving a will stipulating that "the results of the said research shall be made freely available to the peoples of the world . . . and not used or controlled by any group for their own selfish purposes."

Accordingly, this is what Brace Research Institute has done. Operating on an annual budget of less than $100,000 per year, the institute has maintained both an undergraduate and graduate-level program at McGill University in the field of solar and wind energy research, as well as in water desalination. In addition, from 1961–67 the institute initiated

† The fact that wind turbines last longer and cost less in the long run makes little difference to a nation starved for both money and electric power.

and maintained a research laboratory on the island of Barbados, West Indies, to build wind and solar machines. This work is still continuing there, where recent projects have included construction of a special solar-powered cooker for a school in Haiti and the fabrication of several large solar-powered water distillation units as well as installation of various types of wind plants for irrigation and water pumping.

After compilation of engineering data, construction, and field trials of the natural energy machines are completed, Brace disseminates the information as widely as possible in a number of ways. For instance, the institute has available a series of "do-it-yourself" booklets, which detail the construction of solar water heaters, solar water stills, solar cookers, and wind plants. These clearly written and illustrated booklets show in detail how to construct machinery to use the free energy of sun and wind from low-cost materials that can be found in most parts of the world.‡

Brace has built a number of specialized wind plants. One unique design is the Savonius rotor, named after Finnish engineer S. J. Savonius.* The Savonius rotor is of a primitive design—Brace Research Institute's plans call for making it out of 2 45-gallon oil drums cut in half and welded together to form troughs that scoop up the wind and keep the rotor turning. The rotor can operate in areas where wind speeds are low—between 8 and 12 miles per hour. It is designed to pump water where the water supply is no more than 15 feet below ground. Cost is very low, since it can be fabricated using a variety of readily available materials, a few tools, and a welding set. BRI, for instance, built one on Barbados for about $50. A simple wooden frame supports the rotor at a 6-foot height. Its only other parts are a drive shaft and a single-action diaphragm water pump. The rotor can pump 300 Imperial gallons of water per hour under 10 feet of water pressure.

A larger wind generator is the 32-foot-diameter Brace Airscrew Windmill, constructed in cooperation with the Barbados Government as a deep well pump for irrigation. It delivers about 4 horsepower (about 3 kilowatts equivalent in electrical energy) to a water pump in a 15-mile-per-hour wind. At this speed, 130 gallons of water can be pumped per minute from a 100-foot depth.

In 1969, the Barbados Ministry of Overseas Development observed

‡ Available from Brace Research Institute, Macdonald College of McGill University, Ste. Anne de Bellevue 800, Montreal, Quebec, Canada.

* Savonius was commissioned to build three of his plants in the 1920s on the estate of Col. Henry H. Rogers at Southampton, Long Island. It was on learning about this project that Palmer Putnam, mastermind of the famed Grandpa's Knob wind plant, turned his interests to the development of wind power.

of the Brace Airscrew Windmill: "In a region desperately short of intermediate technological developments in the rural areas, it is gratifying to note the development . . . of a simple mill scientifically designed to exploit wind power. . . ." A representative from the Ministry, A. Wilson, called for a long-range effort to improve the prototype machine and incorporate wind power into the agricultural programs of the country.[50]

After the First World War, a number of American inventors and small companies, following in the footsteps of the Europeans, began the development of wind systems for farm use in the American West, where the annual wind regime favored this form of energy. The systems were in widespread use for several decades before the spread of federally subsidized (Rural Electrification Administration) electric power in the hinterlands.

Jacobs Wind Electric Company

One of the most successful manufacturers of homestead-size wind generators was Jacobs Wind Electric Company of Minneapolis, Minnesota, which produced wind-electric power plants from 1928 to 1957. The founder of the company, Marcellus L. Jacobs, experimented with conventional windmills in the early 1920s in an attempt to convert them from water pumpers to electrical generators. He wanted the electricity to power a ranch-based radio station, KGCX (now affiliated with the Mutual Broadcasting System) in Montana. He had no success trying to convert the slow old prairie windmills to electricity generation, so he started from scratch and developed a completely new wind power plant.

His two most significant innovations were special 3-blade propellers and a flyball governor to control the pitch of the plant's blades, allowing them to "feather" in high winds in order to maintain a constant speed to power the generator. The diameter of the plant's blades was about 14 feet; they were made of carefully selected, kiln-dried Sitka spruce, regularly used for aircraft propellers, and were finished with a light coat of aluminum-based paint.

Jacobs, who is still an active engineer and inventor living near Fort Myers, Florida, learned to fly "about the same time we started experimenting with the windmills," he recalls. From the World War I-vintage plane he flew came the inspiration for using an airplane-style propeller instead of the traditional windmill blades.

He concluded that the violent vibrations that were causing problems with his new experimental design stemmed from the 2-blade propeller design. When the blades were in the vertical position, they had no significant centrifugal force to overcome, but in the horizontal position, they had to overcome some 1,500 pounds of centrifugal force. This cycle with each rotation of the blade caused tremendous vibration, he found. By substituting a 3-bladed propeller, the centrifugal force became evenly balanced and the vibration was virtually eliminated. On the basis of these findings, Jacobs patented a 3-blade wind generator propeller in 1928.

Jacobs' major competitor over the coming years was the "Wincharger," made by the company of the same name. The Wincharger proved somewhat unsatisfactory because of the excessive vibration attributed to its 2-blade propeller design.

Jacobs' other major wind-generator innovation, the flyball governor, controlled the pitch of the blades. In effect, the governor turned the blades edgewise into the wind when wind speed rose above 18 miles per hour. This gave positive control of the propeller speed independent of the plant's generator, and protected the generator by preventing uneven load.

The plant's generator, rated at 1,500 watts' capacity, contained an oversized armature weighing 160 pounds, which allowed for a direct power drive from the blades to the generator. No gears were used in the plant, eliminating cold-weather problems with oiling and maintenance. Another "overdesign" feature was the use of more than 50 pounds of copper wire for field coil windings in the generator's frame, which improved the generator's efficiency still further.

The plant was designed to charge batteries made especially for the generators by Willard Battery Company. The batteries were assembled in sets of 16 and were rated to store energy for several days during periods of no winds. Each lead-acid battery was sealed in a glass container and weighed 100 pounds.

Jacobs Wind Electric Company manufactured a single standardized wind plant, designed to deliver a minimum of 400 to 500 kilowatt-hours per month.† The wind plants were supplied in both 32-volt and 110-volt direct-current models. With a full set of batteries, all necessary electrical cables, and a 60-foot tower, they cost just over $1,000 at the factory in the '30s, '40s, and '50s. Direct-current appliances were specially designed and produced for the most popular system—the 32-volt plant. The cost of the appliances was approximately the same as for conventional alternating-current counterparts. A special, heavily insulated

† Many of Jacobs' plants were used in both Arctic and Antarctic climates. Admiral Byrd used one of his generators at "Little America" at the South Pole in the 1930s, where it gave virtually maintenance-free service for more than a decade.

freezer made for use with the plant could keep food frozen for up to 5 days without power.

Use of high-quality materials and heavy-duty design made the wind generators remarkably durable and trouble-free. Records kept on 1,000 plants showed that annual maintenance costs averaged less than $5 per year.[51]

The Israeli Institute of Technology installed a Jacobs wind plant in Eilat, Israel, in 1953, and monitored its performance for 3 years. The researchers estimated that the 110-volt, 3-kilowatt-rated plant cost them about $1,000 per kilowatt to install in Israel. Costs of the power amounted to 15 cents per kilowatt-hour, compared with 18 cents per kilowatt-hour if they had used a diesel engine to achieve the same power—based on high diesel fuel costs at the location. Again, maintenance problems were virtually nonexistent, prompting a report on the survey to conclude that "the prices of wind- and diesel-generated electricity do not take into consideration the maintenance costs which tip the balance even further in favor of wind power plants." The report recommended the development of wind plants at locations in underdeveloped areas, as opposed to Diesel units.[52]

Jacobs sold more than $75 million worth of wind power plants before his company closed its doors in 1957. He states that, without question, the spread of Rural Electrification Administration (REA)-subsidized power facilities signaled the end of his business, though it remained active for more than 2 decades after the REA legislation was passed.‡[53]

Wind Power at Grandpa's Knob

During the Second World War, a massive 1,250-kilowatt wind electrical station was operated at "Grandpa's Knob" in the mountains of central Vermont. The plant had 2 blades made of a new aluminum alloy, with a wheel diameter of 175 feet. Electrical generation was synchronous, and the current was used as backup power to pump water behind a local dam. The wind generator was designed to operate in winds of 17 miles per hour and higher.

The 1,250 kilowatts of power that the wind generator produced during sporadic periods of operation were fed into the lines of Central Vermont Public Service Corporation. The plant was conceived and designed by Palmer C. Putnam, an engineer who had become interested

‡ One of the last major uses for Jacobs' plants was for the cathodic protection of pipelines. The wind generators generated electricity that was used to give the pipelines an electrical charge in order to prevent their electrolytic deterioration. The electrical charge in the earth causes the earth to react with exposed metal, creating corrosion, unless the metal has its own charge—which the wind generators supplied.

in wind power in the early 1930s when he built a house on Cape Cod only to find both the winds and the electric utility rates "surprisingly high."

"It occurred to me," he recalled in his book *Power from the Wind,*[54] "that a windmill to generate alternating current might reduce the power bill, provided the power company would maintain standby service when the wind failed and would also permit me to feed back into its system as dump power the excess energy generated by the windmill."

Putnam never built the wind generator for his Cape Cod home, but by careful research, perseverance, and persuasion, he convinced leading scientists and industrialists to back an experimental project to use wind energy on a large scale as a supplement to commercial electrical power. Putnam found that the best wind generator blade designs were those that had evolved from aeronautical experience gained in World War I and the years following.

Putnam contacted several prestigious aeronautical engineers and experimented with blade designs at a Cape Cod location. For help in selecting blade design for his large-scale power project, he retained the leading U.S. aerodynamics specialist, Dr. Theodor von Karman, director of Guggenheim Aeronautical Laboratory at the California Institute of Technology.

By 1939, Putnam had refined his designs and arranged for General Electric Company to manufacture the electrical equipment for the project. His friend, Vannevar Bush, Massachusetts Institute of Technology's dean of engineering who was to become director of the U. S. Office of Scientific Research and Development during World War II, had referred Putnam to a General Electric vice president in 1937.

Putnam's next step was to convince officers of S. Morgan Smith Company, a York, Pennsylvania, manufacturer of hydraulic turbines, to undertake the wind project as the major engineering contractor. Central Vermont Public Service Corporation provided the site for the wind generator, and agreed to purchase the unit upon satisfactory completion of field trials.

The chief engineer for the project was the head of the Department of Civil and Sanitary Engineering at MIT, Dr. John B. Wilbur. As project manager, Palmer Putnam was able to skillfully coordinate the direction of the project, even though his academic consultants and industrial partners were scattered across the country.

Putnam had specified the basic design of the generating unit in 1937, and S. Morgan Smith Company accepted it in 1939. Putnam recalls that "[the] test unit, which was engineered under Wilbur's direction, was fabricated, erected, and wind-driven for the first time on August 29, 1941, 17 months after the decision to go ahead, and 23 months after the decision to explore the problem of large-scale wind power."[55] What is remarkable is that in today's terms, the "lead time" required to build and put large power plants "on line" is frequently 5 to 8 years. Putnam and his team of scientists, engineers, and businessmen were able to start with a new technological concept and produce commercial power within 2 years!

Twenty minutes after starting the unit, it was generating 700 kilowatts of power, and for the first time in history, synchronous wind-power was fed directly to a utility electric grid. Things went smoothly for about 18 months, until a main bearing in the generator failed in February 1943. The cause of the failure was not directly related to the generation of wind power. The bearings used in Putnam's generator had a highly polished lead surface, lubricated with a film of oil. This oil film failed, destroying a bearing. Technological improvements have in the meantime eliminated the use of these bearings. Today, they would be supplanted with antifriction roller or ball bearings.

Because of wartime shortages, the company was unable to replace the bearings for more than 2 years. During that time, the wind plant's blades were locked in place and subjected to the continuous buffeting of the wind, with accompanying stress on the machinery and blades. In the original assembly of the blades and supporting spar, rivet holes had been drilled and punched in a new aluminum alloy material used in the blade. This proved to be a serious mistake, since cracks developed in the metal around the punched holes.[56]

After the bearing was finally replaced, the wind station operated for less than a month. On the morning of March 26, 1945, cracks widened in the spars of one of the big machine's 8-ton blades, which suddenly gave way and was hurled 750 feet through the air, landing on its tip. The cracking problems in the blades had been noted by project engineers in 1942, when they decided to make final repairs instead of sending the entire structure back to the factory for overhaul. As Putnam pointed out, "the decision was to carry out modifications in the field. Otherwise, it would have been necessary to discontinue the field test program until such time as postwar blades could be fabricated, delivered, and erected; and that choice, as it turned out, would have meant a delay of some 4 years."[57]

Putnam's experiment at Grandpa's Knob proved that wind power could work on a large scale, but the project also proved that numerous changes would have to be made before the project could succeed.

Before and during operation of the Grandpa's Knob wind plant,

Putnam's team conducted an elaborate investigation of the wind power available to a large generator in the Vermont region.

The program included both ecological testing and the erection of specially built wind-measurement devices. The ecological program consisted of evaluating the observable effects of wind on vegetation and trees. It was noted that the best natural indicator of suitable winds for power production was the "flagging" of trees. "A tree is said to be flagged when its branches have been caused by the wind to stretch out to leeward while the trunk is bare on the windward side like a flagpole carrying a banner flapping in the breeze," Putnam explained, noting that "the most important indications of wind in New England are given by flagged balsam and spruce."[58]

A meteorological program was instituted to attempt to more accurately measure the wind's strength. Measuring stations with specially built anemometers were established on 14 mountaintops, and at Grandpa's Knob a 185-foot tower—called "the Christmas Tree"—was fitted out with instruments to measure the wind's structure.

Tremendous variations in the Vermont mountain winds were found in the survey, as well as the damning evidence that the early calculations of available wind power at the Grandpa's Knob site were in error. The measurements indicated that high-velocity winds varied widely in the region. The ecological program identified the following phenomena, according to Putnam: "In one patch 100 yards wide, the high-velocity wind strain has been deflected downward for unknown causes and sears the ridge. . . . The transition from this zone of high wind to the zone of winds permitting nearly normal growth occurs in a matter of yards.[59]

The team had calculated originally that the wind generator would produce 4,400 kilowatt-hours of electricity per kilowatt of installed capacity per year. Instead, it produced only 30 percent of this calculated output.[60] Other sites surveyed showed more favorable wind qualities, but since the machine had already been built at Grandpa's Knob, there was nothing they could do to move it.

The original calculation of available winds at the Grandpa's Knob site indicated an average wind speed of 24 miles per hour. More elaborate measurements, however, showed that winds at the site actually averaged only 16.7 miles per hour. Had this been known in the early stages, the wind machine would have been built at a more suitable site than Grandpa's Knob. Likewise, had more knowledge been available on the stress effects of the materials used in construction, better construction techniques could have been utilized.

Within a few months after the blade failed, S. Morgan Smith

Company decided that—after an expenditure of $1.25 million on the project—they could not justify further expense. However, at the company's request, Putnam's engineers estimated the cost of continuing large-scale wind projects based on the lessons learned at Grandpa's Knob. They concluded that 6 more units, based on the original machine, could be installed at another Vermont location at a cost of $190 per kilowatt. However, it was concluded that the actual worth of the additional 9,000 kilowatts of power to Central Vermont Public Service Corporation would be only $125 per kilowatt.

Putnam found that although the technical problems of mastering the winds for electrical generation had been for the most part solved, the development of wind power on a national scale would require a national wind power survey and government aid to solve the economic problems of fitting wind power into either a local or regional national power grid. He believed that in the 100-to-3,000-kilowatt range, wind generators could realize a profit if produced in factory lots as small as 100 per year.

Subsequent Large-Scale Wind Power Interest

The Federal Power Commission became interested in the Grandpa's Knob experiment during World War II, and commissioned Percy H. Thomas, a senior engineer of the commission, to investigate the potential of wind power production for the entire country. Thomas' survey, *Electric Power from the Wind,* was published in March 1945—ironically, the same month that the blade flew off Putnam's machine at Grandpa's Knob.

The survey was optimistic, and suggested the development of units even larger than Putnam's. Thomas recommended the development of wind turbines in 2 sizes: 2-blade, 7,500-kilowatt unit; and a 3-blade, 6,500-kilowatt machine. The latter would be used in areas where the winds blew from 18 to 21 miles per hour, and the larger wind turbine would be used in very favorable locations, such as the Vermont locations suggested by Putnam.[61]

Thomas estimated that for a production "run" of 10 plants, the cost per kilowatt of rated capacity would be $68 per kilowatt for the large machine and $75 per kilowatt for the smaller one.[62] He suggested that the machines should be constructed near hydroelectric projects, if possible, and pointed out that:

> Operation of wind power stations in combination with hydroelectric developments having large storage reservoirs, thus securing *dependable capacity* [emphasis in original] from the wind would, it is believed, re-

sult in very substantial economies, especially in regions where the "firming" hydro capacity is already available or can be installed at low cost. Actual kwh [kilowatt-hour] cost estimates are possible where the cost of firming hydro and its primary energy generation are known.[63]

Thomas pointed out a number of advantages of a system of wind generators:

A single windpower generator will make an exceedingly irregular contribution of energy to the system hour by hour. However, when the unit has a capacity small in comparison with the total load, no serious inconvenience occurs. . . . Diversity with a large number of units in the system smooths out the supply to the utility very materially.[64]

His survey concluded that more data should be acquired on wind velocity and duration at favorable sites; wind tunnel tests should be conducted on promising blade models; and a full-size advanced generator should be built—equipped with instrumentation to monitor all performance details. "The wind, with all its daily variations, is, as measured by monthly averages, a surprisingly stable and reliable source of energy," he added.[65]

Neither the Federal Power Commission nor private companies acted on Thomas' recommendations, probably due to the ill-timed and well-publicized demise of the Grandpa's Knob facility. Thomas continued his work on wind power, however, and was able to generate considerable interest in other branches of government. In 1951, representatives of the Interior Department and the Federal Power Commission testified in favor of a proposed bill that provided for the government to develop large-scale wind power, and to build some generators designed by Percy Thomas. The bill was introduced in Congress by the chairman of the House Committee on Interior and Insular Affairs, John Murdock of Arizona, who wanted a pilot plant costing between $2.25 million and $2.75 million built on public lands in his state.[66]

Assistant Secretary of the Interior William E. Warne argued that the demonstration plant was needed on the basis of energy conservation:

Wind-driven power generation is attractive also on the basis of conservation of materials, manpower, and natural resources, in addition to water. The operation of a wind-driven generating plant would, of course, involve no fuel transportation problem. There would be no demand on transportation facilities and associated manpower that are characteristics of fuel electric plants. Wind energy is inexhaustible. Its use might save oil that otherwise would be used as fuel in steam plants. Petroleum conservation has long been one of our basic national policies

[sic], and the saving of oil is particularly urgent in times of national emergency such as the present [the Korean War].[67]

The members of Representative Murdock's committee were very skeptical, however, about the economic value of the plant. The construction cost of the wind generator (at 7,500 kilowatts for $2.25 million) was $333 per kilowatt, compared to $140 to $225 per kilowatt for fossil-fueled plants. Secretary Warne argued that the costs of wind power, while higher than for fossil fuel power, would be lower than for hydroelectric power. He estimated the costs of hydroelectric plants at the time at $350 to $600 per kilowatt.

Nonetheless, the committee was not impressed with the prospects of large-scale federally subsidized wind power, and the bill was killed.

Large-Scale Wind Power

Interest in the large-scale production of wind power in America was not renewed until William E. Heronemus, professor of civil engineering at the University of Massachusetts, revived the subject in 1970. In the past 3 years, he has proposed wide-ranging networks of huge wind generators in numerous U.S. locations to provide virtually all the nation's electricity from the winds.[68]

Professor Heronemus, a former naval architect, reminds that he is also "a former farm boy fascinated by windmills." His arguments for wind power today are laced with constant references to the pollution-free nature of wind power, its long history of technological feasibility, and its safety vis-à-vis America's burgeoning nuclear power industry.

While thousands of individual sites may produce sizable quantities of wind power, he has pinpointed as the most promising U.S. locations: throughout the Great Plains; along the eastern foothills of the Rocky Mountains; on the Texas Gulf Coast; the Green and White Mountains of New England; and off the continental shelf of the northeastern United States.

William Heronemus' Design

Professor Heronemus' proposed designs for large wind-power installations are based largely on the work done by Palmer Putnam for the Grandpa's Knob generator. Professor Heronemus' suggestions for wind generators in the Midwest indicate that 2 types of installations may be feasible. One installation would be an enormous 210-foot-diameter wind turbine mounted on a 1,000-foot tower—compared with Putnam's 175-foot-diameter turbine on the 2,000-foot Grandpa's

Knob. He points out that even for such a high tower, 3 large turbines could be supported, as was suggested by Percy Thomas in 1945.

A possible disadvantage of a single massive wind machine is the fact that it must be designed to operate only at high wind speeds, whereas smaller wind generators can be designed to operate at low-to-medium wind speeds. Professor Heronemus offers a solution to this problem: a somewhat shorter—600-foot—tower with 20 smaller 2-bladed turbines mounted around the top. For this arrangement, he envisions the possible emplacement of one such machine per square mile in a 300,000-square-mile area of the Great Plains. He concludes that this vast network of wind generators would represent about 190 million kilowatts of installed electrical generating capacity—roughly half the 387 million kilowatts of installed electrical generating capacity in the United States in 1971.[69]

Offshore Wind Power System

Many early schemes for the use of windmills to power electrical generators suffered from the inevitable Achilles heel: What happens when the wind doesn't blow? Professor Heronemus proposes to remedy the problem by converting the wind's energy into a fuel that can be readily transported and stored for later use.

In January 1973 he testified before an Atomic Energy Commission licensing board, established to hear arguments for and against a proposed nuclear power plant at Shoreham, Long Island, New York.* Professor Heronemus proposed an offshore chain of wind generators as an alternative to the plant. He said that any large corporation could begin to deliver components of such an "Offshore Wind Power System" (OWPS) within 24 months of a program start, since the technology is well understood.[70]

Here's how the system would work for Long Island: Beginning in 1975, the first of 2 networks of wind stations would be emplaced off the New York Shoals, off the Long Island shoreline. The wind stations would be mounted either on towers emplanted in the relatively shallow shoal waters, or on 500-foot-by-200-foot floating platforms. Each station would support 3 wind towers, each of which would in turn house 2 200-foot-diameter wind turbines. The 2-blade wind turbines are based on Putnam's design, as refined by an engineering study performed after World War II by New York University.[71] A total of 640 of these wind stations would be linked together along 16 north-to-south lines. The wind generators would produce electric current for powering electrolyzers, machines that electrically break water down into its

* It was licensed by the AEC in April 1973.

basic components, hydrogen and oxygen. The oxygen would be released to the atmosphere, while the hydrogen gas would be pumped ashore and stored in large tanks for later use in fuel cells.

Professor Heronemus proposes a whole range of wind plants as candidates for the system, including:

1. A floating aluminum and concrete tower 450 feet above the ocean's surface housing 34 60-foot-diameter wind turbines, each turbine geared to a 100-kilowatt generator.
2. A floating tower rising 340 feet above sea level housing 3 200-foot generators, each coupled to a 600-kilowatt generator.
3. The same tower as in No. 2, housing 200-foot-diameter turbines coupled to a 2,000-kilowatt generator.

Professor Heronemus' original proposal calls for coupling the wind wheels to AC power generators, then converting the AC power to DC to feed it into an electrolyzer. It may be possible to bypass this and feed the electrolyzer with DC power directly. Today, however, large-scale DC equipment is not available, and Professor Heronemus designed his system to take advantage of existing mechanical equipment wherever possible.

His cost estimates, using available prices of equipment and material in the United States in 1972, are as follows:

1. Sixty-foot-diameter turbines with 100-kilowatt generators—$10,750, or $517,000 for the whole tower with 34 wind turbines.
2. A 200-foot-diameter turbine with a 600-kilowatt generator—$102,800, or $389,000 for the 3-turbine platform.
3. A 200-foot-diameter turbine with a 2,000-kilowatt generator—$239,000, or $914,000 for the 3-turbine platform.

The electrolyzer station is envisioned as an underwater structure constructed of 24-inch-thick reinforced concrete, which would house equipment for conversion of electricity to hydrogen, as well as the operating crew. From the electrolyzer submarine station, hydrogen gas would be pumped to a series of underwater hydrogen storage tanks, also made of reinforced concrete, where it would be retained until pumped ashore for use.

After being pumped ashore, the fuel is burned in fuel cells. Professor Heronemus originally conceived of the fuel cells as 5-kilowatt units designed for use in individual homes, but he indicates that larger fuel cells—up to 20,000-kilowatt capacity—may be cheaper, due to decreased costs of production. As has been pointed out, an enormous advantage of the fuel cell is that it is an essentially pollution-free device, since water vapor is its single emission.

The large-scale wind power system calls for the wind to generate electricity (AC) which is then converted to DC power. The DC power is used to make hydrogen, which is pumped ashore. Ashore, the hydrogen is converted back into electricity in the fuel cell. Professor Heronemus has proposed a wind power system for all the New England states that would produce 159 billion kilowatt-hours of electricity per year—the amount of electricity he believes will be needed in the New England States between 1976 and 1990—and would cost $22.4 billion to install.

Professor Heronemus believes that power from the offshore system would cost 2.18 cents per kilowatt-hour to produce, which he compares with the 1.66 cents per kilowatt-hour that he estimates it will cost to produce in New England in 1990. In addition, he believes that with further production and development of fuel cells, power costs in the system will be brought much lower, making them even less expensive than the costs of power from a nuclear plant.[72]

As far as aesthetics are concerned, the wind towers proposed by Professor Heronemus would pose no more of an eyesore or threat to shipping than the forest of oil rigs sprouting from the Gulf of Mexico and now being proposed for the offshore Atlantic Ocean.

Professor Heronemus, who served in the Navy from 1938 to 1965, worked in the naval reactor program, where he became convinced of the absolute necessity for uncompromisingly high safety standards in the Navy's submarine reactors. He remembers with sadness and regret the loss of the nuclear submarines *Thresher* and *Scorpion* in the 1960s.

His heightened awareness of quality control in atomic reactor engineering is reflected in these comments on a nuclear plant in Vermont, the "Vermont Yankee" power station in Vernon, Vermont:

> I have visited all portions of the plant; I am not impressed. For example, I would never accept for the U. S. Navy the piping and wiring in the rod drive mechanism enclosure. The complexity of the instrumentation and its cabling between reactor and control room is such that I have not been at all surprised at the recent set of spurious alarm incidents which they have experienced.[73]

Professor Heronemus believes that nuclear plants are not engineered with the quality standards essential for adequate protection of the public. He argues that widespread development of wind power, as well as other natural energy sources, is inevitable because they are ultimately cheaper than atomic plants.

In addition, Professor Heronemus points out that due to the decentralized nature of wind technology, the loss of a single wind turbine would not cripple the electric service area with a huge blackout. For example, the April 1973 blackout affecting 150 miles of Florida's "Gold Coast"—from Fort Pierce down into the Florida Keys—was caused by a single nuclear plant failure, as was also New York City's June 1970 blackout following the outage at Consolidated Edison's Indian Point reactor. "One wind station disabled would mean one thirteen thousandth of the system capacity would be lost!" he says, as opposed to the complete system loss that can be precipitated by a nuclear station failure.

He sardonically outlined for the AEC's licensing board at the Long Island hearing the environmental impact of his proposed Offshore Wind Power System:

Thermal pollution: 0
Sulfur dioxide effluent: 0
Nitrogen oxide effluent: 0
Carbon monoxide effluent: 0
Carbon dioxide effluent: 0
Hydrocarbon effluent: 0
Maximum radioactivity discharge: 0
Number of casks of high-level wastes per year: 0
Probability of a radioactive loss-of-coolant accident: 0
Probability of an emergency core cooling failure: 0[74]

Current American Experience with Wind Power: "The New Pioneers"

Within recent years, the combined pressures of environmental and urban deterioration have caused many people to turn once again to wind power on a homestead scale. One young couple, Robert and Eileen Reines of Albuquerque, New Mexico, have built one of the world's first structures powered almost totally by wind and solar power.

Reines, a true child of the nuclear age, was born at Los Alamos, New Mexico, on the site of the Manhattan Project. His father, who was codiscoverer of the neutrino, an elementary atomic particle, worked on the development of America's first atomic bomb.

Reines returned to the area in the late 1960s as an Air Force lieutenant—after living in Ohio and California—assigned to Kirkland Air Force Base Weapons Laboratory in New Mexico, where he was assigned to monitor atomic tests. After concluding that there were better ways to serve his country—and himself—than working with atomic bombs, he requested permission from the base commanding officer to conduct research on natural energy sources.

Bolstering his arguments with a home-made solar-powered radio, he finally persuaded his superiors, and was given a $25,000 Air Force grant in 1970 to develop solar energy for electricity conversion. Shortly thereafter, he and Eileen bought 30 acres of land in the nearby mountains as a hideaway homesite. In planning the construction of a home at their remote location, they discovered that it would cost more than $1,000 to tap into the local electric utility's grid. This discovery triggered their interest in the possibility of using natural sources of energy.[75]

They pursued the subject, both practically—by experimenting with hand-built model wind generators—and by poring over available technical literature on solar and wind energy. They formed a research foundation in April 1971 to investigate and eventually build natural energy structures. (It is still active and is now called ILS Labs, Inc.—for Integrated Life Support.)

By this time, they had decided to build a dome instead of a conventional rectangular house, and ground was cleared for it in April and May. Robert spent that fall and winter scouring the countryside for old, unused wind generators to salvage for his dome. He found his first in Willard, New Mexico. A rancher sold it to him for $50 plus another $300 for a tower and the costs of transporting it to his site. He spent several hundred dollars more rebuilding the plant's generator, which had rusted, and for new hand-carved 12-foot-diameter redwood blades.

The first wind generator was installed in February 1972. The remainder of the year was spent rebuilding 2 other used wind generators and putting together an elaborate solar collector for use in heating the dome and for domestic hot water. In all, their efforts cost them about $12,000 during the 2-year planning and construction period. They call the dome "Prototype I," and are busy raising funds through ILS Labs to build Prototype II, which will be of improved design—and hopefully half as expensive.

Details of Prototype I

In planning their new home, Reines explains, "we wanted to build a space ship totally powered by the sun and wind, with the same level of living we enjoy in the city." They chose a round structure because it was "revolutionary," and because "the sphere is the strongest structure known to man," in addition to having a good surface-to-volume ratio. An equivalent square structure, he notes, would require more than 1,000 square feet of surface area more than the dome, which would create additional heating and cooling problems. It would require, for instance, 10 times the heat and 5 times the electricity that is required by his hemispheric dome.

Prototype I is heated by 3 solar "black box" collectors, mounted on a hillside near the dome. The collectors trap solar heat, which is then stored for future use in a 3,000-gallon storage tank.† The dome is designed as an efficient "heat trap." Its exterior surface is thin metal, protected by an inside coat of 3 inches of sprayed-on urethane foam insulation. Like a spaceship, it has an airlock—2 doors that seal a 7-foot-long hallway at the dome's entrance. The dome's 18 windows are tiny portholes 8½ inches in diameter, which permit very little heat to escape during winter, although a 6-foot roof skylight has proved to be something of a problem.

The inside temperature is maintained at 65° F in the winter, and the summer temperature sometimes hits 85°. They plan to adapt the winter solar heating unit, which heats by supplying hot water to radiators in the dome, to a solar-powered air cooler in summer. This would involve operating the solar unit at night in the summer to circulate cool water through the system. In the dry New Mexican climate, the circulating water would cool the interior air by evaporation.

The Reineses are confident that, with some adaptation, the system will be able to maintain temperatures between 65° and 75° year-'round in the dome, even though seasonal temperature variations range from 30° below zero in winter to over 100° in summer.

Electricity for the dome is supplied by 3 reconditioned wind generators, including 2 Jacobs wind plants, which are the equivalent of more than 5 kilowatts of installed electrical capacity. The wind generators supply DC electricity, which is converted to AC by an electrical inverter—which Reines claims is 93 percent efficient, assuring minimum power loss. Some electricity is used directly as DC to supply emergency lighting and for experimental purposes.

Storage of electricity is accomplished by a bank of 16 heavy-duty batteries, which store 22,000 watt-hours of current. The electricity is used to supply a full range of appliances, including a hi-fi, tape recorder, television set, refrigerator, electric fan, pumps, mixers, and an electrically operated chemical toilet. They wanted a microwave oven—which uses much less electricity than conventional all-electric ovens—but couldn't afford one for the dome. Consequently, they in-

† Reines uses black copper tubes covered by 2 glass plates to supply the solar heat. The heat is actually "trapped" in a liquid (a water-ethylene glycol solution), which is piped to the storage tank, where it heats the water stored there.

cluded the one item that is dependent on external power supply: a butane cooking stove. It consumes only about $1.30 in fuel every 3 months, and even at that is in the process of being eliminated.

Windworks

An enterprising group of young communards in Wisconsin calling themselves "Windworks" has built a number of small electric wind generators, including a unit designed for Buckminster Fuller.‡ Led by graduate aeronautical engineer Hans Meyer, the group has designed several innovative wind plants that can be built cheaply with easily attainable materials.[76]

One generator, which costs less than $200 to build, offers the following features:

- A 3-blade, 10-foot-diameter wind rotor.
- A 12-foot-high tower.
- Maximum efficiency at low wind speeds—10 to 12 miles per hour—and putting out just over 0.25 horsepower, or 190 watts; capable of developing 6 horsepower, or 4.50 kilowatts, in a 30-mile-per-hour wind.
- No tail vane to turn it into the wind, employing instead a wood, Fiberglas, and cloth "cowling" that looks something like the outer surface of a familiar jet turbine, to orient the generator into the wind and direct air into the blades—in effect, running the plant in "reverse."

The Windworks commune has developed a new light airfoil blade for their generators made of Fiberglas-coated webs of paper. The special honeycomb paper, called "Hexcel," is compressed into a solid block and is cut into the proper propeller contour. It is then stretched out to full length, and the Fiberglas is applied. According to Windworks, the "Hexcel" material was developed for use on NASA's moon vehicles, as pads for their "feet." Three Hexcel blocks—enough for a single wind generator—can be purchased from Windworks for $11.*

John Roesel, Jr., president of Precise Power Corporation in Bradenton, Florida, and developer of the constant-output AC generator referred to earlier, is enthusiastic about developing a combined energy system using the Windworks' Hexcel blade-wind turbine in combination with his generator. He estimates that a complete 5-kilowatt home system would cost around $5,000 and, in selected areas, would supply complete energy requirements.[77]

‡ Fuller financially supports some of their efforts, though their life is Spartan.

* Contact Windworks, Route 3, Mukwonago, Wisconsin 53149.

Solar Wind Company of East Holden, Maine

Solar Wind Company was founded in 1972 by Henry Clews on a strong philosophical conviction: "Do we have the right," Clews questioned, "to steal this limited resource [stored energy] from past and future generations and to use up in a couple of hundred years what it took millions of years to produce; and to pollute our air and water in the process; and to rape our earth by strip mining? We at Solar Wind cannot accept this as a rational approach to power generation."[78]

Clews acknowledges that unlimited power is not available from the wind or any other natural energy source—including solar. "Instead," he notes, "you get moderate amounts of free, nonpolluting, environmentally safe power from an independent source which only costs what you are willing to put into the apparatus for harnessing it. The choice is up to you."

Clews' decision to go into the wind generator sales business grew out of his own difficulties in purchasing one for his Maine home. Unable to find a U.S. supplier, he located an Australian supplier, Quirks, in Sydney, and purchased a wind plant from them in August 1972. Still annoyed at the frustrations of locating a supplier, and at the same time pleased with the Quirks "brushless wind plant" once he finally did install it, Clews arranged to represent the Quirks Company in the United States as the Solar Wind Company, which he founded for that purpose.

The unit he installed for his own home provides all his present electricity requirements, including power for lighting, power tools, appliances, and a water pump. Home refrigeration is now provided by LP gas, but Clews plans to install a second wind generator for operating a freezer and refrigerator.

The 2,000-watt, 115-volt unit is mounted atop a 40-foot steel tower. The generator is fitted with a directional vane and a 12-foot-diameter propeller. It generates 3-phase alternating current by means of an alternator geared to the propeller hub. The only maintenance required is a change of oil every 5 years, according to Clews.

It is cheaper in the long run, Clews maintains, to use an alternator to generate electrical power because it is a more efficient device and will outlast a DC generator. The AC power is rectified to direct current to charge the batteries.

The wind plant begins to turn in 5-mile-per-hour winds, and starts charging the lead-acid batteries when they reach 9 miles per hour. At full output, the propeller spins at only 150 rpm, which is a major factor in the long lifetime of the plant. The generator is geared at a 5-to-1 ratio, and turns at 750 rpm. Clews' batteries are manufactured by Century Storage Lighting Company Ltd., in Australia, and sold through the Quirks Company. Average battery life, according to company literature, is 15 to 20 years.

The installed wind generator, which cost Clews $2,790, included the following accessories and equipment:

- 2,000-watt, 115-volt brushless AC wind plant: $1,475
- 40-foot steel tower: $315
- set of 130 amp/hour, 13-plate glass-lined batteries: $475

Total cost, FOB Australia: $2,265

- Freight, duty, and other charges: $419
- DC-to-AC rotary inverter: $100
- Rectifier and accessories: $106

Clews' wind-powered home is located 15 miles from Maine's Atlantic coast at a site some 350 feet above sea level. Although 2 mountains block some of the wind, he is still able to get winds that average 10 miles per hour. Two to 3 times per week, winds average 15 to 20 miles per hour for periods of about 6 hours each. These high-wind periods are critical to the system's success, since appliances run off the batteries instead of directly off the generator, and the lower prevailing winds couldn't by themselves provide the needed power. Average monthly output is about 110 kilowatt-hours.

In addition to the Quirks wind generators, Solar Wind Company now handles a second and less expensive product line manufactured by Elektro GmbH of Winterthur, Switzerland. February 1973 prices for the highly efficient Elektro generators ranged from $575 for the 300-watt generator to $1,785 for a 6,000-watt unit—only slightly higher than the $1,675 price tag on a 2,000-watt Quirks DC generator. Solar Wind also sells solid-state electronic inverters, several sizes of towers, and the Australian-made batteries for wind plants. Clews is now negotiating with Exide Battery Company in the United States as a possible supplier for his battery storage systems.

Clews stresses that his wind-generating plants are highly sophisticated, built of heavy castings with aircraft manufacturing tolerances, and "not at all suited for somebody to try to build in a garage."[79]

Oregon State University

Research at several U.S. universities is aimed at producing better data on wind power, as well as developing sophisticated production systems for small- and large-scale wind plants.

At the University of Oregon, Dr. Wendell Hewson is directing a research project funded by a consortium of privately owned electric utilities to investigate the possibility of producing electricity from Oregon's high coastal winds. The first year of the $132,000 study was spent analyzing existing Oregon wind records; establishing new wind stations to collect data useful in determining the potential for generating electricity from the winds; and modifying a wind tunnel at Oregon State University to test models for wind flow at potential Oregon sites.

The report of this first year's work estimated the cost of building a 1,500-kilowatt turbine designed by Palmer Putnam on the basis of the Grandpa's Knob experience. Dr. Hewson calculated that such a machine could be produced today for $700 per kilowatt, with a lifetime output of 70,000 kilowatt-hours per installed kilowatt, enabling it to generate electricity (though not at sustained, constant levels) at an over-all cost of 10 mills per kilowatt. However, he did not take into account the pollution-free character of wind power, which might add additional economic incentive.

Dr. Hewson concluded in his study that "the installed cost of $700 per kilowatt is too high, and must be reduced to be really competitive with conventional power sources. The costs should be at least cut in half, although this would still be higher than conventional power sources. Zero fuel costs for wind power tend to even out the over-all cost.

"To be really useful," he continued, "wind power *must* be cheaper than other power sources because of its variability."[80]

The National Science Foundation has considered funding projects in 1973–74 at other universities, including William Heronemus' large-scale wind plant work at the University of Massachusetts; a project at the University of California at Berkeley; work at Oklahoma State University; and a University of Montana proposal involving a series of cars running on tracks and equipped with air foils and generators that would convert wind to electricity to be fed into a utilities grid as a supplementary power source.

Oklahoma State University

Research aimed at improving the efficiency and the design of wind generators has been pursued for more than a decade at Oklahoma State University (OSU). Studies conducted at the school in the 1960s showed that the average wind power in Oklahoma City is 18.5 watts per square foot of area swept by the wind, an amount of energy approximately equal to the solar energy that falls on a square foot of land in Oklahoma, on a yearly average.[81]

A research program under way at OSU since 1961 has led to the development of several types of fuel cells; internal-combustion engines operating on hydrogen fuel (which can be produced by wind generators); and efficient wind generators specifically designed to operate in combination with hydrogen-burning engines or for AC power production.

Since conventional AC generators cannot be attached directly to wind plants because of the differing shaft speed of the wind generator and the AC equipment, the OSU scientists have designed an AC generator that has an output frequency independent of shaft speed. The 10-kilowatt generator they have devised is smaller and lighter than conventional AC wind generators, and they now have a 60-kilowatt prototype machine under construction. They report that the generators produce low-voltage, high-current DC power, which is ideal for hooking a wind turbine to an electrolyzer for hydrogen fuel production. Of course, the generator is also ideal for operation of conventional AC appliances when the wind blows at a sufficient speed to deliver power. Energy storage is in the form of hydrogen, which is burned in a fuel cell also developed at OSU, to produce AC power.

In 1973, the university requested $286,510 from the National Science Foundation for a 2-year engineering study, to conclude with the construction and operation of a 5-kilowatt wind-hydrogen-energy system. OSU electrical engineering professor Jack Allison says that 15-kilowatt AC generators—ideal for family use—complete with electrolyzers and wind generators could be built for $10,000 to $15,000. With mass production of the units, he predicts, "the costs could go down fairly dramatically." An analysis of anticipated installation and operation costs over a 20-year life span indicates that "we're within a factor of 2" of today's electricity costs—i.e., it would now be about twice as expensive as power from conventional sources; and "if electric rates rise much higher, we'll be competitive."[82]

Princeton University

A unique and interesting wind generator design that recalls the days of Dutch windmill sails has been developed by Princeton University aeronautical scientist Tom Sweeney. Sweeney, who worked for NASA designing special aerofoil wings for airplanes, explained that, while working on the sailwing aircraft, "I became intrigued with the idea that the sailwing might make an efficient windmill blade."[83]

He left NASA for Princeton, and in 1968 built a 10-foot wind generator blade with stretched Dacron fabric. When the wind turned the blade, the fabric took on an aerodynamic shape—a reflection of its response to the wind's force. At rest, the fabric was stretched flat by a cable under the "wing." In wind tunnel tests, it withstood wind speeds up to 160 knots.

The great advantage of Sweeney's design is its extremely light weight—a prototype blade for a 25-foot-diameter wind machine weighs 44 pounds! This light weight enables the blade to withstand centrifugal force placed on it much better than the heavier metal blades of more conventional wind-generator designs.

Sweeney is now working under a grant from Fairchild Industries to build a complete 25-foot-blade-diameter wind generator using this principle, which would be rated at 7 kilowatts, a family-size power plant.[84]

Yale University

Under the direction of Everett Barber, Jr., engineering professor at Yale's School of Architecture, a series of experiments are being conducted on the use of wind generators and solar collectors for use in houses, buildings, and new communities. Tests of wind generators have been conducted (in 1973) on the roof of the School of Architecture building.

Barber, who, as noted earlier, has designed a number of efficient solar collectors, has been working with integrated wind/solar systems. In 1972, he applied for a federal grant from the National Bureau of Standards to design a wind/solar system as part of NBS' ongoing program to develop a Modular Integrated Utility System (MIUS).†

He calls his approach a "Natural Utility System," or NUS. Wind power would be used to electrolyze water for hydrogen production; the hydrogen would be used on-site for combustion in steam boilers and gas turbines. In combination with solar collectors for heating and air conditioning, and reuse of wastes for power, all energy needs for the community would be met from natural sources.[85]

Peter Clegg, an English graduate student working with Professor Barber on the project, has designed an "autonomous house," which utilizes a wind generator on the roof as well as a thermal chimney (an approach also employed by Dr. Alvin Miller at the University of Arizona). Though he describes it as "a bit fanciful," the house design is functional, and incorporates a greenhouse for natural solar collection.

Clegg has also designed elaborate, integrated energy systems using natural sources for application to new communities.[86]

† See Chapter 4 on decentralized energy technologies.

Future Uses of Wind Power

The principles of wind power generation are well understood, and the long-term operation of wind generators has been accomplished in many countries. Wind power promises a number of possible future applications:

Small Electric Generators

Small electric generators in the 3-to-15-kilowatt range can be utilized directly to provide power for a house or building, as has been demonstrated repeatedly by an inventive tradition represented by Professor Poul La Cour in the nineteenth century and by Hans Meyer, Robert Reines, Henry Clews, and others today. In combination with solar space heaters—and air conditioners—using flat-plate collectors to supply low-grade heat requirements, wind generators can supply the higher-grade power in the form of electricity to meet the total energy requirements for a house using moderate quantities of energy.

However, the costs of manufacturing the wind plants, procuring several hundred dollars' worth of storage batteries, and erecting towers of 50 feet and higher represent a substantial initial investment. Commercially available units of this size range from around $2,000 to more than $10,000. Minimization of costs by "do-it-yourself" construction can be attained, but to date, no systems have been demonstrated that can duplicate the performance of more expensive commercially available equipment.

To justify the large initial investment in wind power equipment, a system must be erected to give years of dependable service. Wind generators of small size have been built in many countries from low-cost materials, such as surplus military equipment, recycled automotive and machine parts, and the like, but they have not and probably cannot come near the standards set by such entrepreneurs as Marcellus Jacobs (Jacobs Wind Electric Company). To meet high-performance standards, the best materials and techniques must be used in construction, necessitating an approach that does not "cut costs."

In combination with solar heating systems, small-scale wind generators can be used quite efficiently. Again, however, the costs of collecting solar energy and storing it, perhaps in several hundred to a few thousand gallons of water, add additional high initial costs to the system. Robert Reines of ILS Labs is convinced that solar-heated, wind-powered domes can be produced for $6,000 to $7,000 apiece, but only if substantial economies of scale can be reached in mass production of the necessary equipment. He estimates that manufacturing facilities might be able to produce several hundred combined units per week.

The key to the success of such systems lies more in the values of society—specifically those members of society wishing to live in a natural-energy home—than it does in the technological development of the systems. The successful demonstration of wind (and solar) equipment in the twentieth century stands on its own merits. It is not probable purely on economic grounds that these systems will soon prove economically competitive for most locations. Only by carefully engineering natural, low-energy use into the structures for which they are designed to furnish power has their economic worth been demonstrated.

The systems will probably not be "economic" for some years—despite the rising costs of nuclear and conventional fossil fuels. Wind power may not become "competitive" in the United States until the bulk of our fossil fuels have been squandered and the cost of nuclear electricity has risen substantially. The short-sighted economic reasoning employed in the world's decision-making today does not approximate the real costs to society of any form of energy—fossil fuels, nuclear, wind, or solar.‡

Medium-Size Wind Generators

Medium-size wind generators of 15-to-20-kilowatt capacity (perhaps as high as 100-kilowatt) might be used to produce electrical power in favorable locations for small communities. As the needed generating units increase in size, costs decline somewhat, so that the projected investment per person in such a scheme might drop to 20 to 50 percent less than the costs of individual house generating units. The central drawback here is the nature of the wind's energy.

A single, large machine must be situated in an excellent location, one having annual wind velocities averaging more than 20 to 25

‡ Assuming the laws of economic supply and demand were to react overnight to the consequences of a nuclear power plant accident taking the lives of thousands and permanently condemning a few hundred square miles of property, the true value of natural energy might be recognized and orders for wind generators would deluge manufacturers the next day. But this kind of a reaction is not likely. Nor is it likely that the generations of today will mentally poll the generations of tomorrow, as Dave Brower, president of Friends of the Earth, has suggested, asking their opinion of the current exhaustion of precious fossil fuels and the buildup of tens of thousands of years' worth of radioactive by-products from nuclear power plants and weapons. Our generation does not seem immediately interested in getting off the cheap-energy binge with its concomitant environmental abuse.

miles per hour, to be economic. If a hill site is not available to such a community, then a very high tower, perhaps exceeding 100 to 200 feet, might be necessary to extract the necessary energy. This would require additional capital, as well as the availability of specialized personnel and machinery to service the unit.

Smaller wind generators would be more expensive per person, but due to the variations in available wind power in most locations, a community network of wind generators might extract more over-all power than one large machine. Undoubtedly, such a decision would have to be made by a community after conducting a detailed wind survey, based on precise measurement of the wind structure at the desired tower height and a study of available records.

Such wind stations might operate in combination with solar energy equipment to fulfill total energy requirements, but the solar equipment in most places would remain on rooftops, not in central facilities. Only in sunny areas with high solar values would solar collecting equipment be centralized for redistribution to individuals in the system.

Large-Size Wind Generators

Large generators, ranging in size from 100 kilowatts' to several megawatts' capacity, might be connected in an electric utility grid to produce power in a wide area, with storage in batteries, flywheels, water power, compressed air in underground reservoirs, or converted to hydrogen fuel. Perhaps in a large grid, wind could be used as a supplementary power source—as NASA's Ernst Cohn has suggested—with no storage. In each case, the variations in wind flow and strength would be favorably utilized.

The economics of such machines are not precisely known at this time. The initial costs of building large wind-generators may be several times greater per kilowatt of capacity than the costs of building equivalent power stations powered by conventional fuels. Of course, today, the true environmental-societal costs of conventional power are not recorded in the ledger books. Wind power offers society no "Faustian bargain," as does nuclear power; nor does it pollute the air to cause death and disease, as does the combustion of coal and oil.

Technical aspects of building large generators are remarkably well understood. French engineer Louis Vadot, who chaired the sessions on wind power generation at the United Nations' 1961 Rome conference on new energy sources, noted that:

> . . . It is now generally agreed that the design of wind power plants with capacities ranging from the lowest levels to far beyond 100 kilowatts no longer presents any fundamental problems.

> In the field of calculation and theoretical aerodynamics, our present knowledge is sufficient to be usable in a more or less routine way and without need to be on guard against surprises. . . .
>
> In the field of engineering, the record of achievement shows that here too we are on firm ground. In the case of very large machines, there are, of course, still some small imperfections of detail, but it is rare to find any branch of activity in which such a small number of prototypes [100-kilowatt and above] have produced so few failures. This is very encouraging.[87]

Not only is the technical development of electric wind generators encouraging, it is revolutionary. Electric power from the winds offers the twin advantages (particularly in nations with advanced technological capability) of (1) relief from dependence on dwindling fossil fuel reserves or inherently dangerous nuclear power; and (2) totally clean energy. The minimal drawback of higher capital investment for wind plants would be more than repaid by the benefits of inexhaustible energy, safety, and environmental compatibility.

Magnitude of Power from the Wind

Twenty billion kilowatts of available wind power in the world, as predicted by the World Meteorological Organization (WMO) in 1961, is such a significant energy resource that it must not be neglected; although it must be noted that the WMO calculation is 1,000 times higher than estimates made by Palmer Putnam in his book *Power from the Wind,* and has been challenged by the present-day "energy establishment."

A 1964 White House energy study,[88] commissioned 2 years before by then-President John F. Kennedy, concluded that the World Meteorological Organization figure may reflect an "ultimate potential," but suggested no research program to facilitate recovery of the power. Succeeding administrations in the United States have also failed to respond. The federal government has spent little to examine or effect a program to utilize this tremendous energy resource.

One of the reasons has been a misinterpretation of the amount of power that can be economically derived from the winds. The 1964 White House energy study suggested that "economical power generation requires an annual average wind velocity of about 30 miles per hour."[89]

With the advent of advanced equipment, and better understanding of the nature of wind power, this estimate should be revised. Economical power generation from the winds can be achieved from an annual wind regime considerably less than 30 miles per hour—probably on the order of 20 miles per hour for large-scale plants, and less for smaller plants.

Recognition of this vast potential has been slow in coming, but signs of progress are evident. A committee of the National Science Foundation and the National Aeronautics and Space Administration in January 1973 recommended a $610 million federal program to analyze the potential of wind power in the United States, culminating in the construction of a demonstration facility. The committee adopted William Heronemus' estimate of the technical and economic feasibility of developing 159 billion kilowatt-hours of wind power offshore in New England annually by the year 1990. The committee suggested that by the year 2000, less than three decades away, a major U.S. development program in wind power could result in an annual yield of *1.5 trillion kilowatt-hours* of electricity![90]

This amount of power—1.5 trillion kilowatt-hours—was the total electricity consumed in the United States in 1970. With a policy of limited energy growth, wind power might eventually supply all U.S. electric power.

The federal government has been slow in responding to wind power potential. In 1973, however, scientists at NASA-Lewis Research Center in Cleveland, Ohio, announced tentative plans to build a 100-kilowatt wind machine there in 1975. An updated version of the generator may also be emplaced on the island of Culebra off Puerto Rico under a contract with the National Science Foundation.

The project would represent the first phase of a 5-year program to bring wind power to the threshold of widespread commercial application. NASA officials believe that with mass production, electricity generated by wind power will be available at capital costs that a decade from now will be competitive with those of nuclear plants. The aim of the 5-year study is to identify and eliminate problems involved in using wind generators to get electricity. The NASA-Lewis goal is to have a wind system—turbines, generator, tower, and storage—that is cost-effective and commercially viable at the end of the 5-year study.

Alluding to earlier Danish efforts at wind power, Dr. Joseph Savino, technical director of the NASA-Lewis Wind Energy Program, contends that "there's enough energy in the winds to supply all our electrical needs. If the Danes could get 200,000 kilowatts of electricity from the winds back in 1908, we should be able to get the power we need right now."[91]

Implementing a large-scale national program for generating electricity from wind calls for a major commitment. The same is true for the development of wind-powered ships such as the Dyna-Ship.

The federal government's first significant commitment to wind power development came in the fiscal year 1974 budget, in which the Administration requested of Congress only $1.25 million to assess the potential of wind development—one one-thousandth of the $1 billion total for all energy research and development. Such a commitment can hardly be considered serious.[92]

Basil Greenhill has described the pitfalls of *not* developing wind power for shipping, and his analysis holds true for power plants as well.

If the energy crisis forces a shortage of conventional fuels necessary to actually build modern wind-powered ships, he notes,

> a situation in which fuel was virtually unobtainable would, of course, require a much more drastic solution if sea transport was to be maintained. Such a situation would mean virtually a return to a man- and animal-powered, wood-fired industry. There would not be the resources ashore to build a sophisticated vessel of the type I have described. . . . Perhaps the design and experimental work on the development of modern merchant sailing vessels [or power plants] should be begun soon, before the crisis is actually upon us.[93]

Afterword

The Energy Basis for Future Society

For a period of history stretching over almost three centuries, the advanced technological nations of the Western world have been engaged in an energy and resources race that has brought us to the brink of an unknown future, marked by the sudden undercutting of the fossil fuel energy base that supports the currently resource- and energy-rich nations.

Philosophers, theologians, and historians have questioned the various social problems of crime and poverty in the high-energy societies, and have offered many theories about the future of our civilization. Few theories or philosophies, however, have come to grips with the underlying questions of entropy and disorder brought about by the gradual dissipation of the rich products of the Earth, without which the high-energy societies cannot function.

One of the most perceptive historians to question the values of high-energy societies, the American Henry Adams, compared the American love affair with the energy of the great Dynamo to the former religious energy of the Virgin in Christianity, who, he noted, had "acted as the greatest force the Western world ever felt and had drawn man's activities to herself more strongly than any other power, natural or supernatural, had ever done. . . ." This was of enormous consequence to the study of history, Adams observed, because the "historian's business was to follow the track of the energy; to find where it came from and where it went to; its complex source and shifting channels; its values, equivalents, conversions."

Adams was appalled that the track of energy in America led to love of the Dynamo, which precluded human concern for the spirit. He noted that "forty-five years of study had proved to be quite futile for the pursuit of power; one controlled no more force in 1900 than in 1850, although the amount of force controlled by society had enormously increased. . . ."[1]

That moving force in American society proved to be merely mechanical power; the inner force of the human spirit had not risen appreciably.

Adams was a stranger to high-energy society, and would be dismayed to see the development of urban civilization today. Before his death in 1918, he predicted that the acceleration of high-energy use would bring the civilization of the West to an end. He argued that the greater values of human existence were crowded out of man's fascination and worship of electricity, dynamos, steam engines, and automobiles.

Were Adams alive today he might view what appears to be the approaching end of the fossil fuel era as cause for celebration. As the great stocks of fossil fuels decline, people of America and other high-energy societies will have occasion to pause and reflect on the passing era of the all-electric home, jet travel, and the automobile. The reconstruction of society based on sharing and on decentralized living habits in closer harmony with the Earth will not appear as a utopian dream, but as a necessity for survival in many areas.

Though the concept of a *finite* Earth resource has been continually stressed by thinkers in Western societies, no concentrated human actions have been made to prepare for the day when life must revert to a less destructive base. M. King Hubbert of the U. S. Geological Survey, widely respected as the dean of America's energy and resource experts, has predicted for several decades the collapse of the fossil fuel economy. Dr. Hubbert notes:

> It now appears that the period of rapid population and industrial growth that has prevailed during the last few centuries, instead of being the normal order of things and capable of continuance into the indefinite future, is actually one of the most abnormal phases of human history. It represents only a brief transitional episode between two very much longer periods, each characterized by rates of change so slow as to be regarded essentially as a period of nongrowth. It is paradoxical that although the forthcoming period of nongrowth poses no insupera-

> ble physical or biological problems, it will entail a fundamental revision of those aspects of our current economic and social thinking which stem from the assumption that the growth rates which have characterized this temporary period can be permanent.[2]

Making the transition to a lower-energy-based society, in which natural energy forms can serve as the prime movers of a new civilization, will perhaps be history's most challenging experience to Americans and other inhabitants of the high-energy world.

Natural Architecture

The physical architecture of the new society will require the utilization of sophisticated technologies of the present to bridge the gap of history to the styles of the past, when climate reigned supreme. As was noted in the first chapter, it has long been fashionable to consider the form of early human civilization as primitive.

This pervasive myth is best examined in light of the architectural achievements of civilizations of centuries past. What appear primitive in the early structures of human existence have been examined by a number of architects and scientists, and some of the most intriguing findings concern the Indian pueblos in the American Southwest.

Architecture professor Ralph Knowles of the University of Southern California has devoted a number of years to the study of American Indian cultures in the West and Southwest, with particular emphasis on the varieties of architecture practiced by early Indian civilizations. Professor Knowles has found that the structure of the Indian pueblos (villages) in the Southwest was carefully designed to use solar energy, winds, and other climatic elements.

A study of the Acoma tribe's pueblo, which is located atop a 400-foot-high mesa about 50 miles west of Albuquerque, New Mexico, was undertaken by Knowles to ascertain the relationship of the pueblo's construction to the climate of the region. The pueblo was first described by the Spanish Coronado expedition in 1540, and appears to have been continuously inhabited for over a thousand years.[3] An analysis of the two hundred houses that comprise the pueblo was compared to theoretical studies of forms and materials that would have been used at the site to maximize the entry of sun in the houses in the winter but minimize sun entry in the houses in the summer. "It would have been most reasonable if the vertical walls receiving winter sun had a high transmission coefficient and a high heat-storage capacity," Professor Knowles reasoned. "Conversely, the horizontals receiving their maximum energy in the summer should exhibit a low thermal transmission coefficient and a low heat-storage capacity."[4]

A study of the Indians' construction revealed that their choice of materials and method of construction had done exactly that.

Professor Knowles points out in his study that in the desert climate a major construction feature is the ability of walls to store heat, since temperatures drop sharply at night, and sun energy intercepted by the building must be retained for gradual release into the structure (for heating) at night. He found that the use of thick adobe masonry by the Acoma Indians did a superb job of capturing winter heat. In addition, the orientation of the buildings in relation to the position of the summer sun, prevented much of the solar heat from entering the structures, thus keeping them cooler. In fact, the "efficiency" of the structures at the Acoma pueblo brought 50 percent more heat to warm the dwellers in winter than in summer. The study of the Acoma pueblo confirmed Professor Knowles' belief that the Indians deliberately built their structures to use the energy of the sun in a sophisticated fashion. The pueblo, he says, "is an efficient energy system tending to equalize internal energy profiles over the extremes of season and day."[5]

Not only is the Acoma pueblo carefully designed to take advantage of the solar heat in winter for heating, and to repel the heat in summer, but analysis of other Indian settlements shows similar and even advanced forms of design. Professor Knowles studied the design of the pueblo Bonito in Chaco Canyon, New Mexico (in the middle northwest area of the state). Two Indian cultures built structures at that location, the old Bonito pueblo being constructed around A.D. 919, and the new Bonito pueblo constructed between A.D. 1060 and 1080. The new Bonitians expanded the old site, tripling its size.[6] The plan of the pueblo Bonito remained essentially the same during both periods of construction. As shown in the illustration, the pueblo is crescent-shaped, opening to the southeast.

During both periods of construction, both wood and sandstone were used as construction materials—the sandstone ranging up to three feet in thickness in the walls. At the height of its development, the pueblo may have numbered as many as twelve hundred inhabitants, before its fall around A.D. 1130 to a warring tribe.[7]

The pueblo was constructed in such a fashion that the sun in different seasons provided illumination to specific rooms in the tiers of dwellings in the village. Professor Knowles reports that during late afternoons in the summer, from 6 P.M. until sunset, "a gap in the north face of the third floor allows the sunlight to penetrate to the upper terraces and to the eastern arm of the pueblo. Here the last rays of the sun simultaneously intersect two projected corners of a prominently placed section of the upper tier. This special effect would have emphasized

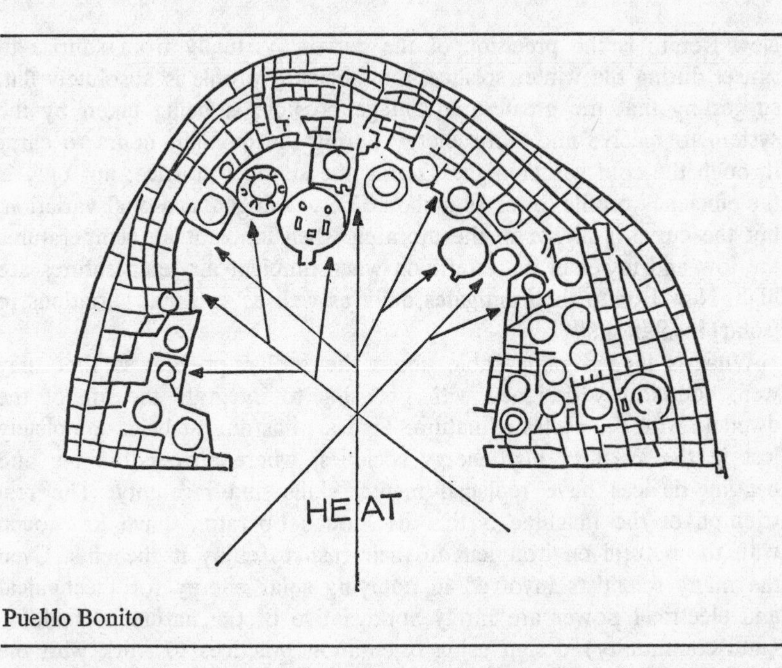

Pueblo Bonito

the limits of this part of the pueblo and would have insured its importance by sharp and detailed contrasts of dark and light."[8]

Similar starting patterns of the solar effect on the pueblo occurred as the winter sun wound its way across the sky. "The azimuth of winter sunset precisely coincides with a line drawn through the extreme limits of the crescent [the over-all pueblo plan]," he says, resulting in the reversal of the summer condition, and placing almost the entire pueblo in shadow.

However, an exception to this was provided by the Indian planners, who left a corner of the crescent unbuilt, providing a clear path for the sunlight. According to Knowles, "this removed section, which provided the special sighting of summer sun at cliff's edge, also allows winter sunset to finally strike the eastern extremity of the upper tier, thus reinforcing its apparent significance. Such special relationships seem remarkable and beyond the possibility of coincidence."[9]

He found that the over-all design of the pueblo provided a form which, like the Acoma pueblo, very efficiently admitted winter sun for heating and repelled summer sun, keeping the structures cool. Both the old Bonito settlement and the new Bonito additions were "efficient energy systems to mitigate seasonal variation in [solar] insolation," he says, adding, "each has a higher efficiency curve in winter than summer [in admitting sun heat]. The truly remarkable fact about

New Bonito is the precision of the curves. Virtually from sunrise to sunset during the winter solstice, the efficiency profile is absolutely flat, suggesting that the greatest advantage possible is being taken by the system to receive and store energy during the daylight hours to carry through the cold winter night. During the summer solstice, not only is the efficiency profile lower as it should be to mitigate seasonal variation, but the curve is higher in the morning when ambient air temperatures are low and lower in the afternoon when ambient air temperatures are high. New Bonito then mitigates daily as well as seasonal variations in [solar] insolation."[10]

What is indeed remarkable about the Indian pueblos is that they were consciously designed with precision to integrate the life of the dwellers with the cycles of nature. That art has almost been completely lost in the Western high-energy societies, where air-conditioning and heating devices have replaced mental skills and ingenuity. The real triumph of the machine is that its human operators have lost touch with the natural environment in their rush to enjoy its benefits. Even the many scientists involved in applying solar energy for mechanical and electrical power are rarely appreciative of the nature of building (and community) design using orientation practices to work with the sun in providing housing.

The ancient Indian civilization of the Southwest applied greater logic to climate design than virtually all of our "progressive" and "learned" urban planners and architects in America today!

Evolution

The evolution of climatic design by humans was learned from observations of the relationship of plants and animals to climate in the natural world. Humans are not equipped with the natural protective mechanisms of many plants and animals, and, in early cultures, humans had to mimic the climatic adaptation of other creatures. The ability of the world's creatures to adapt to climate appears unlimited. Bears hibernate in winter to reduce metabolism and survive the cold and unproductive season. Elephants cool their blood in hot climates by moving their ears, exposing honeycombed blood vessels to the cooling effects of air.

Many desert animals simply burrow into the ground during the hot, sultry days, reappearing at night to search for food. Birds can control their insulation by trapping air bubbles under their feathers; when times warrant, they migrate to more satisfactory climates. Victor Olgyay notes that the insect world outclasses other animals by collective building practices to adapt nests in relation to climate. In temperate areas, ant hills are built on southeast slopes and are elongated on northeast-to-southwest axes to collect the early morning sun. In tropical regions, compass termites build enormous blade-shaped hills that point due north. The various exposures equalize heat distribution, and the mass of earth shields the insects from the beating sun. The towers, he says, reach 400 times their body length, "which translated into human terms would equal 2,400 feet."[11]

Vitruvius and the History of Adaptive Architecture

The history of man's adaptive architecture is largely an unwritten record, as the thousands of years of evolution and adaptation to various climates developed specific styles in different civilizations. In the first chapter, the Greek writer Xenophon is quoted on the construction and orientation of houses. A more complete and fascinating guide to adaptive architecture was written in the time of the Emperor Augustus by the Roman architect Vitruvius.

Vitruvius devoted many pages of his *Ten Books on Architecture* to specific suggestions for siting and orientation of houses, buildings, and cities in regard to the sun, wind, and other climatic factors. He defined architecture as comprised of various characters, including order, symmetry, arrangement, propriety, and economy. On the aspect of propriety, he notes that this is influenced by natural causes [climate], citing as an example the siting of temples in sacred precincts, which should have adequate water sources to heal the sick.

"There will also be natural propriety in using an eastern light for bedrooms and libraries, a western light in winter for baths and winter apartments, and a northern light for picture galleries and other places in which a steady light is needed; for that quarter of the sky grows neither light nor dark with the course of the sun, but remains steady and unshifting all day long."[12]

On city siting: "Such a site will be high, neither misty nor frosty, and in a climate neither hot nor cold, but temperate; further, without marshes in the neighborhood. . . . If the town is on the coast with a southern or western exposure it will not be healthy, because in summer the southern sky grows hot at sunrise and is fiery at noon, while a western exposure grows warm after sunrise, is hot at noon, and at evening all aglow."[13]

On the relationship of wind directions to the location of city streets: ". . . Let the directions of your streets and alleys be laid down on the line of division between the quarters of the two winds. . . . On this principle the disagreeable force of the winds will be shut out from dwellings and lines of houses. For if the streets run full in the face of the winds, their constant blasts rushing in from the open country, and then confined by narrow alleys, will sweep through them with great violence. The lines of houses must therefore be directed away from the quarters from which the winds blow so that as they come in they may strike against the angles of the blocks and their force thus be broken and dispersed."[14]

Vitruvius' guides for construction of cities, temples, the Roman baths, and houses all related to climatic aspects of individual sites in a regional context. He pointed out that each region would foster a different style of construction. "One style of house seems appropriate to build in Egypt, another in Spain, a different kind in Pontus, one still different in Rome, and so on with lands and countries of other characteristics. This is because one part of the earth is directly under the sun's course, another is far away from it, while another lies midway between these two. Hence, as the position of the heaven with regard to a given tract on the earth leads naturally to different characteristics, owing to the inclination of the circle of the zodiac and the course of the sun, it is obvious that designs for houses ought similarly to conform to the nature of the country and to diversities of climate."

In practice, he suggested: "In the north, houses should be entirely roofed over and sheltered as much as possible, not in the open, though having a warm exposure. But on the other hand, where the force of the sun is great in the southern countries that suffer from heat, houses must be built more in the open and with a northern or northeastern exposure. Thus we may amend by art what nature, if left to herself, would mar. In other situations, too, we must make modifications to correspond to the position of the heaven and its effect on climate."[15]

Examples of human adaptation to climate can be found in all cultures of our hemisphere, not just among the Pueblo Indians. In her comprehensive book on the subject, *Native Genius in Anonymous Architecture,* Sibyl Moholy-Nagy notes that the settler "has his own climatology that deals with the four climatic conditions he has to combat: cold, snow, rain and heat. For him this is weather and all there is to climate. His adjustments are geared to these phenomena."[16]

She details many examples of houses and buildings in our hemisphere that make use of site planning, orientation in regard to sun and wind, and the use of construction materials to mitigate the effects of unpleasant climatic conditions and enhance desired conditions.

Houses in tropical zones are classically open, airy structures constructed with light materials so that wind is freely admitted and solar heat is not unduly absorbed in the walls and roof. Moholy-Nagy describes a Haitian house that is admirably adapted to the sultry heat of the island, moderated only by occasional breezes from the mountains:

> The floor-to-ceiling opening admits light by day and air by night, but it is the central dormer that is most admirably adapted as ventilator. Supported by the posts of the porch, it overhangs the house wall and traps each breeze that might come its way, expelling the rising heat admitted through the French door.
>
> The high conductivity of the corrugated tin roof allows for quick cooling off as soon as the sun has set, and it is completely termite proof. The walls are woven screens, joined seam-to-seam where privacy is wanted, and left half open directly under the dormer to increase the upward draft of air.[17]

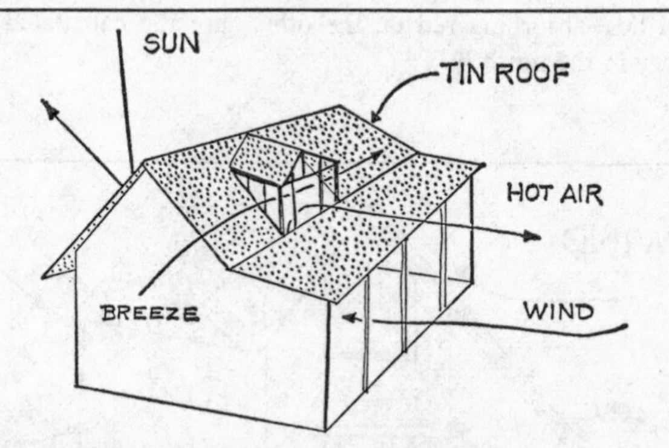

Tropical house: wind- and sun-oriented architecture

She adds that the functionality of the house is a total concept, that "there is no element in this simple settler house that does not serve the functions of human comfort."[18]

In colder climates, houses and other structures were devised to absorb sunlight during the winter season. In New England, the high, slanting roof was built with various breaks to reduce the stress of mounds of snow on the house structure. In other areas, the sloping roof was used in many cases as a barrier to keep snow near the north side of the structure as added insulation (as the Eskimos use ice as insulation in the igloo). Moholy-Nagy decribes the use of snow as insulation, in addition to answering the question: Why are barns red? She describes two barns on the reservation of Fort Klamath, Oregon, built with a long, sloping wall "into the weather" or in the direction of the prevailing winds. The roof ends several feet above ground, and the space between the ground and the sloping roof is filled with bales of hay or alfalfa to serve as a double wall of winter insulation, the accumulating snow (from the winter winds) forming the primary protective layer of insulation, or snow break.

The south walls of the barn are painted red. She explains:

> When farmers started to paint their buildings in the early 19th century there was no "store-bought" paint. Skim milk was mixed with oxide of iron—rust, to wit, scraped from fences and nails—and with lime. The result was a sort of varnish that did not sink into the wood grain but covered it like a skin. . . . The choice of red was by no means accidental. Because red absorbs sunlight, it was a "solar heat" device. The handicap of a region with severe winters has been turned into an advantage. The natural warmth of the animals inside the barn together with the natural provision of a protective snow layer on one side and heat-absorbing red on the other, are the calculated assets of husbandry in the north.[19]

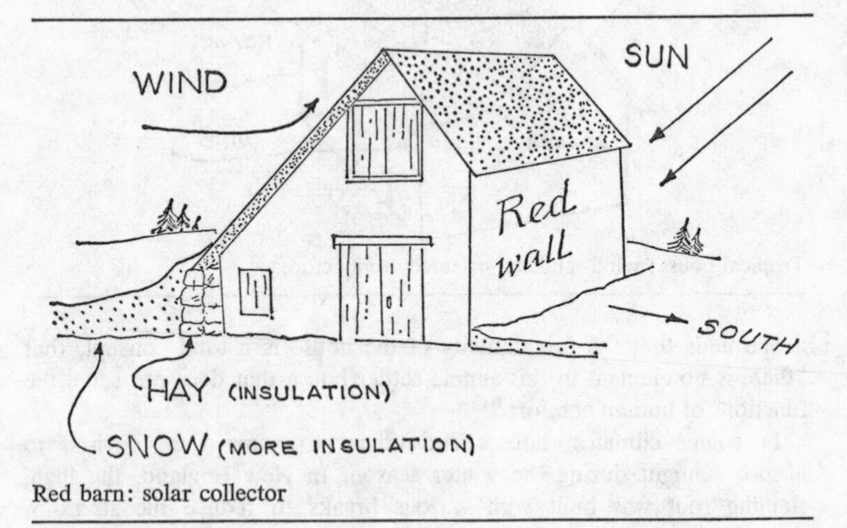

Red barn: solar collector

Before the advent of oil heaters, electricity, and the air-conditioning machines that homogenized the structure and shape of construction and housing, all buildings and homes were designed to use the natural factors of climate. Perhaps the last major attempt in contemporary American civilization to develop the techniques of natural architecture for the population at large (excepting rural farmers and the selected rich who could afford to experiment with natural building techniques as an avocation) was made in the early 1950s by the federal Housing and Home Finance Agency (HHFA).

Architects Victor and Aladar Olgyay developed a comprehensive program for adapting American architecture to climatic considerations under contract with the HHFA at the Massachusetts Institute of Technology. The Olgyay brothers' research culminated in the publication of a manual on climatic design called *Application of Climatic Data to House Design,* which was published by the federal agency in 1953.[20] Although the nation's love affair with industrialized housing and climate control via air conditioning and other energy-intensive mechanical systems overruled the popular acceptance of the Olgyays' work during the 1950s, the manual* has remained one of the best architectural guides for natural design. Additionally, they authored the book *Solar Control and Shading Devices* in 1957,[21] and Victor Olgyay published his comprehensive treatment of the subject, *Design with Climate,* in 1963.[22]

Design with Climate analyzes the relationship of specific houses in various U.S. locations to climatic variables and points out a number of practical ways in which houses, as well as communities, can take advantage of climate, orientation of structures, and native building materials to effect quality housing and community goals, while maximizing the use of solar energy and other climatic variables. Four climatic regions—Minneapolis, New York, Phoenix, and Miami—are studied, and specific recommendations are given for community and individual housing unit development. Olgyay designed for each region prototypes of "balanced houses" that made use of climatic design, and compared their performance in conserving energy to "orthodox" houses, which reflected average houses not designed with climate in mind.

In the temperate climate area of New York, he compared the climatic performance of a square house having a total floor area of 1,225 square feet with a balanced, rectangular house of the same floor area. The rectangular house was designed to maximize the interception of solar energy by incorporating a large south-facing window area to admit winter heat, but equipped with an overhang on the roof to restrict the admittance of solar heat in summer. The house was designed with careful insulation and other techniques to mitigate climatic effects.

The balanced house in New York showed a dramatically improved energy budget (for example, more natural solar energy was used to offset the use of additional fuel) than a conventional house. Laboratory tests of the simulated house model indicated that the balanced house gained more solar heat than a normal house in winter—through solar

* The manual is, unfortunately, out of print.

orientation, reorientation of windows, the rectangular house shape, reduced heat loss through construction techniques, good insulation, and double-glazed windows. (The double-glass windows also reduced heat loss through convection.) Ventilation of appliances and the selective use of stone for walls also improved the energy budget. In total, the balanced house in the New York area reduced winter heat loss by 49 percent over the performance of the conventional model house.

In summer, the balanced house outperformed the conventional house by utilizing shape, arrangement, and window overhangs—in addition to the use of wall shading and improved ventilation techniques—to remove interior air heated by the sun. In all, the improvements in the house energy budget for summer showed a 71 percent higher performance than for a conventional house.[23]

Olgyay's findings are surprising only in a contemporary context, for it must be remembered that until only very recently, all buildings were designed to take maximum advantage of climate; and the savings discovered by redesigning contemporary American houses are simply the result of reapplying ancient architectural wisdom.

The signs of revival of climate-oriented architecture are visible throughout American society, through the efforts of the young, the "technological counterculture," and a small but growing number of engineers and architects disenchanted with the dull, monotonous dimensions of industrialized architecture characterized by generations of Lever Houses and modularized homes. A new renaissance in architecture is on the threshold, heralded by such groups as Sun Mountain Design, a small group of architects and engineers who are currently planning a community using the principles of low-energy housing and planned community design to bring people together with the natural environment. In a proposal for funding to the federal Housing and Urban Development agency, they describe their planned project:

> We would like to emphasize that this is not the usual residential development. We plan to create an environmental community that is designed with nature instead of against it. We will disturb the environment as little as possible, spending less to create an additional value, and hold "zero impact" on the existing ecological balance as a goal.
>
> We intend to use the most advanced concepts of cluster design practical. We plan to design villages that act as a self-contained energy and service system. Cluster housing has it origins in Santa Fe where the compound has historical roots in the defended village or rancho. Even more important, the Indian pueblo is one of the most perfectly designed low technology structures in the world. They store heat by their shape and were created out of the land they stood on. We plan to use tech-

nology where it is appropriate to enhance or preserve the beauty of the land and to benefit the residents.

The time is certainly ripe for a low energy consumption housing system and a development approach that saves the land; the demand is growing.[24]

A rational future architecture will recognize climate once again as the precondition for the science's own existence. In conjunction with the use of natural energy sources, which do not take from the Earth the limited, priceless resources of the fossil fuels or the deadly nuclear materials, a future program of rational architecture will be the hallmark of a secure civilization, which plans not only for the wishes and desires of one generation of humans temporarily caught up in the excesses of a high-energy society, but can accept the steady-state conditions of existence with energy sources based primarily on the sun.

Social Transition

The social transition from a high-energy, fossil-fuel-based society to an intermediate-energy society that uses solar energy as a primary power source—both for agriculture (decentralized) and for localized power generation—cannot be expected to take place overnight, but the seeds of change are evident today.

Biologist John Todd, a founder of the New Alchemy Institute in Woods Hole, Massachusetts, has spent the past several years developing decentralized energy technologies, including small agricultural experiments, and constructing wind generators and solar collectors. He describes the changes he foresees in the energy-short world of the future as people seek new alternatives to the high-energy world of today. His view is not apocalyptic, but is a gentle vision of a postindustrial age:

> . . . . there will evolve a genuine alternative, for the apocalyptic view may cradle within itself the seeds of a social transformation which is accessible to all. If there is little faith in western civilization coping with itself, then the most viable alternative lies at the lowest levels of society . . . the family or small group. If enough people realized that change was primarily their own responsibility and created places of living that were functionally complete microcosms then many planetary stresses would be eased. These microcosms would be ecologically derived communities in which the inhabitants produce their own energy, foods, shelters and communitas and where wastes were considered resources to help restore local environments. Each area would be comprised and shaped by unique peoples and differing climates, soils and resources. They would be highly adaptive and the country could be transformed into a rich matrix of biological and social diversity.

> Change would begin slowly, first in the countryside where several million are already working to become pioneers for the 21st century. In time this change would spread to the cities, the source of a goodly number of environmental woes.[25]

The uses of decentralized, natural-energy technologies in an urban setting are being applied in at least one city—Washington, D.C.—by Community Technology, Inc., a group devoted to social and technological interaction in the Adams-Morgan neighborhood, an inner-city area that has developed town meeting assemblies for dealing with neighborhood development. Community Technology, Inc. is engaged in a series of programs to bring technology and community planning "home" to the residents of the Adams-Morgan area. Their plans are described in a 1973 publication:

> The goals of Community Technology, Inc., and of the projects described in this proposal, are to de-mystify technology, to challenge all of the claimed economies of scale, and to push as far as possible practical demonstrations of high technology in the direct service of human needs and imagination in an urban community.
>
> We propose, beyond the demonstrations, to gather information relating to technology which is both usable by and useful for communities of people—technology which, although sophisticated in concept, is low in impact upon the environment and low also in capital and labor demands. This sort of technology, sometimes called intermediate, is decentralizing or centrifugal in social impact and frugal in resource use. After gathering information, we will concentrate upon the most effective methods of dissemination.
>
> Specific projects immediately planned, in addition to a complete information service, are trout raising in basement-sized areas; vegetable farming in roof-sized areas; use of solar energy on a community scale; use of windmills as an urban energy source; effect of machine tools upon community self-reliance; and redesign of community facilities, including transportation. Additionally, in partnership with every project-demonstration there will be the development of community institutions which best bring citizens and technology into the closest, least dominating, and most liberating relationship.[26]

The implications for society of the use of decentralized energy technologies and community planning have been explored by Murray Bookchin, author of books on anarchism, ecology, and city planning. Bookchin argues that the development of intermediate-level technologies that consume little fuel or natural resources will be the necessary ingredients of a new age of human understanding based on nondestructive attitudes toward the Earth as well as toward ourselves. In an essay written in the 1960s, "Towards a Liberatory Technology," he makes the case for the "technologies for life" that will play a great part in the social interactions of the future:

> A technology for life could play the vital role of integrating one community with another. Rescaled to a revival of crafts and a new conception of material needs, technology could also function as the sinews of confederation. A national division of labor and industrial centralization are dangerous because technology begins to transcend the human scale; it becomes increasingly incomprehensible and lends itself to bureaucratic manipulation. To the extent that a shift away from community control occurs in real material terms (technologically and economically), centralized institutions acquire real power over the lives of men and threaten to become sources of coercion. A technology for life must be *based* on the community; it must be tailored to the community and the regional level. On this level, however, the sharing of factories and resources could actually promote solidarity between community groups; it could serve to confederate them on the basis not only of common spiritual and cultural interests but also of common material needs. Depending upon the resources and uniqueness of regions, a rational, humanistic balance could be struck between autarky, industrial confederation, and a national division of labor.[27]

Conclusion

The response of individuals acting on a community level to the energy and materials shortages of the future will be crucial to survival. A few years ago, when the first pessimistic forecasts of energy shortages and mineral depletion were offered to an unsuspecting Western public, the prospects for decentralization and an end to the high life of excessive resource use seemed far away. The increasing realization that national policies must change to adapt to the new age has become as major a concern to individuals as the changes in lifestyles in their personal lives.

Given the critical shortages of energy and materials in all the economies of the high-energy, high-technology societies of America, Western Europe, and Japan, there appear to be only two courses of action open:

1. The "advanced" nations can continue to pursue the high-growth policies of the past few decades, while devoting only a minimal amount of their Gross National Product to developing and encouraging low-energy technologies (in housing and agriculture) and new, natural uses of energy based on the sun, or
2. The wealthier countries can devote a substantial fraction of their

resources to sweeping research and development efforts to utilize low-energy technologies and develop solar power, wind power, and other natural-energy sources on a great scale.

Continuation of the first policy seems doomed to failure, because the fossil fuels will be quickly depleted, leaving little or no energy with which to build a new technology base that would use natural sources of energy and encourage a decentralized lifestyle for its citizens. Dr. Howard T. Odum has described the pitfalls of the fast-growth policies of the world's developed countries in a recent paper to the Royal Swedish Academy of Science:

> The pattern of urban concentration and the policies of economic growth stimulation that were necessary and successful in energy growth competition periods are soon to shift. There will be a premium against the use of pump priming characteristics since there will be no more unpumped energy to prime. What did work before will no longer work and the opposite becomes the pattern that is economically successful. All this makes sense and is commonplace to those who study various kinds of ecosystems, but the economic advisors will be sorely pressed and lose some confidence until they learn about the steady state and its criteria for economic success. Countries with great costly investments in concentrated economic activity, excessive transportation customs, and subsidies to industrial expansion will have severe stresses. Even now the countries who have not gone so far in rapid successional growth are setting out to do so at the very time when their former more steady state culture is about to begin to become in more favored economic state comparatively. . . .
>
> . . . The terrible possibility that is before us is that there will be the continued insistence on growth with our last energies by the economic advisors that don't understand so that there are no reserves to make a change with, to hold order, and to cushion a period when populations must drop a hundredfold. Disease reduction of man and of his plant production systems could be planetary and sudden if the ratio of population to food and medical systems is pushed to the maximum at a time of falling net energy. At some point the great gaunt towers of nuclear energy installations, oil drilling, and urban cluster will stand empty in the wind for lack of enough level of technology to keep them running. A new cycle of dinosaurs will have passed its way. Man will survive as he reprograms readily to that which the ecosystem needs of him so long as he does not forget who is serving whom. What is done well for the ecosystem is good for man. However, the cultures that say only what is good for man is good for nature will pass and be forgotten like the rest.

There was a famous theory in paleoecology called Orthogenesis which suggested that some of the great animals of the past were part of systems that were locked into evolutionary mechanisms by which the larger ones took over from smaller ones. The mechanisms then became so fixed that they carried the size trend beyond the point of survival, whereupon the species went extinct. Perhaps this is the main question of ecology, economics, and energy. Has the human system frozen its direction into orthogenetic path to cultural crash, or is the great creative activity of the current energy-rich world already sensing the need for change? Are alternatives already being tested by our youth so they will be ready for the gradual transition to a fine steady state that carries the best of our recent cultural evolution into new more miniaturized, more dilute, and more delicate ways of man/nature?[28]

It is likely that many will see the materialistic discomforts of the coming years as threats to their existence. After years of social conditioning to relate only to the value system of the high-energy society of today, many people are unprepared for readjustment to the future. It is encouraging that the possibilities for change have never been more viable in society, and that, from the the troubling times of the present, a new consciousness must emerge, to bring forth a new Enlightenment from an age of enchantment with the Dynamo.

Notes
Chapter 1

1 Friedrich Klemm, *A History of Western Technology*. Cambridge, Massachusetts: MIT Press (1964), p. 17.
2 Harry Elmer Barnes, *An Economic History of the Western World*. New York: Harcourt, Brace (1937), p. 6.
3 Ibid., p. 9.
4 Le Corbusier, *vers une architecture*, pp. 53–55, quoted in Joseph Rykwert, *On Adam's House in Paradise*. New York: The Museum of Modern Art (1972), p. 15.
5 Xenophon, quoted in Marion J. Simon, *Your Solar House*. New York: Simon and Schuster (1947), p. 15.
6 Barnes, loc. cit.
7 Lewis Mumford, *Technics and Human Development*. New York: Harcourt Brace Jovanovich (1967), p. 268.
8 Ibid., p. 238.
9 Barnes, op. cit., p. 67.
10 T. K. Derry and Trevor I. Williams, *A Short History of Technology*. New York and Oxford, England: Oxford University Press (1961), p. 245.
11 Oscar Wilde (1895), quoted in *Petroleum Today* (1972), No. 2.
12 John F. Sandfort, *Heat Engines*. Garden City, New York: Doubleday Anchor Books (1962), p. 3.
13 Mumford, op. cit., p. 286.
14 Derry and Williams, op. cit., p. 39.
15 Francis Bacon, quoted in *A History of Western Technology*, op. cit., p. 174.
16 Sandfort, op. cit., pp. 11–13.
17 Ibid., pp. 20–21.
18 Sam H. Schurr and Bruce C. Netschert, et al., *Energy in the American Economy: 1850–1975*. Baltimore: The Johns Hopkins Press (1960), p. 60. Published for Resources for the Future, Inc.
19 Sandfort, op. cit., pp. 25–42.
20 Ibid., p. 40.
21 Ibid., pp. 42–46
22 Ibid., pp. 48–53.
23 Ibid., pp. xvii–xx.
24 Derry and Williams, op. cit., p. 341.
25 Klemm, op. cit., p. 275.
26 Ibid., pp. 275–76.
27 Sandfort, op. cit., pp. 25–42.
28 Ibid., p. 86.
29 Ibid., pp. 175–85.

30 Isaac Asimov, "In the Game of Energy and Thermodynamics You Can't Even Break Even," *Smithsonian* (August 1970), pp. 4–11.
31 Nicholas Georgescu-Roegen, "Economics and Entropy," *The Ecologist* (July 1972), pp. 13–18.
32 Derry and Williams, op. cit., pp. 372–73.
33 Barnes, op. cit., p. 504.
34 Philip Nobile and John Deedy (eds.), *Complete Ecology Fact Book*. Garden City, New York: Doubleday (1972), p. 10.
35 Gordon Rattray Taylor, *The Doomsday Book*. New York: World (1970), pp. 50–51.
36 Schurr and Netschert, op. cit., p. 36.
37 R. G. Lillard, *The Great Forest*. New York: Knopf (1948), p. 85.
38 Lynn White, Jr., "The Historic Roots of our Ecologic Crisis," *Science*, Vol. 155 (March 10, 1967), pp. 1203–7.
39 Derry and Williams, op. cit., p. 674.
40 Ibid.
41 Ibid., p. 33.
42 Ibid., p. 675.
43 Ibid., p. 668.
44 Jane Jacobs, *The Economy of Cities*. New York: Random House (1969); Vintage Books Edition (1970), pp. 86–90.
45 Allen H. Barton, *Communities in Disaster: A Sociological Analysis of Collective Stress Situations*. Garden City, New York: Doubleday (1969), p. 20.
46 Barry Commoner, *The Closing Circle*. New York: Knopf (1971), p. 46.
47 Norbert Wiener, *The Human Use of Human Beings: Cybernetics and Society*. New York: Avon (1967) (originally published in 1950).
48 Commoner, op. cit., pp. 33–34.
49 *See especially:* Thomas R. Harney and Robert Disch (eds.), *The Dying Generations*. New York: Dell (1971), p. 44.
50 Alexis de Tocqueville, *Democracy in America*. New Rochelle, New York: Arlington House (1965), Vol. II, p. 46.
51 James K. Page, Jr., and Richard Saltonstall, Jr., *Brown-Out and Slow Down: A Citizen's Manual for the Twin Crises of Energy and Transportation*. New York: Walker (1972), pp. 91–92.
52 De Tocqueville, op. cit., p. 328.
53 Page and Saltonstall, op. cit., pp. 92–93.
54 Derry and Williams, op. cit., p. 706.
55 Hyman G. Rickover, "Energy Resources and our Future," a speech presented before the Minnesota State Medical Association, St. Paul, Minnesota (May 14, 1957).
56 Hans H. Landsberg and Sam H. Schurr, *Energy in the United States: Sources, Uses, & Policy Issues,* a Resources for the Future Study. New York: Random House (1968), p. 28.
57 Ibid., p. 29.
58 Schurr and Netschert, op. cit., p. 47, p. 58ff.
59 Ibid., p. 58 (footnote).
60 Landsberg and Schurr, op. cit., p. 33.
61 Schurr and Netschert, op. cit., p. 97.

62 Ibid., p. 63.
63 Richard O'Connor, *The Oil Barons*. Boston: Little, Brown (1971), p. 12.
64 Ibid., pp. 12–13.
65 Ibid., p. 14.
66 Ibid., p. 16.
67 Fred Cottrell, *Energy and Society*. New York: McGraw-Hill (1955), p. 93.
68 Schurr and Netschert, op. cit., p. 96ff.
69 Ibid.
70 Ibid., p. 103.
71 Ibid.
72 Ibid., p. 104.
73 Ibid., p. 115.
74 Ibid.
75 Ibid., p. 116.
76 Klemm, op. cit., p. 342.
77 Ibid.
78 Barnes, op. cit., p. 311.
79 P. W. Kingsford, *Electrical Engineering: A History of the Men and the Ideas*. New York: St. Martin's Press (1969), pp. 131–32.
80 Ibid., p. 132.
81 Derry and Williams, op. cit., pp. 615–16.
82 Siegfried Giedion, *Mechanization Takes Command*. New York: Oxford University Press (1948), pp. 557–58.
83 Derry and Williams, op. cit., p. 635.
84 Sandfort, op. cit., p. 241ff.
85 Landsberg and Schurr, op. cit., p. 55.
86 Giedion, op. cit., p. 559.
87 Schurr and Netschert, op. cit., p. 501.
88 Ibid., pp. 182, 492.
89 Schurr and Netschert, op. cit., p. 70.
90 Reyner Banham, *The Architecture of the Well-tempered Environment*. London: The Architectural Press (1969), p. 65.
91 Derry and Williams, loc. cit.
92 Derry and Williams, op. cit., pp. 612–17.
93 Giedion, op. cit., p. 558.
94 Lawrence Lessing, "The Coming Hydrogen Economy," *Fortune* (November 1972), p. 138ff.
95 Edwin T. Freedly, *Leading Pursuits and Leading Men*. Philadelphia (1854), p. 29, quoted by Giedion in *Mechanization Takes Command,* op. cit.
96 Giedion, op. cit., pp. 156–61.
97 Derry and Williams, op. cit., p. 418.
98 Ibid., p. 416.
99 Ibid.
100 Banham, op. cit., p. 48, quoted from William Gage Snow, *Furnace Heating*. New York, 6th ed. (1923), p. 213.
101 Banham, op. cit., pp. 45–46.

102 Ibid., p. 46, quoted from Bushnell and Orr, *District Heating*. New York (1915), p. 2.
103 Ibid., pp. 51–52.
104 Ibid., pp. 52–53.
105 Ibid., p. 74.
106 Schurr and Netschert, op. cit., p. 511.
107 Ibid., p. 70.
108 Ibid., p. 652.
109 O'Connor, op. cit., pp. 91–93.
110 Schurr and Netschert, op. cit., p. 652.
111 O'Connor, op. cit., pp. 92–93.
112 Schurr and Netschert, op. cit., pp. 652–53.
113 Landsberg and Schurr, op. cit., p. 41.
114 Ibid., p. 33.
115 Ibid., p. 34; Schurr and Netschert, op. cit., p. 91ff.
116 O'Connor, op. cit., p. 87.
117 Ibid., p. 86.
118 Ibid., pp. 86–87.
119 Robert Engler, *The Politics of Oil*. Chicago: University of Chicago Press (1961); second printing (1969), p. 83; *Federal Energy Organization*, a staff analysis prepared at the request of Henry M. Jackson, Chairman, Committee on Interior and Insular Affairs, U. S. Senate. Washington, D.C.: U. S. Government Printing Office (1973), p. 29ff.
120 *Federal Energy Organization*, a staff analysis prepared for U. S. Senate Commitee on Interior and Insular Affairs, Serial No. 93–6 (92–41), Washington, D.C.: U. S. Government Printing Office (1973), p. 29.
121 Ibid., p. 29ff.
122 Ibid.
123 Ibid.
124 Schurr and Netschert, op. cit., chart, p. 182.
125 Landsberg and Schurr, op. cit., p. 52.
126 Ibid.
127 Giedion, op. cit., p. 77.
128 Ibid., p. 78.
129 Cottrell, op. cit., pp. 210–11.
130 John Kenneth Galbraith, *The New Industrial State*. New York: Signet (1967), p. 69.
131 Cottrell, op. cit., p. 212.
132 Landsberg and Schurr, op. cit., p. 37.
133 Ibid., p. 57.
134 Schurr and Netschert, op. cit., p. 118.
135 Ibid.
136 Ibid.
137 Giedion, op. cit., p. 162.
138 Ibid.
139 Schurr and Netschert, op. cit., p. 120.
140 Landsberg and Schurr, op. cit., pp. 41–42.
141 Ibid., p. 43.
142 Ibid., p. 45ff.

143 The National Coal Association, *Bituminous Coal Facts 1972*. Washington, D.C., p. 59.
144 O'Connor, op. cit., p. 405.
145 Ibid., p. 406.
146 Ibid., p. 411.
147 Landsberg and Schurr, op. cit., pp. 52–53.
148 Schurr and Netschert, op. cit., p. 720. *See also* The Federal Power Commission, *The 1970 National Power Survey*. Washington, D.C.: U. S. Government Printing Office (1971), Part I, pp. I-1–I-19.
149 Ritchie Calder, *Living with the Atom*. Chicago: University of Chicago Press (1962), p. 3.
150 Ralph Lapp, *The Logarithmic Century*. Englewood Cliffs, New Jersey: Prentice-Hall (1973), p. 86ff.
151 Calder, op. cit., pp. 11–12.
152 Giedion, op. cit., pp. 616–17, 622.
153 Ibid., pp. 625–27.
154 Banham, op. cit., pp. 172–73.
155 *The Way Things Work: An Illustrated Encyclopedia of Technology*. New York: Simon and Schuster (1967), p. 262.
156 Banham, op. cit., p. 186.
157 Ibid., p. 172.
158 Ibid., p. 222.
159 Ibid., p. 183.
160 Ibid., p. 222, quoted from *Architectural Forum* (November 1950), p. 108.
161 Frank Lloyd Wright, quoted in Peter Blake, *Le Corbusier: Architecture and Form*. Baltimore: Penguin (1960), p. 127.
162 Edison Electric Institute, *The Electric Utility Industry*, 1970–71 Edition.
163 The National Coal Association, op. cit., p. 57.
164 U. S. Bureau of the Census, *Statistical Abstract of the United States, 1972*. Washington, D.C.: U. S. Government Printing Office (1972), p. 505.
165 *Consulting Engineer* (March 1973), p. 218.
166 U. S. Bureau of the Census, op. cit., p. 984.
167 Ronald Ridker, *Population, Resources and Environment*, Vol. III of the report of the U. S. Commission on Population Growth and the American Future. Washington, D.C.: U. S. Government Printing Office (1972).
168 Adapted from Barry Commoner, Michael Corr, and Paul J. Stamler, "The Causes of Pollution," *Environment* (April 1971), pp. 1–19.

Notes
Chapter 2

1 Resources for the Future, Inc., *Annual Report, 1972*. Washington, D.C., p. 12.
2 Ralph Lapp, *The Logarithmic Century*. Englewood Cliffs, New Jersey: Prentice-Hall (1973), p. 75.
3 Luman H. Long (ed.), *The 1972 World Almanac*. New York: Newspaper Enterprise Association (1971), p. 456.
4 Robert Engler, *The Politics of Oil*. Chicago: University of Chicago Press (1961); second printing (1969), p. 37.
5 Ibid.
6 National Petroleum Council, *Guide to National Petroleum Council Report on United States Energy Outlook*. Washington, D.C. (1973), p. 5.
7 Lawrence Rocks and Richard P. Runyon, *The Energy Crisis*. New York: Crown (1972), pp. 15–17.
8 Ralph Lapp, op. cit., p. 39.
9 Ibid.
10 Ibid.
11 Ibid., p. 38.
12 Ibid.
13 National Petroleum Council, *U.S. Energy Outlook*. Washington, D.C. (1972), pp. 18, 29–32.
14 Ibid.
15 Ibid.
16 Ibid.
17 Ibid.
18 Ibid.
19 "Oil Blackmail," an editorial, Washington *Post* (June 14, 1973).
20 David R. Ottaway and Ronald Koven, "Saudis Tie Oil to U.S. Policy on Israel," Washington *Post* (April 19, 1973).
21 Remarks by William E. Simon, Administrator, Federal Energy Office, before the National Academy of Sciences, Washington, D.C., January 29, 1974.
22 "World Bank Fears Monetary Collapse," *Weekly Energy Report*, Washington, D.C., January 14, 1974, p. 1.
23 Hans H. Landsberg, Leonard L. Fischman, and Joseph L. Fisher, *Resources in America's Future*. Baltimore: The Johns Hopkins Press (1963), p. 399. Published for Resources for the Future, Inc.
24 National Petroleum Council, *Guide to NPC Report*, op. cit., p. 17.
25 Ibid.
26 Rocks and Runyon, op. cit., pp. 37, 38.
27 National Petroleum Council, *Guide to NPC Report*, op. cit., p. 5.
28 Ibid., p. 18.

29 National Petroleum Council, *U.S. Energy Outlook,* op. cit., pp. 27–28; Edison Electric Institute, *The Electric Utility Industry,* 1970–71 Edition.
30 National Petroleum Council, *U.S. Energy Outlook,* op. cit., pp. 18, 29–32.
31 Lapp, op. cit., p. 67.
32 Ibid., p. 34ff.
33 Ibid.
34 National Petroleum Council, *U.S. Energy Outlook,* op. cit., p. 35.
35 Earl Cook, "The Role of Energy in Our Society," paper presented to the American Association for the Advancement of Science Annual Meeting, Washington, D.C. (December 29, 1972).
36 *Congressional Record,* June 4, 1971. Washington, D.C.: U. S. Government Printing Office, p. H 4725.
37 A. B. Makhijani and A. J. Lichtenberg, "An Assessment of Energy and Materials Utilization in the U.S.A.," Memorandum No. ERL-M310 (September 22, 1971). Berkeley, California: College of Engineering, University of California.
38 G. Tyler Miller, Jr., *Replenish the Earth: A Primer in Human Ecology*. Belmont, California: Wadsworth (1972), pp. 122–23.
39 Victor Paschkis, "Cutting Back Energy Consumption . . . Why? How?" American Society of Mechanical Engineers Annual Meeting, New York (November 26–30, 1972).
40 Lapp, *The Logarithmic Century,* op. cit., p. 244.
41 Sam H. Schurr, Bruce C. Netschert, et al., *Energy in the American Economy: 1850–1975*. Baltimore: The Johns Hopkins Press (1960). Published for Resources for the Future, Inc.
42 Ibid., pp. 144–64.
43 Ibid., p. 174.
44 Ibid., p. 181.
45 National Economic Research Associates, Inc., *Fuels for the Electric Utility Industry*. New York: Edison Electric Institute (1972), p. 59. The chart in the text is adapted from "Energy Demand Studies: An Analysis and Appraisal," Committee on Internal and Insular Affairs of the U. S. House of Representatives, September 1972, p. 69, U. S. Government Printing Office, Washington, 1972.
46 Ibid., p. 68; Luman H. Long (ed.), op. cit., p. 122.
47 National Economic Research Associates, Inc., op. cit., pp. 61–62.
48 Earl Cook, "The Flow of Energy in an Industrial Society," *Scientific American* (September 1971), pp. 135–44.
49 Federal Power Commission, *The 1970 National Power Survey*. Washington, D.C.: U. S. Government Printing Office (1971), Part I, pp. 1, 3, 8.
50 Cook, "The Role of Energy in Our Society," op. cit.
51 U. S. House of Representatives Committee on Interior and Insular Affairs, "Energy 'Demand' Studies: An Analysis and Appraisal," a staff study, September 1972. Washington, D.C.: U. S. Government Printing Office (1972).

52 Interview with Chester Kylstra and Jesse Boyles, University of Florida, Gainesville, Florida (August 1973).
53 Donella H. Meadows, Dennis L. Meadows, Jørgen Randers, and William W. Behrens III, *The Limits to Growth.* New York: Universe (1972).
54 Peter Glaser, "The Case for Solar Energy," Annual Meeting of the Society for Social Responsibility in Science. London (September 3, 1972).
55 Thomas S. Lovering, "Mineral Resources from the Land," *Resources and Man,* a study of the National Academy of Sciences/National Research Council. San Francisco: Freeman (1969), pp. 122–23.
56 Ibid., p. 119.
57 "The Scramble for Resources," *Business Week* (June 30, 1973), pp. 56–63.
58 Ibid.
59 Miller, op. cit., p. 110.
60 *See also* Preston Cloud, "Mineral Resources from the Sea" and William E. Ricker, "Food from the Sea" in *Resources and Man,* op. cit.
61 National Petroleum Council, *Guide to NPC Report,* op. cit., p. 27.
62 Jerome Kohl, "Energy and the Environment in North Carolina: Yesterday, Today, and Tomorrow," a speech presented at North Carolina Conference on the Environment, Raleigh, North Carolina (August 30, 1972).
63 James Walker, "U.S. Said Hooked on 'Heroin of Technology,'" Tampa (Florida) *Tribune* (March 15, 1973).
64 Joint Committee on Atomic Energy Hearings, *Environmental Effects of Producing Electrical Power,* January 29, 1970. Washington, D.C.: U. S. Government Printing Office (1970), p. 3.
65 Jerome Weingart, "Surviving the Energy Crunch," *Environmental Quality Magazine* (January 1973), pp. 29–33, 67.
66 Council on Environmental Quality, *Environmental Quality—1970.* Washington, D.C.: U. S. Government Printing Office (1970), p. 5.
67 Miller, op. cit., pp. 54–55.
68 Nicholas Georgescu-Roegen, *The Entropy Law and the Economic Process.* Cambridge, Massachusetts: Harvard University Press (1971), p. 280.
69 Ibid.
70 Ibid., p. 281.
71 Robert F. Mueller, "Thermodynamics of Environmental Degradation," paper presented at the Annual Meeting of the American Geophysical Union, Washington, D.C. (1971), p. 1.
72 Environmental Policy Division, Legislative Reference Service, Library of Congress, *The Economy, Energy, and the Environment,* background study prepared for the Joint Economic Committee, U. S. Congress, September 1, 1970. Washington, D.C.: U. S. Government Printing Office (1970), pp. 55–56.
73 Ibid.
74 National Wildlife Federation, *Conservation News* (September 15, 1970), p. 7.

75 R. T. Dewling, Statement Before the Subcommittee on Air and Water Pollution of the U. S. Senate Committee on Public Works, Machias, Maine (September 8, 1970).
76 Julian McCaull, "The Black Tide," *Environment* (November 1969), p. 11.
77 Malcolm F. Baldwin, "Public Policy on Oil—an Ecological Perspective," *Ecology Law Quarterly*, Vol. I, No. 2 (Spring 1971), p. 265.
78 Resources for the Future, Inc., in cooperation with MIT Environmental Laboratory, *Energy Research Needs*, a report to the National Science Foundation. Washington, D.C. (1971), IX-51.
79 Ibid.
80 Ibid., p. IX-53.
81 Study of Critical Environmental Problems (SCEP), *Man's Impact on the Global Environment: Assessment and Recommendations for Action*. Cambridge, Massachusetts: The MIT Press (1970), p. 267.
82 Testimony of Barbara M. Heller before the House Committee on Merchant Marine and Fisheries, Washington, D.C., July 12, 1973.
83 Ibid.
84 McCaull, loc. cit.
85 Max Blumer, Testimony before the Subcommittee on Air and Water Pollution of the U. S. Senate Committee on Public Works, Machias, Maine (September 8, 1970).
86 Resources for the Future, Inc., in cooperation with MIT Environmental Laboratory, op. cit., p. IX-61.
87 Ibid., pp. IX-62, IX-63.
88 McCaull, loc. cit.
89 Robert W. Fri, Acting Administrator of the Environmental Protection Agency, in an address to the Aerospace Industries Association, Williamsburg, Virginia (May 23, 1973).
90 Baldwin, op. cit., p. 266.
91 Heller, loc. cit.
92 Ibid.
93 James A. Fay and James J. MacKenzie, "Cold Cargo," *Environment*, Vol. 14, No. 9 (November 1972), p. 21ff.
94 Timothy H. Ingram, "Peril of the Month: Gas Supertankers," *The Washington Monthly* (February 1973), p. 7ff.
95 Ibid.
96 Ibid., pp. 10–11.
97 Fay and MacKenzie, op. cit., p. 29.
98 M. C. Brown, "Pneumoconiosis in Coal Miners," *Mining Congress Journal* (August 1965).
99 National Safety Council, *Accident Facts*, 1967 Edition. Chicago, p. 26.
100 Resources for the Future, Inc., in cooperation with MIT Environmental Laboratory, op. cit., p. X-22; "The Coal Industry Makes a Dramatic Comeback," *Business Week* (November 4, 1972), p. 55.
101 John V. Conti, "Safety Underground: Coal-Mine Study Shows Record Can Be Improved When Firms Really Try," *The Wall Street Journal* (January 18, 1973).
102 Ibid.

103 J. Davitt McAteer, *Coal Mining Health and Safety in West Virginia.* Morgantown, West Virginia (published privately, July 1970), p. 661.
104 American Chemical Society, *Cleaning our Environment: The Chemical Basis for Action.* Washington, D.C. (1969), p. 145.
105 McAteer, op. cit., p. 660.
106 Environmental Policy Division, Legislative Reference Service, Library of Congress, op. cit., p. 46.
107 Edwin Cubbison and Louise C. Dunlap, *Stripping the Land for Coal—Only the Beginning.* Washington, D.C.: COALition Against Strip Mining (February 1972), p. 2.
108 Ibid., p. 1.
109 Ibid.
110 Ken Hechler, "Strip Mining: A Clear and Present Danger," *Not Man Apart,* Vol. I, No. 7 (July 1971), p. 1.
111 Ibid.
112 Malcolm F. Baldwin, *The Southwest Energy Complex: A Policy Evaluation.* Washington, D.C.: The Conservation Foundation (1973), p. 12.
113 E. A. Nephew, "Healing Wounds," *Environment,* Vol. 14, No. 1 (January/February 1972), p. 12ff.
114 Ibid.
115 Ibid.
116 "The Coal Industry Makes a Dramatic Comeback," *Business Week* (November 4, 1972), p. 54.
117 Nephew, loc. cit.
118 Jane Stein, "Coal Is Cheap, Hated, Abundant, Filthy, Needed," *Smithsonian* (February 1973), p. 19ff.
119 Hechler, loc. cit.
120 Council on Environmental Quality, *Environmental Quality—1972.* Washington, D.C.: U. S. Government Printing Office (1972), p. 6.
121 Ibid.
122 Office of Science and Technology, cited in *Engineering for the Resolution of the Energy-Environment Dilemma,* Committee on Power Plant Siting, National Academy of Engineering, Washington, D.C. (1972), p. 39.
123 National Science Foundation and U. S. Senate Committee on Interior and Insular Affairs, *Summary Report of the Cornell Workshop on Energy and the Environment (February 22–24, 1972)* Washington, D.C.: U. S. Government Printing Office (1972), p. 45.
124 Ibid., pp. 45–46.
125 Ibid., p. 46.
126 Federal Power Commission, op. cit., p. I-11-7.
127 National Science Foundation and U. S. Senate Committee on Interior and Insular Affairs, loc. cit.
128 Ibid., p. 47.
129 Baldwin, *The Southwest Energy Complex,* op. cit., p. 23.
130 Ibid., p. 17.
131 Ibid., pp. 8–14.
132 Ibid., p. 4.
133 Ibid., p. 15.

134 Ibid., p. 17.
135 Ibid., p. 22.
136 Ibid., pp. 23–24.
137 Ibid., p. 24.
138 Council on Environmental Quality, op. cit., p. 111.
139 Ibid.
140 "Sierra Club Takes EPA to Court on Air Quality," *Weekly Energy Report*, Vol. I, No. 24 (July 23, 1973), p. 1.
141 "Study Finds SO_2 Technology Not Ready," *Weekly Energy Report*, Vol. I, No. 27 (August 13, 1973), p. 1.
142 Baldwin, *The Southwest Energy Complex*, op. cit., p. 19.
143 Resources for the Future, Inc., in cooperation with MIT Environmental Laboratory, op. cit., p. IX-19.
144 Ibid.
145 Ibid.
146 Wilson Clark, *U.S. Energy Use: Selected Environmental and Social Policy Issues*. Washington, D.C.: Environmental Resources, Inc. (1970), p. 44.
147 John Clark, "Heat Pollution," *National Parks Magazine* (December 1969).
148 Resources for the Future, Inc., in cooperation with MIT Environmental Laboratory, op. cit., p. IX-25.
149 Ibid., p. IX–34.
150 Dean E. Abrahamson, *Environmental Cost of Electric Power*. New York: Scientists' Institute for Public Information (1970), pp. 8–9.
151 Ibid.
152 Resources for the Future, Inc., in cooperation with MIT Environmental Laboratory, op. cit., p. IX-35.
153 Abrahamson, op. cit., p. 9.
154 Ibid., p. 10.
155 Council on Environmental Quality, *Environmental Quality—1970*. Washington, D.C.: U. S. Government Printing Office (1970), p. 68.
156 Lester Lees, et al., *Smog: A Report to the People*. Pasadena, California: Environmental Quality Laboratory, California Institute of Technology (1972), p. 17.
157 Council on Environmental Quality, *Environmental Quality—1972*, op. cit., pp. 111–12.
158 Lees, op. cit., pp. 52–68.
159 Jude Wanniski, "How the Clean Air Rules Were Set," *The Wall Street Journal* (May 29, 1973).
160 Jeffrey A. Perlman, "Clean-Air Car Devices May Prove Harmful to Health," *The Wall Street Journal* (June 19, 1973).
161 Ibid.
162 "Automakers' Clean Air Last Tango Proceeds," *Rational Transportation* (April 1973), p. 5.
163 Fri, loc. cit.
164 Department of Transportation News Release No. 56–73 (August 13, 1973).
165 Study of Man's Impact on Climate (SMIC), *Inadvertent Climate Modification*. Cambridge, Massachusetts: The MIT Press (1971), p. 12.

166 Abraham H. Oort, "The Energy Cycle of the Earth," *Scientific American* (September 1970), pp. 54–63.
167 Study of Man's Impact on Climate (SMIC), op. cit., p. 170.
168 Ibid., p. 70.
169 Ibid., p. 169.
170 Ibid., pp. 58–59.
171 William W. Kellogg, "Climate, Change and the Influence of Man's Activities on the Global Environment," *Symposium on Energy Resources and the Environment, Kyoto, Japan, July 11, 1972, Mitre Report 72–190*. McLean, Virginia: The Mitre Corporation (1972), p. 65.
172 Ibid., pp. 65–66.
173 Study of Man's Impact on Climate (SMIC), op. cit., p. 10.
174 John Holdren and Philip Herrera, *Energy*. San Francisco and New York: Sierra Club (1971), p. 97.
175 Study of Man's Impact on Climate (SMIC), op. cit., p. 165.
176 Council on Environmental Quality, *Environmental Quality—1970*, op. cit., p. 103.
177 David Howell, "It Was an Act of God—with a Pinch of Salt," *Environmental Action*, May 12, 1973.
178 Study of Man's Impact on Climate (SMIC), op. cit., p. 60.
179 Kenneth S. Davis and John A. Day, *Water: The Mirror of Science*. Garden City, New York: Doubleday (1961), p. 173.
180 National Science Foundation and U. S. Senate Committee on Interior and Insular Affairs, op. cit., p. 54.
181 Council on Environmental Quality, *Environmental Quality—1972*, op. cit., p. 11.
182 Ibid., p. 16.
183 James Rathlesberger (ed.), *Nixon and the Environment*. New York: Taurus Communications (1972), p. 97.
184 *Energy Digest* (July 31, 1973), p. 337.
185 Committee on Interior and Insular Affairs, U. S. Senate, *A Review of the Energy Resources of the Public Lands, Based on Studies Sponsored by the Public Land Law Review Commission*, Serial No 92–5. Washington, D.C.: U. S. Government Printing Office (1971), p. 3.
186 J. H. Dales, "Land, Water and Ownership," *Canadian Journal of Economics* (November 1968), reprinted in Robert Dorfman and Nancy Dorfman (eds.), *Economics of the Environment*. New York: Norton (1972), p. 176.
187 Howard T. Odum, *Environment, Power, and Society*. New York: Wiley (1971), p. 46.
188 Ariel Lugo, et al., "Models for Planning and Research for the South Florida Environmental Study." Gainesville, Florida: University of Florida Center for Aquatic Sciences (1971).
189 Odum, op. cit., p. 125.
190 Council on Environmental Quality, *Environmental Quality—1972*, op. cit., p. 57.
191 Commission on Population Growth and the American Future, *Population, Resources and the Environment*. Washington, D.C.: U. S.

Government Printing Office, Vol. III (1972), p. 375.
192 U. S. Bureau of the Census, *Statistical Abstract of the United States, 1971*. Washington, D.C.: U. S. Government Printing Office (1971), p. 584.

Notes
Chapter 3

1 Bruce C. Netschert, testimony during hearings on "Competitive Aspects of the Energy Industry" before the Subcommittee on Antitrust and Monopoly of the U. S. Senate Judiciary Committee (May 5–6, 1970).
2 "House Democrat Says 11 Big Firms Avoided Paying Any Federal Income Tax Last Year," *The Wall Street Journal* (August 2, 1973).
3 Charles R. Ross, testimony during hearings on "Competitive Aspects of the Energy Industry" before the Subcommittee on Anti-Trust and Monopoly of the U. S. Judiciary Committee (May 5–6, 1970).
4 Philip M. Stern, *The Rape of the Taxpayer*. New York: Random House (1973), p. 229.
5 Ibid., pp. 228–251.
6 James Ridgeway, *The Last Play*. New York: Dutton (1973), pp. 4–5.
7 U. S. Bureau of the Census, *The Statistical Abstract of the U.S.—1971*. Washington, D.C.: U. S. Government Printing Office (1972), reprinted commercially as *The American Almanac*. New York: Grosset & Dunlap, Inc. (1972), p. 757.
8 "You and the Commercial," Columbia Broadcasting System Television Network News Special (April 26, 1973).
9 U. S. Bureau of the Census, op. cit., p. 760.
10 "You and the Commercial," loc. cit.
11 U. S. Bureau of the Census, loc. cit.
12 Lee Metcalf, remarks in *The Congressional Record,* January 26, 1973. Washington, D.C.: U. S. Government Printing Office, pp. E 428–E 429.
13 Ibid.
14 James K. Page, Jr., and Richard Saltonstall, Jr., *Brown-Out and Slow Down,* New York: Walker (1972), p. 53.
15 Ibid.
16 Ibid., p. 54.
17 Ibid., p. 56.
18 Eric Hirst, *Electric Utility Advertising and the Environment,* ORNL Report NSF-EP-18, Oak Ridge, Tennessee: Oak Ridge National Laboratory.
19 David M. Rosenbaum, *The Structure Is the Policy*. McLean, Virginia: The Mitre Corporation (June 1972), p. 49.
20 Ibid.
21 "TVA's Influence on Electric Rates," printed in *Competitive Aspects of the Energy Industry,* hearings before the Subcommittee on Antitrust and Monopoly, Senate Committee on the Judiciary (May 5–6, 1970). Washington, D.C.: U. S. Government Printing Office (1971), pp. 32–44.

22 Ibid., reprint of pamphlet on the TVA.
23 Rosenbaum, op. cit., p. 51.
24 "The Wrong Man for the FPC," Washington *Post* editorial (June 7, 1973).
25 "Oilman Supporter of Nixon Screened Appointee to FPC," Washington *Post* (June 27, 1973).
26 "FPC Official Ordered Gas Data Burned," Washington *Post* (June 10, 1973).
27 Ibid.
28 Edwin Vennard, *The Electric Power Business.* New York: McGraw-Hill (1962), p. 77.
29 Rosenbaum, loc. cit.
30 Vic Reinemer, "The Relation of Power Needs to Population Growth, Economic and Social Costs," address before American Medical Association Congress on Environmental Health, Chicago (April 30, 1973).
31 *Promotion at What Cost?,* presentations at Michigan Public Service Commission Energy Hearings, Case U-4100, Detroit Model Neighborhood Area-Social Planning Project (1972).
32 Ibid., p. 19.
33 Ibid., p. 20.
34 Ibid., pp. 75–77.
35 Ibid., p. 48.
36 "Added Generation Falls Far Short of Industry's Bogey," *Electrical World* (June 15, 1973), pp. 83–85.
37 David B. Large (ed.), *Hidden Waste, Potentials for Energy Conservation.* Washington, D.C.: The Conservation Foundation (1973), p. 96.
38 Office of Emergency Preparedness, Executive Office of the President, *The Potential for Energy Conservation.* Washington, D.C.: U. S. Government Printing Office (1972), p. F-23.
39 Brief for National Intervenors, Sierra Club and SHOCK before District of Columbia Public Service Commission; Case No. 568, Washington, D.C. (1972), p. 30.
40 Ibid., p. 16.
41 Stanford Sesser, "Unreddy Kilowatt?—Critics Say Utilities Worsen Power Problems by Pushing Electricity," *The Wall Street Journal* (July 27, 1972).
42 *Weekly Energy Report* (July 16, 1973), p. 1.
43 *Info.* New York: Atomic Industrial Forum (June 1973), p. 5.
44 Sesser, loc. cit.
45 R. D. Doctor, K. P. Anderson, et al., *California's Electricity Quandary: Vol. III, Slowing the Growth Rate.* Santa Monica, California: Rand Corporation (September 1972), p. 5.
46 Ibid., pp. xiv–xv.
47 Ibid., p. vii.
48 "New Energy Law Proposed in California," *Electrical World* (July 1, 1973), pp. 23–24.
49 Stanford Research Institute, *Patterns of Energy Consumption in the United States,* prepared for Office of Science and Technology, Executive Office of the President. Washington, D.C.: U. S. Government Printing Office (1972), p. 7.

50 Ibid., p. 83.
51 Office of Emergency Preparedness, op. cit., p. E-2.
52 Ibid., p. E-12.
53 Ibid., p. E-8.
54 S. A. Schilling, R. J. Bengstrom, and J. H. Lindholm, Jr., "Reclamation and Recycling: An Economic Overview," *Battelle Research Outlook,* Vol. 3, No. 3 (1971).
55 William D. Smith, "Alcoa Unveils New Process," New York *Times* (January 12, 1973).
56 Roy J. Harris, Jr., "Aluminum Makers Are on Defensive Again," *The Wall Street Journal* (June 4, 1973).
57 Franklin Wallich, *The American Worker.* New York: Ballantine (1972).
58 Stanford Research Institute, op. cit., p. 120.
59 Ibid., p. 121.
60 Ibid., p. 119.
61 DuPont Company press release, March 19, 1973, Wilmington, Delaware.
62 Stanford Research Institute, op. cit., pp. E-16, E-17.
63 Large, loc. cit.
64 DuPont Company, loc. cit.
65 Large, loc. cit.
66 Doctor, Anderson, et al., op. cit., p. 88.
67 Arjun B. Makhijani, "Energy and Well Being," *Environment* (June 1972), pp. 10–18.
68 "Aluminum Makers are on Defensive Again," loc. cit.
69 Bruce M. Hannon, "Bottles, Cans, Energy," *Environment,* Vol. 14, No. 2 (March 1972), pp. 11–21.
70 Philip Nobile and John Deedy (eds.), *Complete Ecology Fact Book.* Garden City, New York: Doubleday (1972), p. 358.
71 George W. Pierson, *The Moving American.* New York: Knopf (1973), p. 93.
72 U. S. Department of Transportation, *Summary of National Transportation Statistics.* Washington, D.C.: U. S. Department of Transportation (December 1972).
73 Eric Hirst, *Energy Intensiveness of Passenger and Freight Transport Modes, 1950–1970,* ORNL Report NSF-EP-44. Oak Ridge, Tennessee: Oak Ridge National Laboratory (April 1973).
74 Ibid., loc. cit.
75 Eric Hirst, *Energy Consumption for Transportation in the U.S.,* ORNL Report NSF-EP-15. Oak Ridge, Tennessee, Oak Ridge National Laboratory (March 1972), p. 25.
76 U. S. Department of Commerce, op. cit., p. 545.
77 Nobile and Deedy, op. cit., p. 385.
78 Hirst, *Energy Intensiveness of Passenger and Freight Transport Modes, 1950–1970,* loc. cit.
79 "Energy Requirements for the Movement of Intercity Freight to Association of American Railroads." Columbus, Ohio: Battelle Columbus Laboratories (December 15, 1972).
80 Saltonstall and Page, op. cit., p. 130.

81 William E. Mooz, "Energy in the Transportation Sector," address at Florida Governor's Conference on Energy Supply and Use (March 1973).
82 R. Stephen Berry and Margaret F. Fels, "The Production and Consumption of Automobiles," to Illinois Institute for Environmental Quality (July 1972).
83 Peter Chapman, "No Overdrafts in the Energy Economy," *New Scientist* (May 17, 1973), pp. 408–10.
84 Berry and Fels, loc. cit.
85 W. E. Fraize and J. K. Dukowicz, "Transportation Issues and Environmental Issues." McLean, Virginia: The Mitre Corporation (1972).
86 Ibid.
87 Paul Friedlander, "The Stirling Air Engine," New York *Times* (May 13, 1973).
88 Council on Environmental Quality, Executive Office of the President, *Energy and the Environment: Electric Power.* Washington, D.C.: U. S. Government Printing Office (August 1973), p. 30.
89 Office of Emergency Preparedness, op. cit., p. C-9.
90 U. S. Department of Transportation press release (August 13, 1973).
91 Eric Hirst, "Transportation Energy Conservation," submitted for hearings before the House of Representatives Committee on Government Operations and Committee on Science and Astronautics (July 1973).
92 Joint Economic Committee, U. S. Congress, *The Economics of Federal Subsidy Programs; Part 6, Transportation Subsidies.* Washington, D.C.: U. S. Government Printing Office (1973), p. 732.
93 Ibid., p. 839.
94 Leslie A. White, *The Evolution of Culture.* New York: McGraw-Hill (1959), p. 281.
95 Frederick J. Smith, *Ecological Perspectives,* Vol. III of the report of the Commission on Population Growth and the American Future. Washington, D.C.: U. S. Government Printing Office (1972), pp. 289–98.
96 H. J. Maidenberg, "Do We Need More Grain Deals?" New York *Times* (July 8, 1973).
97 Michael J. Perelman, "Farming with Petroleum," *Environment* (October 1972), pp. 8–13.
98 Ibid.
99 U. S. Department of Commerce, op. cit., p. 596.
100 Ibid., p. 586.
101 Ridgeway, op. cit., pp. 345–50.
102 Perelman, loc. cit.
103 U. S. Department of Commerce, op. cit., p. 600.
104 Edison Electric Institute, "Questions and Answers About the Electric Utility Industry," 1970–71 Edition. New York: Edison Electric Institute (1970).
105 Barry Commoner, Michael Corr, Paul Stamler, "The Causes of Pollution," *Environment* (April 1971), pp. 2–19
106 Ibid., p. 150.
107 Ibid., pp. 150–51.

108 Denzel E. Ferguson, "The New Evolution," *Environment* (July/August 1972), pp. 30–35.
109 Ibid.
110 Ibid.
111 Perelman, loc. cit.
112 Robert van den Bosch, "The Cost of Poisons," *Environment* (September 1972), pp. 18–31.
113 David Pimental, "Realities of a Pesticide Ban," *Environment* (March 1973), pp. 18–29.
114 U. S. Department of Commerce, op. cit., p. 606.
115 Maidenberg, loc. cit.
116 René Dubos, "Humanizing the Earth," B. Y. Morrison Memorial Lecture presented at annual meeting of American Association for the Advancement of Science, Washington, D.C. (December 29, 1972).
117 Robert Rodale, "Living on the Land in China Today," *Organic Gardening and Farming*, Vol. 20, No. 5 (May 1973), p. 36.
118 U. S. Department of Commerce, op. cit., p. 584.
119 Commoner, Corr and Stamler, loc. cit.
120 Ibid.
121 Commoner, *The Closing Circle,* op. cit., p. 156.
122 Eric Hirst, "Energy Costs of Food in the United States," draft report. Oak Ridge, Tennessee: Oak Ridge National Laboratory (1973).
123 Jim Hightower, *Hard Tomatoes, Hard Times.* Washington, D.C.: Agribusiness Accountability Project (1972), p. i.
124 Ibid., p. 55.
125 Ibid., p. 64.
126 National Bureau of Standards and Executive Office of Consumer Affairs, "Seven Ways to Reduce Fuel Consumption in Household Heating." Washington, D.C.: U. S. Government Printing Office (1971).
127 National Bureau of Standards and Executive Office of Consumer Affairs, "Eleven Ways to Reduce Energy Consumption and Increase Comfort in Household Cooling." Washington, D.C.: U. S. Government Printing Office (1971).
128 "A Guide for the Use of FHA Interim Revision No. 51A." New York: National Mineral Wool Insulation Institute (1971).
129 U. S. Department of Housing and Urban Development, "Minimum Property Standards for Multifamily Housing," General Revision No. M-21. Washington, D.C.: U. S. Government Printing Office (June 1972).
130 Large, op. cit., pp. 22–23.
131 "Impact of Improved Thermal Performance in Conserving Energy." New York: National Mineral Wool Insulation Association, Inc. (1972).
132 "Home Sweet Electric Mobile Home." New York: Electric Energy Association (1972).
133 "Residential Energy Consumption: Phase I Report," Department of Housing and Urban Development Report No. HUD-HAI-1. Columbia, Maryland: Hittman Associates (March 1972), p. II-62.
134 Federal Power Commission, *The 1970 National Power Survey*, Part I. Washington, D.C.: U. S. Government Printing Office (1971), pp. I-3–I-8.

135 A. B. Makhijani and A. J. Lichtenberg, "An Assessment of Residential Energy Utilization in the U.S.A." Berkeley, California: University of California (1972), Table III.
136 Ibid.
137 D. G. Harvey and John Kudrick, "Minimization of Residential Energy Consumption," preliminary copy. Columbia, Maryland: Hittman Associates (1972).
138 John Neely, thesis in energy utilization. Cambridge, Massachusetts: Harvard University (1972), pp. 111–12.
139 Ibid.
140 Ken Bolyard and Alejandro Martinez, "A Study of the Electrical Energy Use of Current Television Receivers and Some Implications of This Data." Cambridge, Massachusetts (1972).
141 Doug Garr, "Air Conditioning for 1973: More Cool, Less Current," *Popular Science* (May 1973), pp. 79–81, 172.
142 Michael Corr and Dan MacLeod, "Getting It Together," *Environment* (November 1972), pp. 2–9, 45.
143 Dan Kennedy, "Energy Conservation . . . Some Wheat, Some Chaff," *AGA [American Gas Association] Monthly* (February 1973), pp. 4–7.
144 "New Proposal to Conserve Gas," *New York Journal of Commerce* (December 1972) (undated clip).
145 National Bureau of Standards and Executive Office of Consumer Affairs, "Eleven Ways to Reduce Consumption . . . ," loc. cit.
146 Frank Lloyd Wright, *The Future of Architecture*. New York: Horizon Press, Bramhall House edition (1953), pp. 156–57.
147 "The Tallest Skyscraper," *Time* (June 11, 1973), pp. 54, 59.
148 Ibid.
149 Richard G. Stein, "A Matter of Design," *Environment* (October 1972), pp. 17–29.
150 Ibid.
151 Ibid.
152 Advertisement, L-O-F, *Business Week* (December 23, 1972).
153 "New Skyscraper Announced for Midtown Manhattan," Miami *Herald* (July 30, 1973).
154 Peter Blake, *Le Corbusier—Architecture and Form*. Baltimore: Penguin (1960), p. 105.
155 Fred S. Dubin, "Energy Conservation Through Building Design and a Wiser Use of Electricity," American Public Power Association Conference, San Francisco (June 26, 1972).
156 *Roundtable on Energy Conservation in Public Buildings,* U. S. General Services Administration, Washington, D.C., July, 1972, 65pp.
157 "Buildings That Save a Watt and More," *Progressive Architecture* (special issue, October 1971), p. 106.
158 Richard G. Stein, "A-E Team Can Short-Circuit Major Crisis," *Consulting Engineer* (March 1973), pp. 122–26.
159 Leslie Larson, *Lighting and Its Design*. New York: Whitney Library of Design (1964).
160 Leslie Larson, "IES Illumination Levels," a statement. New York (1972).

161 Ralf Kelman, "Good Morning Sunshine," *Win* (January 15, 1973), pp. 16–17.
162 Stein, "A-E Team Can Short-Circuit Major Crisis," loc. cit.
163 Ralph Nevins and Preston McNall, Jr., "Proposed Classification of Human Response to the Environment," *ASHRAE* [*American Society of Heating, Refrigeration and Air-Conditioning Engineers*] *Journal* (June 1971).
164 Lester T. Avery, *ASHRAE Journal* (June 1971), p. 12.
165 *Roundtable on Energy Conservation in Public Buildings,* loc. cit.
166 Sital Daryanani, "HVAC System Design for Saving Energy," *ASHRAE Journal* (February 1973).
167 *Roundtable Conservation in Public Buildings,* loc. cit.
168 Stein, "A-E Team Can Short-Circuit Major Crisis," loc. cit.
169 Ibid.
170 David Sage, Inc., "Report of Study of Energy Consumption for Heating," prepared for Mechanical Contracting Industry of Long Island (November 1972).

Notes
Chapter 4

1 Environmental Policy Division, Congressional Research Service, Library of Congress, *Energy—The Ultimate Resource,* prepared for the Task Force on Energy of the Subcommittee on Science, Research and Development of the U. S. House of Representatives Committee on Science and Astronautics. Washington, D.C.: U. S. Government Printing Office (1971), p. 118.

2 Frank J. Prial, "Now Look Who's Feeling the Heat," New York *Times* (July 26, 1972).

3 Mike Toner, "Nuclear Power Unit Went First in Failure's Chain of Events," *Miami Herald* (April 4, 1973).

4 William Steigelman and Hugh Spencer, "Improved Urban Energy Use," *Proceedings of Effective Energy Utilization Symposium.* Philadelphia, Pennsylvania: Drexel University (June 8–9, 1972), p. 249.

5 Urban C. Lehner, "Many Steam Heat Companies Bemoan Loss of Customers, New Air Standards," *The Wall Street Journal* (January 10, 1973).

6 U. S. Delegation to the Soviet Union, *National Bureau of Standards Report Number 10–184.* Washington, D.C.: U. S. Department of Commerce (May 1, 1970), p. 205.

7 "District Heating by Underground Network is Clean and Cheap," *Sweden Now,* Vol. 6, No. 3 (1972), pp. 42–43.

8 Lee Metcalf, speech printed in *Congressional Record.* Washington, D.C.: U. S. Government Printing Office (February 27, 1973), pp. S 3397–S 3402.

9 L. B. Young, *Power Over People.* New York: Oxford University Press (1973), p. 27.

10 Ibid.

11 Resources for the Future, Inc., in cooperation with MIT Environmental Laboratory, *Energy Research Needs,* a report to the National Science Foundation. Washington, D.C. (1971), p. V-105.

12 John R. Free, "Cryogenic Power Lines, Cool Aid for our Energy Crisis," *Popular Science* (October 1972), pp. 69–71, 130.

13 Lawrence Rocks and Richard P. Runyon, *The Energy Crisis.* New York: Crown (1972), p. 97.

14 Young, op. cit., p. 32.

15 Arthur R. Kantrowitz, Statement Before U. S. Senate Committee on Interior and Insular Affairs, Subcommittee on Minerals, Materials and Fuels (December 18, 1969).

16 International Atomic Energy Agency, "MHD Electrical Power Generation," *1972 Status Report of the International Atomic Energy Agency.* Vienna (1972).

17 U. S. Department of the Interior, Press Release No. 1174-71 (August 25, 1971).
18 Office of Science and Technology, Executive Office of the President, *MHD for Central Station Power Generation: A Plan for Action.* Washington, D.C.: U. S. Government Printing Office (June 1969).
19 Subcommittee on Science, Research and Development, U. S. House of Representatives Science and Astronautics Committee, *Serial M, Briefings of the Task Force on Energy.* Washington, D.C.: U. S. Government Printing Office (1971).
20 John B. Dicks, statement on MHD Production of Electric Energy Before Subcommittee on Department of Interior and Related Agencies Appropriations, U. S. Senate Appropriations Committee (May 1973).
21 Ibid.
22 Kantrowitz, loc. cit.
23 "Environmental Pollution Control Through MHD Power Generation." Everett, Massachusetts: Avco Everett Research Laboratory (September 1970).
24 Dicks, loc. cit.
25 Terri Aaronson, "The Black Box," *Environment* (December 1971), pp. 10–18.
26 Ibid.
27 Ibid.
28 "Three on Apollo," *Bee-Hive,* United Aircraft publication (Fall 1969), pp. 10–15.
29 Office of Science and Technology, Executive Office of the President, *Energy R&D and National Progress.* Washington, D.C.: U. S. Government Printing Office (1964), p. 294.
30 Ernst M. Cohn, "A Survey of U.S. Activities Relating to Fuel Cells," Presentation at Fourth Joint Meeting, UJNR Energy Panel, Tokyo, (Fall 1972).
31 Roger Benedict, "Little Black Box: Fuel Cell, Long Seen as Electricity Source, Moves Ahead in Tests," *The Wall Street Journal* (May 19, 1971).
32 Ibid.
33 Interview with William H. Podolny, chief engineer, Advanced Power Systems, United Aircraft Corporation, Pratt & Whitney Division, East Hartford, Connecticut (Fall 1972).
34 Thomas H. Maugh III, "Fuel Cells: Dispersed Generation of Electricity," *Science,* Vol. 178, No. 4067 (December 22, 1972), pp. 1273–74b.
35 Ibid.
36 S. Baron, "Fuel Cell Power Generation and Gasification," paper presented at the American Public Power Association Annual Conference, San Francisco (June 26–28, 1972).
37 Ibid.
38 William H. Podolny, "Fuel Cell Powerplants for Rural Electrification," United Aircraft Corporation, Pratt & Whitney Division, East Hartford, Connecticut (December 1971).
39 Ibid.
40 Baron, loc. cit.

41 C. E. Heath, Letter, *Science,* Vol. 180, No. 4086, pp. 543–44 (May 11, 1973).
42 Maugh, loc. cit.
43 Discussions with Lester Shapiro, Engelhard Minerals and Chemicals Corporation Washington office (Fall 1972).
44 *Twenty Year Advance Program: 1973–1992*. New York: Consolidated Edison Company (March 1973).
45 "EEI Looks Toward Larger Fuel Cells," *Electrical World* (February 1, 1973), pp. 32–33.
46 News item, *Electrical World* (May 15, 1973).
47 Benedict, loc. cit.
48 *Energy Conversion Digest,* Vol. 9, No. 7.
49 Maugh, loc. cit.
50 Baron, loc. cit.
51 *Direct Conversion of Energy*. Washington, D.C.: Division of Technical Information, U. S. Atomic Energy Commission (1969), p. 25.
52 John F. Sandfort, *Heat Engines*. Garden City, New York: Doubleday Anchor Books (1962), p. 256.
53 L. O. Tomlinson, "Comparing Combined Cycle Plants," *Gas Turbine International* (November/December 1972), pp. 20–28.
54 William D. Metz, "Power Gas and Combined Cycles: Clean Power from Fossil Fuels," *Science,* Vol. 179, No. 4068 (January 5, 1973), pp. 54–56.
55 "The Combined Cycle Gains New Friends," *Business Week* (March 11, 1972), pp. 44–45.
56 "The Fluorocarbon Turbine—A Newcomer on the Prime Mover Scene," *Power* (April 1971), p. 61.
57 Tomlinson, loc. cit.
58 Metz, loc. cit.
59 *Twenty Year Advance Program: 1973–1992,* loc. cit.
60 Fred S. Dubin, *Total Energy*. New York: Educational Facilities Laboratories, Inc. (1970), pp. 9–10.
61 Extrapolated for 1974; as of 1972, there were 550 Total Energy plants in the United States *See* Paul R. Achenbach and J. B. Coble, "Site Analysis for the Applications of Total Energy Systems to Housing Developments," Seventh Intersociety Energy Conversion Engineering Conference, San Diego (September 25–29, 1972).
62 *The Potential for Energy Conservation,* op. cit.
63 Dubin, op. cit.
64 Dubin, op. cit., p. 37.
65 Michael Baybak, "Total Energy—The Gas Industry's Newest Weapon in the Battle of the Fuels," *House and Home* (October 1966), reprint.
66 Peter Dirksen, Jr., "Gas Total Energy Case History: Worcester, Massachusetts Science Center," *ASHRAE* [*American Society of Heating, Refrigeration and Air-Conditioning Engineers*] *Journal* (April 1973), pp. 39–45.
67 "Total Energy and Pneumatic Waste Collection Demonstrations," Office of the Assistant Secretary for Research and Technology, U. S. Department of Housing and Urban Development (May 1972); Paul R.

Achenbach and J. B. Coble, "Description of Equipment and Instrumentation for a Field Study of a Total Energy System in an Apartment Development," Seventh Intersociety Energy Conversion Engineering Conference, San Diego (September 25–29, 1972).
68 Achenbach and Coble, "Site Analysis for the Applications of Total Energy Systems . . . ," loc. cit.
69 *Modular Integrated Utility System: Program Description,* Office of Research and Technology, U. S. Department of Housing and Urban Development (December 1972).
70 A. M. Smith and Jack Wolfe, *The Total Utility Plant Concept.* San Antonio, Texas: Southwest Research Institute (1972).
71 *Modular Integrated Utility System: Program Description,* loc. cit.
72 "Do-It-Yourself Power Catches On," *Business Week* (November 30, 1968), pp. 62–64.
73 C. Girard Davidson, "Report to the City of New York's Consumer Council on Consolidated Edison in Regard to Reliability of Service, Adequacy of Future Power Supply, and Rates—Rough Draft." New York (January 10, 1968).
74 Ibid., p. 38.
75 Ibid.
76 "Suggestions for Economical Operation of Your Heat Pump System," instruction sheet prepared for heat pump customers by Commercial Customer Department, Potomac Electric Power Company (March 19, 1968).
77 "Energy Conservation with Heat Actuated Heat Pumps," American Gas Association Position Paper (June 21, 1972), p. 11.
78 *Central Air-Conditioning Progress,* Vol. I, No. 2, General Electric Corporation (1972).
79 Eric Hirst and John C. Moyers, "Potential for Energy Conservation Through Increased Efficiency of Use," testimony submitted to the U. S. Senate Committee on Interior and Insular Affairs (March 1973), p. 15.
80 Interviews with air-conditioning sales and contracting firms, Fort Myers, Florida (May 1973).
81 Hirst and Moyers, loc. cit.
82 Dan Kennedy, "Energy Conservation: The Potential of the Heat-Actuated Heat Pump," *A.G.A.* [*American Gas Association*] *Monthly* (April 1973), pp. 18–25.
83 Ibid.
84 Ibid.

Notes
Chapter 5

1 The National Coal Association, *Bituminous Coal Facts 1972*. Washington, D.C. (1972), p. 7.
2 Michael Fortune, "Synthetic Fuels from Coal," unpublished paper prepared for Energy Policy Project, Washington, D.C. (1972), p. 1ff.
3 Ibid., pp. 2–3, quoted from Stewart, E. G. *Town Gas—Its Manufacture and Distribution*. London: Science Museum (1958).
4 Ibid., p. 3.
5 Resources for the Future, Inc., in cooperation with MIT Environmental Laboratory, *Energy Research Needs,* a report to the National Science Foundation. Washington, D.C. (1971), p. V-31.
6 Ibid.
7 William D. Metz, "Power Gas and Combined Cycles: Clean Power from Fossil Fuels," *Science,* Vol. 179, No. 4068 (January 5, 1973), p. 56ff.
8 Resources for the Future, Inc., et al., op. cit., pp. III-40–III-41.
9 National Economic Research Associates, Inc. *Fuels for the Electric Utility Industry, 1971–1985*. New York: Edison Electric Institute (1972), pp. 119–20.
10 Ibid., pp. 122–23.
11 Allen L. Hammond, William D. Metz, and Thomas H. Maugh III, *Energy and the Future*. Washington, D.C.: American Association for the Advancement of Science (1973), p. 11ff.
12 Metz, op. cit., pp. 54ff.
13 Ibid.
14 Fortune, op. cit., p. 15.
15 Ibid.
16 Ibid.
17 "Clean Power from Dirty Fuels," *Scientific American,* Vol. 227, No. 4 (October 1972), pp. 26–35.
18 National Petroleum Council, *U.S. Energy Outlook: An Interim Report*. Washington, D.C. (1972), pp. 44, 17.
19 Lawrence Rocks and Richard P. Runyon, *The Energy Crisis*. New York: Crown (1972), p. 41.
20 National Petroleum Council, op. cit., p. 37.
21 Ibid.
22 American Chemical Society, *Environmental Science and Technology,* Vol. 5 (December 1972), p. 1183.
23 Ibid.
24 Jane Stein, "Coal Is Cheap, Hated, Abundant, Filthy, Needed," *Smithsonian* (February 1973), p. 18ff.
25 Ibid.

26 Ibid.
27 American Chemical Society, loc. cit.
28 Stein, loc. cit.
29 *Weekly Energy Report,* Vol. I, No. 26 (August 6, 1973), p. 2.
30 Ibid.
31 American Chemical Society, loc. cit.
32 National Economic Research Associates, Inc., op. cit., p. 137.
33 Resources for the Future, Inc., et al., op. cit., p. IV-38.
34 National Economic Research Associates, Inc., op. cit., p. 140.
35 Fortune, op. cit., p. 32.
36 National Economic Research Associates, Inc., op. cit., p. 139.
37 Resources for the Future, Inc., et al., op. cit., p. IV-46.
38 National Economic Research Associates, Inc., op. cit., p. 138.
39 Resources for the Future, Inc., et al., op. cit., pp. IV-54–IV-55.
40 Fortune, op. cit., p. 42.
41 Philip H. Abelson, editorial, *Science,* Vol. 180, No. 4091 (June 15, 1973), p. 1127.
42 Resources for the Future, Inc., et al., op. cit., p. IV-10.
43 Ibid.
44 Resources for the Future, Inc., et al., op. cit., p. IV-8.
45 "The Facts on Oil Shale," *Not Man Apart,* Vol. I, No. 11 (November 1971), pp. 10–11.
46 Resources for the Future, Inc., et al., op. cit., pp. IV-8–IV-9.
47 National Economic Research Associates, Inc., op. cit., p. 35.
48 "The Facts on Oil Shale," op. cit., p. 10.
49 Ibid., p. 11.
50 Ibid., p. 10.
51 Ibid., p. 11.
52 *Weekly Energy Report,* loc. cit.
53 National Economic Research Associates, Inc., loc. cit.
54 Resources for the Future, Inc., et al., op. cit., p. IV-25.
55 National Economic Research Associates, Inc., loc. cit.
56 Resources for the Future, Inc., et al., loc. cit.
57 National Petroleum Council, op. cit., p. 58.
58 Resources for the Future, Inc., et al., op. cit., p. IV-25ff.
59 "FPC: 30 Nuclear Plants Delayed," *Info,* No. 30 (March 1973), p. 4.
60 Ibid.
61 Robert Williams, "A Primer on Nuclear Reactors and the LMFBR Commitment." Washington, D.C.: privately published (1973).
62 Ibid.
63 Ibid.
64 Ralph Lapp, *The Logarithmic Century*. Englewood Cliffs, New Jersey: Prentice-Hall (1973), p. 94ff.
65 Jack Schubert and Ralph Lapp, *Radiation: What It Is and How It Affects You*. New York: Viking (1957), pp. 20–22.
66 Ibid., p. 14.
67 Ibid., Chapter 6.
68 *The Safety of Nuclear Power Reactors* (*Light Water-Cooled*) *And Related Facilities, Wash-1250*. Washington, D.C.: Atomic Energy Commission (July 1973), p. I-28.

69 Ernest J. Sternglass, *Low-Level Radiation.* New York: Ballantine (1971), p. 142.
70 John W. Gofman and Arthur R. Tamplin, *Poisoned Power.* Emmaus, Pennsylvania: Rodale Press (1971), p. 97.
71 Sternglass, loc. cit.
72 Gofman and Tamplin, op. cit., pp. 133–34.
73 *The Safety of Nuclear Power Reactors,* op. cit., Chapter 4.
74 Thomas O'Toole, "Radiation Usage High; Panel Warns of Deaths," Washington *Post* (November 16, 1972).
75 *The Safety of Nuclear Power Reactors,* op. cit., p. 3.
76 John Gofman, Mike Gravel, and Wilson Clark, *The Case for a Nuclear Moratorium.* Environmental Action Foundation, Washington, D.C. (1973).
77 Ibid., p. 13ff.
78 Alvin M. Weinberg, "Nuclear Energy—18 Years After," speech presented before American Public Power Association, San Francisco (June 27, 1972).
79 *U. S. Government Organizational Manual,* 1971–72. Washington, D.C.: U. S. Government Printing Office (1971), p. 392.
80 Wilson Clark, "History of U.S. Atomic Energy Programs." Washington, D.C. (1971).
81 *Theoretical Possibilities and Consequences of Major Accidents in Large Nuclear Power Plants, Wash-740.* Washington, D.C.: U. S. Atomic Energy Commission (1957), pp. 1–6.
82 PL 88-703, quoted in Sheldon Novick, *The Careless Atom.* New York: Delta (1970), p. 71.
83 Sheldon Novick, *The Careless Atom.* New York: Delta (1970), p. 71.
84 Ibid., p. 63.
85 "AEC Details Near-Failures in Power Plants Before '65," Miami *Herald* (June 27, 1973).
86 Novick, op. cit., p. 75.
87 H. Peter Metzger, *The Atomic Establishment.* New York: Simon and Schuster (1972), p. 37.
88 *Info,* No. 33 (June 1973), p. 5.
89 Novick, op. cit., pp. 1–5.
90 Ibid., pp. 7–10.
91 *The Safety of Nuclear Power Reactors,* op. cit., pp. 6–28.
92 Ibid., pp. 6–29.
93 Novick, op. cit., p. 11.
94 *The Safety of Nuclear Power Reactors,* loc. cit.
95 Ibid., p. ii.
96 Robert Gillette, "Nuclear Safety: AEC Report Makes the Best of It," *Science,* Vol. 179, No. 4071 (January 26, 1973), pp. 360–63.
97 Ibid.
98 Letter to Lester Rogers, AEC's director of regulatory standards, from EPA's director of federal activities, Sheldon Meyers (February 16, 1973).
99 Ibid.
100 *The Safety of Nuclear Power Reactors,* op. cit., pp. 3–4.

101 W. K. Ergen (chm.), *Emergency Core Cooling: Report of Advisory Task Force on Power Reactor Emergency Cooling*. Washington, D.C.: U. S. Atomic Energy Commission (1967).
102 Ibid., pp. 47–48.
103 Ibid.
104 Wilson Clark, "Nuclear Plants Are Unsafe," *Not Man Apart* (July 1971), pp. 8, 23.
105 Tom Alexander, "The Big Blowup over Nuclear Blowdowns," *Fortune*, (May 1973), p. 216ff.
106 "World's First Series of Destructive Safety Tests," *Energy Digest* (October 4, 1972), p. 225.
107 "Reactor Core Cooling Hearings," *Energy Digest* (March 28, 1972), pp. 59–64.
108 Ibid.
109 Quoted in "Concluding Statement, Safety Phase, ECCS Hearings," Consolidated National Intervenors, AEC Docket No. RM-50-1. Washington, D.C. (March 15, 1973), p. 4.41.
110 Ibid., p. 1.4.
111 Ibid., p. 2.7.
112 *Technical Report on Densification of Light Water Reactor Fuels*, report of Regulatory Staff, U. S. Atomic Energy Commission. Washington, D.C. (November 1972).
113 John Holdren and Philip Herrera, *Energy*, San Francisco and New York Sierra Club (1971).
114 Dr. George L. Weil, *Nuclear Energy: Promises, Promises*. Washington, D.C.: George L. Weil, publisher (1972).
115 Ibid.
116 Anthony Ripley, "Political Shifts Threaten Growth of Atomic Power," New York *Times* (November 25, 1972).
117 "Added Generation Falls Far Short of Industry's Bogey," *Electrical World* (June 15, 1973), p. 83.
118 David Bird, "Con Edison Scores Makers of Nuclear Generating Plants for 'Glorious Promises,'" New York *Times* (November 19, 1972).
119 Thomas Ehrich, "Atomic Lemons: Breakdowns and Errors in Operation Plague Nuclear Power Plants," *The Wall Street Journal* (May 3, 1973).
120 Ibid.
121 Paul G. Edwards, "Vepco Fined for A-Plant Infractions," Washington *Post* (May 18, 1973).
122 Carl W. Houston, "Present and Future Safety in the Construction of Nuclear Power Stations." Johnson City, Tennessee, privately published (January 5, 1971).
123 "GE Control Rod Problems Spread," *Weekly Energy Report* (July 30, 1973), p. 1.
124 *The Safety of Nuclear Power Reactors*, op. cit., pp. 8–23.
125 Ralph Lapp, "Nuclear Energy: How Soon, How Safe," *Consulting Engineer* (March 1973), pp. 156–63.
126 "Geological Survey Report," *Weekly Energy Report* (May 14, 1973), p. 5.
127 Williams, op. cit., p. 8.

128 Ibid.

129 *Energy Research Policy Alternatives,* Senate Interior Committee Hearing, Serial 92-30 (June 7, 1972), Washington, D.C.: U. S. Government Printing Office (1972), p. 502.

130 Donald P. Geesaman, "Plutonium and Public Health." Boulder, Colorado: University of Colorado (April 19, 1970).

131 A. B. Long, "Plutonium Inhalation: The Burden of Negligible Consequence," *Nuclear News* (June 1971), pp. 69–73.

132 Edward Teller, paper presented to American Nuclear Society, *Nuclear News* (August 1967).

133 Quoted in Richard Curtis and Elizabeth Hogan, *Perils of the Peaceful Atom.* Garden City, New York: Doubleday (1969), p. 14.

134 Richard Curtis and Elizabeth Hogan, *Perils of the Peaceful Atom.* Garden City, New York: Doubleday (1969), p. 162.

135 Ibid., p. 155.

136 Quoted in *Environmental Impact Statement for Liquid Metal Fast Breeder Reactor* (draft). Washington, D.C.: U. S. Atomic Energy Commission (July 1971).

137 George C. Wilson, "AEC Told to Preview Impact of Reactors on Environment," Washington *Post* (June 13, 1973).

138 "Calvert Cliffs Decision," *Council on Environmental Quality Monitor,* Vol. 1, No. 8 (1971).

139 *Liquid Metal Fast Breeder Reactor Demonstration Plant,* Hearings Before the Joint Committee on Atomic Energy, September 1972. Washington, D.C.: U. S. Government Printing Office (1972), p. 32.

140 Amory B. Lovins, "The Case Against the Fast Breeder Reactor," *Bulletin of the Atomic Scientists* (March 1973), pp. 29–35.

141 Ibid.

142 *Environmental Impact Statement for Liquid Metal Fast Breeder Reactor,* op. cit., p. 28.

143 Dr. Thomas B. Cochran, "An Economic and Environmental Analysis of an Early U. S. Commitment to the Liquid Metal Fast Breeder Reactor," Washington, D.C., Resources for the Future, Inc., 1972, p. 127. (Now in book form, *The Liquid Metal Fast Breeder Reactor,* Resources for the Future, Inc., 1974.)

144 Ibid., p. 128.

145 Ibid., p. 129.

146 Ibid., p. 131.

147 Ibid., p. 42k.

148 Ibid., p. 131.

149 Ibid., p. 20.

150 Ibid., p. 132.

151 Ibid., p. 135.

152 *The Safety of Nuclear Power Reactors,* op. cit., pp. 4–31.

153 Ibid., pp. 4–34.

154 Richard Lewis, *The Nuclear Power Rebellion,* New York: Viking (1972), pp. 270–71.

155 *Environmental Survey of the Nuclear Fuel Cycle,* U. S. Atomic Energy Commission Fuels and Materials Directorate of Licensing (November 1972).

156 "'Mad Bombers' Now Eying Reactor Plutonium," *Energy Digest*
171 William Hambleton, remarks at annual meeting of American Asso-
(January 17, 1972), p. 4.
157 Victor Gilinsky, "Bombs and Electricity," *Environment,* Vol. 14, No. 7 (September 1972), pp. 11–17.
158 Ibid.
159 Ibid.
160 Ibid.
161 Timothy H. Ingram, "Nuclear Hijacking: Now Within the Grasp of Any Bright Lunatic," *Washington Monthly* (January 1973), pp. 21–28.
162 Ibid.
163 Metzger, op. cit., p. 223.
164 M. King Hubbert, "Energy Resources," *Resources and Man.* San Francisco: Freeman (1969), p. 233.
165 Ibid., p. 235.
166 Wilson Clark, "AEC May Put 'Hot' Liquid in Rocks," *Environmental Action* (October 3, 1970).
167 "An Evaluation of the Concept of Storing Radioactive Wastes in Bedrock Below the Savannah River Plant Site," Report by Committee on Radioactive Waste Management, National Academy of Sciences, Washington, D.C. (1972).
168 *The Safety of Nuclear Power Reactors,* op. cit., pp. 4–81.
169 Richard S. Lewis, *The Nuclear Power Rebellion.* New York: Viking (1972), p. 154.
170 Ibid., p. 167.
ciation for the Advancement of Science, Philadelphia (December 29, 1971).
172 *The Safety of Nuclear Power Reactors,* op. cit., pp. 4–86.
173 Metzger, loc. cit.
174 M. S. General Accounting Office, *Progress and Problems in Programs for Managing High-Level Wastes* (January 29, 1971).
175 Ibid.
176 Lee Dye, "Thousands Periled by Nuclear Waste," Los Angeles *Times* (July 5, 1973).
177 William Hines, "Sun Eyed as Atomic Dump," Chicago *Sun-Times* (January 1972).
178 *Energy Digest,* first issue (February 1972), p. 40.
179 *See* National Wildlife Federation, *Nuclear Plants and You: A Citizens Handbook,* Washington, D.C. (1971).
180 *Congressional Record* (March 14, 1973), p. S 4573.
181 *Congressional Record* (July 2, 1973), pp. E 4331–E 4334.
182 Allen V. Kneese, "Benefit-Cost Analysis and Unscheduled Events in the Nuclear Fuel Cycle," comments submitted to AEC Directorate of Licensing on Environmental Survey of Nuclear Fuel Cycle: Power Reactor Licensing Rule Making, U. S. Atomic Energy Commission (November 1972).
183 G. J. Mischke, "The Search for Fusion Power," *Naval Research Reviews* (April 1971), p. 1ff.

184 Tom Alexander, "The Hot New Promise of Thermonuclear Power," reprint from *Fortune* (June 1970).
185 Los Alamos Scientific Laboratory, *Central Station Power Generation by Laser-Driven Fusion, LA-4858-MS,* Vol. I. Los Alamos, New Mexico (February 1972), p. 1ff.
186 Alexander, loc. cit.; Resources for the Future, Inc., et al., op. cit., p. VI-109.
187 Holdren and Herrera, op. cit., p. 107.
188 T. K. Fowler, "Fusion Power—How Soon?," paper prepared for the Atomic Energy Commission (June 19, 1972), p. 9.
189 Herman Postma, "Power from Fusion," *Energy Research and Development,* Hearings before the Subcommittee on Science, Research, and Development of the Committee on Science and Astronautics, U. S. House of Representatives, May 9, 10, 11, 23, 24, 25, and 30, 1972. Washington, D.C.: U. S. Government Printing Office (1972), p. 613.
190 Holdren and Herrera, op. cit., p. 107ff.
191 Alexander, loc. cit.
192 Holdren and Herrera, op. cit., p. 108.
193 Alexander, loc. cit.
194 Ibid.
195 Victor Cohn, "Nuclear Fusion Breakthrough Seen," Washington *Post* (December 2, 1972).
196 Untitled paragraph under "Nuclear Developments," *Electrical World* (January 1, 1973), p. 29.
197 "Fusion: Electricity by 1993 with Funding, Say Scientists; But Congress Will Continue to Favor Fast Breeder," *Energy Digest,* Vol. I, No. 15 (November 26, 1971), p. 171.
198 Ibid.
199 William D. Metz, "Laser Fusion: A New Approach to Thermonuclear Power," *Science,* Vol. 177, No. 4055 (September 29, 1972), p. 1180.
200 Ibid.
201 *Heat, Light and Sound,* Curtis Publishing Company, New York, 1968, p. 139.
202 Metz, "Laser Fusion," loc. cit.
203 Ibid.
204 Ibid.
205 Ibid.
206 Ibid.
207 Alexander, loc. cit.
208 Ibid.
209 R. F. Post, "Fusion Power," paper presented to National Academy of Sciences Symposium on Energy for the Future (April 26, 1971).
210 Alexander, loc. cit.
211 Ibid.
212 Resources for the Future, Inc., et al., op. cit., p. VI-114.
213 "Fusion: Electricity by 1993 with Funding . . . ," loc. cit.
214 Ibid.
215 Fowler, loc. cit.
216 Postma, op. cit., p. 615.

217 Resources for the Future, Inc., et al., op. cit., p. VI-111; Postma, op. cit., p. 617.
218 Postma, op. cit., p. 618.
219 Ibid.
220 Ibid.
221 Ibid., Table II, p. 617.
222 Ibid.
223 Ibid.
224 Ibid.
225 Walter J. Hickel, *Geothermal Energy,* University of Alaska, College, Alaska, 1972.
226 AEC Staff Paper on Geothermal Energy (July 17, 1972).
227 Interview with Dr. Tsvi Meidav and John Banwell, United Nations geothermal energy consultants, United Nations Headquarters, New York (August 1972).
228 Ibid.
229 Allen L. Hammond, "Geothermal Energy: An Emerging Major Resource," *Science,* Vol. 177, No. 4053 (September 15, 1972), pp. 978–80.
230 John Banwell and Tsvi Meidav, "Geothermal Energy for the Future," paper presented at the 138th Annual Meeting of the American Association for the Advancement of Science, Philadelphia (December 1971).
231 Ibid.
232 John Banwell, United Nations, New York (February 1973).
233 Hickel, op. cit., p. 14.
234 Banwell, loc. cit.
235 C. P. Gilmore, "Hot New Prospects for Power from the Earth," *Popular Science* (August 1972), pp. 56–60, 126.
236 Earl C. Gottschalk, Jr., "Earth Power: Steam Below Ground Seen Giving Big Boost to U.S. Energy Supply," *The Wall Street Journal* (March 20, 1973).
237 Banwell and Meidav, "Geothermal Energy for the Future," loc. cit.
238 Martin Goldsmith, *Geothermal Resources in California Potentials and Problems.* Pasadena, California: California Institute of Technology (1972).
239 Gottschalk, loc. cit.
240 Committee on Interior and Insular Affairs, U. S. Senate, *Conservation of Energy,* Washington, D.C.: U. S. Government Printing Office (1972), pp. 87–90.
241 Meidav and Banwell, interview, loc. cit.
242 Ibid.
243 Banwell and Meidav, "Geothermal Energy for the Future," loc. cit.
244 Ibid.
245 Ibid. and interview, loc. cit.
246 Committee on Interior and Insular Affairs, U. S. Senate, op. cit., p. 88.
247 Goldsmith, loc. cit.
248 Meidav, interview, loc. cit.
249 Hammond, loc. cit.
250 Meidav, interview, loc. cit.
251 Banwell and Meidav, "Geothermal Energy for the Future," loc. cit.

252 Ibid.
253 Committee on Interior and Insular Affairs, U. S. Senate, loc. cit.
254 Banwell and Meidav, "Geothermal Energy for the Future," loc. cit.
255 "The Development of Dry Geothermal Energy Sources," Atomic Energy Commission staff paper, with author's interpolation of temperatures and depths. Washington, D.C. (1972).
256 *Energy Research and Development,* Hearings before the Subcommittee on Science, Research, and Development of the Committee on Science and Astronautics. Washington, D.C.: U. S. Government Printing Office (1972), p. 343; Gilmore, loc. cit.
257 Gilmore, loc. cit.
258 "Interior's Geothermal Regulations Inadequate," *Sierra Club Bulletin,* Vol. 58, No. 2 (February 1973), p. 19.
259 "Geothermal Energy and the Environment," summary of transactions of meeting held at United Nations Headquarters, New York (May 22, 1972), p. 12.
260 "Interior's Geothermal Regulations Inadequate," loc. cit.
261 "APPA Warns Interior its Geothermal Leasing Proposal Lacks Anti-trust Safeguards," *Public Power Weekly Newsletter* (November 26, 1971), p. 3.
262 Banwell, interview, loc. cit.
263 "Geothermal Energy and the Environment," loc. cit.
264 Goldsmith, op. cit., p. 31.
265 Ibid., p. 32.
266 Ibid., pp. 32–33.
267 Gilmore, loc. cit.
268 "Geothermal Energy and the Environment," op. cit., pp. 15–16.
269 Hickel, op. cit., pp. 71–77.
270 Ibid., p. 6.
271 Edward P. Clancy, *The Tides.* Garden City, New York: Doubleday (1968), p. 136.
272 Ibid., pp. 136–37; National Petroleum Council, op. cit., p. 57.
273 F. L. Lawton, "Tidal Power in the Bay of Fundy," *Tidal Power,* proceedings of an international conference on the utilization of tidal power held May 24–29, 1970, at the Atlantic Industrial Research Institute, Nova Scotia Technical College, Halifax, Nova Scotia. New York: Plenum (1972) p. 3.
274 National Petroleum Council, loc. cit.
275 Lawton, op. cit., p. 4.
276 Ibid.
277 Hubbert, op. cit., Table, p. 213.
278 Clancy, op. cit., p. 140ff.
279 Ibid., pp. 144–45.
280 L. B. Bernshtein, "Kislaya Guba Experimental Tidal Power Plant and Problem of the Use of Tidal Energy," *Tidal Power,* proceedings of an international conference on the utilization of tidal power held May 24–29, 1970, at the Atlantic Industrial Research Institute, Nova Scotia Technical College, Halifax, Nova Scotia. New York: Plenum (1972), pp. 217–18.
281 Ibid., p. 233.

282 Hubbert, op. cit., p. 213.
283 Luman H. Long (ed.), *The 1972 World Almanac*. New York: Newspaper Enterprise Association (1971), p. 456.
284 Hubbert, loc. cit.
285 National Petroleum Council, op. cit., p. 59.
286 Ibid.
287 Hubbert, loc. cit.
288 National Petroleum Council, op. cit., p. 58.
289 Clancy, op. cit., p. 147ff.
290 D. H. Waller, "Environmental Effects of Tidal Power Development," *Tidal Power*, proceedings of an international conference on the utilization of tidal power, held May 24–29, 1970, at the Atlantic Industrial Research Institute, Nova Scotia Technical College, Halifax, Nova Scotia. New York: Plenum (1972), p. 627ff.
291 Holdren and Herrera, op. cit., p. 54.
292 Federal Power Commission, *The 1970 National Power Survey*, Part I. Washington, D.C.: U. S. Government Printing Office (1971), p. I-7-1.
293 Ibid., p. I-7-21.
294 Ibid.
295 Ibid., pp. I-7-21–I-7-22.
296 Ibid., p. I-7-23.
297 Holdren and Herrera, loc. cit.
298 Ibid., pp. 54–55.
299 Boyce Richardson, "Running Amok at James Bay," *Sierra Club Bulletin*, Vol. 57, No. 10 (December 1972), p. 5ff.
300 Ibid., p. 9.
301 Ibid., p. 6.
302 Ibid., p. 10.
303 Rocks and Runyon, op. cit., p. 56.
304 Henry van der Schalie, "Control in Egypt and the Sudan," *The Careless Technology*. Garden City, New York: Natural History Press (1972).
305 M. Taghi Farvar, introduction to "Schistosomiasis, The Disease of Slowed-Down Waters," *The Careless Technology*, ibid.
306 Hubbert, op. cit., p. 209.
307 Holdren and Herrera, op. cit., p. 55.
308 Wilson Clark, "U.S. Energy Use," unpublished manuscript (1970), pp. 42–43.
309 Federal Power Commission, loc. cit.
310 Holdren and Herrera, op. cit., pp. 55–56.
311 Lawrence Lessing, "The Master Fuel of a New Age," *Fortune* (May 1961), p. 152ff.
312 *Chemical & Engineering News* (September 25, 1972), p. 32.
313 Lawrence W. Jones, *The Hydrogen Fuel Economy: An Early Retrospective*. Ann Arbor, Michigan: University of Michigan (November 1972), p. 6.
314 Lawrence Lessing, "The Coming Age of Hydrogen," *Fortune* (November 1972), p. 138.
315 Hammond, Metz, and Maugh, op. cit., p. 121.

316 Wilson Clark, "Hydrogen May Emerge as the Master Fuel to Power a Clean-Air Future," *Smithsonian* (August 1972).
317 Derek P. Gregory, "The Hydrogen Economy," *Scientific American* (January 1973).
318 Hammond, Metz, and Maugh, loc. cit.
319 "Hydrogen: Likely Fuel of the Future," *Chemical & Engineering News* (June 26, 1972), p. 15.
320 Gregory, loc. cit.
321 Cesare Marchetti, "Hydrogen, Master-Key to the Energy Market," *Euro Spectra*, Vol. X, No. 4 (December 1971), p. 123.
322 Lessing, "The Coming Age of Hydrogen," loc. cit.
323 Ibid., p. 142.
324 Jones, op. cit., p. 7.
325 Lawrence Jones, "Liquid Hydrogen as a Fuel for the Future," *Science* (October 22, 1972).
326 Lessing, "The Coming Age of Hydrogen," op. cit., p. 142.
327 Jones, op. cit., pp. 4–5.
328 Kurt H. Weil, "The Hydrogen I.C. Engine—Its Origins and Future in the Emerging Energy-Transportation-Environment System," Seventh Intersociety Energy Conversion Engineering Conference, 1972; Conference Proceedings, American Chemical Society (1972), p. 1357.
329 Jones, op. cit., pp. 9–10.
330 Clark, op. cit.
331 Ibid.
332 Ibid.
333 Jones, *The Hydrogen Fuel Economy: An Early Retrospective*, loc. cit.
334 Ibid., p. 10.
335 Ibid.
336 Clark, op. cit.
337 Ibid.
338 Gregory, loc. cit.
339 "Hydrogen: Likely Fuel of the Future," op. cit., p. 16.
340 Lessing, "The Coming Age of Hydrogen," op. cit., p. 141.
341 Marchetti, op. cit., p. 118.
342 Hammond, Metz, and Maugh, op. cit., p. 119.
343 Marchetti, op. cit., p. 117.
344 Lessing, "The Coming Age of Hydrogen," op. cit., p. 140.
345 Ibid.
346 Jones, *The Hydrogen Fuel Economy: An Early Retrospective*, loc. cit.
347 Gregory, loc. cit.
348 Ibid.
349 "Nixon Energy Office Adds $, Shifts $, to Total $1 Billion," *Energy Digest*, Vol. III, No. 19 (October 17, 1973), p. 1.

Notes
Chapter 6

1 Farrington Daniels, *Direct Use of the Sun's Energy*. New Haven, Connecticut: Yale University Press (1964).
2 Ibid., p. 23.
3 Watson, Lyall, *Supernature*. Garden City, New York: Anchor Press/Doubleday (1973).
4 Donald Culross Peattie, *Flowering Earth*. New York: Viking (1965).
5 David M. Gates, "The Flow of Energy in the Biosphere," *Scientific American* (September 1971), pp. 89–90.
6 Miller, G. Tyler, *Energetics, Kinetics, and Life*. Belmont, California: Wadsworth (1971), p. 239ff.
7 Jerome Weingart, "Everything You've Always Wanted to Know about Solar Energy but Were Never Charged Up to Ask," *Environmental Quality* (December 1972), p. 40.
8 Ibid.
9 Aden Baker Meinel and Marjorie Pettit Meinel, *Power for the People*. Tucson, Arizona: privately published (1970), p. 6.
10 Hans Rau, *Solar Energy*. New York: Macmillan (1964), p. 43.
11 D. S. Halacy, Jr., *The Coming Age of Solar Energy*. New York: Harper & Row (1963), pp. 197–98.
12 Meinel and Meinel, op. cit., p. 11.
13 Rau, op. cit., p. 44.
14 Ibid., p. 45.
15 A. S. E. Ackermann, quoted in C. G. Abbott, *The Sun and the Welfare of Man*, New York, Smithsonian Institution (1929), pp. 224–25.
16 Ibid., p. 210.
17 Rau, op. cit., p. 46.
18 Abbot, op. cit., p. 211.
19 Halacy, op. cit., p. 203.
20 Abbot, op. cit., p. 212.
21 Ibid., p. 216.
22 Daniels, op. cit., p. 10.
23 Abbot, op. cit., p. 222.
24 Ibid., p. 224.
25 Ibid., pp. 224–25.
26 Meinel and Meinel, op. cit., p. 42.
27 Abbot, op. cit., pp. 245–46.
28 Ibid.
29 Ibid., p. 47.

30 E. R. Hagemann, "R. H. Goddard and Solar Power," *Solar Energy*, Vol. 6, No. 2 (1962), p. 52.
31 Ibid., pp. 47–54.
32 Harold L. Alt, "Sun Effect and the Design of Solar Heaters," American Society of Heating and Ventilating Engineers *Transactions* (1935), pp. 131–48.
33 F. A. Brooks, "Solar Energy and Its Use for Heating Water in California," *Bulletin 602*, University of California College of Agriculture (November 1936).
34 Weingart, op. cit., pp. 39–43.
35 A. M. Zarem and Duane D. Erway (eds.), *Introduction to the Utilization of Solar Energy*, University of California Engineering and Sciences Extension Series. New York: McGraw-Hill (1963), p. 299.
36 "Applied Solar Energy: The Amazing Solartek Panel." Opa Locka, Florida: Solar Products Corporation (no date).
37 John I. Yellott, "Japan Applies Solar Energy," *Mechanical Engineering* (March 1960), pp. 49–51.
38 Harry Tabor, "Solar Energy Research in Israel," memorandum, National Physical Laboratory, Hebrew University (December 1972).
39 Halacy, op. cit., p. 204.
40 John I. Yellott, "Solar Energy: Its Use and Control," *Heating Piping & Air Conditioning* (October 1966), p. 137.
41 D. Proctor, "The Use of Waste Heat in a Solar Still," *Solar Energy*, Vol. 14, No. 4 (March 1973), p. 433.
42 J. A. Eibling, S. G. Talbert, and G. O. G. Löf, "Solar Stills for Community Use—Digest of Technology," *Solar Energy*, Vol. 13, No. 2 (May 1971), p. 276.
43 Felix Trombe, Leon Gion, Claude Royere, and Jean François Robert, "First Results Obtained with the 1000 KW Solar Furnace," *Solar Energy*, Vol. 15, No. 1 (May 1973), p. 67.
44 Halacy, op. cit., p. 71.
45 Energy Study Group, Office of Science and Technology, *Energy R&D and National Progress*. Washington, D.C.: U. S. Government Printing Office (1964).
46 Ibid., p. 284.
47 *Resources for Freedom*, Report of the President's Materials Policy Commission, Vol. IV, *The Promise of Technology*. Washington, D.C.: U. S. Government Printing Office (1952).
48 "Solar and Atomic Energy—A Survey," Bureau of Business and Economic Research, University of Maryland, College Park, Maryland, Vol. 12, No. 4 (March 1959).
49 Personal correspondence from John H. Cover (March 1972).
50 Ad Hoc Panel on Solar Cell Efficiency, *Solar Cells: Outlook for Improved Efficiency*, National Academy of Sciences, Washington, D.C. (1972).
51 "Nixon Administration Asks 20 Per Cent Increase in Energy R&D," *Energy Digest*, Vol. III, No. 2 (January 30, 1973), front page; and Warren H. Donnelly, "Federal Funds for Research and Development for Civil Power, Fiscal Years 1966 to 1972." Congressional Research Service, Library of Congress (August 30, 1971).

52 Martin Wolf, "Cost Goals for Silicon Solar Arrays for Large Scale Terrestrial Applications," from Ninth IEEE Photovoltaic Specialists' Conference, Catalogue #72 CHO 613-0-ED, Silver Spring, Maryland, Institute of Electrical and Electronic Engineers, 1972, p. 346.
53 Wilson Clark, "How to Harness Sun Power and Avoid Pollution," *Smithsonian,* (November 1971).
54 William R. Cherry, "The Generation of Pollution Free Electrical Power from Solar Energy," NASA-Goddard Space Flight Center No. X-760-71-135 (March 1971).
55 Interview with A. I. Mlavsky, Tyco Labs, Waltham, Massachusetts (August 1972).
56 Ibid.
57 C. G. Currin, K. S. Ling, E. L. Ralph, W. A. Smith, and R. J. Stirn, "Feasibility of Low Cost Silicon Solar Cells," Institute of Electrical and Electronics Engineers Photovoltaic Specialists Conference, Silver Spring, Maryland (May 2, 1972).
58 E. L. Ralph, "A Commercial Solar Cell Array Design," Solar Energy Society Conference, Greenbelt, Maryland (May 10, 1971).
59 Ibid.
60 Ad Hoc Panel on Solar Cell Efficiency, loc. cit.
61 Ibid., p. 17.
62 Peter Glaser, "Power from the Sun: Its Future," *Science,* Vol. 162, No. 3856 (November 1968), pp. 857–61.
63 Peter Glaser, interview.
64 "Satellite Solar Power Station: An Option for Power Generation," testimony prepared for briefings before the Task Force on Energy of the Subcommittee on Science, Research, and Development, Committee on Science and Astronautics, U. S. House of Representatives (March 1972), p. 146ff.
65 Ibid., pp. 146ff.
66 E. L. Ralph and F. Benning, "The Role of Solar Cell Technology in the Satellite Solar Power Station," paper presented at Institute of Electrical and Electronics Engineers Photovoltaic Specialists Conference, Silver Spring, Maryland (May 2, 1972).
67 Vincent Falcone, Jr., "Atmospheric Attenuation of Microwave Power," *Journal of Microwave Power,* Vol. 5, No. 4 (December 1970), p. 274.
68 National Science Foundation, National Aeronautics and Space Administration (NSF/NASA), Solar Energy Panel, *Solar Energy as a National Energy Resource,* College Park, Maryland: University of Maryland (1972).
69 Peter Glaser, John Mockovciak, and John Yerkes, interview, Washington, D.C. (March 1973).
70 Martin Wolf, loc. cit.
71 Terri Aaronson, "Mystery," *Environment,* Vol. 12, No. 4 (May 1970), p. 10.
72 Meinel and Meinel, op. cit., p. 125.
73 National Science Foundation/National Aeronautics and Space Administration (NSF/NASA) Solar Energy Panel, op. cit.
74 Cherry, loc. cit.
75 Meinel and Meinel, op. cit., pp. 211–26.

76 "Studies of Thermal Conversion of Solar Energy for Electrical Power Production," research proposal submitted to the National Science Foundation by the University of Arizona, Tucson, Arizona (1971).
77 Meinel and Meinel, op. cit., p. 218.
78 Aden Baker Meinel and Marjorie Pettit Meinel, "Physics Looks at Solar Energy," *Physics Today* (February 1972), pp. 44–50
79 Meinel and Meinel, *Power for People*, op. cit., p. 242.
80 Ibid.
81 Report No. ATR-73(7283-01)-1, prepared by the Aerospace Corporation, El Segundo, California, for the National Science Foundation (November 15, 1972).
82 *The Minnesota/Honeywell Solar Concept*, Minneapolis, Minnesota: Honeywell, Inc. (1972).
83 George P. Miller, Introduction to *Solar Energy Research*, staff report, Committee on Science and Astronautics, U. S. House of Representatives (January 1973).
84 Meinel and Meinel, "Physics Looks at Solar Energy," loc. cit.
85 William J. D. Escher, interview (December 1972).
86 Alvin F. Hildebrandt, Gregory M. Haas, et al., "Large-Scale Concentration and Conversion of Solar Energy," *ES*, Vol. 53, No. 7 (July 1972), p. 684ff. *See also Symposium on Energy, Resources and the Environment*, Proceedings of the Fifth Meeting of the Symposium Committee, February 22, 23, 24, and 25, McLean, Virginia: The Mitre Corporation (1972).
87 N. C. Ford and J. W. Kane, "Solar Energy and Public Power." Amherst, Massachusetts: University of Massachusetts (April 1971).
88 Marjorie Pettit Meinel and Aden Baker Meinel, testimony before the Task Force on Energy of the Subcommittee on Science, Research, and Development of the U. S. House of Representatives Committee on Science and Astronautics (March 6, 1972).
89 Report No. ATR-73(7283-01)-1, op. cit.
90 Thomas Maugh III, "Solar Energy: The Largest Resource," *Science*, Vol. 177, No. 4054 (September 22, 1972), pp. 1088–93.
91 Clarence Zener, "Solar Sea Power," *Physics Today* (January 1973).
92 James H. Anderson, Jr., "Economic Power and Water from Solar Energy," Paper No. 72-WA/Sol.-2, presented at the American Society of Mechanical Engineers Winter Annual Meeting, New York, (November 26–30, 1972).
93 Ibid.
94 Ibid.
95 Ibid.
96 H. E. Karig, "Thermal Power Systems Using Ocean Temperature Gradients as Source of Energy," presented at the American Society of Mechanical Engineers Winter Annual Meeting, New York (November 26–30, 1972).
97 P. G. Wybro and C. S. Chen, "Application of Sea Thermal Energy for Underwater Habitats," presented at the American Society of Mechanical Engineers Winter Annual Meeting, New York (November 26–30, 1972).

98 "National Network of Pollution-Free Power Sources," research proposal to the National Science Foundation from the University of Massachusetts, Amherst, Massachusetts (April 1971).
99 Ibid.
100 C. Zener and Carnegie Mellon University research team, presentation to U. S. Department of Commerce (January 1973).
101 James H. Anderson, Jr., "Economic Power and Water from Solar Energy," op. cit.
102 Irwin Goodwin, "Caribbean Being Harnessed," Washington *Post* (October 25, 1970).
103 Ibid.
104 Zener, "Solar Sea Power," op. cit.
105 "PELE Energy Laboratory Experiments," Research Proposal to the National Science Foundation from the University of Hawaii Center for Engineering Research, Honolulu, Hawaii (1972), pp. 19–21.
106 Harry Tabor, "Large Area Solar Collectors (Solar Ponds) for Power Production," Paper No. S/47, Vol. 4, *UN Conference on New Sources of Energy, Rome, 1961*. New York: United Nations (1961).
107 Tabor, "Solar Energy Research in Israel," loc. cit.
108 NSF/NASA Solar Energy Panel, loc. cit.
109 T. R. Schneider, "Substitute Natural Gas from Organic Materials," paper presented at American Society of Mechanical Engineers Annual Winter Meeting, New York (November 26–30, 1972).
110 New Energy Forms Task Group of the National Petroleum Council's Committee on U.S. Energy Outlook, *U.S. Energy Outlook, 1972*. Washington, D. C.: National Petroleum Council (1972).
111 George Szego, *The Energy Plantation: A Cost-Effective Means of Providing All the U.S. Energy and Power Needs by Utilizing Solar Energy*. Warrenton, Virginia: InterTechnology Corporation (February 1972).
112 A. W. Fisher, Jr., "Economic Aspects of Algae as a Potential Fuel," *Solar Energy Research*. Madison, Wisconsin: University of Wisconsin Press (1961), pp. 185–89.
113 Eugene Rabinowitch, "Photochemical Utilization of Light Energy," *Solar Energy Research*. Madison, Wisconsin: University of Wisconsin Press (1961), pp. 193–202.
114 Ibid., p. 194.
115 "Hydrogen Production by Photosynthesis and Hydrogenase Activity—An Energy Source," research proposal submitted to the National Science Foundation by Case Western Reserve University, Cleveland (1972).
116 An Enquiry into Biological Energy Conversion," Proceedings of a workshop held by the National Science Foundation at Oak Ridge National Laboratory (October 12–14, 1972).
117 Ibid.
118 Howard T. Odum, *Environment, Power and Society*. New York: Wiley-Interscience Press (1971), pp. 125–28.
119 Ibid., pp. 127–28.
120 Bill Kovach, "Garbage Smothered Cities Face Crisis in Five Years," New York *Times* (June 10, 1973).

121 "Council on Environmental Quality Says Resource Recovery from Solid Waste Is Now Feasible—Does Not Plan Aid to Communities," *Energy Digest*, Vol. III, No. 9 (May 15, 1973), p. 227.
122 "Problems and Opportunities in Management of Combustible Solid Wastes," International Research and Technology Corporation, report prepared for Environmental Protection Agency Contract No. 68–03–0060 (October 1972).
123 "Council on Environmental Quality . . . ," loc. cit.
124 Thomas H. Maugh III, "Fuel from Wastes: A Minor Energy Source," *Science* (November 10, 1972), pp. 599–602.
125 Ibid.
126 Interview with Ram Bux Singh, *The Mother Earth News* (November 1972), pp. 6–11.
127 Interview with Martin Wolf, Philadelphia (December 1972).
128 Interview with Ram Bux Singh, loc. cit.
129 Ibid.
130 "An Enquiry into Biological Energy Conversion," loc. cit.
131 Clarence Oswald, letter to Senator Mike Gravel (February 14, 1972).
132 W. J. Oswald and C. G. Golueke, "Solar Power Via a Botanical Process," *Mechanical Engineering* (February 1964), pp. 40–43.
133 Larry L. Anderson, "Energy Potential from Organic Wastes: A Review of the Quantities and Sources," Bureau of Mines Information Circular No. 8549, Washington, D. C.: U. S. Bureau of Mines (1972).
134 Ibid.
135 Gary Soucie, "How You Gonna Keep It Down on the Farm?" *Audubon* (September 1972), pp. 113–15.
136 Ibid.; *see also* Barry Commoner, *The Closing Circle*. New York: Knopf (1971), p. 82.
137 "Chicken Power," Maine *Times* (May 19, 1972).
138 "Problems and Opportunities in Management of Combustible Solid Wastes," op. cit., pp. 157–76.
139 Jonathan Allen, "Sewage Farming," *Environment*, Vol. 15, No. 3 (April 1973), p. 36ff.
140 David Brand, "Battle of Survival: World Scientists Fight the Protein Shortage with Diverse Weapons," *The Wall Street Journal* (February 7, 1973).
141 "Problems and Opportunities in Management of Combustible Solid Wastes," op. cit., pp. 186–209.
142 Donald E. Carr, *The Deadly Feast of Life*. Garden City, New York: Doubleday (1971), p. 141.
143 NSF/NASA Solar Energy Panel, op. cit., p. 5.

Notes
Chapter 7

1 Stanford Research Institute, "Patterns of Energy Conservation in the U.S." Office of Science and Technology, Executive Office of the President, Washington, D.C.: U. S. Government Printing Office (January 1972), p. 8.
2 Ibid.
3 "Space Heating with Solar Energy," *Heating and Ventilating* (September 1950), pp. 88–90.
4 Maria Telkes, "A Review of Solar House Heating," *Heating and Ventilating* (September 1949).
5 Maria Telkes and Eleanor Raymond, "Storing Solar Heat in Chemicals—A Report on the Dover House," *Heating and Ventilating* (November 1949), pp. 80–86.
6 Aladar Olgyay, "Design Criteria of Solar Heated Houses," Paper No. S/93, Vol. 5, *UN Conference on New Sources of Energy, Rome, 1961*. New York: United Nations (1961).
7 George O. G. Löf, "The Heating and Cooling of Buildings with Solar Energy," *Introduction to the Utilization of Solar Energy*. New York: McGraw-Hill (1963).
8 Maria Telkes "Solar Heat Storage," Paper No. 64 WA/Sol-9, American Society of Mechanical Engineers Winter Annual Meeting, New York (November 29–December 4, 1964).
9 Philip Sporn and E. R. Ambrose, "Solar Heat Pump Beats Handicaps," *Heating, Piping and Air Conditioning* (March 1956).
10 E. R. Ambrose, *Heat Pumps and Electric Heating*. New York: Wiley (1956).
11 E. B. Penrod and K. V. Prasanna, "Will Solar Energy be the Heat Source for Tomorrow's Heat Pump?" *Heating, Piping and Air Conditioning* (May 1960), pp. 118–26.
12 R. C. Jordan and J. L. Threlkeld, "Design and Economics of Solar Energy Heat Pump Systems," *Heating, Piping and Air Conditioning* (ASHVE Journal Section) (1954), pp. 1–9.
13 Ibid.
14 F. H. Bridgers, D. C. Paxton, and R. W. Haines, "Performance of a Solar Heated Office Building," *Heating, Piping and Air Conditioning* (November 1957), pp. 165–70.
15 Löf, loc. cit.
16 Masanosuke Yanagimachi, "Report on 2½ Years Experimental Living in Yanagimachi Solar House II," *UN Conference on New Sources of Energy, Rome, 1961*. New York: United Nations (1961).
17 C. D. Engebretson, "MIT Solar House IV," *UN Conference*, ibid.

18 Resources for the Future, Inc., in cooperation with MIT Environmental Laboratory, *Energy Research Needs,* a report to the National Science Foundation, Washington, D.C. (1971), p. VIII-27.
19 Löf, loc. cit.
20 George O. G. Löf, remarks at Arthur D. Little Conference on Solar Climate Control, Fort Lauderdale, Florida (March 1973).
21 George O. G. Löf et al., "Design and Performance of Domestic Heating System Employing Solar Heated Air—The Colorado Solar House," Paper No. S/114, *UN Conference,* op. cit.
22 Harry Thomason, "Solar Space Heating, Water Heating, Cooling in the Thomason Home," Paper No. S/3, *UN Conference,* op. cit.
23 Maria Telkes, interview, Newark, Delaware (October 1972).
24 "Energy Conservation with Solar Climate Control," Technical Memorandum, Arthur D. Little, Inc., Cambridge, Massachusetts (April 6, 1972).
25 R. A. Tybout and George O. G. Löf, "Solar House Heating," *Natural Resources Journal,* Vol. 10, No. 2 (1970).
26 George O. G. Löf and Richard A. Tybout, "Cost of House Heating with Solar Energy;" *Solar Energy,* Vol. 14 (1973), pp. 253–78.
27 Jerome Weingart, remarks at National Science Foundation Conference on Solar Heating and Cooling, Washington, D.C. (March 1973).
28 Richard Caputo, telephone interview (April 1973).
29 *Practical Applications of Solar Energy,* Leaflet No. 83. Gainesville, Florida: College of Engineering, University of Florida (November 1956).
30 Erich A. Farber, "The Direct Use of Solar Energy to Operate Refrigeration and Air-Conditioning Systems," Technical Progress Report No. 15. Gainesville, Florida: College of Engineering, University of Florida (November 1965).
31 Löf, "The Heating and Cooling of Buildings with Solar Energy," loc. cit.
32 Ron Alward, interview, New York (November 1972).
33 R. K. Swartman and V. H. Ha, "Performance of a Solar Refrigeration System Using Ammonia-Sodium Thiocyanate," Paper No. 72-WA/Sol.-3, American Society of Mechanical Engineers Winter Annual Meeting, New York (November 26–30, 1972).
34 Löf, "The Heating and Cooling of Buildings with Solar Energy," loc. cit.
35 E. M. Barber, Jr., interviews and correspondence, New Haven, Connecticut (1972–73).
36 Wolf von Eckardt, "Attractive, Feasible Solar Architecture," Washington *Post* (May 19, 1973).
37 Peter E. Glaser, "The Case of Solar Energy," speech presented at Annual Meeting of the Society for Social Responsibility in Science, London (September 3, 1972); available from Arthur D. Little, Inc., Cambridge, Massachusetts.
38 "Energy Conservation with Solar-Natural Gas Climate Control," Arthur D. Little proposal to American Gas Association, Arthur D. Little, Inc., Cambridge, Massachusetts (January 14, 1972).

39 D. Elliott Wilbur, remarks at Arthur D. Little Conference on Solar Climate Control, Fort Lauderdale, Florida (March 1973).
40 "Solar Climate Control Could Save U.S. Millions of Barrels of Oil per Year," press release. Cambridge, Massachusetts: Arthur D. Little, Inc. (May 10, 1973).
41 K. W. Böer, interview, Newark, Delaware (October 1972).
42 K. W. Böer, "Direct Solar Energy Conversion for Terrestrial Use," Institute of Energy Conversion, University of Delaware, Newark, Delaware; and "Future Large Scale Terrestrial Use of Solar Energy," Institute of Energy Conversion, University of Delaware, Newark, Delaware (undated).
43 National Science Foundation/National Aeronautics and Space Administration (NSF/NASA) Solar Energy Panel, *Solar Energy as a National Energy Resource*. College Park, Maryland: University of Maryland (December 1972), pp. 13–21.
44 Press Release No. 733–122, National Science Foundation (March 6, 1973).
45 Harold Hay, interviews and correspondence, Los Angeles, California and Washington, D.C. (1972–73).
46 Steve Baer, remarks at U. S. Solar Energy Society meeting, University of Florida, Gainesville, Florida (April 1972).
47 "Biosphere," poster, Biotechnic Press, Albuquerque, New Mexico, 87125, 1973.

Notes
Chapter 8

1 Farrington Daniels, *The Direct Use of Solar Energy*. New Haven, Connecticut: Yale University Press (1964), p. 23.

2 Abraham H. Oort, "The Energy Cycle of the Earth," *Scientific American* (September 1970), pp. 59–61; Vernon C. Finch et al., *The Elements of Geography*. New York: McGraw-Hill (1957), pp. 46–57.

3 Ibid. (*The Elements of Geography*), p. 46.

4 E. W. Golding, *The Generation of Electricity by Wind Power*. New York: Philosophical Library (1955), p. 100.

5 Henry Clews, "Electric Power from the Wind." East Holden, Maine: Henry Clews, publisher (1973), p. 5.

6 "New Sources of Energy: Proceedings of the Conference, Rome, August 21–31, 1961," Vol. VII. New York (1964).

7 T. K. Derry and Trevor I. Williams, *The Oxford Short History of Technology*. New York: Oxford University Press (1961), pp. 367–68.

8 Basil Greenhill, "The Sailing Ship in a Fuel Crisis," *The Ecologist* (September 1972).

9 James Carlill, "Wind Power," reprinted from *Edinburgh Review* (October, 1918), in *The Annual Report of the Smithsonian Institution* (1918), Washington, D.C.: U. S. Government Printing Office (1920), pp. 147–56.

10 Greenhill, loc. cit.

11 Ibid.

12 Ibid.

13 James McCawley, "Return of the Tall Ships," *Rudder* (1971), pp. 29–31, 70–72.

14 Vitruvius, *The Ten Books on Architecture*. New York: Dover (1960).

15 Golding, op. cit.

16 Derry and Williams, op. cit., p. 254.

17 Golding, op. cit.

18 Derry and Williams, op. cit., p. 258.

19 Sam H. Schurr and Bruce C. Netschert, *Energy in the American Economy, 1850–1975*. Baltimore, Maryland: The Johns Hopkins Press (1960), p. 486.

20 *Windmills in Foreign Countries*. Special Consular Reports, Vol. 31, U. S. Department of Commerce and Labor. Washington, D.C.: U. S. Government Printing Office (1904), p. 17.

21 Ibid., pp. 75–76.

22 "Disappearing Windmills," New York *Times* (December 21, 1972).

23 "Letters to the Editor," New York *Times* (January 3, 1973).

24 Golding, op. cit.

25 Ibid.

26 Ibid., p. 23.
27 Clews, op. cit.
28 Ibid.
29 Interview with John Roesel, Jr., Bradenton, Florida (April 1973).
30 Golding, op. cit., p. 232.
31 Ibid., p. 223.
32 David Byrd, "Batteries May Make Pump Storage Unnecessary," New York *Times* (May 30, 1972).
33 Material received from Stal Laval Company (1972).
34 "The Windup Car," *Environment* (June 1970).
35 Julian McCall, "A Lift for the Auto," *Environment* (December 1971).
36 Discussion with John Roesel, Jr. (April 1973).
37 Ibid.
38 *UN Conference, Rome, 1961*, op. cit., pp. 376–77.
39 Discussion with Ernst Cohn (February 1973).
40 *UN Conference, Rome, 1961*, op. cit., p. 377.
41 Ibid., p. 376.
42 *Windmills in Foreign Countries*, op. cit., p. 266.
43 *UN Conference, Rome, 1961*, op. cit., pp. 229–40.
44 Palmer C. Putnam, *Power from the Wind*. New York: D. Van Nostrand (1948), p. 105.
45 Ing Narcisco Levy, "Current State of Windpower Research in the Soviet Union." Brace Research Institute Technical Report, T56, G. T. Ward (ed.) (September 1968).
46 *UN Conference, Rome, 1961*, op. cit., pp. 201–6.
47 Ibid.
48 Golding, op. cit.
49 *UN Conference, Rome, 1961*, op. cit., pp. 241–47.
50 Discussions held in 1972–73 with representatives of Brace Research Institute, McGill University, Montreal, Quebec, Canada.
51 Discussion with Marcellus Jacobs, Fort Myers, Florida (1973).
52 *UN Conference, Rome, 1961*, op. cit., pp. 326–34.
53 Discussion with Marcellus Jacobs, Fort Myers, Florida (1973).
54 Putnam, op. cit.
55 Ibid.
56 Ibid.
57 Ibid.
58 Ibid.
59 Ibid.
60 Ibid.
61 Percy H. Thomas, "Electric Power from the Wind: A Survey." Washington, D.C.: Federal Power Commission (March 1945).
62 Ibid.
63 Ibid.
64 Ibid.
65 Ibid.
66 "Production of Power by Means of Wind-Driven Generator," Hearing Before Committee on Interior and Insular Affairs, U. S. House of Rep-

resentatives (H.R. 4286). Washington, D.C.: U. S. Government Printing Office (September 1951).
67 Ibid.
68 Discussions with William E. Heronemus, University of Massachusetts, Amherst, Massachusetts (October 1972 and January 1973).
69 Ibid.
70 William E. Heronemus, "Power from the Offshore Winds," paper presented at annual meeting of Marine Technology Society, Washington, D.C. (October 1972).
71 Interview with Heronemus, October 1972 and January 1973, op. cit.
72 Ibid.
73 Testimony presented by William E. Heronemus before Atomic Safety and Licensing Board Hearing on Shoreham Nuclear Power Plant (January 1973).
74 Ibid.
75 Interviews with Robert Reines (1971–73). *See also* John Dreyfuss, "Unique Dome Home Harnesses Sun and Wind," Los Angeles *Times* (January 1, 1973).
76 Interview with Ben Wolf, "Windworks" (October 1972) and subsequent correspondence.
77 Interview with John Roesel, Jr., Bradenton, Florida (April 1973).
78 Telephone interviews with Henry Clews, East Holden, Maine (1972–73). *See also* Clews, op. cit.
79 Clews, loc. cit.
80 Correspondence with Wendell Hewson, Corvallis, Oregon (1972).
81 Oklahoma State University Research Proposal to National Science Foundation to develop wind power (August 18, 1972).
82 Telephone interview with Jack Allison, Oklahoma State University (February 1973).
83 "Can We Harness Pollution-free Electric Power from Windmills?" Stephen Kidd and Douglas Garr, *Popular Science* (November 1972), pp. 70–72.
84 Ibid.
85 Interviews with Everett Barber, Jr. (1972–73).
86 Correspondence with Peter Clegg (January 1973).
87 *UN Conference, Rome, 1961*, op. cit., pp. 198–99.
88 "Energy Research and Development and National Progress," Executive Office of Science and Technology, Washington, D.C. (1964), p. 334.
89 Ibid.
90 National Science Foundation/National Aeronautics and Space Administration (NSF/NASA) Solar Energy Panel, *Solar Energy as a National Energy Resource*. College Park, Maryland: University of Maryland (December 1972), pp. 65–68.
91 Wilson Clark, "Interest in Wind Is Picking Up as Fuels Dwindle," *Smithsonian* (November 1973), p. 70ff.
92 "Nixon Energy Office Adds $, Shifts $, to Total $1 Billion," *Energy Digest*, Vol. III, No. 19 (October 17, 1973), p. 1.
93 Greenhill, loc. cit.

Notes
Afterword

1 Henry Adams, *The Education of Henry Adams,* Sentry Edition. Boston: Houghton Mifflin (1961), p. 389.
2 M. King Hubbert, "Energy Resources," *Resources and Man.* San Francisco: Freeman (1969), p. 239.
3 Ralph Knowles, *Energy and Form.* Cambridge, Massachusetts: MIT Press (1974).
4 Ibid.
5 Ibid.
6 Ibid.
7 Ibid.
8 Ibid.
9 Ibid.
10 Ibid.
11 Victor Olgyay (with assistance from Aladar Olgyay), *Design with Climate.* Princeton, New Jersey: Princeton University Press (1963), p. 2.
12 Vitruvius, *The Ten Books on Architecture* (Morris H. Morgan, trans.), originally published Cambridge, Massachusetts: Harvard University Press (1914); republished New York: Dover (1960), pp. 15–16.
13 Ibid., p. 17.
14 Ibid., p. 27.
15 Ibid., p. 170.
16 Sibyl Moholy-Nagy, *Native Genius in Anonymous Architecture.* New York: Horizon (1957), p. 83.
17 Ibid., p. 96.
18 Ibid.
19 Ibid., p. 90.
20 Victor Olgyay and Aladar Olgyay, *Application of Climatic Data to House Design.* Washington, D.C.: Housing and Home Finance Agency (1953).
21 Aladar Olgyay and Victor Olgyay, *Solar Control and Shading Devices.* Princeton, New Jersey: Princeton University Press (1957).
22 Victor Olgyay, *Design with Climate.* Princeton, New Jersey: Princeton University Press (1963).
23 Ibid., p. 137.
24 "Arroyo Barranca Project," project proposal, Sun Mountain Design Ltd., Santa Fe, New Mexico (1973).
25 John Todd, "Limiting Growth . . . A Libertarian Alternative." Woods Hole, Massachusetts: New Alchemy Institute (1973).
26 "A Proposal for Community Technology." Washington, D.C.: Community Technology, Inc. (1973).

27 Murray Bookchin, "Towards a Liberatory Technology," *Post-Scarcity Anarchism*. Berkeley, California: Ramparts (1971).
28 Howard T. Odum, "Energy, Ecology, and Economics," invited paper prepared for the Royal Swedish Academy of Sciences (1973).

INDEX